COASTAL PROCESSES

D1388651

The world's coastlines, dividing land from sea, are geological environments unique in their composition and the physical processes affecting them. Humans have been building structures throughout history at these dynamically active intersections of land and the oceans. Although coastlines were initially used for naval and commercial purposes, more recently recreation and tourism have increased activity in the coastal zones dramatically. Shoreline development is now causing a significant conflict with natural coastal processes.

This text on coastal engineering will help the reader understand these coastal processes and develop strategies to cope effectively with shoreline erosion. The book is organized into four parts: (1) an overview of coastal engineering using case studies to illustrate problems; (2) a consideration of the hydrodynamics of the coastal zone reviewing storm surges, water waves, and low-frequency motions within the nearshore and surf zone; (3) a discussion of coastal responses, including equilibrium beach profiles and sediment transport; and (4) a presentation of applications such as erosion mitigation, beach nourishment, coastal armoring, tidal inlets, and shoreline management.

Students, practicing engineers, and researchers in coastal engineering and coastal oceanography will find this book an invaluable resource for understanding the mechanisms of erosion and designing shoreline structures.

Robert G. Dean is Graduate Research Professor, Department of Civil and Coastal Engineering, University of Florida. Professor Dean is a member of the National Academy of Engineering and serves as President of the Coastal Engineering Research Council.

Robert A. Dalrymple is E. C. Davis Professor of Civil and Environmental Engineering and founding director of the Center for Applied Coastal Research at the University of Delaware.

Professors Dean and Dalrymple are also authors of the well-known *Water Wave Mechanics for Engineers and Scientists*.

COASTAL PROCESSES

with Engineering Applications

ROBERT G. DEAN
University of Florida

ROBERT A. DALRYMPLE
University of Delaware

PUBLISHED BY THE PRESS SYNDICATE OF THE UNIVERSITY OF CAMBRIDGE
The Pitt Building, Trumpington Street, Cambridge, United Kingdom

CAMBRIDGE UNIVERSITY PRESS
The Edinburgh Building, Cambridge CB2 2RU, UK
40 West 20th Street, New York NY 10011–4211, USA
477 Williamstown Road, Port Melbourne, VIC 3207, Australia
Ruiz de Alarcón 13, 28014 Madrid, Spain
Dock House, The Waterfront, Cape Town 8001, South Africa

http://www.cambridge.org

First published 2002
First paperback edition 2004

Typefaces Times Ten 10/13 pt. and Helvetica Neue   *System* LaTeX2$_\varepsilon$ [TB]

A catalogue record for this book is available from the British Library

Library of Congress Cataloguing in Publication Data
Dean, Robert G. (Robert George), 1930–
    Coastal processes / Robert G. Dean, Robert A. Dalrymple.
       p.   cm.
    Includes bibliographical references and indexes.
    ISBN 0 521 49535 0 hardback
    1. Coastal engineering. I. Dalrymple, Robert A., 1945– II. Title.
  TC205 .D43 2001
  627′.58 – dc21   00-067434

ISBN 0 521 49535 0 hardback
ISBN 0 521 60275 0 paperback

Contents

Preface

This book is written for graduate students, researchers, and practitioners in the fields of coastal engineering, nearshore oceanography, and marine geology. Although the treatment in many chapters is rather mathematical, it is hoped that our message does not get swamped by the delivery.

The book, which deals primarily with sandy coastlines, is divided into four parts. The first, Introduction to Coastal Processes, provides an overview of the problems of coastal engineering based on examples and a geological perspective of the field. Part 2, Hydrodynamics of the Coastal Zone, reviews storm surges, water waves, and low-frequency motions within the nearshore and surf zone. The third part, Coastal Response, discusses the equilibrium beach profile and sediment transport. Finally, the last section, Shoreline Modification and Analysis, covers aspects of erosion mitigation such as beach nourishment and coastal armoring, tidal inlets, and shoreline management.

We have attempted to include much of the important work in the field, but, given a book with such a broad scope, we have been forced to omit (or overlook) a considerable amount of the literature. An attempt has been made to reference those contributions that clarified the physics of the processes or provided a model for engineering applications. Nevertheless, the book is biased toward our own experiences, which means that much of our work and many U.S. examples are presented. To our colleagues and friends whose work we have used, thanks, and to those whom we have egregiously omitted, our apologies.

The field of coastal engineering is changing rapidly. Perforce, this book is a snapshot of the field (albeit with a long exposure, when one considers how long it took us to write the book!), and many parts of it are subject to becoming outmoded soon. The reader is cautioned to review the recent literature before drawing conclusions. The bulk of the literature in the field of coastal engineering appears in such journals as *Coastal Engineering; Journal of Waterway, Port, Coastal and Ocean Engineering; Coastal Engineering Journal*; and the *Journal of Geophysical Research*, as well as a variety of conference proceedings. Chief among these conferences is the biennial International Conference on Coastal Engineering, which is hosted by different countries around the world. The reader is referred to these original sources to provide a fuller explication of the field.

Robert G. Dean
Department of Civil and Coastal Engineering
University of Florida

Robert A. Dalrymple
Center for Applied Coastal Research
Department of Civil and Environmental Engineering
University of Delaware

Acknowledgments

We are pleased to acknowledge our host institutions, the University of Florida and the University of Delaware, which have supported us and our colleagues and students over the years. We appreciate having the opportunity to spend our lives working in an exciting field with interesting coworkers and with numerous intriguing problems yet to be solved.

We also thank our funding sources – primarily, the Sea Grant Program of the U.S. Department of Commerce, but also the Army Research Office and the Office of Naval Research (RAD).

PART ONE

INTRODUCTION TO COASTAL PROCESSES

CHAPTER ONE

Overview

In Egypt, the construction of the Aswan High Dam and others on the Nile River has caused extreme erosion problems at the Nile Delta, where whole villages have disappeared as the shoreline has retreated at rates of 30 to 50 m/yr! Before construction of the dams for flood control, irrigation, and water supply, the Nile delivered about 20 million metric tons of sediment annually to the Mediterranean Sea. This sediment supply resulted in two large deltas (Damietta and Rosetta*), which extend 50 km into the sea. As each of the dams on the Nile was completed, the reservoir behind the dam began to capture a significant portion of the annual riverine sediment load.

To combat the ensuing erosion, large coastal structures have been placed along the shoreline near the river mouths to limit further shoreline retreat. Nevertheless, erosion is continuing in water depths below the base of the structures, and storm waves are attacking the coast with increasing intensity.

Surprisingly, the length of shoreline affected by the erosion is relatively short. Further, field measurements conducted by the Alexandria Coastal Research Institute show that, farther from the river mouths, the shoreline continues to advance in response to the prior era of abundant sediment supply and delta building.

1.1 INTRODUCTION

The world's coastlines, dividing land from sea, are geological environments unique in their composition and the physical processes affecting them. Many of these coastlines have beaches composed of loose sediments such as gravel, sand, or mud that are constantly acted upon by waves, currents, and winds, reshaping them continuously. However, despite the different wave climates that exist around the world and the variations in coastline composition, the nature and behavior of beaches are often very similar.

---

* The famed Rosetta stone, which led to the deciphering of hieroglyphics, was found in 1799 in the town of Rosetta. Inscribed on this black basaltic stone, which now resides in the British Museum in London, were three versions of a 196 B.C. decree set forth in hieroglyphics, demotic writing (the everyday script), and Greek. Twenty-three years later, Jean-Francois Champollion was able to translate the hieroglyphics for the first time in 1500 years!

3

Waves gather their energy and momentum from winds blowing over possibly huge expanses of uninterrupted ocean, yet much of this accumulated energy is dissipated within the fairly narrow surf zone. The breaking of the waves within this zone is responsible for the transformation of organized wave motion into chaotic turbulence, which mobilizes and suspends the sediments composing the beach. Also, the breaking waves create nearshore currents that flow along the shoreline and in the cross-shore direction. These currents can transport large quantities of sediment in both directions in volumes as large as hundreds of thousands of cubic meters of sand per year in some places.

At this dynamically active intersection of land and the oceans, humans have been building structures throughout history. Ports and harbors have always served as bases for naval forces and as commercial egresses to upland trade routes or major centers of civilization. More recently, as recreation and tourism at the shoreline have become more important economically, coastal development, taking the form of homes and businesses, has increased to such an extent that over 50 percent of the U.S. population now lives within 50 miles of the coastline. In 1995, it was estimated (IIPLR 1995) that over three trillion U.S. dollars of insured property was located adjacent to the U.S. Atlantic and Gulf shorelines alone. This shoreline development is causing an increasingly important conflict with the natural coastal processes.

There are many historical examples of engineering works that have interfered with sediment transport processes, causing severe beach erosion and associated structural damage or, conversely, large accumulations of sand that have rendered some facilities useless. In addition to the ports, human interactions adversely impacting the shoreline have included navigational channels and jetties, groin fields and seawalls, dam construction on rivers that reduces sand supply to the coast, sand mining of beaches and river beds that supply sand to these beaches, and hydrocarbon and groundwater extraction, inducing local ground subsidence and associated inundation and erosion.

During the past several decades, increasing emphasis has been placed on the coastal zone owing to the rapid development of this region and the hazardous effects and costs of short- and long-term natural processes. Episodic and cyclical events such as hurricanes along the East Coast of the United States and El Niño on the West, monsoons in the Bay of Bengal, and severe storms on the North Sea have caused loss of life and widespread damage, and governments and taxpayers have become concerned about the costs resulting from inappropriate construction practices. As a result, many countries that have expended immense amounts of money to protect shorelines are reexamining their policies. In the United States, most coastal states (including those along the Great Lakes) have initiated or are in the process of initiating controls on the types and locations of coastal structures. These restrictions may require that the structure be able to withstand a rare storm event, such as a 100-year hurricane, and that the structure be set back from the shoreline 30 or so times the annual shoreline recession rate. In addition to the impact of storm events, concern exists over the long-term effects of mean sea level rise and the possibility of an increase in the rate of rise in the coming decades caused by greenhouse gases.

Over the last 50 years, coastal engineering has become a profession in its own right with the objective of understanding coastal processes and developing strategies to cope effectively with shoreline erosion. With a more sophisticated and knowledgeable approach to coastal processes, coastal engineers can design effective coastal protection and mitigation schemes and avoid the mistakes of the past. Also, a greater knowledge of sediment transport mechanics at beaches may permit the development of novel means to mitigate erosion problems. With the population pressure on the shoreline and the threat of sea level rise and coastal storms, the need for coastal engineerings and research into coastal processes is certain to increase (NRC 1999).

The best understanding of coastal processes, including the nearshore flows and the resulting sediment transport, and the ability to transform this understanding into effective engineering measures require the following:

A blend of analytical capability,

An interest in the workings of nature,

The ability to interpret many complex and sometimes apparently conflicting pieces of evidence, and

Experience gained from studying a variety of shorelines and working with many coastal projects.

We say this in part because, at this time, the mathematical and statistical equations governing the behavior of the sand and water at the shoreline are not yet fully known, thus precluding our ability to make models for precise long-term predictions about the coastal zone. In this sense, the field of coastal processes is still as much an art as a science and requires a good intuitive grasp of the processes that occur in the coastal zone. In fact, the best computer and physical modelers are those who are skeptical about their results and continuously compare their models with well-documented case studies and field experiments. Further, much more research is necessary to improve our abilities to make predictions of coastal behavior, particularly in response to man-induced perturbations. However, despite the rudimentary state of our knowledge, the student of the shoreline will find the beauty and dynamic nature of the coast rewarding along with the numerous secrets that Mother Nature still guards so zealously.

In this vein, you will notice that many times we take two approaches to problems: one is the macroscale, which utilizes conservation laws or heuristic arguments that provide reasonable solutions, and the other is the microscale, which involves examining the detailed physics of the process. Presently, the macroscale approach is more useful to the coastal engineer, for the detailed physics of coastal processes are still being unraveled by coastal researchers; however, the day may not be too far off when the macroscale approach is replaced by, or merged, with the microscale one.

We have organized this book into four parts so that it will be coherent and useful to readers with different backgrounds and interests. Part One provides a general description of coastal processes with emphasis on long-term forces and responses. This section can be read with little or no background in mathematics and points out the ever-present forces tending to cause equilibrium both in profile and in planform. When natural or human-induced changes occur, a new set of forces is induced to reestablish equilibrium consistent with these changes. Familiarization with this material will assist in developing an overall comprehension of, and intuitive feel for, large-scale, long-term dynamics. Also reviewed are the coastal landforms found in the coastal zone and their causative forces. Many times the geometry of these forms contains information that can assist in characterizing the dominant waves, currents, and other important forces. Part Two develops the theories representing the forces in the nearshore region with a focus on longshore and cross-shore hydrodynamic processes. This part is relatively mathematical and may be of more value to the reader with a strong research background. For the less mathematically inclined reader, the equations may be largely ignored. Although Part Three, dealing with the response of the shoreline to the forcing, is strongly recommended for those concerned with the construction of coastal engineering works or remedial erosion measures, this material is not absolutely necessary as a prerequisite to Part Four, which is directed toward engineering applications. The emphasis in Part Four is twofold: (1) presentation of techniques to predict the impact that a project may have, and (2) discussion of various methods employed to mitigate beach erosion.

In writing this book, the challenge has been to assist students of the coasts to sharpen their abilities to interpret coastal phenomena, to predict the consequences of a given coastal project, and to incorporate this knowledge into the design process but at the same time to provide the level of scientific detail to satisfy a researcher in the field. We hope we have met the challenge. The measure of our success, of course, will be the degree to which the book can be applied to improving existing coastal projects and beneficially guiding decisions and design of future coastal construction.

1.2 SOME TERMINOLOGY OF THE COASTS

1.2.1 DESCRIPTIVE TERMS

It is a remarkable fact that beaches around the world are quite similar in composition and shape. The beach profile, which is a cross section of the beach taken perpendicular to the shoreline, is generally composed of four sections: the offshore, the nearshore, the beach, and the coast, as shown in Figure 1.1. The sand making up this profile

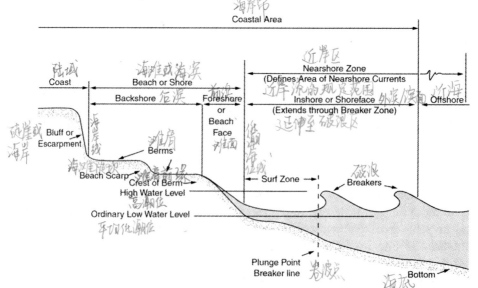

Figure 1.1 Beach profile terminology (adapted from the *Shore Protection Manual* 1984).

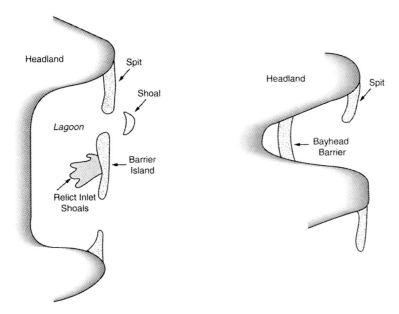

Figure 1.2 Shoreline planform terminology.

is shaped by waves coming from the offshore and breaking in the nearshore zone, where sandbars may exist. The *foreshore*, or *swash* zone, is the region of the profile that is alternately wet or dry as the waves rush up this steep portion of the profile. The dry beach may have one or more *berms*, which are horizontal sections of the profile, and *scarps*, which are near-vertical cuts caused by wave action during higher water levels perhaps associated with a storm. The landward portion of the beach may have sand dunes created by winds blowing sand off the beach into these features (aided by the sand-trapping capability of beach grass and other vegetation) or a bluff or a cliff (particularly on elevated eroding shorelines).

In planform (looking down on the coast as in an aerial photograph), the shoreline may have several interesting features. In Figure 1.2, the coast is fronted by a *barrier island* with tidal inlets transecting it at various locations. This situation occurs in numerous locations around the world, and a large concentration of barrier islands is found in North America. The inlets provide a means of water flow between the ocean and the lagoon system behind the islands. Often, old depositional features associated with relict (closed) inlets can be found, as shown in the middle of the barrier island. A *baymouth barrier* is a sandy feature that closes off a bay, whereas a *spit* is a depositional feature that grows from a headland or other prominent feature.

Tidal inlets play a major role in the sand budget of many shorelines, for these inlets affect the longshore transport of sand along the coast by capturing a significant portion of it and hence removing it from the active beach system. The size and shape of inlets are the result of a balance between the sand that is carried into them by the waves and the scouring ability of tidal flows that course through them daily. Some of their common features are shown in Figure 1.3. The two most important features of a tidal inlet are the ebb and flood tidal shoals, which may be very voluminous features that began when the inlet was created; tidal shoals can increase to tremendous sizes,

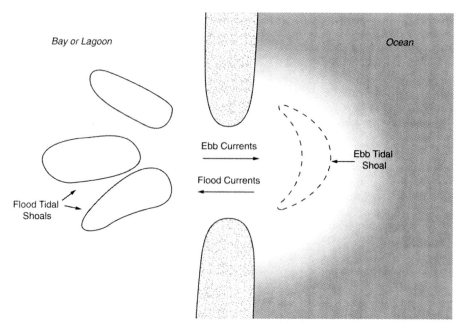

Figure 1.3   Tidal inlet terminology.

often containing amounts of sand equivalent to many years' worth of the annual gross transport of sand along the shoreline. These shoals are created by the ebb and flood inlet currents, which jet out sand transported into the inlet by waves. (Chapter 13 will discuss tidal inlets in detail.)

Estuaries differ from inlet-bay systems in that a river provides a considerable amount of fresh water to the bay system, leading to the formation of strong salinity differences often characterized by density fronts. The water chemistry of estuaries is more complicated than that of a more homogeneous inlet-bay system, and phenomena such as the flocculation of fine sediment occur. Estuaries are a subject unto themselves, and other texts can provide the reader with more information.

1.2.2 TRANSPORT PROCESSES

The beach profile and planform shapes discussed in the preceding section are a result of the action of waves and currents at the shoreline. The waves not only suspend the sediments but give rise to nearshore currents that carry the suspended sediment alongshore or cross-shore. As will be discussed in Chapter 5, a *longshore current* is driven by waves breaking obliquely to the shoreline and flows in a direction corresponding to the wave direction. Often, this current turns seaward and becomes a *rip* current, taking sediment (and hapless swimmers) offshore.

The sediment carried by the waves and currents is referred to as the *littoral drift*, and the amount of sediment moved along the coast is the *littoral transport*, or longshore sediment transport, which is usually measured in units such as cubic meters per year or cubic yards per year (see the appendix to this chapter for conversion among different units). As the wave environment changes during the year, the transport can change directions; however, at most coastlines there is a dominant direction of sediment transport. *Downdrift* refers to a direction coincident with this dominant transport direction, whereas *updrift* is the opposite direction.

The *cross-shore transport*, which is caused by wave- or wind-induced mean cross-shore flows, is largely responsible for the existence of sandbars and other beach profile changes. These changes can be slow, on the order of years in duration, or they can occur rapidly during storms with time scales on the order of hours.

1.3 EXAMPLES OF COASTAL ENGINEERING PROJECTS

An unfortunate legacy of construction at the shoreline is the large number of projects built without the requisite historical data for the site or appropriate knowledge about coastal processes. These deficiencies, in part, are due to the relatively recent development of our understanding of coastal behavior and the difficulty of obtaining wave and sediment transport data. Additionally, concern over the effect of a coastal project on adjacent beaches has only become important in the last several decades as more and more people use beaches for dwellings, recreation, and industry.

Here we will describe several examples to illustrate the broad range of problems in the field and, we hope, to whet your appetite for the opportunities and challenges ahead. Many of the problems presented will be addressed in detail in later chapters of this book. Some problems illustrate pitfalls that can be encountered, whereas others pose general or specific coastal engineering concerns.

A common thread in most of the problems is that there are often never enough data available to assess the problem accurately. Either the data (say for waves or for littoral transport) do not exist, or the length of the data record is too short to draw reliable conclusions. To counter this recurring problem, a major tenet of coastal engineering should be *to design flexibility wherever possible into every project to correct for unknown parameters and poorly estimated factors and to allow for fine-tuning of the project afterwards.* This necessary flexibility requires that the coastal engineer plan carefully and creatively to enable project performance to be improved, if necessary, depending on the subsequent interaction of the project with the environment.

1.3.1 BEACH NOURISHMENT

Beach nourishment, or beach fill, is the placement of large quantities of sand on an eroding beach to advance the shoreline seaward of its present location. Beach nourishment is one of the more common methods for erosion mitigation because this approach usually does not involve the construction of permanent structures. This erosion control technique is a way of setting the beach system back in time to when the shoreline was more advanced seaward. Central among the engineering questions to be addressed are: What is the additional beach width that results for a given volume of added sand? What is the lifetime of the project? What are the amounts of turbidity and biological disruption to be caused during placement of the fill material? What are the advantages of using coarser but possibly more expensive sand?

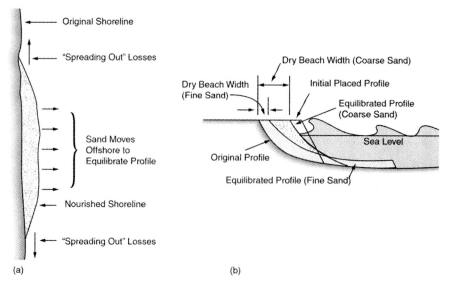

Figure 1.4 Beach nourishment showing plan view and profile (from Dean and Abramian 1993). (a) Plan view showing "spreading out" losses and sand moving offshore to equilibrate profile. (b) Elevation view showing original profile, initial placed profile, and adjusted profiles that would result from nourishment project with coarse and fine sands.

When a beach nourishment project is constructed, there are two known processes that diminish the expected additional beach width, each operating on a different time scale (see Figure 1.4). Equilibration of the on–offshore beach profile from the arbitrary shape created by placing sand on the beach to the natural equilibrium shape created by the environment occurs on a shorter time scale of months or years and includes a transfer of sand from the dry beach and the shallow constructed portions of the profile to the offshore to form an equilibrium profile. The second process is a result of the planform perturbation created by the fill that results in the sand spreading out in the alongshore directions. For reasonably long projects, this time scale is on the order of several years to decades.

The gradual transport of the beach nourishment sand away from the placement area results in a diminution of the beach width in the region of interest. Eventually the beach will erode back to its original position, for the beach fill has not removed the cause(s) of erosion but only provided new sand to be eroded. (Beach fill is the *only* erosion mitigation scheme that involves adding new sand to the coastline.) As the beach recedes to its original position, another beach fill (renourishment) may be required. In fact, for long-term protection of a beach using beach nourishment, a plan for periodic renourishment must be developed.

Generally, the material for beach nourishment projects is obtained by dredging from an offshore *borrow area*, although land sources are also used. The material obtained is frequently finer and more poorly sorted than that naturally present on the beach, which may reduce the effectiveness of the project. It will be shown in our studies of equilibrium beach profiles later in the book that, if sand finer than the native is used, a much greater volume of fill may be required to yield a desired beach width. Also, the longevity of the project will be shown to depend critically on the project length. Doubling the project length quadruples the project life. The wave height is an even more significant determinant. If two identical projects are constructed in areas in which the wave heights differ by a factor of two, the life of the project in the higher wave area will only be 19 percent of that in which the lower waves prevail. (Chapter 11 discusses this more fully.)

1.3.2 EFFECTS OF NAVIGATIONAL ENTRANCES – A GENERIC PROBLEM

Navigational entrances include natural inlets or waterways that have been modified to provide deeper and more stable channels as well as entrances that have been created for navigational purposes. Frequently, one or two jetties are constructed to reduce the infilling of the channel by littoral transport and to provide wave sheltering of transiting vessels.

Modified or constructed inlets have a very significant potential to interfere with natural sediment transport processes and thereby cause disruption to the adjacent beaches, as shown in Figure 1.5. The effects are usually greater where the net long-shore sediment transport is large.

Coastal engineers are often asked to assess and remedy the impact of navigational entrances. Quite simply, this requires reinstating the natural sediment transport processes around the entrance; however, our track record in accomplishing this objective has been less than impressive. In past decades this reinstatement of the transport was not accorded high priority, but with the increasing recognition of the effects of navigational entrances on neighboring beaches, the escalating values of shorefront property, and the increased storm damage potential resulting from the additional

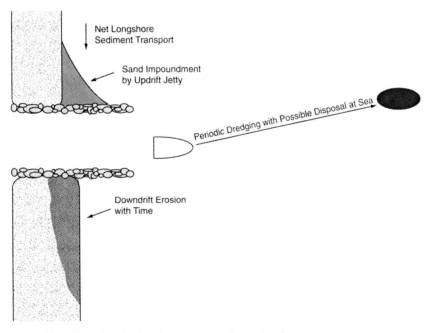

Figure 1.5   Effect of navigational entrance on adjacent beaches.

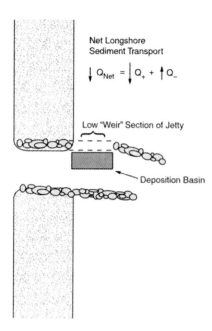

Figure 1.6 Characteristics of a weir jetty. Weir jetty acts as a sand flow control ideally allowing only the net transport to pass over.

erosion, new efforts are being undertaken to counter the adverse impacts of these entrances.

Reinstating the natural net longshore sediment transport is neither an inexpensive nor a straightforward task. One approach is to install a sand bypassing plant, located on the updrift side of the inlet, that pumps sand from there to the downdrift side. Another approach is to construct a low section into the updrift jetty across which sand flows into a designated *deposition basin*, where it remains until removed by a floating dredge for downdrift placement (see Figure 1.6). Innovative approaches, including jet pumps that involve no moving parts, have been used in the last decade or so. Often, however, these installations have not been completely effective as the result of various operational problems (one notable exception is Indian River Inlet, Delaware).

The erosional impact of coastal inlets on downdrift beaches is so widespread and of such magnitude that it will likely remain high on the coastal engineer's menu for many decades.

1.3.3 PONCE de LEON INLET, FLORIDA – A WEIR JETTY SYSTEM

Ponce de Leon Inlet is located on the northeast Florida coast immediately south of Daytona Beach. The inlet was modified in 1970 by the addition of a weir constructed in the updrift jetty and a deposition basin located immediately adjacent to the weir to accommodate the net longshore sediment transport that would pass over the top of the weir. Figure 1.6 presents the general layout of the jetty system and the weir section.

This was the fourth weir jetty to be constructed in the world, and serious early difficulties were encountered. The weir section experienced significant structural problems, the valuable updrift beach appeared to erode as a result of the new jetty system (possibly due to too much sand passing over the weir), and the inlet channel migrated into the deposition basin, removing any sand that reached it. Public pressure was so great that a decision was made first to place a low stone revetment adjacent to the weir to repair the structural problems. However, this did not reduce the excessive overpassing of sand, and eventually the entire weir section was filled in with stone.

1.3.4 PORT ORFORD, OREGON

The success of many coastal engineering projects is improved by an understanding of all the coastal processes that affect the project site. Also in many cases, a very useful consideration is that of equilibrium – both in planform and profile. Many times the design elements will perturb the natural system, which will then evolve toward a new equilibrium. Knowledge of the balance of forces affecting the original equilibrium and the manner in which the design will disturb those forces and the resulting new equilibrium can help avoid surprises and provide guidance toward the most effective design.

Port Orford is located in a small indentation on the Pacific Coast. Such features are known as *crenulate bays*, hooked bays, half-heart bays, or by a variety of other terms. Their equilibrium half-heart planforms are due to the existence of a beach between erosion-resistant updrift and downdrift features and a predominant angle of wave attack. In the case of Port Orford, sketched in Figure 1.7(a), the predominant wave direction is from the northwest. The updrift (north) rock ledge extends as a submerged feature to the south. A pier is located in the northern, generally sheltered, portion of the embayment.

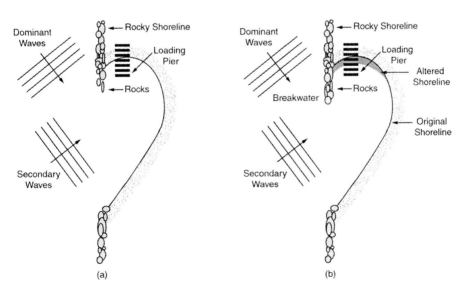

Figure 1.7 General characteristics at Port Orford, Oregon. (a) Original situation, and (b) modified situation (after Giles and Chatham 1974).

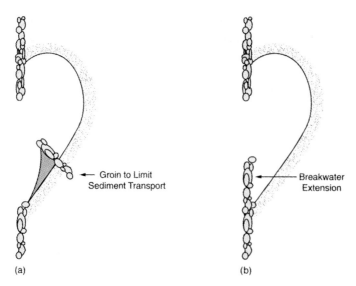

Figure 1.8 Two of the proposals for Port Orford, Oregon (after Giles and Chatham 1974).

Although the loading pier received substantial sheltering from the northwest waves, an undesirable level of wave agitation from more westerly and southwesterly waves led to a desire to increase the sheltering at the loading pier. In 1965, a decision was made to construct a protective structure on the rocky ledge (see Figure 1.7(b)).

At the risk of showing excellent hindsight, it should have been recognized in the design considerations that the embayment was in planform equilibrium with the updrift control provided by the rocky ledge. Extending this feature had the effect of setting into play forces that advanced the shoreline southward, thereby causing shoaling at the pier. This sand was drawn predominantly from the south. The pier could have been extended, but this would have placed the loading vessels in the same relative position with respect to the new control as the original and hence would have been without much overall benefit. The chain of events described in the preceding paragraph did occur and could have been anticipated. Is there a design that could have avoided these problems? In a restudy, the Waterways Experiment Station (Giles and Chatham 1974) recommended various options with differing lengths of sand control structures to restrict sand transport from the south, as shown in Figure 1.8.

This example illustrates pitfalls that can be avoided by understanding the forces that are in balance in the natural system and the consequences of interacting with this force balance.

1.3.5 EFFECTS OF SHALLOW-WATER PROFILE DEEPENING

The amount of sediment transport as a function of water depth is a consideration in many coastal engineering problems. The example discussed in the next paragraph presents one design option that was considered for loading liquefied natural gas (LNG) tankers on the north coast of Sumatra and illustrates the trade-off between

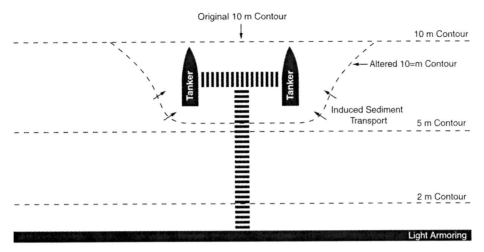

Figure 1.9   Induced sediment transport as a result of profile deepening to accommodate a terminal.

initial capital cost and subsequent maintenance costs as well as the need for knowledge of sediment transport mechanics.

This design option entailed a pier extending from shore with a T head along which the ships would berth for loading of the LNG product. Because the natural bottom slope in this area is reasonably mild and the tankers required 10-m depth, a pier extended to that depth contour would be long and costly. An alternative, presented in Figure 1.9, was to dredge a deepened area into shallower water. The latter option would clearly incur a smaller capital cost; however, the maintenance dredging could be greater and involve costly delays following rapid depositional events such as might be caused by storms. It is clear that, without maintenance dredging, the deepened area will, over time, tend to reestablish the original profile owing to the deposition of sediment. However, the time scale at which this occurs is poorly known and extremely relevant.

Given the lack of reliable calculation procedures, field measurements are frequently employed and can be quite effective in predicting the behavior of a design. Techniques include monitoring and interpreting the deposition in test pits, installing suspended sediment traps, or constructing temporary groins to impound the littoral transport. Field techniques are treated more fully in Chapter 6.

1.3.6 STORM IMPACT ZONES

In recognition of the vulnerability of the coastal area and the consequences of damage and loss of life due to a major storm, many coastal states have developed, or are developing, programs governing coastal construction. Generally these programs establish setback lines relative to the active beach that prohibit construction seaward of these lines, regulate the elevations and types of construction landward of this line, or both. There is a precedent through the Flood Insurance Program of the U.S. Federal Government to relate these requirements to the effects of the predicted 100-year storm characteristics of that area. Thus, a need exists to be able to predict the erosion associated with such events. The problem can be posed as follows: given

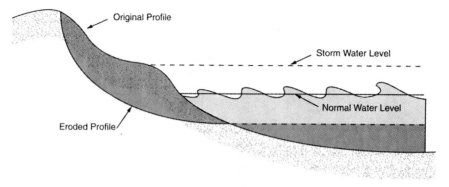

Figure 1.10 Profile erosion due to storm tides and waves.

the time-varying storm tide and waves associated with a 100-year storm, determine the time-varying beach profile; Figure 1.10 presents an example. To address this problem, sediment transport and continuity equations are necessary, and, at present, are solved through numerical modeling (Chapter 10). This problem is complicated further in cases in which the storm surge elevation exceeds the land elevation and *overwash*, a landward transport of water and sand, occurs. We will come back to this topic of shoreline management in Chapter 14.

1.3.7 OFFSHORE BREAKWATERS

An offshore breakwater is generally a shore-parallel structure composed of rocks and is designed to reduce the wave energy shoreward of the structure. These breakwaters can be designed as emergent or submerged (see Figure 1.11 for an emergent break-water). In Japan, in excess of 3000 such structures have been constructed, whereas less than 100 have been built in the United States. In addition to the wave sheltering of the region shoreward of the breakwater, wave diffraction occurs at the breakwater tips, and thus the waves are turned inward toward the centerline of the breakwater. This combination of wave sheltering and diffraction causes sand to deposit behind the breakwater. Breakwaters constructed in areas with a strong net longshore sediment transport can trap sand from this system and cause downdrift erosion. The degree of

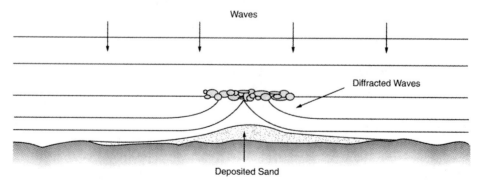

Figure 1.11 Sand deposited as a result of offshore breakwater.

interaction of the breakwater with the natural system depends, as you might expect, on the breakwater crest elevation, alongshore length, and separation distance from shore. Some structures have been proposed to hold sand placed in conjunction with a beach nourishment project, letting the ambient sand transport pass without interference. Whether this can be accomplished is still under debate by coastal engineers and geologists.

1.3.8 EFFECTS OF GROINS – IMPERIAL BEACH, CALIFORNIA

Groins are shore-perpendicular structures designed to trap sand from the littoral system or to hold sand placed in conjunction with a beach nourishment project. They are installed either as a single groin or multiply in a *groin field*. As might be expected, groins generally function best in areas in which there is a strong longshore sediment transport (see Figure 1.12).

Two groins were constructed between 1959 and 1963 to stabilize an eroding shoreline at Imperial Beach, California. Subsequently, the beach eroded as fast or faster than the natural beach had before groin construction, which sometimes happens with the improper use of groins. Observations showed that circulation cells formed within the groin compartments and caused seaward-directed currents adjacent to the groins. The Waterways Experiment Station (Curran and Chatham 1977) conducted a model study of the site that recommended several options, including the construction of offshore breakwaters and sills that would limit the wave energy and the offshore transport.

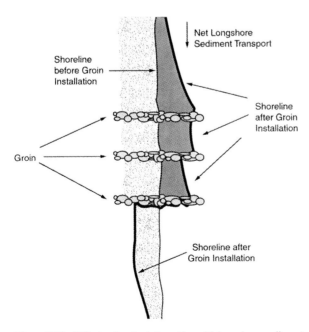

Figure 1.12   Effects of groins interacting with longshore sediment transport.

In retrospect, in addition to the occurrence of the circulation cells, the stability and longevity of sand within the groin compartment was reduced by the near-normal wave incidence at the site and the finer-than-native sand that had been placed on the beach. This fine sand would equilibrate to a beach profile with a much smaller dry sand beach width, and some of this equilibration process may have been interpreted as sand lost from the system.

1.3.9 RECREATIONAL BEACHES

In areas with favorable climates, the coastal zone is often developed for recreational purposes. In particular, in the Caribbean and along the Mediterranean coastline of Spain recreational beaches are being constructed at a rapid pace. Many of these places have high-quality natural beaches, whereas at others, rocky or marshy shorelines may exist. Also, beaches may be too steep (because of coarse sand) or too energetic for use by the beach-going public.

Constructed recreational beaches usually require retention structures to prevent the movement of sand along the beach. If only shore-perpendicular structures are placed, there is the possibility that changes in wave direction will cause fill sand to be lost around the tips of the structures. Also, if the fill sand is finer than the native, it will move offshore and again be lost around the structure tips. In these cases, T groins, sometimes with sills connecting the ends, can be used (see Figure 1.13). This design also has the advantage of providing a control on the amount of wave energy at the created beach, thus reducing the wave hazard to swimmers.

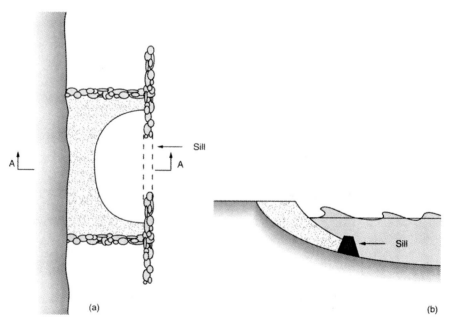

Figure 1.13 T groins for coastal protection and beach construction, including an offshore sill. (a) Planform; (b) Profile through Section A–A.

1.3.10 COASTAL ARMORING

At eroding shorelines, one sure means of preventing land loss is by coastal armoring, usually with seawalls or revetments. The effects of armoring on the adjacent beaches and the offshore profile are not well understood, and the concern over future sea level rise and the consequent shoreline erosion has mobilized groups favoring letting natural processes occur in general and opposing armoring in particular. Claims have been made that armoring causes the offshore profile to steepen and the adjacent beaches to erode. Many times a coastal engineer must attempt, in a diagnostic mode, to unravel shoreline changes to separate the natural behavior from the impact of armoring or other human interventions. In cases where armoring would be acted upon by the waves daily and the littoral transport rates are low, armoring definitely impacts the adjacent (predominantly downdrift) beaches. However, in cases where the armoring would be required to provide protection only rarely (say, if it were located within or behind the dune system), it may be regarded as an insurance policy – out of sight until needed but there to provide protection during a major storm.

The crux of the debate about armoring is the value of the upland and the potential for loss of life during storms. If the cost of armoring, including consideration of the impacts on adjacent beaches, is lower than the benefits to be accrued (both in the short and long term), then armoring or other shoreline protection schemes are appropriate.

REFERENCES

Curran, C.R., and C.E. Chatham, "Imperial Beach, California; Design of Structures for Beach Erosion Control," U.S. Army Corps of Engineers, Waterways Experiment Station, Rpt. H-77-15, Vicksburg, MS, 1977.

Dean, R.G., and J. Abramian, "Rational Techniques for Evaluating the Potential of Sands for Beach Nourishment," U.S. Army Corps of Engineers, Waterways Experiment Station, Rpt. DRP-93-2, Vicksburg, MS, 1993.

Giles, M.L., and C.E. Chatham, "Remedial Plans for Prevention of Harbor Shoaling, Port Orford, Oregon," U.S. Army Corps of Engineers, Waterways Experiment Station, Tech. Rpt. H-74-4, Vicksburg, MS, 1974.

Insurance Institute for Property Loss Reduction (IIPLR), "Coastal Exposure and Community Protection: Hurricane Andrew's Legacy," Boston, 1995.

National Research Council (NRC), *Meeting Research and Educational Needs in Coastal Engineering*, Washington, DC, National Academy Press, 1999.

U.S. Army Corps of Engineers, *Shore Protection Manual*, Coastal Engineering. Research Center, Washington, DC: U.S. Government Printing Office, 3 Vols., 1984.

EXERCISES

1.1 On the basis of your experience, list and discuss methods of coastal erosion control for a section of eroding sandy beach. Can you think of something that has not been tried before?

1.2   What are the various kinds of coastlines that you are aware of beyond sandy beaches?

1.3   Discuss in detail any actual coastal erosion mitigation scheme that you are familiar with. What is its purpose? How well has it accomplished the goal?

APPENDIX: USEFUL UNITS

The coastal engineer will encounter both metric and so-called English customary units, that is, foot-pound-second system. Because both systems of units are used in practice, this book will use both freely. If an original figure to be used in this book is in English customary units, it will be left as such. However, if there is a choice between the two systems, the metric system will be chosen. A few useful conversions are presented in Table 1.1. An example, to convert 15 cubic meters to the equivalent amount in cubic yards: $15(1.307) = 19.605$ cu yd.

Table 1.1   Some Useful Conversions between the Metric and English Systems

Conversion	Conversion Factor (Multiply by)
Feet to meters	0.305
Meters to feet	3.28
Cubic yards to cubic meters	0.765
Cubic meters to cubic yards	1.307
Cubic yards per foot to cubic meters per meter	2.51
Cubic meters per meter to cubic yards per foot	0.398
Pounds to newtons	4.45
Newtons to pounds	0.225

CHAPTER TWO

Sediment Characteristics

On the southeastern shore of the Chesapeake Bay, parts of the shoreline are receding at very high rates of up to 10 m per year! Many of the shorelines in this region consist of narrow sand beaches backed by small bluffs several meters in height. The bluff, however, comprises considerably finer material than the beach and does not contain much sand-sized or coarser material. At Taylor Island, Maryland, for example, the mean diameter of the beach sand is greater than 0.5 mm, the bluff material is 99 percent silt, and the sand fraction has a mean diameter of 0.12 mm. Dalrymple et al. (1986) hypothesized that the high erosion rate there occurs because the erosion of the bluff yields only fine materials, which are easily transported away by the waves and tides, and contributes very little sand to the beach. Had the upland been composed of coarser material, bluff erosion would have produced more sand, resulting in a wider beach that would have slowed the bluff erosion.

2.1 INTRODUCTION

A beach can be composed of a wide variety of materials of many sizes and shapes. Remarkably, however, there is a very narrow range over which most beach materials vary, both in composition and size. Along the U.S. East Coast, for example, despite the variation in exposure to waves and the composition of the upland, the mean grain size of most of the beaches varies by a factor of only five (U.S. Army *Shore Protection Manual* 1973). Most beaches there are made of silica sand, although a significant shell fraction is found in Florida. Why are all of these beaches similar in characteristics? In fact, why are beaches around the world so similar? Why is sand basically the same size everywhere? Perhaps another question is more germane: Why are there beaches at all?

This chapter will not answer these questions because it only addresses the characteristics of sand. However, by the time you finish the remaining chapters, these questions should have been answered.

2.2 SAND COMPOSITION

Most sand is *terrigenous*, that is, a byproduct of the weathering of rocks; therefore, its composition reflects the nature of its origin. In most locations, the erosion of granitic mountains and the subsequent transport of the erosion products to the coastline by rivers have led to a very significant fraction (around 70 percent) of the beach sand being composed of quartz, and about 20 percent consisting of feldspar. These materials are very hard and resist the abrasion encountered on the trip from the mountains to the beaches, whereas most of the other less-resistant minerals have been totally abraded away. Eroding coastal headlands and the onshore transport of offshore sediments are other sources of these materials.

Other minerals exist in sand as well: hornblende, garnet, magnetite, ilmenite, and tourmaline, to name a few. They can often be seen concentrated as black layers that sometimes exist on or within the beach.* These minerals are referred to as heavy minerals because they are more dense than quartz. To quantify the difference, the *specific gravity* is used, which is the density of the mineral divided by the density of fresh water. For the heavy minerals, the specific gravity is often much greater than 2.87, whereas the specific gravity of quartz is 2.65. (To obtain the true weight of these minerals, it is necessary to multiply the specific gravity by the weight of water per unit volume; thus, for example, the weight of a solid cubic meter of granite would be 2.65×9810 N, or about 26,000 N.)

In locations where the local rocks consist of materials other than quartz, the sand composition can be radically different than described in the previous section. At Point Reyes, California, the sand has a significant fraction of jade as manifested by its characteristic green color. In Hawaii, another green sand occurs, but the contributing mineral is olivine, which is another semiprecious stone created by cooling lava. The Hawaiian islands can also boast of black sand beaches composed of basaltic sands resulting from the erosion of volcanic rock or, in some cases, the direct flow of molten lava into the ocean.

In the Bahamas, an interesting sand called aragonite is created when calcium carbonate precipitates out of cold waters flowing over the warm Bahamian banks. These *oolitic* sands consist of well-rounded grains usually containing a biological "seed" in the center that initiated the precipitation process. Because there are no sources for granitic sands on the Bahamian islands, this soft oolitic sand is not abraded rapidly and forms beaches. Sands created by chemical processes are referred to as *authigenic* sands.

In tropical regions, the production of sand by biological activity can overwhelm the production of sand by weathering or precipitation. *Biogenous* sands can be a result of the abrasion of seashells or the destruction of coral reefs. The south Florida shoreline, far from the mountains of Georgia, the nearest terrigenous source of sand, has a very high shell content ranging from less than 5 to 40 percent.

2.3 GRAIN SIZES

Sand has many different sizes, which can be readily seen by examining a handful. Figure 2.1 shows a photograph of a typical sand sample with its variety of sizes and shapes. To quantify the sizes of the sand within the sample, we resort to statistical

---

* Komar (1989) reviews the various physical mechanisms that lead to the concentration of these minerals.

Figure 2.1 Photograph of beach sand.

measures. The most obvious is a mean (or average) diameter of the sand, which is usually measured in millimeters. The typical sizes of sand on U.S. beaches are between 0.15 and 2 mm (mean diameter).

Not all the sediment composing a beach is sand. Geologists have developed size classifications to determine what qualifies as sand, what is gravel, and so on. One of the more popular classifications is the Wentworth scale, which classifies sediment by size (in millimeters) based on powers of two, as shown in Table 2.1. On the Wentworth scale, granular particle between 0.0625 and 2 mm is considered to be sand. Finer materials are primarily silts and clays, whereas larger sediments can be pebbles and cobbles. In numerous locations, cobbles are the primary sizes forming the beach as, for example, along sections of Chesil Beach, England.

Because the Wentworth sand size classification depends on powers of two, Krumbein (1936) introduced the *phi scale* as an alternative measure of size. The phi (ϕ) size is related to the grain size by

$$\phi = -\log_2 d, \tag{2.1}$$

such that $2^{-\phi} = d$, where d is measured in millimeters. (An equivalent mathematical form, using natural logarithms, is $\phi = -\ln d / \ln 2 = -\log_{10} d / \log_{10} 2$.) The use of the phi scale is widespread, particularly in the coastal geology literature, because it leads to a convenient display of sand size distributions, as discussed in the remainder of this section. A disadvantage of the phi scale is that larger phi values correspond to smaller sand sizes brought about by the introduction of the minus sign in Eq. (2.1). As examples, a phi size of 3.5 indicates a very fine sand (0.088 mm), a phi of 1.0 means a medium–coarse sand (0.5 mm), and a sand with $\phi = -5$ is a large pebble (32 mm).*

To determine the range of sizes present in a sand sample, a size analysis must be done. *Sieving* the sand is the most common means of finding the range of sizes in

---

* Krumbein put in the minus sign because most sands are smaller than 2 mm, and, without it, all sand sizes would be carrying around a minus sign as excess baggage.

Table 2.1   Wentworth Scale of Sediment Size Classification

Wentworth Scale Size Description		Phi Units (ϕ)	Grain Diameter d(mm)	U.S. Standard Sieve Size	Unified Soil Classification (USC)	
Boulder		−8	256		Cobble	
Cobble			76.2	3 in		
		−6	64.0		Coarse	Gravel
			19.0	3/4 in		Gravel
Pebble		−2.25	4.76	No. 4	Fine	
		−2	4.0		Coarse	
Granular		−1	2.0	No. 10		
Sand	Very coarse	0	1.0	2.0		
	Coarse	1	0.5		Medium	Sand
	Medium	1.25	0.42	No. 40		Sand
		2	0.25			
	Fine	2.32	0.20	No. 100		
		3	0.125	No. 140	Fine	
	Very Fine	3.76	0.074	No. 200		
		4	0.0625			
Silt		8	0.00391		Silt or Clay	
Clay		12	0.00024			
Colloid						

the sample. Usually, sieves, which are pans with a wire screen of a standard mesh size serving as the bottom, are used; see Table 2.1 for the sieve classifications. The sieves are arranged in a stack such that the coarser sieves are at the top and the finer at the bottom. The sample is placed in the top sieve, and the sieves are shaken until the sand has fallen as far as possible through the stack; different sized fractions are trapped by the sieves of varying sizes. Machines, such as a Roto-tap, are used for this step. The weight of sand caught by each sieve is then determined, and the percentage of the total weight of the sample passing through the sieve is determined.

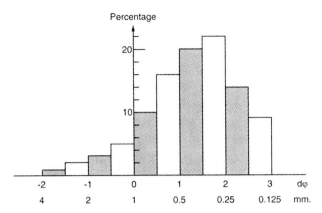

Figure 2.2 Example of sand size histogram.

Sediment size data are usually presented in several forms. One can plot a histogram of sediment sizes, such as presented in Figure 2.2, where the ordinate represents the sample percentage by weight between the two sieve sizes bracketing that value. This provides an experimental representation of the distribution of sand sizes present. A second way, used most often, is the cumulative size distribution, illustrated in Figure 2.3, which shows "percent coarser," and thus the value at a particular diameter represents the total sample percentage by weight that is coarser than that diameter. By custom, this plot is usually presented with the logarithm of the diameter on the abscissa (decreasing to the right).

In many cases, the distribution of sand has been shown almost to obey a lognormal probability law; thus, if normal probability paper is used for the cumulative percentage passing and the phi scale is used for the sand size, a straight line will result (e.g., Otto 1939). The log-normal probability density function is given by $f(\phi)$,

$$f(\phi) = \frac{1}{\sigma_\phi \sqrt{2\pi}} e^{-\frac{(\phi - \mu_\phi)^2}{2\sigma_\phi^2}} \qquad (2.2)$$

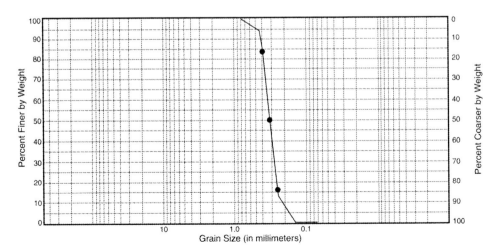

Figure 2.3 Example of cumulative sand size distribution.

where μ_ϕ is the mean grain size in phi units, and σ_ϕ is the standard deviation in size. The probability that a sand size is coarser than a given size ϕ_g is

$$P[\phi < \phi_g] = \int_{-\infty}^{\phi_g} f(\phi)\, d\phi \tag{2.3}$$

It can be shown from this formula (Problem 2.2) that the probability that a grain size is greater than the mean grain size is one-half, whereas the probability that a grain size is greater than zero size ($\phi \to \infty$) is, of course, one.

For distributions that are not log-normal, the Gram–Charlier distribution, which includes skewness and kurtosis corrections, has been proposed (Inman 1952).

2.3.1 STATISTICS OF THE SAND SIZES

The sand size distribution contains a considerable amount of information concerning the sand sample; however, this distribution is primarily used to obtain numerical measures of the sample that can convey nearly the same amount of information, or at least provide sufficient information, for most problems.

One common measure of sand sample is d_{50} (or ϕ_{50}), which is the *median size*. This sand size can be obtained directly from the cumulative distribution curve, for it is the size for which half the sample by weight is coarser and half is smaller. According to normal probability theory, 68 percent of all sizes will lie within plus or minus one standard deviation of the mean. Therefore, the phi sizes ϕ_{84} and ϕ_{16}, corresponding to $\phi_{(50\,\pm\,68/2)}$, should play a major role in describing a sediment. Otto (1939) and Inman (1952) proposed that *mean* diameter be defined as

$$M_{d\phi} = \frac{(\phi_{84} + \phi_{16})}{2} \tag{2.4}$$

Other measures of the mean diameter have been proposed. Folk and Ward (1957), who examined sand samples dominated by both large and small sizes, proposed the following measure for bimodal distributions:

$$M_{d\phi} = \frac{(\phi_{84} + \phi_{50} + \phi_{16})}{3} \tag{2.5}$$

The differences between the two definitions are small for distributions approaching a log-normal distribution. For a sand with a symmetrical size distribution, the mean and the median sizes are the same.

The *sorting* of a sand sample refers to the range of sizes present. A perfectly sorted sample would contain sand of all the same diameter, whereas a poorly sorted sand contains a wide range of sizes. A numerical measure of the sorting is the standard deviation, σ_ϕ, which is defined as

$$\sigma_\phi = \frac{(\phi_{84} - \phi_{16})}{2} \tag{2.6}$$

A well-sorted sample is a poorly *graded* sample, whereas distributions with a wide range of sizes are termed well graded or poorly sorted. A perfectly sorted sand (homogeneous in size) would have the same values for ϕ_{84} and ϕ_{16}; therefore, $\sigma_\phi = 0$.

For realistic sand size distributions on a beach, a $\sigma_\phi \leq 0.5$ is considered well sorted, whereas a sample with $\sigma_\phi \geq 1$ is considered poorly sorted.

Another measure of the distribution is the *skewness*, which occurs when the sediment size distribution is not symmetrical. The skewness is

$$\alpha_\phi = \frac{M_{d\phi} - \phi_{50}}{\sigma_\phi} \qquad (2.7)$$

with $M_{d\phi}$ based on Eq. (2.4). A negative skewness indicates that the distribution is skewed to smaller phi sizes (larger grain sizes). Duane (1964) showed that negative skewness is an indicator of an erosive environment, for the finer materials have been winnowed out by the action of currents or waves. On the other hand, a depositional environment would likely have a positive skewness.

A final measure determines the peakedness of the size distribution – the *kurtosis*. Inman (1952) proposed

$$\beta_\phi = \frac{(\phi_{16} - \phi_5) + (\phi_{95} - \phi_{84})}{2\sigma_\phi} \qquad (2.8)$$

For a normal distribution, $\beta_\phi = 0.65$. If the distribution is spread out wider than the normal distribution (that is, a larger range of sizes), the kurtosis will be less than 0.65.

EXAMPLE

Figure 2.3 shows an actual sand sample from the west coast of Florida. Let us examine the statistical measures of the sample. (Note that the grain size decreases from left to right along the abscissa.)

Sediment Parameters from Figure 2.3

% coarser	d (mm)	ϕ
16	0.40	1.32
50	0.32	1.64
84	0.27	1.89

The median grain size d_{50} from the table is 0.32 mm or 1.64ϕ. The mean grain size is

$$M_{d\phi} = \frac{(\phi_{84} + \phi_{16})}{2} = \frac{1.89 + 1.32}{2} = 1.61\phi. \qquad (2.9)$$

The Folk and Ward expression gives 1.62 ϕ. The sorting is $\sigma_\phi = (\phi_{84} - \phi_{16})/2 = 0.285\phi$, which indicates the sample is quite well sorted, and the skewness is $\alpha_\phi = (M_{d\phi} - \phi_{50})/\sigma_\phi = -0.081$, which means larger grain sizes are slightly more prevalent.

2.3.2 SPATIAL AND TEMPORAL VARIATIONS IN SAND SIZE

The statistical properties of the sand vary across the beach profile, vertically into the beach, and also along the beach. In Figure 2.4, as we move in the cross-shore direction from offshore (right) to onshore (left) on a variety of beaches, we see that the offshore sands are often finer than the sand in the nearshore region, which is more

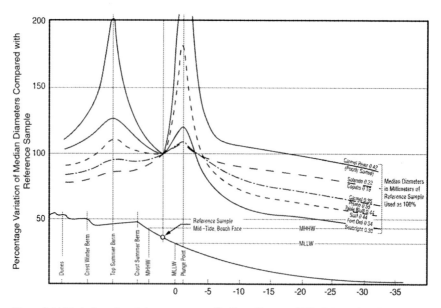

Figure 2.4   Variation of sand sizes across a profile (from Bascom 1951. © American Geophysical Union).

dynamic owing to the influence of the shoaling and breaking waves. At the location of the breaker line, denoted Plunge Point in the figure, where the turbulence levels are the highest, the grain size reaches a maximum. Across the surf zone, the sand is smaller until the swash region (where the waves rush up and down across the beach face), where again the size increases. The size variation across the dry beach can vary due to the action of winds, which can winnow out the finer sand fractions, and the infrequent occurrence of storm waves. The tops of unvegetated sand dunes often have a larger size, again caused by wind winnowing, which results in "armoring" the tops of the dunes with coarser sand (called a lag deposit). The sorting of the sand varies along with the mean diameter as the sand often gets well-sorted in regions of high turbulence.

Variations along the beach can be due to a variety of nonuniform processes. Beach cusping, which is the scalloping of the beach face by the waves, as discussed in Chapter 9, is generally accompanied by the collection of coarse materials at the horns of the cusps.* Variations in wave energy attacking the beach can mean a variation in sand size and beach face slope. Bascom (1951) showed that there is a decrease in sand diameter with decrease in exposure to waves and a decrease in beach face slope with decreasing sand size (Figure 2.5).

Temporal variations occur on many different time scales from days to seasons to years. Annual variations can be seen readily, particularly on some New England and California beaches, which are covered with sand in the summer, but, when the wave climate becomes more severe in the winter, the sand is stripped from the beaches, leaving behind a cobble beach. The sand spends the winter offshore in sand bars and

---

* Useful to know on the jade-rich Pt. Reyes, California, beaches.

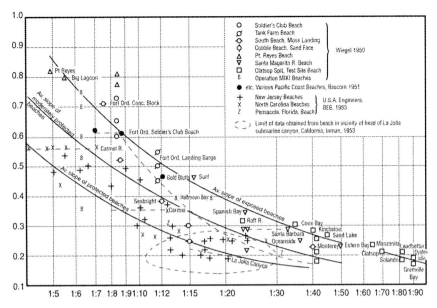

Figure 2.5 Grain size and beach face slope (from Weigel 1965).

is returned to the foreshore by milder wave action the following summer, once again covering the cobble.

2.4 SHAPE

The shape of sand grains affects their behavior in the marine environment. A flat grain would certainly settle in water differently than a spherical grain. Also, the shape is a measure of age. More rounded sand grains are likely to have undergone the grinding action of waves for a longer time compared with more angular grains.

Several measures have been proposed to describe the shapes of grains. A popular one is the Zingg (1935) classification based on the measures of the three major axes of a grain (denoted as a, b, and c) in order of decreasing size. If all three of these axes are the same, the grain is spherical. Other shapes occur, depending on the ratios of b/a and c/a, as shown in Figure 2.6.

Another measure is the Corey shape factor C_0 (Corey 1949), which is defined as $C_0 = c/\sqrt{ab}$. The maximum of the Corey shape factor is unity, corresponding to a sphere, and the minimum is zero, a disk. Typical values are on the order of 0.7.

Measures of sand shape, however, are not used in coastal engineering computations at this time, for the uncertainty in predicting the transport of sand is so great as to preclude the use of more sophisticated parameters in transport theories.

2.5 POROSITY

The shape of the sand grains affects the packing of sand in a beach. Between each sand grain and its neighbors are spaces that allow water to percolate through the sand and that permit organisms to live within it.

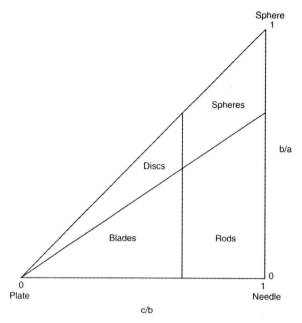

Figure 2.6 Zingg shape classification.

The porosity p is defined as the volume of voids per unit volume of sample. For sand, p is usually between 0.3 and 0.4. It may vary along a beach owing to changes in sand sorting or sizes. Freshly deposited sand has a high value of porosity, which is subsequently reduced by waves and currents, rearranging the grains into a more stable arrangement. Typically, the upper foreshore is very porous, and the limit of wave uprush is an area of high porosity. This region allows the wave uprush to sink into the beach and flow back offshore through the sand. This loss of water volume means that the wave backwash is less capable of carrying sand offshore, thus providing a mechanism for accretion of sand on the beach face.

To determine the actual weight of a given volume of sand, the porosity is used. For example, the weight of V cubic meters of sand with density ρ_s is

$$W = \rho_s g \, (1 - p) V$$

The porosity term is necessary to remove the void volume.

2.6 FALL VELOCITY

An important hydrodynamic characteristic of a sand particle is its fall velocity w, which is the maximum speed attained by a falling particle under the action of gravity (in other words, the terminal velocity). The sizes of suspended sand in the water column, for example, will depend on the fall velocity, for a large sand grain, which falls rapidly, will be less likely to be suspended in the water column when compared with a finer sand grain.

The fall velocity can be calculated theoretically in still water from a force balance on a single falling grain, where the relevant forces are the weight of the grain W, the buoyancy force F_B provided by the water, and the fluid drag F_D that the falling particle experiences. The force balance for the particle (with positive forces acting downwards) is

$$W - F_B - F_D = \rho_s g \left(\frac{\pi d^3}{6}\right) - \rho g \left(\frac{\pi d^3}{6}\right) - \rho C_D \left(\frac{\pi d^2}{8}\right) w^2 = 0, \qquad (2.10)$$

where ρ_s and ρ are the densities of sand and water, $(\pi d^3/6)$ is the volume of the assumed spherical grain of diameter d, and C_D is the drag coefficient of the falling grain, known to be a function of the Reynolds number, which is a dimensionless parameter defined in this context as $\mathbf{Re} = wd/v$. The kinematic viscosity of the water v is related to the dynamic viscosity μ by the density, that is, $v = \mu/\rho$. Most textbooks on fluid mechanics have tables and charts of the viscosity of water as a function of temperature. As a representative value, $v = 1.0 \times 10^{-6}$ m^2/s for fresh water at 20°C. Because $\rho = 1000$ kg/m^3 for fresh water and 1035 kg/m^3 for salt water, $\mu \simeq 1.0 \times 10^{-3}$ N s/m^2.

Solving the force balance in Eq. (2.10) leads to an expression for the fall velocity:

$$w = \sqrt{\frac{4(\rho_s - \rho)gd}{3\rho C_D}}. \qquad (2.11)$$

The drag coefficient for spheres has been obtained analytically for very small Reynolds numbers. For Reynolds numbers less than unity, Stokes (1851) obtained $C_D = 24/\mathbf{Re}$. Substituting his result into the fall velocity equation (2.11), we obtain

$$w = \frac{(\rho_s - \rho)gd^2}{18\mu}, \qquad (2.12)$$

which is referred to as Stokes law. From this relationship, which is restricted to slowly falling grains, we see that the fall velocity increases with the density of the sand grain and its size. Increasing the water temperature decreases the viscosity, and thus the sand falls faster in warmer water.

The Oseen correction, for slightly larger Reynolds numbers, and including inertial effects, gives

$$C_D = \frac{24}{\mathbf{Re}} \left(1 + \frac{3\mathbf{Re}}{16}\right).$$

For larger Reynolds numbers, but less than 100, Olson (1961) gives the approximation

$$C_D = \frac{24}{\mathbf{Re}} \left(1 + \frac{3\mathbf{Re}}{16}\right)^{1/2}. \qquad (2.13)$$

As an alternative to this theoretical approach, which is restricted to small Reynolds numbers anyway, numerous charts of fall velocity versus grain size are available based on both theoretical and empirical data. Rouse (1937; see also Vanoni 1975) shows in Figure 2.7 the fall velocity of spherical grains in both air and water as a function of size and temperature.

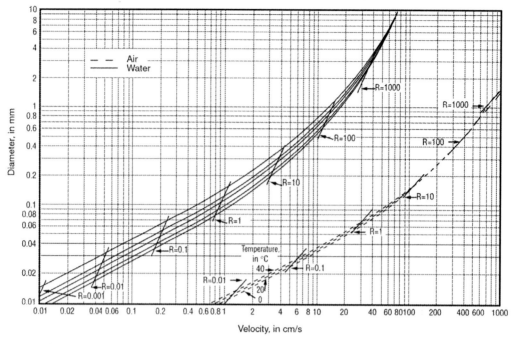

Figure 2.7 Fall velocity of spherical grains as a function of diameter and water temperature (reprinted, by permission, from Rouse 1937).

In the surf zone, the water is not still; there are high levels of turbulence, wave-induced velocities and accelerations, and mean currents. In addition, millions of sand grains are suspended by the wave action. Therefore, the results above, which apply for a single grain in still water, serve as guidance; in nature, the fall velocity will be dictated by the additional effects of fluid motions, sediment concentration, and sediment characteristics, including shape. Nielsen (1992) provides an overview of these effects in his Chapter 4.

REFERENCES

Bascom, W., "Relationship Between Sand Size and Beach Face Slope," *Trans. Am. Geophys. Union*, 32, 866–874, 1951.

Corey, A.T., "Influence of Shape in the Fall Velocity of Sand Grains," M.S. Thesis, A&M College, Colorado, 1949.

Dalrymple, R.A., R.B. Biggs, R.G. Dean, and H. Wang, "Bluff Recession Rates in Chesapeake Bay, *J. Waterway, Port, Coastal, and Ocean Eng.*, ASCE, 112, 1, 164–168, 1986.

Duane, D.B., "Significance of Skewness in Recent Sediments, West Pamlico Sound, North Carolina," *J. Sed. Petrology*, 34, 4, 864–874, 1964.

Folk, R.L., and W.C. Ward, "Brazos River Bar, a Study of the Significance of Grain Size Parameters," *J. Sed. Petrology*, 27, 3–27, 1957.

Inman, D.L., "Measures for Describing the Size Distribution of Sediments," *J. Sed. Petrology*, 22, 3, 125–145, 1952.

Komar, P.D., "Physical Processes of Waves and Currents and the Formation of Marine Placers," *Rev. Aquatic Sci.*, 1, 3, 393–423, 1989.

Krumbein, W.C., "Applications of Logarithmic Moments to Size Frequency Distribution of Sediments," *J. Sed. Petrology*, 6, 1, 35–47, 1936.

Nielsen, P., *Coastal Bottom Boundary Layers and Sediment Transport*, Singapore: World Scientific Press, 324 pp., 1992.

Olson, R., *Essentials of Engineering Fluid Mechanics*, Scranton, PA: International Textbooks, 404 pp., 1961.

Otto, G.H., "A Modified Logarithmic Probability Graph for the Interpretation of the Mechanical Analysis of Sediments," *J. Sed. Petrology*, 9, 62–76, 1939.

Richardson, J.F., and W.N. Zaki, "Sedimentation and Fluidization, Part 1. *Trans. Institution of Chemical Engineers*, 32, 35–53, 1954.

Rouse, H., "Nomogram for the Settling Velocity of Spheres," Division of Geology and Geography Exhibit D, Report of the Commission on Sedimentation, 1936–1937, Washington, DC: National Research Council, 57–64, 1937.

Stokes, G.G., "On the Effect of the Internal Friction of Fluids on the Motion of Pendulums," *Trans. Cambridge Phil. Soc.*, 9, 8, 1851.

U.S. Army, Corps of Engineers, Coastal Engineering Research Center, *Shore Protection Manual*, 3 Vols., 1973.

Vanoni, V.A., ed., *Sedimentation Engineering*, New York: American Society of Civil Engineers, 745 pp., 1975.

Wiegel, R.L., *Oceanographical Engineering*, Englewood Cliffs, NJ: Prentice–Hall, 531 pp., 1965.

Zingg, T., "Beitrag zur Schotteranalyse," *Schweiz. Min. u. Pet. Mitt.*, 15, 39–140, 1935.

EXERCISES

2.1 Discuss the relative merits of the fall velocity versus the grain size as a descriptor of sediment behavior in the nearshore zone.

2.2 Show for the case of a log-normal distribution that the probability a sand size is coarser than a given size, ϕ_g, can be integrated to yield

$$P[\phi < \phi_g] = \int_{-\infty}^{\phi_g} f(\phi)\,d\phi = \frac{1}{2}\left[1 + \mathrm{erf}\left(\frac{\phi_g - \mu_\phi}{\sqrt{2}\sigma_\phi}\right)\right],\qquad (2.14)$$

where erf (z) is the *error function* of argument z given by

$$\mathrm{erf}(z) = \frac{2}{\sqrt{\pi}}\int_0^z e^{-x^2}dx$$

(If ϕ_g is smaller than μ_ϕ, note that $\mathrm{erf}(-z) = -\mathrm{erf}(z)$. Show that the probability of $\phi_g = \mu_\phi$ is $\frac{1}{2}$.

2.3 Show that $\mu_\phi = \int_{-\infty}^{\infty} \phi f(\phi)\,d\phi$ and $\sigma_\phi = \int_{-\infty}^{\infty} (\phi - \mu)^2 f(\phi)\,d\phi$.

2.4 For a sand sample with a mean grain size $\mu_\phi = 1$ and standard deviation of 0.5, what are the probabilities of grain sizes coarser than 0.5 ϕ units, 1ϕ units, and 1.5ϕ units? What is the significance of these sizes?

2.5 An analysis of a sand sample shows that $d_{84} = 1.6\phi$, $d_{50} = 1.5\phi$, and $d_{16} = 1.2\phi$. Is this sediment well-sorted? What is the mean diameter?

2.6 Find the fall velocities associated with the sand sizes for the Example in the chapter. Use a temperature of $20°C$ and Figure 2.7 for the first estimate. Then compute the fall velocity using Eqs. (2.11) and (2.13).

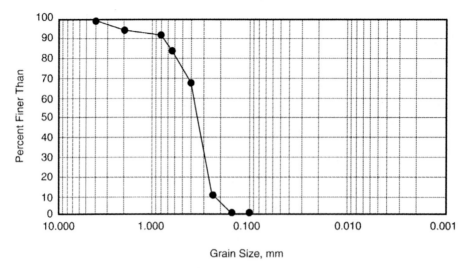

Figure 2.8   Sand size distribution for Problem 2.7.

2.7   From the cumulative sand size distribution shown in Figure 2.8,

    (a) compute median, mean, skewness, and kurtosis;

    (b) show the actual sand size distribution and compare it with a normal distribution having the same mean and standard deviation;

    (c) plot the distribution of the fall velocity versus percent passing.

2.8   A sand grain settling in a fluid with a high concentration of like sand grains falls more slowly than if it were the only settling grain (Richardson and Zaki 1954). Explain why.

2.9   Explain the effect of temperature on the fall velocity of sand. How much variation during a year might there be on the fall velocity associated with a sand grain with a 0.4 mm diameter?

2.10   Equations (2.4) and (2.6) are estimates of the mean sediment size and the sorting based on the phi size. Show, based on these equations, that the equivalent expressions for the geometric grain size are

$$\overline{d} = \sqrt{d_{16}d_{84}}$$

$$\sigma_\phi = -\log_2\left(\sqrt{\frac{d_{84}}{d_{16}}}\right) = \log_2\sqrt{\frac{d_{16}}{d_{84}}}$$

CHAPTER THREE

Long-Term Processes

Natural and anthropogenic effects combine to result in the maximum erosional stress on the barrier islands located near the mouth of the Mississippi River. These islands include Grand Isle, Timbalier, and Isle Dernieres to the west of the active delta and the Chandeleur Islands to the east. Natural effects include the subsidence resulting from the weight of the delta and the soft unconsolidated underlying muds. However, in this instance, the anthropogenic effects probably dominate and include additional subsidence resulting from withdrawal of hydrocarbons and the reduction of sediment supply from the river by the construction of upstream impoundments and jetties that direct the riverine sediment offshore to deep water. Additionally, in some areas, inlets and jetties have disturbed local sediment transport pathways. The total relative rise in sea level in the region of the Mississippi River Delta is on the order of 1 cm per year – eight times the worldwide rate. The resulting erosion rate from the combined effects is so great that entire islands can move and vanish within periods of decades. See, for example, Nummedal and Cuomo (1984) for an historical perspective on some of this erosion.

3.1 INTRODUCTION

The processes that shape shorelines can be examined with many different time scales. The beach changes constantly under the action of the individual waves that suspend the beach sediment and move it about; however, it is the erosion on the scale of hours and days that is responsible for the cumulative damage by a storm. Finally, it is the continued erosion (or accretion) over months and years that is important for the long-term shoreline recession (or accretion) of uplands and the performance of coastal structures.

An understanding of long-term coastal processes on the order of hundreds and thousands of years is important because it provides a background for interpreting the pervasive forces that have resulted in the shaping of the shorelines; by presumption, many of these forces are still active, albeit perhaps to a lesser degree than in the past. The processes discussed here are the rise in sea level relative to the land and the concept of an equilibrium profile. These two important long-term phenomena

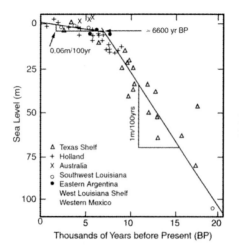

Figure 3.1 Historical sea level rise curve (adapted from Shepard 1963).

(which vary spatially) result in the ultimate characteristics of a particular section of shoreline. An understanding of these processes and their effects also provides a framework for the interpretation or prediction of the results of modern-day processes.

3.2 RELATIVE SEA LEVEL CHANGE

A long-term geological process of utmost importance to a shoreline is relative sea level change, which can occur as the result of a change in water volume of the oceans or the subsidence or emergence of the land by geologic processes. As will be discussed later, relative sea level change causes the shoreline to be out of equilibrium with the sea level and brings into play processes that tend to restore equilibrium. These processes can cause the shoreline either to erode or accrete.

Geologists have determined from the geological record that the earth experiences very low frequency cycles of cooling and warming in which the greatest cooling results in ice ages. At least four major ice ages have occurred in the last 300,000 years (the Quaternary age) according to Fairbridge (1961). These cycles appear to be caused by changes in the solar irradiance due to variations in the orbit and tilt of the earth as it revolves around the sun. Milankovitch showed, beginning in 1920, that the tilt of the earth has a 41,000-year cycle and the precession of the earth has a 22,000-year cycle, which affects the distance between the sun and earth. By computing the historical insolation of the northern latitudes, Milankovitch found that the times of low insolation corresponded to the times when the ice ages were assumed to occur.* The subsequent growth and melting of glaciers give rise to a change in sea level, which is called glacial *eustacy*, or a glacially induced change in sea level.

The last such ice age occurred about 20,000 years ago. Figure 3.1 presents results from Shepard (1963) showing that, at its most extreme, the sea level was 120 m below its present level. Large ice sheets covered much of North America and northern Europe. The areas occupied by the continental shelves of today were largely dry, and the river valleys and canyons extended a substantial distance across these shelves to the sea. Also, because of the lower sea level, the gradients of the rivers were greatly increased relative to present values. During periods of high rainfall, these increased

* Imbrie and Imbrie (1979) present an historical account of the unraveling of the ice age mystery.

gradients caused higher velocities, and large quantities of sediment were transported from mountains onto the continental shelves. As the glaciers retreated and released vast quantities of water to the oceans, sea level rose quickly until about 6000 years ago, when the rate of sea level rise decreased dramatically. Since that time, sea level has increased several meters. This more recent sea level rise and subsequent shoreline recession is referred to as the *Holocene transgression*.

Establishing historical sea level rise curves has not been an easy task. Some of the evidence comes from marine geologists, who have discovered submarine terraces created by the erosion of waves on erstwhile subaerial (above water) shorelines. Radiocarbon dating of sediment cores taken of these submerged terraces gives the approximate date when each terrace was at sea level. Also, marsh deposits found below present sea level in some places can be carbon dated. These old marshes were formed at sea level originally.

In contrast to the global sea level rise caused by the increased amount of water in the oceans due to the melting of the glaciers that is still ongoing, there are local causes of sea level change. Because some of the local sea level change is due to the rise in water level and some is due to the sinking or rising of the land, the net sea level change is often referred to the *relative* sea level rise. In this way, only the net change in sea level is important, not what is rising with respect to what.

The local causes of a relative sea level change are numerous. There is *tecto-eustacy* due to the change in relative sea level caused by land movement. This is particularly ongoing today in the high northern latitudes where there is still a rebound of the land in response to the reduction of glacial loading. In parts of Alaska, relative sea levels are falling, not rising, at a rate of 1.2 cm/yr. However, the land is sinking in other locations, including a good portion of the U.S. eastern seaboard.

Steric-eustacy refers to the change in sea level due to the warming of the water and its subsequent expansion. This is a significant contributor to the present concerns about global warming due to the greenhouse effect. One degree (Celsius) of uniform increase in seawater temperature has been calculated to result in a 2-m sea level rise. (During the ice ages it has been estimated that the earth's average temperature was only 5°C colder than present.) On the basis of numerous studies, the National Academy of Sciences concluded that the atmospheric concentrations of carbon dioxide will double during the next 100 years owing to the combustion of fossil fuels (yielding up in a relatively short time much of the carbon dioxide that has been tied up in coal and oil over millions of years). The increased concentrations of gases, such as carbon dioxide and nitrous oxide, absorb the infrared radiation emanating from the warm surface of the earth that normally cools it. The trapped radiant energy then heats the atmosphere, and predictions of 1–4.5°C warming of the atmosphere within the next 100 years have been made (Climate Research Board 1979, 1982).

Other factors contributing to local relative sea level rise include consolidation of the land. This is occurring, as noted before, at the mouth of the Mississippi River, where the soft muds deposited by the river are consolidating, causing the land to sink with respect to an absolute datum. An anthropogenic cause of relative sea level rise is the pumping of fresh water from coastal aquifers, which has led to the compaction of these aquifers, resulting in sinking of the land. Oil and other hydrocarbon extraction also results in land subsidence as, for example, in Houston, Texas, and Terminal Island, California.

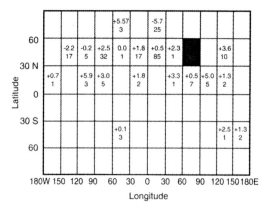

Figure 3.2 Characteristics of tide gauge data by 30° longitude and latitude sectors. The lower values in each sector represent the number of tide gages, and the upper values represent the linear long-term relative sea level change determined from the gauges in millimeters per year (after Pirazzoli 1986).

3.2.1 MEASUREMENT OF SEA LEVEL RISE

Tide gages have been installed in numerous locations around the world primarily to provide water depth information for navigation. At a few sites, there are over 100 years of data on the tides, which permits the extraction of the more recent relative sea level rise signal. San Francisco, is the location of the longest U.S. gage record, extending over 100 years, whereas the tidal record for Rotterdam, the Netherlands, extends back over 400 years.

Numerous investigators have examined these historical tide gage records on a worldwide basis to determine if it is possible to extract recent global sea level rise data. One of the major problems facing these investigators has been the paucity of the gages. Some highly populated areas of the world have a very dense network of tide gages, whereas less-populated areas and the islands in the middle of the oceans have very few gages. Pirazzoli (1986) examined 1178 tide gage records stored at the Permanent Service for Mean Sea Level (PSMSL) located at the Proudman Oceanographic Laboratory, Bidston Observatory, United Kingdom.* Fully 97 percent of the gages examined by Pirazzoli were located in the northern hemisphere. Further, a requirement for a valid tide record is that the gage be stably situated for a period of 40 years. Figure 3.2 shows some of the estimates of relative sea level rise based on these gages.

Sturges (1987) has pointed out the difficulty in isolating sea level rise information from tide gages by hypothesizing that a coherence between gages with no phase lag would indicate a sea level rise signal, whereas a coherence with a phase lag would mean some atmospheric or hydrodynamic effect was responsible. Comparing gages from San Francisco and Honolulu, Sturges observed that there was coherence at periods of 5 to 10 years with a phase lag. He concluded considerable natural "contamination" of the tidal record occurs for time spans of up to 50 years, making it difficult to discern the eustatic rate.

If we examine some individual gages to look at short-term trends in sea level, it is obvious that the levels have been changing at a fairly slow rate over the past century. The best estimate of this worldwide or eustatic rise in sea level as determined from

---

* See Wordworth (1991) for more information on the PSMSL, which maintains a data base of monthly and annual mean sea level information from over 1600 tide gages worldwide.

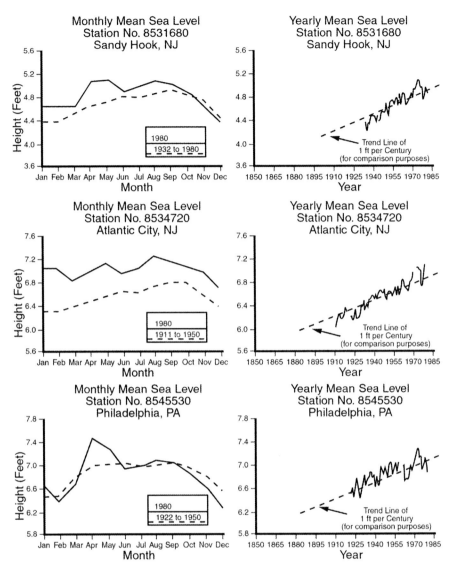

Figure 3.3 Monthly mean sea level variations and long-term sea level data at East Coast locations based on tide gage data (adapted from Hicks, Debaugh, and Hickman 1983).

tide gage records is approximately 11 cm per century. Figures 3.3, 3.4, and 3.5 present, in the right-hand panels, data from some of the more long-term tide gages on the East Coast of the United States, Gulf of Mexico, and Alaska. Each data point on these plots represents the average tide gage recordings over a 1-year period. Note that, in addition to the upward trend of the data, considerable fluctuations can be observed around the trend line.

During the course of a year, the sea level varies (Marmer 1952). On the basis of monthly tidal averages, an obvious seasonal effect on the tides is observable. See Figures 3.3, 3.4, and 3.5 again and examine the left-hand sides of the figures, which show the mean seasonal variations for two different time spans. This seasonal variation, which can be up to 20 cm, is not simply attributable to the variation in snow and ice levels in the higher latitudes. Some of the annual variability is due to changes

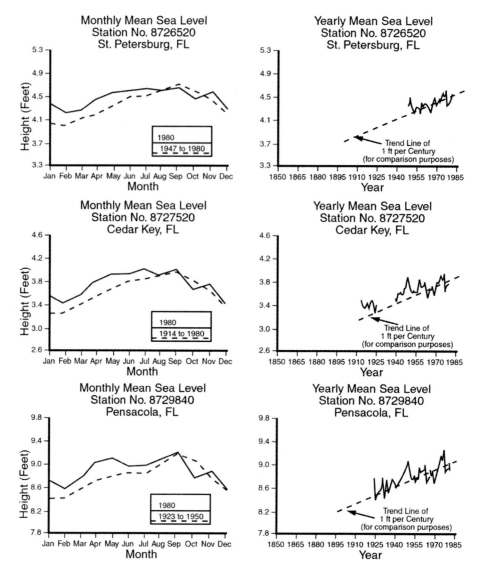

Figure 3.4 Monthly mean sea level variations and long-term sea level data at three Florida west coast locations based on tide gage data (adapted from Hicks et al. 1983).

in regional barometric pressure and changes in local oceanic currents, which, with their Coriolis-induced surface tilts, change sea level along the coastline. Additionally, the ocean water temperature changes during the year, yielding a steric effect; river runoff has a seasonal character, changing the local salinity; and the winds change seasonally, which can also have a significant effect.

For most of the data in Figures 3.3 and 3.4, the average sea level rise rate is approximately 30 cm per century. This is undoubtedly the combined result of the eustatic rate (11 cm per century) and local effects. The Alaskan tide gage data, Figure 3.5, show a rapid sea level lowering due to glacial rebound at these higher latitudes. Figures 3.6 and 3.7 present average sea level variations for the Gulf Coast and United States, respectively. It can be seen that these trends are somewhat less than 30 cm/century.

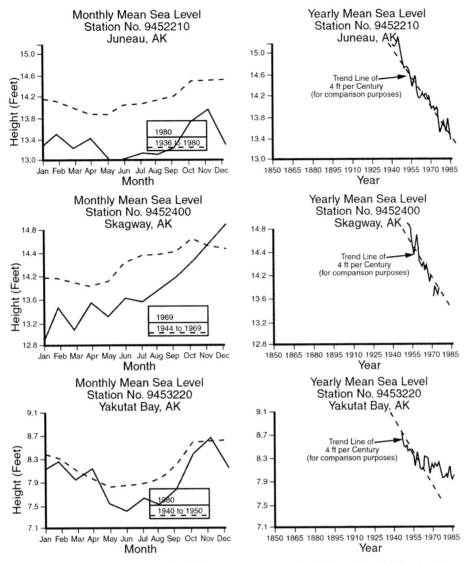

Figure 3.5 Monthly mean sea level variations and long-term sea level data at three Alaskan locations where glacial rebound is evident based on tide gage data (adapted from Hicks et al. 1983).

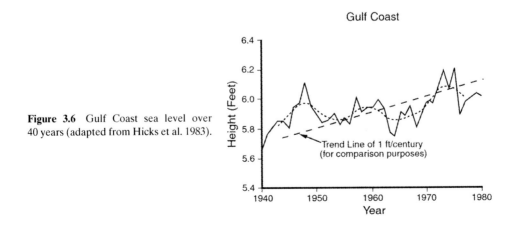

Figure 3.6 Gulf Coast sea level over 40 years (adapted from Hicks et al. 1983).

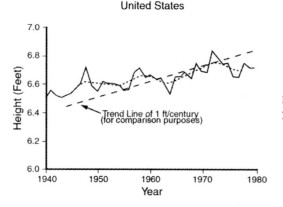

Figure 3.7 United States sea level over 40 years (adapted from Hicks et al. 1983).

3.2.2 FUTURE SEA LEVEL TRENDS

Future trends of sea level change are impossible to predict with high accuracy. Hoffman, Keyes, and Titus (1983) and Hoffman, Wells, and Titus (1986), based on the warming trends, developed different scenarios for sea level rise to the year 2100 by using predictions from numerical climate models. These, as well as those of the Revelle committee of the National Research Council (1983), are shown in Figure 3.8. Clearly, the high scenario, which involves the detachment of the Ross ice sheet from the Antarctic continent, would have disastrous impacts on the world. (However, the developers of these projections indicate that they entail substantial uncertainty.) If these estimates are close to being correct, there will be very large associated responses of the shoreline and an imperative for a revised strategy relative to development along the shoreline and stabilization measures.

3.3 EQUILIBRIUM BEACH PROFILE

A beach profile is the shape that would result if one could cut a vertical slice through the beach and look at it from the side. It is typically depicted as a line displaying the

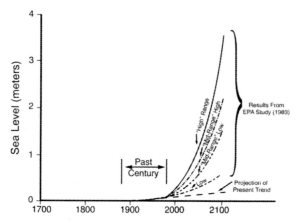

Figure 3.8 Temporal estimates (by year) of sea level rise projections developed by the EPA (Hoffman et al. 1983).

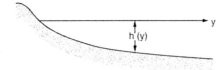

Figure 3.9 Schematic of an equilibrium profile.

water depth and land elevation versus distance as measured from the beach berm or dune to some offshore point (Figure 3.9). The shape of the beach profile is a result of the natural forces acting on the sand making up the beach. In the absence of wave activity, the beach sand would have a linear profile with a slope corresponding to the sediment's angle of repose; this angle measures the steepest slope at which sand can be piled (about 32°). This slope reflects the balance of the force of gravity, tending to roll the sand grains down the slope, and the support forces provided by neighboring grains in the sand pile. On a real beach acted upon by waves, the profile is generally concave upwards with slopes much less than the angle of repose. This is because other forces (constructive and destructive) affect the stability of the sand grains. The concept of destructive forces, tending to flatten the profile, is most readily apparent when the effects of a severe storm that may include elevated water levels and larger than normal wave heights are observed. During this erosional event, sediment is taken from the beach face and transported seaward, where it is deposited as an offshore bar. The constructive forces are evident after the storm and slowly act to move the sand back onto the beach, actually steepening it against the gravitational forces.

In Chapter 7, the concept of an equilibrium beach profile will be discussed in more detail; here we note that, for sediment of a given size, there will be a unique beach profile in equilibrium with the given wave and tide characteristics of the beach; that is, the constructive and destructive forces on the sand grains are in balance. If any of the wave and tide conditions are altered, such as an increased water level or a change in wave height or period, a new equilibrium profile will exist, and the previous profile will evolve toward the new equilibrium shape.

A few of the known empirical relationships between equilibrium profile shape and sediment, wave, and tide characteristics are as follows:

- *Sediment size* – Coarse sediments are associated with steep equilibrium beach slopes. These sediments seem to be able to withstand the destructive forces more easily than finer sands, or the constructive forces affect them to a greater extent. This effect was noted by the Beach Erosion Board (1933), which found that steeper New Jersey beaches were characterized by larger grain sizes. Bascom (1951) showed that the beach face slope for California beaches was correlated positively with grain size (see Figure 2.5).
- *Wave height* – Increasing wave heights cause milder slopes probably because increased wave heights mean greater destructive forces and only with a milder beach slope can the forces reach equilibrium. Also, higher waves result in a wider surf zone, spreading the destructive forces over a wider region and into deeper water.
- *Wave period* – Increasing wave periods tend to cause sediment to be transported shoreward and the mean shoreline to advance seaward such that the average beach slope is steeper than for shorter-period waves.

- *Water level* – An increasing tide or water level causes sediment to be transported seaward. The increased water level requires a new equilibrium profile but one lifted vertically and moved shoreward. This means shoreline recession and an offshore transport of sediment must take place.

3.3.1 DIMENSIONLESS PARAMETERS

The individual effects of wave and sediment characteristics on the equilibrium beach profile can be described conveniently by combinations of these characteristics rather than by treating them separately. Besides lumping together the effects of several beach or wave characteristics, these parameters are also dimensionless, meaning that, regardless of the dimensional units used, the parameters have the same value.

The first parameter we present is the so-called Dean number.[*] This parameter is

$$D \equiv \frac{H_b}{wT}, \qquad (3.1)$$

where H_b is the breaking wave height, T is the wave period, and w is the sediment fall velocity. A second dimensionless parameter is the *surf similarity parameter* defined by Battjes (1974) as

$$\zeta \equiv \frac{\tan \beta}{\sqrt{H_0/L_0}}, \qquad (3.2)$$

where $\tan \beta$ is the beach slope and H_0 and L_0 are the deep-water wave height and length, respectively.[†] The third parameter is a type of Froude number defined in terms of the breaking wave height H_b and the fall velocity w:

$$F = \frac{w}{\sqrt{g H_b}}, \qquad (3.3)$$

where g is the acceleration of gravity.

In considering these three parameters, it is clear that the first and third are more convenient for application because these depend only on the wave and sediment characteristics, whereas the second parameter requires knowledge of the beach profile itself. Additionally, the surf similarity parameter requires that a single beach slope be identified, which makes comparison of results for different beaches difficult because one may choose to define the effective beach slope as that of the beach face, whereas another may choose the average slope out to the breaker line, and so on.

3.3.2 EXPECTED BEACH PROFILE RESPONSES

Armed with our knowledge of equilibrium profiles, it is relatively easy to predict qualitatively the effects of changes in wave and sediment characteristics, or, given an initial condition, the evolution of a particular beach profile.

---

[*] See, for example, Suh and Dalrymple (1987) for the naming convention. The first author of this text disclaims any part in it.

[†] Wave parameters will be discussed more fully in the next chapter.

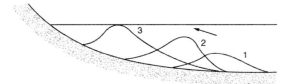

Figure 3.10   Ridge and runnel system forming bars.

If an initial beach profile is much steeper than the equilibrium profile, equilibrium can only come about by nature's somehow providing the extra material to build up the offshore depths. This material could come from the longshore transport of material caused by oblique wave incidence and the longshore current, or the material could be the result of erosion of the shoreline with the resulting offshore sediment transport providing the milder beach profile.

If, on the other hand, the initial profile is much milder than the equilibrium profile, there are several possibilities that could bring about an equilibrium beach profile. One of these is the longshore sediment transport, which removes the sand necessary to form the profile. The second would be the result of onshore sediment transport and a seaward movement of the shoreline to the point where the beach profile had steepened to achieve the equilibrium value. The third possibility is a result of onshore transport, which occurs in a nonuniform manner, known as a "ridge and runnel" system. These shoreward moving ridges may grow to such an elevation that they become emergent, thereby forming barrier islands with lagoons trapped behind them such that a new equilibrium beach profile is formed. The shoreward progression of a ridge and runnel system is shown in Figure 3.10.

3.4 CLASSIFICATION OF SHORELINES

3.4.1 INTRODUCTION

The shapes and characteristics of the coastlines of the world are extraordinarily diverse as a result of the local environment and the original geological processes leading to their development and subsequent fate. In the United States, the rocky shorelines of the coast of Maine differ greatly from the sandy beaches of Florida, which even differ from one side of the state to the other.

The classification of shorelines is of value because shorelines often respond in a generic way to the natural forces that bear upon them. Thus, a study of shorelines by type (or classification) provides a basis for understanding shoreline responses to different types of forces.

3.4.2 SHORELINE CLASSIFICATION SCHEMES

Some of the more famous classification schemes are those of Johnson (1919) and Shepard (1963), although that of Valentine (1952) is widely used. To provide a background for coastal classification, Johnson's classification is provided here in Table 3.1. The first two major categories are shorelines of submergence and emergence, which are the same ones we will use.

Table 3.1   Shoreline Classification Scheme of D.W. Johnson

I. Shorelines of submergence		
	Ria	Drowned river valleys
	Fjord	Drowned glacial troughs
II. Shorelines of emergence		Coastal plain shoreline
III. Neutral shorelines		
	Delta	
	Alluvial plain	
	Outwash plain	
	Volcanoes	
	Coral reef	
	Fault	
IV. Compound shorelines		Combination of preceding shorelines

Source: Johnson 1919.

The classification scheme we have selected is based on an attempt to understand the response of sandy shorelines to natural and anthropogenic (human-related) forces better. The dynamical classification scheme is based on the concept of an equilibrium beach profile and submerging or emerging coastlines. The advantage of this classification is that the ongoing mechanisms for such coastlines may be apparent from the classification.

3.4.3 SHORELINES OF SUBMERGENCE

A relative sea level rise, which results from the rising of the mean sea level, land subsidence, or both, creates a shoreline of submergence. We will discuss this shoreline response first in terms of the beach profile and then in terms of the shoreline planform.

3.4.3.1 Beach Profile

We consider the idealized situation in which the increased water level comes to rest against a relatively consolidated material that is fairly steep. In fact, we will assume that the water level intersection is at a steep rock cliff, as shown in Figure 3.11. The surf zone in this case will be very narrow owing to the steep slope; hence, the energy of the incoming waves will be dissipated within a very narrow surf zone. A great deal of turbulence will be generated within the surf zone that has a substantial capacity to dislodge large pieces of rock, usually by eroding weak seams. Once this rock has been dislodged, it serves as an abrasive or grinding agent to loosen additional rock and to reduce neighboring rocks to smaller sizes. As a result, the profile will become less steep, and the material that has been dislodged will be deposited below the water level at the base of the cliff. As storms occur, the increased water level and waves reach higher on the cliff, causing more erosion. As this process continues, the slope of the beach profile becomes milder, as shown in Figure 3.11(b). Owing to

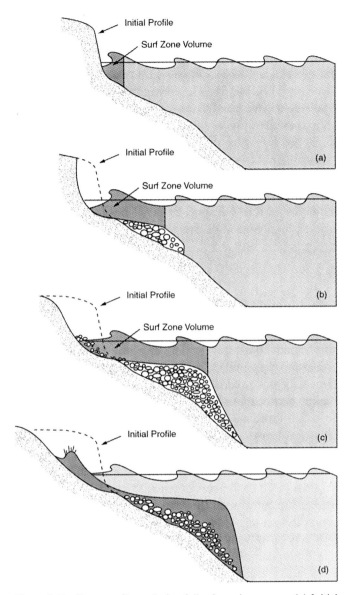

Figure 3.11 Shore profile evolution following submergence. (a) Initial profile following submergence of shoreline; (b) large fragments caused by erosion; (c) later Stages – Surf zone volume increases, particles are ground to finer and finer sizes; (d) equilibrium profile produced with formation of sand beach and dune.

the milder slope, the waves now break much farther offshore, and their energy is dissipated within a much greater water volume, resulting in lesser turbulence and a lesser capacity for transporting material offshore.

The same processes that eroded the cliffs abrade the eroded material, grinding it finer and finer and eventually reducing it to sand or silt. As will be shown in Section 8.4, during periods of storms, the finer material is carried offshore, leaving only the coarser material, which then is rolled and tumbled together, interacting to produce still finer material. With the return of milder wave activity, the sand-sized fine sediment moves ashore and forms a wider beach, whereas the silt-sized material is left offshore. The action of the wind (the *aeolian processes*) now comes into play and transports some of this material landward. Grasses with a reasonable tolerance for salinity will grow and cause windblown particles of sand to be trapped by their blades and deposited on the beach. The grass grows still higher through this sand, trapping more sand in a bootstrap operation of dune growth. The roots of the grass also stabilize the sand.

With the passage of time, an equilibrium profile will be approached such that waves are dissipated over a broad, mildly sloping surface. At this stage there is a long-term balance between the destructive forces that tend to transport sediment offshore and the constructive forces that transport material onshore. In addition, because of the more gradual dissipation of wave energy, the rate of sand abrasion is reduced greatly.

3.4.3.2 Shore Planform

We now view the beach in a three-dimensional sense, taking into account the along-shore transport of sand, or what is called the *littoral transport*. First, we should distinguish subaerial erosion from subaqueous (underwater) erosion. In particular, subaerial weathering and erosion tend to result in a land surface that is highly irregular and is characterized by river valleys, canyons, and tributaries. An initially planar surface at the coastline would soon become incised by erosional features that deliver the eroded sediment to the shoreline. In stark contrast, the opposite occurs for subaqueous erosion. These processes tend to reduce and eventually eliminate the original irregularities in the bottom, as shown in Figure 3.12. For example, if a positive bathymetric feature (such as a shoal) exists beneath the water, it will be subjected to erosional processes due both to the convergence and dissipation of wave energy on the feature as well as the abrasion of the feature by sediments of various sizes being transported along the bottom. A negative bathymetric feature (such as a channel or canyon), which may have been the result of subaerial erosion before the relative sea level rise, will be filled ultimately by sediment transported in the nearshore zone being trapped in the feature.

Let us examine the shoreline of Figure 3.12 after a relative sea level rise of 10 m. The resulting drowned shoreline is shown in Figure 3.13 and is characterized by protruding headlands and embayments. The coast of the state of Maine and the Norwegian fjord coastline are two examples of irregular coastlines caused by drowning.

The two dominant processes tending to modify the initially irregular subaqueous planform are the waves and currents. In addition, the natural resistance of the coastal materials to erosion will play a substantial role. The waves will generally approach the shoreline at an angle, and the direction may vary seasonally and with storms. Owing to wave refraction, the wave energy will be concentrated on the headlands rather than the embayments. Thus, more erosion products will tend to be developed at the headlands, and the waves and currents along the shoreline will transport this material in a preferential direction. Initially, there will be no substantial beaches

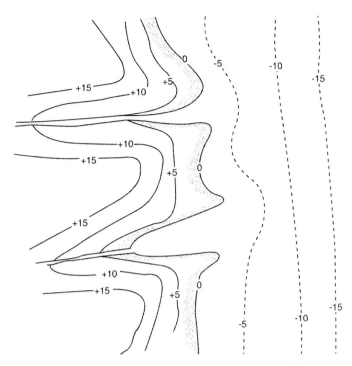

Figure 3.12   Differences in subaerial and subaqueous weathering.

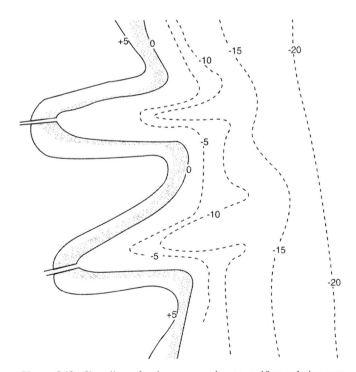

Figure 3.13 Shoreline of submergence due to a 10-m relative sea level rise.

on the entire shoreline at the water boundary of the submerged terrain. The waves approaching the shoreline will produce the energy necessary to abrade the erosion products to sand and to transport the resulting material.

As the material is reduced to finer and finer sizes, it is transported by the waves and currents acting at that location. In the very early stages of this process, because of the high wave energy levels acting at the headland and the quite limited sediment supply, no beaches form at the base of the headland; rather, the finer fractions are transported in the directions of the dominant nearshore currents and deposited underwater at an elevation consistent with the supply and the transporting capability of the waves and currents. (The transporting wave-related forces are much lower on sediment deposited at a substantial depth than for sediment deposited at or near the mean water line.) We will focus on several generic planform features produced in this erosion process.

Spits

A *spit* is formed when the dominant waves and currents carry the sediment into an elongated subaerial depositional feature extending away from an eroding headland (or another source of sand). A *simple spit* is shown in Figure 3.14(a), where the supply of material produced at a headland is sufficient to be deposited in a feature more or less parallel to the shoreline. Sand transported along the trunk of the spit is deposited at its end in deeper water, permitting the spit to grow longer. If the sand is produced by the headland at a slower rate, there would be a tendency for this material to be transported farther into the embayment, forming a *bayside beach* rather than a spit, or, for even a slower sand production rate, a *bayhead beach* only.

Spits can grow in a variety of shapes, depending on the wave climate, but they are due primarily to a dominant wave (and thus sediment transport) direction. The sediments are carried from the source downdrift into deep water. The continued transport of sediment allows the filling of the deeper water and the growth of the spit. The growth rate of the spit is related to the sediment transport rate and the depth of water into which the spit is growing.

Recurved spits occur as the wave climate changes for a time from the conditions forming the spit to waves arriving from the downdrift direction. No sediment is now supplied to the spit, and the waves from the new direction tend to cause transport of the material at the *distal* (terminal) end of the spit into the embayment, as shown in Figure 3.14(b). If this process is repeated more than once, hooked features may be left in the spit in a sequential manner much as shown in Figure 3.15(a). This type of spit is referred to as a *compound recurved spit*. After a spit is formed, changes sometimes occur to the more protected parts of the spit causing these to be recurved back even further, and this particular type is called a *complex spit* (see Figure 3.15(b)).

The highest elevations on a spit are an indicator of the vertical reach of the waves at the time of spit formation; that is, the sediment is cast upwards to an elevation equal to the upper run-up limit of the waves. For various reasons during periods of deposition, this upper limit may vary, resulting in features that are called *beach ridges*. The difference in elevation between the crest and swale of a beach ridge may vary substantially, ranging from tens of centimeters to perhaps one meter. A series of beach ridges is sometimes referred to as a "beach plane" or "Chenier plane."

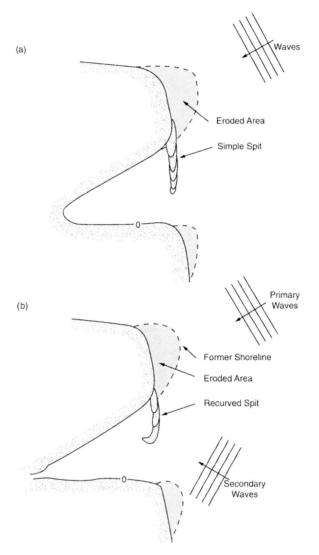

Figure 3.14 Formation of simple and recurved spits. (a) Simple spit formation; and (b) recurved spit formation (one cause).

The growth of spits may be limited by the increase of currents at an inlet formed by the spit. For example, if spits form in two directions from adjacent headlands and grow toward the center of the embayment, eventually the tidal currents in the inlet (flowing between the ocean and the embayment during a tidal cycle) will become stronger and stronger because of the decreased width of the inlet cross section. Finally, the currents force a balance with the littoral drift of materials into the inlet, and a final equilibrium inlet cross section results. These currents may also contribute to the recurving of the spit ends (see Figure 3.16(b)). If the tidal currents are not strong enough to keep the sand out of the inlet, the inlet may close, permitting a *baymouth barrier* to form, as shown in Figure 3.16(a).*

* In Chapter 13, we will examine the methods to predict whether or not an inlet will remain open.

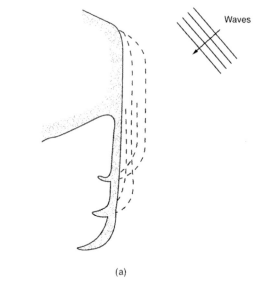

(a)

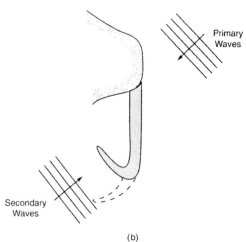

Secondary
Waves

(b)

Figure 3.15 Development of compound recurved and complex spits. (a) Compound recurved spit; (b) complex spit.

One evidence of this mechanism of spit and barrier island generation, given by Johnson (1919), is that if the headlands provide a source for material and the spit or barrier island is broken by inlets, we would expect, because inlets are a sink for littoral materials as evidenced by the ebb and flood shoals on the bay and ocean sides of the inlet, that near the headland the inlets would be widely spaced. On the other hand, at the distal end of the barrier island chain the inlets would become more closely spaced because the supply of sediment would be much less this far from the source area. This variation in inlet spacing appears to hold for the barrier islands on the Delmarva Peninsula, for example.

Sometimes other sandy features may form in a headland–embayment system such as spitlike deposits on the shorelines of the embayment called *midbay barriers* or *midbay bars* instead of spits and barrier islands.

If a hill should become isolated as an island because of submergence, special depositional features may form as it erodes. One possibility is for the material to erode relatively fast and to deposit more or less parallel to the dominant direction

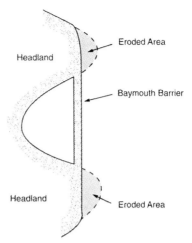

Figure 3.16 Example of baymouth barriers.

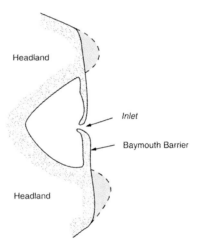

of the waves. In this case, the feature is called a *winged headland*. In cases where the erosion occurs at a lesser rate relative to the transporting capacity of the waves, the material may be transported around both sides of the hill, meeting and enclosing a small lagoon.

Tombolos

In addition to the local deposition at offshore islands, neighboring shorelines may be affected. The most common example is that of a *tombolo* formation, which is a depositional feature resulting in a subaerial connection between the offshore island and the shoreline, as shown in Figure 3.17(a). This depositional feature is the result of wave sheltering of the island and the modification of the wave crests in such a way that they are curved inward toward the sheltered area behind the island. The tombolo* may grow from the shore toward the island, or, if there is sufficient sand on the island, from the island toward the shore.

---

* The word *tombolo* comes from Italian and referred originally to sand dunes as well as to what we now call tombolos.

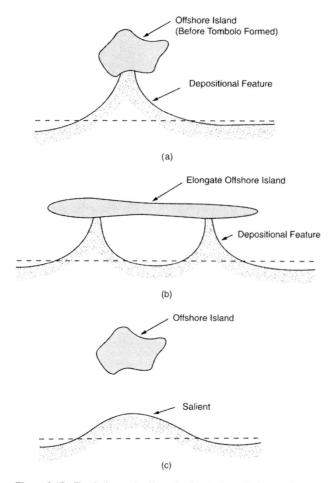

Figure 3.17 Tombolos and salients in the shelter of offshore islands. The dotted line indicates the shoreline in the absence of the island. (a) Tombolo formed in the lee of an offshore island. (b) Double tombolo formed in the lee of elongate offshore island. (c) Depositional feature in lee of offshore island that is too far offshore to form tombolo.

In some cases, when the island is generally elongate and long compared with the separation distance from the shoreline, a double tombolo may form, as shown in Figure 3.17(b). Triple tombolos are also known to exist.

If the offshore feature is located too far offshore of the coastline, a true tombolo will not form, but a shoreline protuberance, called a *salient*, can occur, as shown in Figure 3.17(c).

Barrier Islands

Barrier islands occur along a substantial portion of the world's shorelines (13 percent). They are most prevalent in the midlatitudes and in areas with a small tidal range. The East and Gulf Coasts of the United States contain more than 4300 km of these islands.

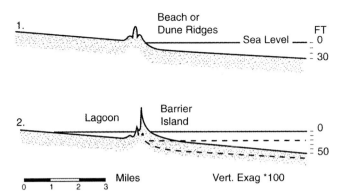

Figure 3.18 Formation of barrier islands by submergence (from Hoyt 1967. Reproduced with permission of the Geological Society of America, Boulder, Colorado, USA. Copyright © 1967 Geological Society of America).

Numerous theories have been advanced for the development of barrier islands. For shorelines of submergence, in addition to the barrier–spit model given in the preceding paragraphs, the theory of Hoyt (1967) may apply. He postulated that barrier islands are the result of sand dunes drowned in place by a relative rise in sea level (see Figure 3.18). His theory is based on the examination of sediment cores taken behind barrier islands in the state of Georgia (U.S.A.) that indicate the absence of salt water species of vegetation, implying that the lagoon formed after the barrier.

Once barriers are formed along a shoreline, they respond to the rising sea level in several modes. The first is by growing upwards and landwards with the sea level rise. This migration process requires a significant onshore transport of sand, which comes about through several means:

1. Flood tidal shoals built landward of the island by inlets that become part of the island when the inlet migrates or closes.
2. *Overwash* that occurs during storms when the waves wash sand directly over the island; often the overwash occurs at low spots in the dunes, and the flow and sand are locally confined, leading to overwash fans, which are similar in planform to river deltas.
3. Aeolian transport of sand by onshore winds.

This model may have permitted the continuous existence of barrier islands along outwash shorelines like the U.S. East Coast over the last 6000 years of sea level rise. Another response of barrier islands is to be eroded or drowned in place. This implies that there is a lack of sand supply or that the relative sea level rise occurred so rapidly that the landward migration processes could not keep pace.

The subject of barrier island migration has been hotly debated. Shoreline communities on barrier islands have attempted various erosion mitigation schemes to forestall the erosion process, whereas critics argue that these measures interfere with the ongoing migration process and are ultimately futile efforts. Also, development on barriers has been discouraged in the United States by governmental actions so as not to impede natural processes.

Inlets

The morphology of tidal entrances differs greatly from one inlet to the next. An ability to interpret these features is important, for their geometry represents the combined signature of the sediment supply or transport characteristics, the wave characteristics, and the tidal flow. Our ability to interpret these features is not perfect and, even recently, misinterpretations have resulted in significant design errors in the construction of inlet training structures.

Generally, inlets tend to migrate in the direction of the longshore sediment transport. The migration process can be demonstrated by referring to the barrier island–inlet system shown in Figure 3.19(a). With the dominant sediment transport from the top of the figure toward the bottom, the material tends to be deposited on the updrift shoulder of the entrance channel. This may be considered to result in a momentary decrease in the channel cross-sectional area, thereby increasing the inlet current velocity and putting erosional pressure on both the updrift and downdrift banks of the entrance channel. Because the downdrift bank has limited (if any) sediment supply,

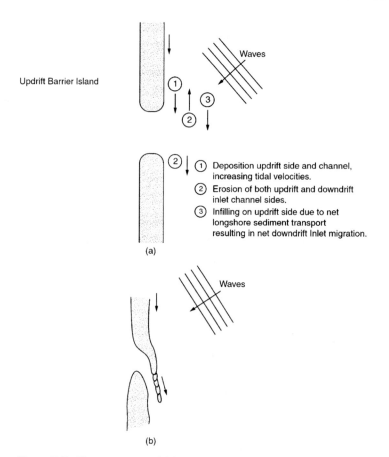

Figure 3.19 Two responses of inlets to net longshore sediment transport. (a) Inlet migration in downdrift direction; (b) overlapping by updrift barrier island as a result of net longshore sediment transport.

it will erode without an accompanying deposition, whereas the material eroded on the updrift bank will be replaced by additional deposition associated with the longshore sediment transport. In this manner, there may be a gradual and progressive migration of the inlet in the downdrift direction. In some cases, the overall result of such a process will be a cycle by which the inlet migrates from near the updrift end of a baymouth barrier to near the downdrift end, after which a breakthrough occurs during a storm at a weak point near the updrift end of the bay. This new inlet is more hydraulically efficient than the old one, which then closes. This breakthrough process causes a significant amount of sand to bypass the inlet in one fell swoop, causing substantial changes to the downdrift shoreline. The process then repeats with a recurrence period of perhaps decades.

A similar type of behavior may occur at inlets with the exception that there may not be a migration of the inlet channel but rather the updrift deposition may occur as an overlap to the downdrift shoreline, as shown in Figure 3.19(b). Usually this downdrift overlap will continue to a point at which the channel is deflected to such a degree that the hydraulics of the inlet system becomes relatively inefficient. During a large freshwater discharge event caused perhaps by a severe storm, there may be a breakthrough that may leave the depositional feature stranded as a small barrier island. The feature may either erode and eventually attach to the downdrift shoreline or it may survive due to the protection provided by the growth of another spit feature seaward of the previously active spit.

Another recognizable characteristic of natural inlets is a relative offset between the updrift and downdrift shorelines. Generally one might expect that, due to the source of material being located on the updrift shoreline, this side of the inlet would be displaced farther seaward than that of the downdrift shoreline and this is sometimes the case as indicated in the previous paragraph. However, an offset, as shown in Figure 3.20(a), is by no means consistently reliable as an indicator of the direction

(a)

Figure 3.20 Examples of inlets with updrift and downdrift offsets. (a) Inlet with updrift offset; (b) inlet with downdrift offset.

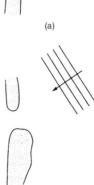

(b)

Figure 3.21   Townsends Inlet, New Jersey. The net littoral drift is from north (top) to south.

of longshore sediment transport. For example, Figure 3.21 presents a plan view of
Townsends Inlet, New Jersey, where there is a downdrift offset.

Hayes, Goldsmith, and Hobbs (1970) have carried out systematic studies of inlets
that exhibit downdrift offsets and have found that they are predominantly the result
of the sheltering effect of the ebb tidal shoal, which is formed by sediments being
jetted offshore of the inlet by the ebb tidal currents. These shoals tend to be more
dominant on the updrift side simply because of their nearness to the source of sediment supply. In some cases these ebb tidal shoals may contain many millions of cubic
meters of sediment. In areas where there are substantial reversals in the direction
of longshore sediment transport, this and other mechanisms, which will be discussed
in greater detail in Chapter 11, lead to a trapping of sediment in the lee of the ebb
tidal shoal such that the waves from the dominant direction are greatly reduced and
thereby have a limited sediment-transporting capacity in these regions. This results

in a progradation, or building, of the downdrift shoreline seaward such that there is a downdrift offset at beaches where this process occurs.

Most natural inlets migrate in the downdrift direction. Some inlets, however, may migrate in the updrift direction. One of these is Redfish Pass at the north end of Captiva Island, Florida. This inlet has been studied by Walton and Dean (1976), who determined the mechanism for the updrift migration to be basically the same as that just described for downdrift offsets. Because of the sheltering influence of the ebb tidal shoal, sediment is transported from the downdrift beaches back toward the inlet, where it is deposited and exerts a tendency toward updrift migration.

Many of the forces and processes discussed here for natural inlets will also be applicable to the case of modified or stabilized inlets. Our emphasis has been the geomorphology of natural features to provide the necessary background for discussions of both natural and modified inlets. In Chapter 13, we will also discuss the factors important for the long-term survivability of tidal inlets.

Cuspate Features

Cuspate features encompass a range of geomorphological forms that occur on outer coasts or along elongate bays or river channels. Examples of those occurring along outer coasts include Cape Canaveral, Florida, which is a fairly large-scale cuspate feature, and many of the other capes along the East Coast, including Cape Hatteras, Cape Lookout, and so forth. In many of these cases, the depositional history of the cuspate feature can be inferred from the character of the beach ridges embedded in the feature. Moreover, in some cases, where these depositional features are truncated, the evolution of the cuspate feature may still be evident to some degree.

One of the more interesting cuspate features is that of a so-called cuspate foreland, which tends to occur along elongate bays. In many cases these cusps have substantial subaqueous deposits located off the points of the cusps. The most beautiful set of these is on Nantucket Island, Massachusetts, as seen in Figure 3.22. These features have been studied by Rosen (1975), who concluded that they are both erosional and depositional and caused by reversals in wave direction that result in the erosion of the embayments and deposition on the points of the cusps (see Figure 3.23). Other mechanisms that have been mentioned in the past are eddies that form in the lee of the points and cause sediment to be transported back to the points, as the process indicated in Figure 3.23 shows, or seiches (long-period standing waves; Wilson 1972) that have nodes or antinodes at which the points can form. If this were the case, one would probably expect to find the cusp points in an elongated bay located at the same position on the two sides of the bay. However, this is not generally true, as shown in Figure 3.24, for Santa Rosa Sound, Florida. One possible mechanism for the occurrence of these arcuate cuspate features is related to the longshore sediment transport characteristics. This will be discussed more fully in Chapter 8.

3.4.3.3 Summary

Shorelines of submergence tend to result in initially irregular shorelines owing to the effects of subaerial weathering. Usually the resulting shore profile is steeper

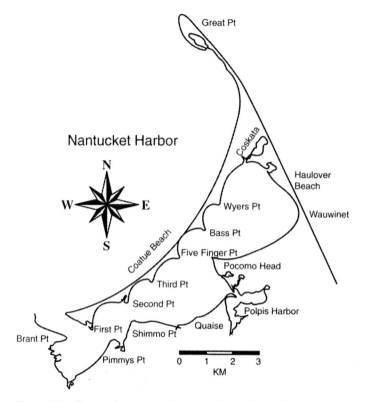

Figure 3.22 Cuspate features on Nantucket Island, Massachusetts.

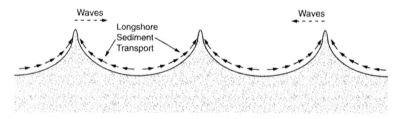

Figure 3.23 Rosen's interpretations of cuspate features on Nantucket Island, Massachusetts.

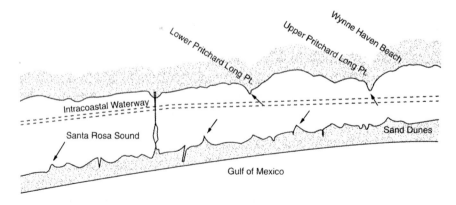

Figure 3.24 Cuspate features in Santa Rosa Sound, Florida.

than equilibrium. Wave and current energy, acting over geological time, result in an approach of the shore profile and planform to equilibrium. This process generally involves erosion of the promontories and deposition in depressions and across (at least partially) embayments to produce smooth contours. Resulting features may include a variety of spits, beach ridges, and barrier islands. Barriers may extend from adjacent headlands constricting the flow into the embayment to such an extent that the ebb and flood tidal flows increase sufficiently to make further deposition impossible, resulting in a viable natural inlet. The geometrical and migrational characteristics of inlets contain information relating to the local currents and sediment transport characteristics; however, our present ability to interpret this information has not been developed to a reliable level.

3.4.4 SHORELINES OF EMERGENCE

A lowering of sea level or an increase of the land elevations due to tectonic changes leads to a relative sea level drop and the appearance of a shoreline of emergence. As discussed earlier in this chapter, the topography due to subaerial weathering and submarine processes results in quite different three-dimensional forms. In particular, submarine weathering and transport process result in smooth contours that are more or less parallel to the shoreline. Therefore, straight shorelines are characteristic of shorelines of emergence. The resulting beach profile is also one that is out of equilibrium and has an excess of sand in the profile.

3.4.4.1 Barrier Island Development

We have discussed the formation of barrier islands for shorelines of submergence resulting from spit development, possibly from an eroding headland, and as caused by the drowning of sand dunes (Hoyt's theory). Shorelines of emergence also set the stage for barrier island development because of the excess of sand in the offshore profile. We will discuss two more explanations here: those advanced by de Beaumont and Gilbert. At the outset it is reasonable to expect that none of the individual theories can explain the formation of all barrier islands; however, a discussion of the processes associated with the various theories is helpful in interpreting the causes of barrier island formation at different locations.

de Beaumont Theory (1845)

The process underlying this theory is simply the reestablishment of the equilibrium beach profile on a mildly sloping beach through onshore sediment transport and the trapping of a lagoon between the resulting island and the shoreline, as shown in Figure 3.25(b). In this case, following the emergence phase, the dominantly onshore sediment transport processes reachieve equilibrium, resulting in an erosion zone below the initial beach profile. This sand is transported shoreward and deposited into a barrier feature to form a barrier island, and a lagoon is trapped behind

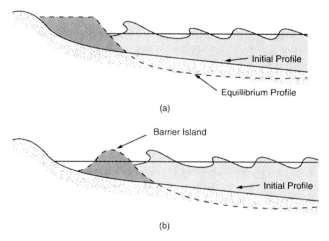

Figure 3.25 Two possible evolutionary sequences due to an excess of sand in profile. (a) Onshore transport to achieve equilibrium profiles, resulting in shoreline advancement; (b) onshore transport to achieve equilibrium profiles, resulting in barrier island formation.

the barrier island. From the standpoint of the wave processes, it is easy to imagine the vertical growth of the island up to the limit of wave uprush, which, of course, can vary substantially depending on tidal elevations and various wave conditions. The aeolian processes serve to transport the unvegetated sand farther landward where it is trapped by dune grasses, and the deposition there then proceeds vertically. The rate of landward transport of sand by wind depends on several factors, including the strength of the wind, the size and moisture content of the sediment, and the width of the unvegetated sand feature; therefore, the wider the berm (which is at about the limit of wave uprush vertically), the greater the landward transport of sediment toward the growing dune region. From several examples in nature (and laboratory wave tank tests), it is evident that the mechanism of the de Beaumont theory is operative in at least some cases.

Gilbert Theory (1885)

The fundamental mechanism of the de Beaumont theory is cross-shore sediment transport imbalances. The process underlying the Gilbert theory is longshore sediment transport that forms barrier islands progressively as spits extend in a downdrift direction from features such as headlands or inlets. The theory is equally valid for shorelines of emergence and submergence.

As in the case of the de Beaumont theory, there are numerous examples in nature from which it is clear that the Gilbert theory describes the development and growth of small barrier islands quite adequately. An example of this is the southern end of Captiva Island, Florida, where under natural conditions the net southerly longshore sediment transport intermittently forms a spit across Blind Pass, deflecting the channel to the south. As the channel becomes less and less hydraulically efficient, the conditions at the northern end of the spit become more conducive to tidal

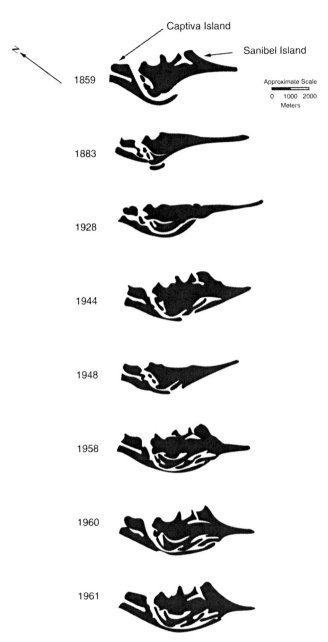

Figure 3.26 Cyclical spit development from south end of Captiva Island and overlapping of the north end of Sanibel Island, Florida.

breakthrough and the formation of a new inlet, which then results in the start of a new cycle with the formation of a new barrier island gulfward of that established earlier. Figure 3.26 shows a series of charts that demonstrate the building of successive barrier islands extending southward from Captiva Island and overlapping Sanibel Island.

Summary of Barrier Island Models

Three different mechanisms for the formation of barrier islands have been discussed. One is cross-shore sediment transport associated with a profile having an "excess of sediment." The barriers are due to the construction of an offshore bar from the material residing on the ocean bottom, which then continues to build into a subaerial feature that continually grows upward primarily through the influence of the wind. The second mechanism, attributed to Gilbert, is longshore sediment transport. The barrier results from the development of spits, which are occasionally breached to form barrier islands. Finally, the model of Hoyt, presented earlier, assumes the submergence of ridge-like coastal features. In actuality, all three theories are likely to be valid but apply at different locations or times. Swift (1968) concluded that the three models are not mutually exclusive, but they do indicate different sources of sand. The important and prevailing aspect of barrier formation is that a substantial source of sand must be available.

3.4.5 OTHER COASTAL TYPES

In addition to the shorelines of submergence and emergence, there are other types of shorelines as shown in Table 3.1. We will confine our analysis to shorelines with beaches.

3.4.5.1 Deltaic Shorelines

A coastline in the vicinity of a large sediment-laden river, such as the Mississippi River, is dominated by the presence of a delta at the mouth of the river because sediment is provided to the coast at a rate greater than the waves can remove it. Shepard and Wanless (1971) described the deltas as cuspate, arcuate, and digitate (birdfoot), depending on the planform of the delta (see Figure 3.27). When the

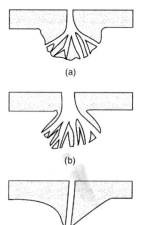

(a)

(b)

(c)

Figure 3.27 Schematic forms of river deltas. (a) Arcuate; (b) digitate; (c) cuspate.

supply of sediment is greater than the erosive forces on the delta and the river has divided into a number of distributaries, the delta is arcuate. An old abandoned delta left behind as another portion of the delta becomes more active or the supply of sediment in the river is reduced becomes digitate and then disappears. A cuspate delta occurs when the river does not have distributaries.

Deltaic shorelines change in planform often as they erode, switch channels, consolidate, and sink, particularly if there is a decreasing supply of sediment from the river. The Mississippi River Delta and the Louisiana shoreline serve as an example.

3.4.5.2 Biogenous Coastlines

Pacific atolls are created when volcanic islands are surrounded by coral reefs, which grow upwards with sea level. Destruction of the coral by storms, fin fish, starfish, and disease creates a calcareous beach sand.

Worm reefs, created through the cementing of sand grains by the Sabellariad worm, provide a noneroding bottom for an otherwise sandy shoreline. These reefs may afford a reasonable amount of coastal protection to the beach during storms.

3.4.6 SUMMARY

Coastlines may be characterized as coastlines of submergence or emergence. The coastline of submergence has an irregular shoreline and a profile steeper than equilibrium. The predominant sediment transport direction is offshore or alongshore. Eroding headlands yield spits, barrier islands, and inlets. Emergent shorelines are much straighter and milder in offshore slope than submerged shorelines, and they are characterized by an excess of sand in the profile. The predominant onshore or alongshore sediment transport permits the growth of barrier islands.

REFERENCES

Bascom, W.N., "The Relationship Between Sand Size and Beach Face Slope," *Trans. Am. Geophys. Union*, 32, 6, 866–874, 1951.

Battjes, J.A., "Surf Similarity," *Proc. 14th Intl. Conf. Coastal Engr.*, ASCE, 466–480, Copenhagen, 1974.

Beach Erosion Board, Interim Report, U.S. Army Corps of Engineers, 1933.

Climate Research Board, *Carbon Dioxide and Climate: A Scientific Assessment*, Washington, DC: National Academy of Science Press, 1979.

Climate Research Board, *Carbon Dioxide: A Second Assessment*, Washington, DC: National Academy of Science Press, 1982.

de Beaumont, L.E., *Leçons de Geologie Pratique*, 7me Leçon–Levees de Sables et Galets. Paris, 1845.

Fairbridge, R.W., "Eustatic Changes in Sea Level," *Physics and Chem. of the Earth*, 4, 99–185, 1961.

Gilbert, G.K., *Lake Bonneville*, U.S. Geological Survey Monog., 1890.

Hayes, M.O., V. Goldsmith, and C.H. Hobbs, III, "Offset Coastal Inlets," *Proc. 12th Intl. Conf. Coastal Engr.*, Washington, DC: ASCE, 1187–1200, 1970.

Hicks, S.D., H.A. Debaugh, Jr, and L.E. Hickman, *Sea Level Variations for the United States, 1855–1980.* Rockville, MD: National Ocean Survey, 1983.

Hoffman, J.S., D. Keyes, and J.G. Titus, "Projecting Future Sea Level Rise; Methodology, Estimates to the Year 2100, and Research Needs," Washington, DC: U.S. Environmental Protection Agency, 121 pp., 1983.

Hoffman, J.S., J.B. Wells, and J.G. Titus, "Future Global Warming and Sea Level Rise," in *Iceland Coastal and River Symposium '85*, G. Sigbjarnarson, ed., Reykjavik, Iceland: National Energy Authority, 1986.

Hoyt, J.H., "Barrier Island Formation," *Bull. Geol. Soc. America*, 78, 1125–1136, 1967.

Imbrie, J., and K.P. Imbrie, *Ice Ages: Solving the Mystery*, NJ: Short Hills, Enslow Publishers, 1979.

Johnson, D.W., *Shore Processes and Shoreline Development*, New York: Wiley, 584 pp., 1919. Reprinted by Hafner Pub. Co., New York, 1972.

Marmer, H.A., "Changes in Sea Level Determined from Tide Observations," *Proc. 2nd Intl. Conf. Coastal Engr.*, Reston, VA: ASCE, 1952.

National Research Council, *Changing Climate*, Washington, DC: National Academy Press, 1983.

Nummedal, D., and R.F. Cuomo, "Shoreline Evolution Along the Northern Coast of the Gulf of Mexico," *Shore & Beach*, 52, 4, 11–17, 1984.

Pirazzoli, P.A., "Secular Trends of Relative Sea Level (RSL) Changes Indicated by Tide Gauge Records," *J. Coastal Res.*, Spec. Issue 1, 1–26, 1986.

Rosen, P.S., "Origin and Processes of Cuspate Spit Shorelines," in *Estuarine Research*, 2, L.E. Cronin, ed., New York: Academic Press, 77–92, 1975.

Shepard, F.P., *Submarine Geology*, 2nd ed., New York: Harper and Row, 557 pp., 1963.

Shepard, F.P., and H.R. Wanless, *Our Changing Coastlines*, New York: McGraw–Hill, 579 pp., 1971.

Sturges, W., "Large-Scale Coherence of Sea Level at Very Low Frequencies," *J. Phys. Oceanog.*, 17, 11, 2084–2094, 1987.

Suh, K.D., and R.A. Dalrymple, "Offshore Breakwaters in Laboratory and Field," *J. Waterway, Port, Coastal, and Ocean Eng.*, Reston, VA: ASCE, 113, 2, 105–121, 1987.

Swift, D., "Coastal Erosion and Transgressive Stratigraphy," *J. Geol.*, 76, 1968.

Valentine, H., "Die Küsten der Erde," *Petermanns Geog. Mitt.*, Ergänzungsheft, 246, 1952.

Walton, T.L., and R.G. Dean, "Use of Outer Bars of Inlets as Sources of Beach Nourishment Material," *Shore and Beach*, 44, 2, 13–19, 1976.

Wilson, B.S., "Seiches," in *Advances in Hydroscience*, Ven Te Chow, ed., 8, New York: Academic Press, 1972.

Wordworth, P.L., "The Permanent Service for Mean Sea Level and the Global Sea Level Observing System," *J. Coastal Res.*, 7, 699–710, 1991.

EXERCISES

3.1 Consider the section of chart shown in Figure 3.28. In which direction is the longshore sediment transport? Discuss your reasoning in detail.

3.2 What are the two dominant hydrodynamic processes for sediment accumulation behind an offshore island?

3.3 What are the two dominant processes that result in longshore sediment transport in the surf zone?

3.4 On the basis of the deposition pattern in the lee of the detached breakwater shown in Figure 3.29, determine the dominant wave direction. Explain your answer.

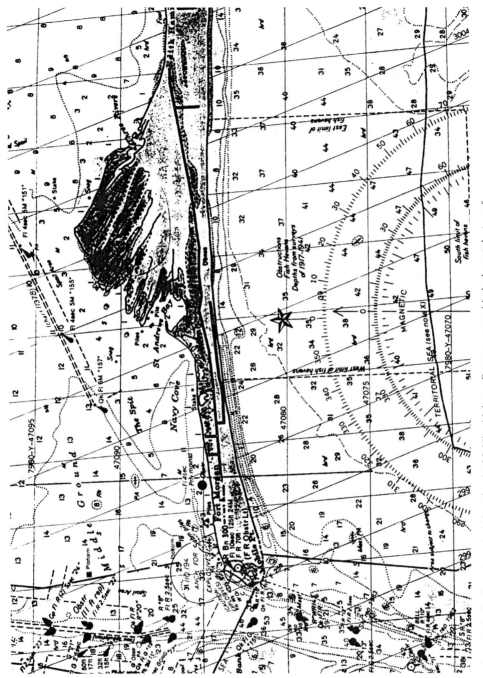

Figure 3.28 Section of nautical chart for the Mobile, Alabama Bay entrance with the bay at the top of the figure and the Gulf of Mexico at the bottom.

67

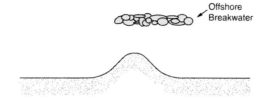

Figure 3.29 Detached breakwater for Problem 3.4.

3.5 Sea level is rising at a rate of 30 cm/century with consequences for barrier islands.

   (a) Consider first the two-dimensional case of a barrier island that maintains its geometry relative to sea level by overwash of sand and landward migration without change of cross section (see Figure 3.30(a)). Suppose that no sand is carried offshore. If the slope on the plane "base" is 1:100, what distance will the barrier retreat in one century?

   (b) Suppose, as a second case (Figure 3.30(b)), that, as the barrier retreats, a portion of the sand is transported offshore and no additional sand is introduced into the system. What effects will this have on the barrier with time? Compare the retreat rate with that determined in **(a)**.

   (c) As a third case (Figure 3.30(c)), suppose that, as the barrier retreats, sand is eroded from the bottom and incorporated into the island. What effects will occur with time? Compare the retreat rate with that computed in **(a)**.

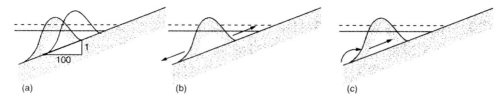

Figure 3.30 Barrier islands for Problem 3.5. (a) Sand volume in barrier unit remains constant; (b) some land lost offshore; (c) some sand is incorporated into the island as it migrates landward.

3.6 A pair of spits is developing east and west from a headland that faces south. A causeway (solid land fill) is later constructed to an offshore island (see Figure 3.31). Discuss, using sketches where helpful, the effects you would expect this construction to have on future spit development.

3.7 Consider the initial planform shown in Figure 3.32. The waves are from the southwest, and the headland is composed of erodible sand and very durable rock. Describe and sketch the resulting planforms that you might expect:

   (a) Just after all of the sand has been eroded.

   (b) A considerable time after all of the sand has been eroded without any significant additional material supplied by the headland.

3.8 The shoreline response is shown for each of the three offshore bathymetries in Figure 3.33. Suggest one or more dominant cause(s) for each of the three cases and discuss how each case differs from the others. Then discuss the probable effect on the shoreline of a storm that includes increased water levels and wave heights.

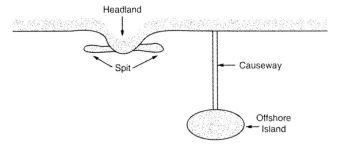

Figure 3.31 Spit and causeway for Problem 3.6.

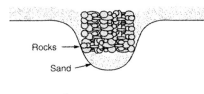

Figure 3.32 Eroding headland for Problem 3.7.

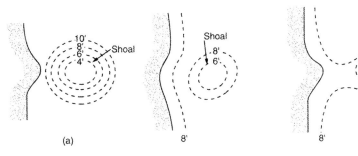

(a)

Figure 3.33 Three offshore shoals.

PART TWO

HYDRODYNAMICS OF THE COASTAL ZONE

CHAPTER FOUR

Tides and Storm Surges

Most waves are formed by relatively steady winds blowing over the water surface, but, in some cases, waves can be formed by unusual meteorological events. As an example, squall line surges are caused by the movement of a relatively small atmospheric pressure perturbation at the approximate speed of the waves being generated. This results in energy being transferred to the same waves continuously, causing them to build up at a relatively rapid rate in a type of resonance. Examples include an event in 1954 that caused seven fatalities along the Lake Michigan shoreline (Ewing, Press, and Donn 1954) and one in 1929 that also caused fatalities (Proudman 1929, 1952). Some surprising characteristics of these types of waves are the small number of substantial waves produced (one to three), the height of the waves (1 to 3 m), and the small magnitude of the pressure change required to cause this phenomena (1 to 3 mbar).

On the evening of July 3, 1992, at Daytona Beach, Florida, a freak wave about 2 m high surged up the beach on an otherwise calm night. Driving on the beach is allowed in this area, and several cars were carried up the beach by the wave and jammed under the boardwalk. A park ranger driving down the beach was overtaken by the wave, which floated his jeep up the beach on the uprush and then back down; the vehicle was then set down near where it had been picked up. The ranger continued on his drive. The initial speculations about the cause of this surprising event were numerous and far-ranging, including waves produced by slumping of sediment on the continental shelf, a meteorite impact, and other more esoteric and unlikely causes. Observers soon reported a sudden increase in wind speed occurred with the change from a warm and humid evening to cool and windy conditions. Later, radar images demonstrated that a squall originated offshore of Georgia and moved south along first the Georgia and later the Florida coast at a speed that corresponded approximately to that of a long free wave at the water depth associated with the offshore distance of the squall line propagation path (see Thieke, Dean, and Garcia 1994; Churchill, Houston, and Bond 1995; Sallenger et al. 1995). A similar wave occurred along the west coast of Florida 3 years later with a wave height of over 3 m from Tampa Bay to Naples, Florida (Paxton and Sobien 1998).

4.1 INTRODUCTION

The coastline responds to various forcing mechanisms that provide the energy and momentum to drive the littoral processes. In addition to the long-term mechanisms discussed in the previous chapter, there are short-term forces, including storm, wave, current, and wind effects. This and the following chapter focus on these forces, which act over periods of seconds, hours, and days rather than years and centuries. Relative sea level rise (discussed in Chapter 3) results in a slowly changing water level near the shoreline with rates of rise measured in millimeters per year. More rapidly changing sea levels are caused by astronomical tides (with rates of water level rise measured in meters per hour), storm surges (meters per hour), and waves (meters per second). This chapter will cover the slower water level changes (tides and surges), and the next will deal with waves and wave-induced phenomena such as longshore and rip currents.

4.2 ASTRONOMICAL TIDES

The daily rise and fall of the tides is caused by the gravitational attractions of the moon and sun acting on water particles on the surface of the earth. A convenient model for examining the mechanisms of the tides is the *equilibrium theory of tides* proposed by George H. Darwin (1898), the son of the famous biologist Charles Darwin. We note that the first modern theory of tides was developed by Sir Isaac Newton as a means of proving his law of gravitation.

For the sake of convenience, we will imagine that the earth is perfectly spherical and totally covered with a uniformly thick layer of water. Thus, we can neglect the influence of the continents on the tides.

Now, there are two possible mechanisms for tides on our imaginary earth. The first is the rotation of the earth about its axis and the resulting centrifugal force on the water particles. This would cause the water particles near the equator to bulge outwards. However, as observers on this world at a fixed location, we would see no variation in water level during a day. The tidal bulges would vary with latitude but not temporally at a fixed location. Thus, we need another mechanism to explain the tides.

A second mechanism is the gravitational attractions of the moon and the sun. To determine these effects we use Newton's equation for gravitational attraction:

$$F = \frac{Gm_1m_2}{r^2} \tag{4.1}$$

Here, r is the distance between the centers of the two bodies with masses m_1 and m_2, and G is the gravitational constant, which is 6.6×10^{-11} m^{2}N/kg^2. We take the earth–moon system first, although the methodology is exactly the same for the earth–sun system. The mutual attraction, earth on the moon and vice versa, must be balanced by a centrifugal force (otherwise the two bodies would have a net attraction for each other with dire consequences for us), which results in a rotating system (like a weightlifter's dumbbell) with an axis of rotation actually located within the earth. The period of rotation for the combined earth–moon system is 27.3 days. The centrifugal acceleration of this rotation implies that there is a *uniform* force on all the water particles that is directed away from the moon.

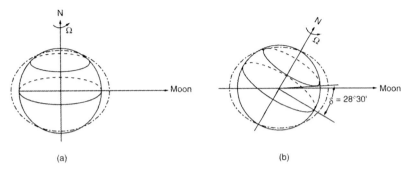

(a)   (b)

Figure 4.1 Astronomical tidal bulges (reprinted, with permission, from Dean 1966). (a) Moon declination $= 0°$; (b) moon declination $\delta = 28°30'$.

The forces acting on a single particle must be balanced. These forces are the gravitational attraction of the earth, which is the same for all the water particles, the gravitational attraction of the moon, which is stronger for those particles on the side of the earth facing the moon, and the centrifugal acceleration about the common axis. If these are added for all the particles of water on our hypothetical planet, it can be shown (e.g., Dean 1966) that there is a bulge in our sphere of water on the side of the earth facing the moon where the moon's gravitational attraction is greater, and there is also a bulge on the side away from the moon (where the centrifugal force is the more dominant force). These tidal bulges are illustrated in Figure 4.1.

To determine the tides at a particular location, we now imagine the earth as rotating through the tidal bulges. The types of tides we obtain will depend on the *declination* of the earth with respect to the earth–moon rotational plane; this declination, relative to a line joining the centers of the earth and the moon, ranges from 0 to 28.5°. If the declination of the earth is zero, a point located on the earth's surface rotating through the two tidal bulges during a day will experience two equal high and low tides per day. If, on the other hand, the earth's axis is tilted, it is possible to have two unequal high and two unequal low tides during the day. These types of tides are *semidiurnal*; however, in the second case, there is a *diurnal inequality* in tidal elevations. Alternatively, depending on the latitude, the site can experience a single high and low tide per day, as shown in Figure 4.2, which is referred to as a *diurnal tide*.

The sun, despite its immense mass, creates a tide less than half that of the moon owing to its much larger distance from the earth (Eq. (4.1)). However, the sun's influence is important because the solar tide can add to or subtract from the lunar tide. The two tidal effects add during the time of the new moon, when both the moon and the sun are on the same side of the earth, and during the full moon, when again they are aligned but on opposite sides of the earth. The coincidence of the solar and lunar tidal components creates the high *spring tides*. During the first and last quarters of the moon's phase, the tidal components tend to cancel, resulting in *neap tides*. Because there are two tidal bulges on the earth, and the moon rotates about the earth in 27.3 days, the timing between spring tides is approximately 14 days (27.3/2).

In addition to the variation in the tides that are a result of the relative positions of the moon and the sun (the fortnightly inequality), there are effects caused by the varying distances between the moon and the earth (the parallactic inequality), which

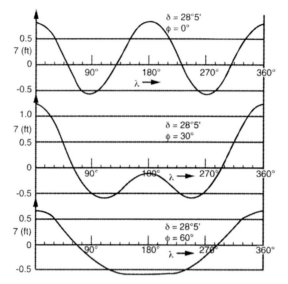

Figure 4.2   Tidal ranges (reprinted, with permission, from Dean 1966).

has a period of about 27.55 days, and the varying distance between the earth and the sun. Examples of these tides are the *perigean tides*, which are semidiurnal tides that occur monthly when the moon is closest to the earth while traveling its elliptical orbit, and *tropic tides*, which are diurnal tides occurring twice monthly caused by the extremes of the moon's declination. Doodson (1921) developed the amplitude and phase for 396 different tidal components. Some of the more important of these are shown in Table 4.1. In principle, by superimposing these tidal components, the tides at a given site could be calculated theoretically for many years. However, the presence of the continents and other local effects alter the relative sizes of the various tidal components for a given site. Figure 4.3 shows some of the possible types of tides experienced during a month at different locations around the world.

Land masses and offshore bathymetry alter the relative sizes of the tidal components; consequently, the local tides must be determined by measurements. Typically, a tide recorder is installed for a given length of time (in the United States this

Table 4.1   Important Components of the Astronomical Tides

Tidal Type	Symbol	Period (Solar Hour)	Relative Amplitude	Description
Semidiurnal	M_2	12.42	100.0	Principal lunar tide
	S_2	12.00	46.6	Principal solar tide
	N_2	12.66	19.1	Monthly variation in lunar distance
	K_2	11.97	12.7	Changes in declination of sun and moon
Diurnal	K_1	23.93	58.4	Solar–lunar constituent
	O_1	25.82	41.5	Principal lunar diurnal constituent
	P_1	24.07	19.3	Principal solar diurnal constituent
Longer	M_f	327.86	17.2	Moon's fortnightly constituent

Source: Doodson 1921.

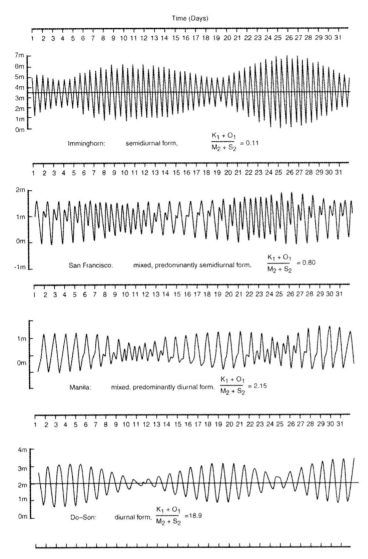

Figure 4.3 Types of possible tides (from Defant 1958. © Springer-Verlag; used with permission).

function is carried out by the National Ocean Survey). The longest significant periodicity in the astronomical tidal record is 18.61 years; however, it is not necessary to keep gauges in for that period of time to obtain reliable data. The recorded data are then analyzed for the various tidal components.

Several tidal datums are used by hydrographic charting agencies to reference the offshore bathymetry because the water level of the ocean surface is variable. The National Ocean Survey uses mean sea level (MSL), and the U.S. Army Corps of Engineers uses mean low water, whereas the British Hydrographic Office uses the lowest astronomical tide* as a chart datum. Other tidal datums in use are the mean

---

* The lowest astronomical tide is defined as the lowest tide that can be created by astronomical forcing over 18.61 years. Nonastronomical forcing, such as excessive winds, could possibly result in lower tides.

higher high water (the mean of the higher tides in sites with diurnal inequalities), mean lower low water, and the national geodetic vertical datum (NGVD). The NGVD is a fixed U.S. reference datum based on the 1929 MSL that avoids the problem of annual variations in MSL; however, this means that the NGVD may have no relationship to the local MSL.

Wood (1978) examined the importance of the tidal extremes; that is, when there is a coincidence of the various tidal components (such as the spring and the perigean tides) to make extremely large tides. He referred to these as the *perigean spring tides*. Zetler and Flick (1985) determined the maximum annual astronomical tides for various sites on the California coastline for the years 1983–2000. The interannual variation in these tides is about 7 percent. These tides in and of themselves are not dangerous, but their occurrence simultaneously with storms can lead to catastrophic effects. Wood has found more than 100 historical coastal floodings (over a 341-year history) when storms occurred at the same time as the perigean spring tide.

Some of the measured tide is not astronomically induced, because storms, atmospheric pressure disturbances, and other mechanisms can lead to short-term changes in local water levels. The El Niño (Philander 1990) is an important phenomenon on the U.S. West Coast that occurs periodically and creates elevated sea levels, owing to the collection of warm water near the coast and the change in wind patterns to create more onshore winds. In 1982–83 and 1997–98, increased shoreline erosion occurred in California because of the coincidence of storms and the El Niño elevated sea levels.

4.3 STORM SURGES

Hurricanes, typhoons, monsoons, and tropical storms consist of large wind fields driven by pressure gradients from a central low pressure and temperature gradients in the atmosphere. These storms can have winds exceeding 200 km/hr over large areas. The winds create storm surges by blowing the ocean water up against the coastline. For some sites, either with shallow offshore bathymetry, such as the Gulf of Mexico, or, where the coastline forms a funnel, such as offshore of Bangladesh, which lies on the delta of the Ganges River, the storms can cause immense flooding and the loss of many thousands of lives.

A storm surge can create high water levels lasting for several days. The record high storm surge in the United States occurred in 1969 when Hurricane Camille struck the coast of Louisiana with winds exceeding 270 km/hr and killed 256 people. The maximum storm surge was estimated to be almost 7 m at Pass Christian, Mississippi. Fortunately, this was a fast-moving storm, and this high water level lasted only several hours. The more recent Hurricane Hugo (occurring in 1989) caused storm surges in excess of 6 m in McClellandville, South Carolina, which is located near Bull's Bay, a broad shallow bay on the Atlantic Coast of South Carolina.

The destructiveness of a storm surge depends on its magnitude and duration, the associated wind-driven waves, and its coincidence with the astronomical tides. A storm intensity scale used in the United States for hurricanes is the Saffir–Simpson Scale (Saffir 1977, Simpson 1979), which is shown in Table 4.2. Hurricanes are placed

Table 4.2 Saffir–Simpson Scale for Hurricane Intensity

Scale	Pressure	Winds	Surge	Damage
1	≤980 mbar	74–95 mph	4–5 ft	Minimal
2	965–979	96–110	6–8	Moderate
3	945–964	111–130	9–12	Extensive
4	920–944	131–155	13–18	Extreme
5	<920	>155	>18	Catastrophic

in one of five classes (Category 5 is the most destructive), depending on the storm's central pressure, maximum wind speed, and surge elevation.

Some of the more destructive storms on the U.S. East Coast have been associated with northeasters, which can last for days. The famous Ash Wednesday storm of 1962 persisted over five high tides, creating immense flooding and storm damage for the Middle Atlantic states.*

Several attempts have been made to develop an equivalent to the Saffir–Simpson scale for northeasters. Halsey (1986) developed a scale based on the damage created by the storm rather than difficult-to-measure storm parameters. A key parameter in her criterion is storm duration (based on the integer number of tidal cycles). Another scale was developed by Dolan and Davis (1992). They chose the square of the wave height as a measure of wave energy and the duration of the storm (t_d) as a parameter. By using cluster analysis to examine more than 1300 northeasters that created waves greater than 5 ft at Cape Hatteras, NC, Dolan and Davis found that the storms could be classified as shown in Table 4.3.

Recently, Kriebel et al. (1997) developed a northeaster *risk index* based on the potential for the northeaster to erode the beach or dune. By examining the response of a beach representative of the Delaware shoreline to a combination of wave height (H), surge elevation (S) (both in feet), and duration (hours), they found empirically that the following parameter is a good predictor of shoreline recession:

$$I = HS\,(t_d/12)^{0.3},$$

where the storm duration t_d is divided by 12 hours to get the number of tidal cycles, as in Halsey's scale. They then scaled this parameter to calibrate it to the March 1962 storm taken as a Class 5 storm. The final risk index $\mathcal{I}$, with a range of 0 to 5, is

$$\mathcal{I} = 5I/400 \tag{4.2}$$

Although this procedure is general and can be applied to other locales subject to northeasters, this particular $\mathcal{I}$ is calibrated and valid only for the Delaware coast.

---

* The U.S. record for the most damage incurred by a storm (up to 1999) goes to Hurricane Andrew: $20 billion! This Saffir–Simpson Category 5 storm made landfall just south of Miami on August 24, 1992, and destroyed numerous structures mostly by the action of the wind. The damage total tops the previous record of $7 billion caused by Hurricane Hugo (Class 4 when it made landfall near Charleston, South Carolina, on September 22, 1989) by a factor of three (Schmidt and Clark 1993). More information on Hurricane Hugo can be found in a special issue of *Shore and Beach*, Vol. 58, No. 4, October 1990.

Table 4.3 Dolan–Davis Scale for Northeasters

Scale	$H^2 t_d$	Damage
1	$\leq 771\ \text{ft}^2\ \text{hr}$	Weak
2	771–1760	Moderate
3	1760–10,000	Significant
4	10,000–25,000	Severe
5	>25,000	Extreme

Many remedies are available to protect lives and property from the storm surge. In many locations, adequate forecasting of these storms can prevent major loss of life by evacuation. However, the abandoned property could be destroyed. Large engineering works can in many instances result in saving both lives and property. The Galveston seawall, built in 1903, was designed to protect that city from devastating hurricane surges (see Chapter 12, Section 12.7).

Storm surge barriers in rivers and estuaries serve to prevent the surge from reaching low-lying areas. These barriers are found in many locations, and the largest is the Osterschelde barrier in the Netherlands near the border with Belgium. This massive structure, built across the mouth of the Osterschelde estuary, has gates that permit the normal tides to flow in and out of the estuary; however, when a large storm surge is forecast, the gates are closed. Surge barriers are presently being planned for the inlets to the lagoon surrounding Venice, Italy.

4.3.1 STORM SURGE COMPONENTS

A storm surge consists of several components, arising from the barometric pressure reduction in the low-pressure storm, the wind stress, the Coriolis force, and the wave set-up. Each of these components will be discussed in the following paragraphs.

The *barometric tide* is the response of the coastal waters to the low pressure at the center of the storm. At the site of the storm, the water is drawn up into the low-pressure region, or, it can be viewed as being pushed up into the low-pressure region by the surrounding high pressure air. Figure 4.4 shows a schematic of this barometric

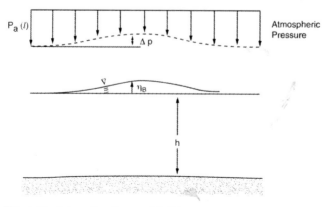

Figure 4.4 Schematic of barometric tide.

response mechanism. However, this is not the major cause of storm surges, for its magnitude, which is on the order of tens of centimeters, is too small.

We can roughly model this phenomenon by using hydrostatic principles. For this purpose, we assume that the water and the storm are stationary. Then, from hydrostatics, the pressure $p(x, z)$ along the ocean bottom $(z = -h)$ must be constant. At a large distance $(x = \ell)$ away from the storm, the bottom pressure is

$$p(\ell, -h) = \gamma h + p_a(\ell), \tag{4.3}$$

where γ is the specific weight of sea water, $\approx 10{,}000\,\text{N/m}^3$, h is the water depth, and p_a is the local atmospheric pressure at ℓ. This bottom pressure must be the same as found under the center of the storm, $p(0, -h)$; otherwise, water would accelerate according to the pressure difference between these points, which is contrary to our hydrostatic assumption. The bottom pressure at the center of the storm is

$$p(0, -h) = \gamma(h + \eta_{\text{B}}) + p_a(\ell) - \Delta p, \tag{4.4}$$

where $h + \eta_{\text{B}}$ is the total water depth there, and η_{B} is the barometric response to the low pressure at the center of the storm (measured in terms of a pressure difference Δp from the ambient pressure p_a. Meteorologists in the United States usually measure Δp in inches of mercury, a holdover from the use of mercury barometers, or in millibars, which are related by 1 mbar $= 0.0295$ in. Hg.) Now, by equating Eqs. (4.3) and (4.4) for pressure, η_{B} is found:

$$\eta_{\text{B}} = \frac{\Delta p}{\gamma} \tag{4.5}$$

As a rule of thumb, $\eta_{\text{B}} = 1.04 \Delta p$, when Δp is measured in millibars and η_{B} has the units of centimeters.

A simple dynamic analysis of the storm surge (Dean and Dalrymple 1991, Chapter 5) shows that the surge can be strongly amplified if the low-pressure system moves exactly at the speed of a shallow water wave, thereby creating a type of resonance. This is the mechanism for the squall line surge discussed in the preface to this chapter.

The *wind stress tide* is created by the frictional drag of the wind blowing over the water. The coupling between the two is through the wind stress, which is the horizontal force per unit area exerted on the water surface. Wind stress cannot be determined theoretically, and numerous wind stress determination experiments have been conducted. The empirical formula for wind stress is

$$\tau_s = \rho c_f W^2,$$

where ρ is the density of the water (*not* the air), W is the wind speed (usually measured at an elevation of 10 m in meters per second), and c_f is a dimensionless friction coefficient. Typical values for c_f are around 1.2×10^{-6} to 3.4×10^{-6}, with the higher values for greater wind speeds, because of the increased roughness of the water surface. (Van Dorn 1953, among others, presents a relationship based on field measurements for c_f versus wind speeds.)

To determine the magnitude of the wind stress tide, we examine the forces operating on a control volume consisting of a column of water of length Δx and unit

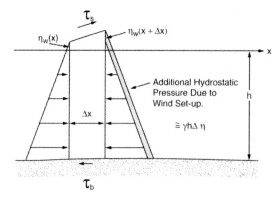

Figure 4.5   Schematic of wind stress force balance.

width with height $h + \eta_w$, as shown in Figure 4.5. The wind stress τ_s acting on its surface creates a force $\tau_s \Delta x$ acting in the wind direction. The hydrostatic forces act on both sides of the element; however, on the right side of the column, the water level is increased by an amount $\Delta \eta_w$ as the result of wind stress. In addition, at the bottom, there is a bottom shear stress τ_b, which, by convention, acts in the direction shown and is caused by the action of the flowing water in contact with the bottom. The force balance on the column is

$$\frac{1}{2}\rho g(h + \eta_w)^2 - \frac{1}{2}\rho g(h + \eta_w + \Delta \eta_w)^2 + \tau_s \Delta x - \tau_b \Delta x = 0 \qquad (4.6)$$

Neglecting the assumed small product, $(\Delta \eta_w)^2$, we obtain

$$\frac{\Delta \eta_w}{\Delta x} = \frac{(\tau_s - \tau_b)}{\rho g(h + \eta_w)} \qquad (4.7)$$

As Δx becomes small, we can replace the left-hand side with a derivative, yielding a differential equation to solve for η_w:

$$\frac{d\eta_w}{dx} = \frac{(\tau_s - \tau_b)}{\rho g(h + \eta_w)} \qquad (4.8)$$

This simple one-dimensional representation of the wind stress contribution to the storm surge includes many important features of the problem. As shown in Eq. (4.8), the water surface slope is related to the wind stress acting on the water surface. The greater the wind stress, the greater the water surface slope, which corresponds to your own experience of blowing across a hot cup of coffee. Also, as the water depth becomes shallow, the water surface slope is larger for the same wind stress owing to the $(h + \eta_w)$ term in the denominator. This accounts for the large storm surges in the Gulf of Mexico, which has a very broad, shallow continental shelf when compared with Atlantic and Pacific coast shorelines.

    Simple expressions for the wind stress tide can be obtained from this equation, but remember that it is only a steady-state, one-dimensional equation and is only useful for getting general ideas about the magnitude of the storm surge. For example, solving Eq. (4.8) for a constant continental shelf depth h and uniform and constant

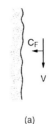

(a)

Figure 4.6 Schematic of Coriolis tide. (a) Plan view showing coastal current, V, and associated Coriolis force (C_F); (b) profile showing Coriolis induced storm tide, η_c.

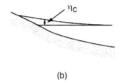

(b)

wind stress, we obtain

$$\eta_w = h\left(\sqrt{1 + \frac{A_s x}{\ell}} - 1\right), \quad \text{where } A_s = \frac{2n\tau_s \ell}{\rho g h^2}, \tag{4.9}$$

x is located at the edge of the continental shelf pointing shoreward, and the shelf width is ℓ. The factor n has been introduced here to account for the bottom shear stress. It is formally $n = 1 - \tau_b/\tau_s$, which usually has values of about 1.15–1.30. It is greater than unity, for the bottom shear stress is negative because the water is usually flowing offshore at the bottom. The dimensionless parameter A_s is the ratio of the wind shear stress force applied to the shelf region to the hydrostatic force. Larger values of A_s lead to larger storm surges. Dean and Dalrymple (1991) also present the case of a shoreline with a uniform sloping offshore profile.

The *Coriolis tide* η_c occurs when the storm forces strong currents to flow along the shoreline. If, for example, in the northern hemisphere the wind-induced coastal current is flowing southward along the U.S. East Coast, as in Figure 4.6, the Coriolis force due to the rotation of the earth can only be balanced by a hydrostatic variation in the water surface. The governing equation for this case is

$$\frac{\partial \eta_c}{\partial x} = \frac{fV}{g}, \tag{4.10}$$

where f is the Coriolis parameter, which is equal to $2\Omega \sin \phi$, where Ω is the angular rotation rate of the earth (7.272×10^{-5} radians/s),* ϕ is the latitude, and V is the magnitude of the depth averaged current. This storm surge component can be important for large currents, but it can also act to reduce a storm surge when the current is flowing in the opposite direction.

Wave setup is a phenomenon that occurs primarily within the wave breaking zone and results in a superelevation of the water level η_{su}. The detailed explanation

---

* This value corresponds to one rotation of the earth (2π radians) in 24 hours.

of wave setup will be presented in the next chapter; here, suffice it to say that the setup is due to the transfer of wave momentum from the breaking waves to the water column. A simple expression for a beach profile that is monotonically decreasing in the shoreward direction is

$$\eta_{su} = -\frac{H_b^2}{16h_b} + \mathcal{K}(h_b - h), \quad \mathcal{K} = \frac{3\kappa^2/8}{1 + 3\kappa^2/8}, \tag{4.11}$$

which applies within the surf zone. The κ is determined from an assumption that within the surf zone the wave height and the local water depth are related by $H = \kappa h$. The parameter H_b is the breaking wave height at the offshore edge of the surf zone given by $H_b = \kappa h_b$. The value of the setup at the still water line is obtained as follows by setting $h = 0$ and using $\kappa = 0.78$:

$$\eta_{su} = 0.188 H_b \tag{4.12}$$

The maximum value of the setup is found by setting the total water depth equal to zero or $h = -(\eta_{su})_{max}$

$$(\eta_{su})_{max} = \frac{\eta_{su}(h = 0)}{1 - \mathcal{K}} \approx 0.23 H_b \tag{4.13}$$

As an example, for hurricane waves 5 m in height, the maximum wave setup would be on the order of 1 m, which is a substantial portion of a storm surge on the order of 2–6 m.

Another measure of η_{su} was made by Guza and Thornton (1981), who carried out extensive measurements of setup and wave climate on a California beach. Their results showed that, for realistic wave climates, an adequate measure of the waves for use in setup calculations is the offshore significant wave height H_{so}, which is defined as the average of the one-third highest waves. They found that the wave setup could be written as

$$(\eta_{su})_{max} = 0.17 H_{so} \tag{4.14}$$

The sum of these four components – the barometric tide, the wind stress tide, the Coriolis tide, and the wave setup – provides the total storm surge. In general, the major contributor to the surge is the wind stress component. However, remember that the results developed here were for steady-state conditions; dynamical effects can be significant.

4.3.2 PREDICTION OF THE STORM SURGE

People model storm surges in detail for a variety of reasons, including most, if not all, of the preceding components for real shorelines as a function of time. One possible reason is the prediction of the 50- or 100-year storm surge at a coastal site for the design water levels for coastal structures or the establishment of hazards and insurance rates for coastal communities. Another purpose, involving real-time modeling, is for hazard mitigation and public safety. Alternatively, the hindcasting of a given historical storm surge may be carried out to determine the nature and extent of the surge or to calibrate and verify a "new" surge model. The last problem is far easier than the first two, for often data concerning the storm parameters and the wind fields can be obtained. Owing to the lack of long-term water level records or any records at all, statistical surge information is usually unavailable. For actual surge occurrences, often not much data are available except from a few established tide gauge sites and site-specific evidence such as high water levels inside buildings, elevation of wave damage, and other indicators of storm water level.

For any surge model, an adequate representation of a wind field is necessary because the spatial extent of the wind and pressure fields associated with a storm are needed as input. Further, the path of the storm and its correct forward speed are necessary. One simple model that has been used is the irrotational vortex from hydrodynamics, which can be characterized by a pressure difference between the central pressure of the vortex and the surrounding atmospheric pressure.

One simple model of the storm surge is the bathystrophic storm surge developed by Freeman, Baer, and Jung (1957). This quasi-static model involves solving reduced equations of motion for the storm surge induced by the wind stress and the Coriolis tide. The governing equations are

$$g(h + \eta)\frac{\partial \eta}{\partial x} = (h + \eta)fV + \frac{\tau_{sx}}{\rho} \tag{4.15}$$

$$\frac{\partial V}{\partial t} = \frac{\tau_{sy} - \tau_{by}}{\rho(h + \eta)}, \tag{4.16}$$

where x and y are directed onshore and alongshore, respectively. The first equation balances the hydrostatic forces of the surge in the x direction with the Coriolis force induced by flow V in the y direction and the wind shear stress component in the x direction. The second equation balances the inertial force in the y direction with the surface and bottom shear stresses in the y direction. This quasi-static model neglects the onshore flow U, which is taken as zero. These two equations can be solved readily by finite difference methods to yield a time history of the storm surge elevation as a function of the changing wind (stress) field of the storm.

The National Weather Service uses the SLOSH (Sea, Lake, and Overland Surges from Hurricanes) model (Jelesnianski, Chen, and Shaffer 1992) operationally to predict storm surges. The numerical SLOSH model runs on a grid system. Convective accelerations are neglected, but some nonlinearities are included (particularly in shallow water). Comparisons with historical storms show that the model is accurate to within ±20 percent.

The Federal Emergency Management Administration (FEMA) of the U.S. Government contracted for the development of a more elaborate "standard" model (Tetra Tech 1981) for the calculation of storm surges for flood insurance purposes. This model uses rectangular grids to represent the offshore geometry with a depth defined at the center of each grid. The model's parameterized hurricane wind field is defined by five parameters: Δp, the central pressure deficit; R, the radius from the eye of the storm to the region of maximum winds; V_F, the forward translation speed of the storm; θ, the translation direction of the storm; and, finally, a parameter positioning the hurricane at some time during its history.

The hurricane is started when it is well seaward of the coastal area of interest and the finite difference forms of the equations of motion and continuity are applied, advancing the hurricane at each time step. These equations of motion include the barometric pressure, surface (wind) and bottom stresses, and the Coriolis forcing directly. Neither the FEMA model nor any of the other models have included the effect of wave setup explicitly, although its importance is recognized and is likely to be included soon.

To determine the 100-year storm surge correctly for a given coastal site, several very important steps must be carried out. First, an adequate historical database of storm parameters is necessary. This requires examining historical storm tracks and meteorological observations. Ideally, a good joint probability distribution of central pressures and storm sizes is available. Then, a coupled wind and surge model is used to generate many realizations of hurricanes. Each of these storms will have a random or, for some parameters, a correlated selection of parameters from the historical statistical database. The resulting storm surges are then ranked by magnitude, and their historical return periods are determined.

REFERENCES

Churchill, D.D., S.H. Houston, and N.A. Bond, "The Daytona Beach Wave of 3–4 July, 1992: A Shallow Water Gravity Wave Forced by a Propagating Squall Line," *Bull. Amer. Meteor. Soc.*, 76, 21–32, 1995.

Darwin, G.H., *The Tides and Kindred Phenomena in the Solar System*, New York: Houghton–Mifflin, 1898.

Dean, R.G., "Tides and Harmonic Analysis," in *Estuary and Coastline Hydrodynamics*, A.T. Ippen, ed., New York: McGraw–Hill Book Co., 1966.

Dean, R.G., and R.A. Dalrymple, *Water Wave Mechanics for Engineers and Scientists*, Singapore: World Scientific Press, 353 pp., 1991.

Defant, A., *Ebb and Flow; The Tides of Earth, Air, and Water*, Ann Arbor: University of Michigan Press, 121 pp., 1958.

Dolan, R., and R.E. Davis, "An Intensity Scale for Atlantic Coast Northeast Storms," *J. Coastal Res.*, 8, 4, 840–853, 1992.

Doodson, A.T., "The Harmonic Development of the Tide-generating Potential," *Proc. Roy. Soc. London*, A, 100, 1921.

Ewing, M., F. Press, and W.L. Donn, "An explanation of the Lake Michigan wave of 26 June 1954," *Science*, 120, 684–686, 1954.

Freeman, J.C., L. Baer, and G.H. Jung, "The Bathystrophic Storm Tide," *J. Mar. Res.*, 16, 1, 1957.

Guza, R.T., and E.B. Thornton, "Wave Set-up on a Natural Beach," *J. Geophys. Res.*, 96, 4133–4137, 1981.

Halsey, S.D., "Proposed Classification Scale for Major Northeast Storms: East Coast USA, Based on Extent of Damage," *Geological Society of America*, Abstracts with Programs (Northeastern Section) 18, 21, 1986.

Jelesnianski, C.P., J. Chen, and W.A. Shaffer, "SLOSH: Sea, Lake, and Overland Surges from Hurricanes," National Oceanic and Atmospheric Administration (NOAA) Technical Rpt. NWS 48, 71 pp., April 1992.

Kriebel, D., R. Dalrymple, A. Pratt, and V. Sakovich, "A Shoreline Risk Index for Northeasters," *Proc. Conf. Natural Disaster Reduction*, ASCE, 251–252, 1997.

Paxton, C.H., and D.A. Sobien, "Resonant Interaction Between an Atmospheric Gravity Wave and Shallow Water Wave Along Florida's West Coast," *Bull. Amer. Meteor. Soc.*, 79, 12, 2727–2732, 1998.

Philander, S.G.H. *El Niño, La Niña, and the Southern Oscillation*, San Diego: Academic Press, 289 pp., 1990.

Proudman, J., "The Effects on the Sea of Changes in Atmospheric Pressure," *Geophysical Supplement, Monthly Notices of Royal Astronomical Society*, London, 2, pp. 197–209, 1929.

Proudman, J., *Dynamical Oceanography*, New York: Wiley, pp. 295–300, 1952.

Saffir, H.S., "Design and Construction Requirements for Hurricane Resistant Construction," New York, ASCE, Preprint No. 2830, 20 pp., 1977.

Sallenger, A.H., Jr., J.H. List, G. Elfenbaum, R.P. Stumpf, and M. Hansen, "Large Wave at Daytona Beach Florida, Explained as a Squall Line Surge," *J. Coastal Res.*, 11, 1383–1388, 1995.

Schmidt, D.V., and R.R. Clark, "Impacts of Hurricane Andrew on the Beaches of Florida," *Proc. 1993 Natl. Conf. Beach Pres. Technology*, Florida Shore and Beach Pres. Assoc., Talahassee, 279–308, 1993.

Simpson, R.H., "A Proposed Scale for Ranking Hurricanes by Intensity," *Minutes of the Eighth NOAA, NWS Hurricane Conference*, Miami, 1979.

Tetra Tech, Coastal Flooding Storm Surge Model, Washington, DC: Federal Emergency Management Administration, 1981.

Thieke, R.J., R.G. Dean, and A.W. Garcia, "Daytona Beach 'Large Wave' Event of 3 July 1992," *Proc. 2nd Intl. Symp. Ocean Wave Meas. and Analysis*, ASCE, New Orleans, 45–60, 1994.

Van Dorn, W.C., "Wind Stress on an Artificial Pond," *J. Mar. Res.*, 12, 1953.

Wood, F.J., *The Strategic Role of Perigean Spring Tides in Nautical History and North American Coastal Flooding, 1635–1976*, U.S. Dept. Commerce, National Oceanic and Atmospheric Administration, 538 pp., 1978.

Zetler, B.D., and R.E. Flick, "Predicted Extreme High Tides for California: 1983–2000," *J. Waterway, Port, Coastal, and Ocean Engr.*, 111, 4, 758–765, 1985.

EXERCISES

4.1 Using the tidal constituents in Table 4.1, calculate and plot tidal charts similar to those of Figure 4.3 for a month. Take different random phases for the constituents.

4.2 Consider a continental shelf of uniform depth $h = 5$ m and a width of 60 km. Considering a wind stress coefficient $C_f = 3.0 \times 10^{-6}$, calculate the storm surges at the shoreline for the lower-limit wind speeds for each of the Saffir–Simpson scales.

4.3 Given a storm duration of 24 hr, a maximum surge height of 2 m, and a wave height of 3 m, determine the risk of a storm for the Delaware shoreline. If a second storm occurred almost immediately thereafter, determine how to calculate the risk to the shoreline given the first storm.

CHAPTER FIVE

Waves and Wave-Induced Hydrodynamics

Freak waves are waves with heights that far exceed the average. Freak waves in storms have been blamed for damages to shipping and coastal structures and for loss of life. The reasons for these waves likely include wave reflection from shorelines or currents, wave focusing by refraction, and wave–current interaction.

Lighthouses, built to warn navigators of shallow water and the presence of land, are often the target of such waves. For example, at the Unst Light in Scotland, a door 70 m above sea level was stove in by waves, and at the Flannan Light, a mystery has grown up about the disappearance of three lighthouse keepers during a storm in 1900 presumably by a freak wave, for a lifeboat, fixed at 70 m above the water, had been torn from its mounts. A poem by Wilfred W. Gibson has fueled the legend of a supernatural event.

The (unofficial) world's record for a water wave appears to be the earthquake and landslide-created wave in Lituya Bay, Alaska (Miller 1960). A wave created by a large landslide into the north arm of the bay sheared off trees on the side of a mountain over 525 m above sea level!

5.1 INTRODUCTION

Waves are the prime movers for the littoral processes at the shoreline. For the most part, they are generated by the action of the wind over water but also by moving objects such as passing boats and ships. These waves transport the energy imparted to them over vast distances, for dissipative effects, such as viscosity, play only a small role. Waves are almost always present at coastal sites owing to the vastness of the ocean's surface area, which serves as a generating site for waves, and the relative smallness of the surf zone, that thin ribbon of area around the ocean basins where the waves break and the wind-derived energy is dissipated.

Energies dissipated within the surf zone can be quite large. The energy of a wave is related to the square of its height. Often it is measured in terms of energy per unit water surface area, but it could also be the energy per unit wave length, or energy flux per unit width of crest. The first definition will serve to be more useful for our purposes. If we define the wave height as H, the energy per unit surface area is

$$E = \frac{1}{8}\rho g H^2,$$

where the water density ρ and the acceleration of gravity g are important parameters. (Gravity is, in fact, the restoring mechanism for the waves, for it is constantly trying to smooth the water surface into a flat plane, serving a role similar to that of the tension within the head of a drum.) For a 1-m high wave, the energy per unit area is approximately 1250 N-m/m^2, or 1250 J/m^2. If this wave has a 6-s period (that is, there are 10 waves per minute), then about 4000 watts enter the surf zone per meter of shoreline.* For waves that are 2 m in height, the rate at which the energy enters the surf zone goes up by a factor of four. Now think about the power in a 6-m high wave driven by 200 km/hr winds. That plenty of wave energy is available to make major changes in the shoreline should not be in doubt.

The dynamics and kinematics of water waves are discussed in several textbooks, including the authors' highly recommended text (Dean and Dalrymple 1991), Wiegel (1964), and for the more advanced reader Mei (1983). Here, a brief overview is provided. The generation of waves by wind is a topic unto itself and is not discussed here. A difficult, but excellent, text on the subject is Phillips' (1980) *The Dynamics of the Upper Ocean*.

5.2 WATER WAVE MECHANICS

The shape, velocity, and the associated water motions of a single water wave train are very complex; they are even more so in a realistic sea state when numerous waves of different sizes, frequencies, and propagation directions are present. The simplest theory to describe the wave motion is the Airy wave theory, often referred to as the *linear wave theory*, because of its simplifying assumptions. This theory is predicated on the following conditions: an incompressible fluid (a good assumption), irrotational fluid motion (implying that there is no viscosity in the water, which would seem to be a bad assumption, but works out fairly well), an impermeable flat bottom (not too true in nature), and very small amplitude waves (not a good assumption, but again it seems to work pretty well unless the waves become large).

The simplest form of a wave is given by the linear wave theory (Airy 1845), illustrated in Figure 5.1, which shows a wave assumed to be propagating in the positive x direction (left to right in the figure). The equation governing the displacement of the water surface $\eta(x, t)$ from the mean water level is

$$\eta(x, t) = \frac{H}{2} \cos k(x - Ct) = \frac{H}{2} \cos(kx - \sigma t) \tag{5.1}$$

Because the wave motion is assumed to be periodic (repeating identically every wavelength) in the wave direction ($+x$), the factor k, the *wave number*, is used to ensure that the cosine (a periodic mathematical function) repeats itself over a distance L, the wavelength. This forces the definition of k to be $k = 2\pi/L$. For periodicity in time, which requires that the wave repeat itself every T seconds, we have $\sigma = 2\pi/T$, where T denotes the wave period. We refer to σ as the *angular frequency* of the waves. Finally, C is the speed at which the wave form travels, $C = L/T = \sigma/k$. The C is referred to as the wave *celerity*, from the Latin. As a requirement of the Airy

---

* If we could capture 100 percent of this energy, we could power 40 100-watt lightbulbs.

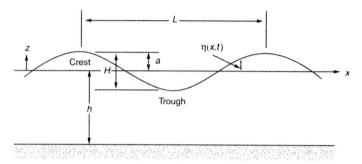

Figure 5.1   Schematic of a water wave (from Dean and Dalrymple 1991).

wave theory, the wavelength and period of the wave are related to the water depth by the *dispersion relationship*

$$\sigma^2 = gk \tanh kh \tag{5.2}$$

The "dispersion" meant by this relationship is the frequency dispersion of the waves: longer period waves travel faster than shorter period waves, and thus if one started out with a packet of waves of different frequencies and then allowed them to propagate (which is easily done by throwing a rock into a pond, for example), then at some distance away, the longer period waves would arrive first and the shorter ones last. This frequency dispersion applies also to waves arriving at a shoreline from a distant storm.

The dispersion relationship can be rewritten using the definitions of angular frequency and wave number

$$L = \frac{g}{2\pi} T^2 \tanh kh = L_0 \tanh kh, \tag{5.3}$$

defining the deep water wavelength L_0, which is dependent only on the square of the wave period. This equation shows that the wavelength monotonically decreases as the water depth decreases owing to the behavior of the $\tanh kh$ function, which increases linearly with small kh but then becomes asymptotic to unity in deep water (taken arbitrarily as $kh > \pi$ or, equivalently, $h > L/2$).

The dispersion relationship is difficult to solve because the wavelength appears in the argument of the hyperbolic tangent. This means that iterative numerical methods, such as Newton–Raphson methods, or approximate means are used to solve for wavelength; see, for example, Eckart (1951), Nielsen (1983), or Newman (1990). One convenient approximate method is due to Fenton and McKee (1989), which has a maximum error less than 1.7 percent, which is well within most design criteria. Their relationship is

$$L = L_0 \left\{ \tanh\left[\left(2\pi \sqrt{(h/g)}/T\right)^{3/2} \right] \right\}^{2/3} \tag{5.4}$$

EXAMPLE

A wave train is observed to have a wave period of 5 s, and the water depth is 3.05 m. What is the wavelength L?

First, we calculate the deep water wavelength $L_0 = gT^2/2\pi = 9.81\,(5)^2/6.283 = 39.03$ m. The exact solution of Eq. (5.3) is 25.10 m. The approximation of Eq. (5.4) gives 25.48 m, which is in error by 1.49 percent.

The waveform discussed above is a progressive wave, propagating in the positive x direction with speed C. Near vertical walls, where the incident waves are reflected from the wall, there is a superposition of waves, which can be illustrated by simply adding the waveform for a wave traveling in the positive x direction to one going in the opposite direction:

$$\eta(x,t) = \frac{H}{2}\cos(kx + \sigma t)$$

The resulting standing wave is

$$\eta(x,t) = H\cos kx \cos\sigma t,$$

which has an amplitude equal to twice the amplitude of the incident wave. By subtracting, instead of adding, we have a standing wave with different phases involving sines rather than cosines. Examining the equation for the standing wave, we find that the water surface displacement is always a maximum at values of kx equal to $n\pi$, where $n = 0, 1, 2, \ldots$. These points are referred to as *antinodal* positions. Nodes (zero water surface displacement) occur at values of $kx = (2n - 1)\pi/2$, for $n = 1, 2, 3, \ldots$, or $x = L/4, 3L/4, \ldots$.

Under the progressive wave, Eq. (5.1), the water particles move in elliptical orbits, which can be decomposed into the horizontal and vertical velocity components u and w as follows:

$$u(x,z,t) = \frac{H\sigma}{2}\frac{\cosh k(h+z)}{\sinh kh}\cos(kx - \sigma t) \tag{5.5}$$

$$w(x,z,t) = \frac{H\sigma}{2}\frac{\sinh k(h+z)}{\sinh kh}\sin(kx - \sigma t) \tag{5.6}$$

At the horizontal seabed, the vertical velocity is zero, and the horizontal velocity is

$$u(x, -h, t) = u_b \cos(kx - \sigma t),$$

where

$$u_b = \frac{H\sigma}{2\sinh kh} = A\sigma$$

The parameter A is half the orbital excursion of the water particle over the wave period and has been shown to be related to the size of sand ripples formed on the bottom.

The associated pressure within the wave is given by

$$p(x,z,t) = -\rho gz + \rho g\frac{H}{2}\frac{\cosh k(h+z)}{\cosh kh}\cos(kx - \sigma t)$$

$$= -\rho gz + \rho g\left(\frac{\cosh k(h+z)}{\cosh kh}\right)\eta(x,t) \tag{5.7}$$

Table 5.1 Asymptotic Forms of the Hyperbolic Functions

Function	Large kh ($>\pi$)	Small kh ($<\pi/10$)
$\cosh kh$	$\frac{1}{2}e^{kh}$	1
$\sinh kh$	$\frac{1}{2}e^{kh}$	kh
$\tanh kh$	1	kh

in which the first term on the right-hand side is the hydrostatic pressure component, and the second is the dynamic (wave-induced) pressure component. The dynamic pressure is largest under the wave crest, for the water column is higher at that location and at a minimum under the wave trough.

Finally, the propagation speed of the wave energy turns out to be different than the propagation speed of the waveform owing to the dispersive nature of the waves. This is expressed as the group velocity C_g, which is defined as

$$C_g = nC,$$

where

$$n = \frac{1}{2}\left(1 + \frac{2kh}{\sinh 2kh}\right) \tag{5.8}$$

The rate at which the waves carry energy, or the energy flux per unit width, is defined as $\mathcal{F} = EC_g$. Both C_g and n are functions of the water depth; n is 0.5 in deep water and unity in shallow water.

To simplify equations involving hyperbolic functions in shallow and deep water, asymptotic representations of the functions can be introduced for relative water depths of $h/L < 1/20$ and $h/L > 1/2$, respectively. These asymptotes are shown in Table 5.1.

5.2.1 OTHER WAVE THEORIES

The Airy theory is called the linear theory because nonlinear terms in such equations as the Bernoulli equation were omitted. The measure of the nonlinearity is generally the wave steepness ka, and the properties of the linear theory involve ka to the first power. For this text, the linear wave theory is sufficient; however, the tremendous body of work on nonlinear wave theories is summarized here.

Higher order theories (including terms of order $(ka)^n$, where n is the order of the theory) have been developed for periodic and nonperiodic waves. For periodic waves, for example, the Stokes theory (Stokes 1847, Fenton 1985, fifth order), and the numerical Stream Function wave theory (Dean 1965, Dalrymple 1974) show that, because of nonlinearity, larger waves travel faster and wave properties are usually more pronounced at the crest than at the trough of the wave, causing, for instance, the wave crests to be more peaked than linear waves. The Stream Function theory, which allows any order, is solved numerically, and the code is available from the authors.

For variable water depths, including deep water, approximations have been made to the governing equations to permit solutions over complicated bathymetries. A major step in this effort was the development of the mild-slope equation by Berkhoff (1972), which is valid for linear waves. Modifications for nonlinear waves and the effects of mean currents were made by Booij (1981) and Kirby and Dalrymple (1984).

In shallow water, the ratio of the water depth to the wavelength is very small ($kh < \pi/10$ or, equivalently, as before, $h/L < 1/20$). Taking advantage of this, wave theories have been developed for both periodic and nonperiodic waves. The earliest was the solitary wave theory of Russell (1844), who noticed these waves being created by horse-drawn barges in canals. The wave form is

$$\eta(x, t) = H \operatorname{sech}^2 \left(\sqrt{\frac{3H}{4h^3}} (x - Ct) \right) \tag{5.9}$$

This unusual wave decreases monotonically in height from its crest position in both directions, approaching the still water level asymptotically. The speed of the wave is

$$C = \sqrt{gh} \left(1 + \frac{H}{2h} \right) \tag{5.10}$$

In the very shallow water of the surf zone on a mildly sloping beach, waves behave almost as solitary waves. Munk (1949) discusses this thoroughly.

For periodic shallow water waves, the analytic cnoidal wave theory is sometimes used (Korteweg and deVries 1895). This theory has, as a long wave limit, the solitary wave, and as a short wave limit, the linear wave theory.

A more general theory that incorporates all of the shallow water wave theories is derived from the Boussinesq equations (see Dingemans 1997 for a detailed derivation of the various forms of the Boussinesq theory). For variable depth and propagation in the x direction, equations for the depth averaged velocity u and the free surface elevation can be written as (Peregrine 1967)

$$\frac{\partial u}{\partial t} + u \frac{\partial u}{\partial x} = -g \frac{\partial \eta}{\partial x} + \frac{h}{2} \frac{\partial^2}{\partial x^2} \left(h \frac{\partial u}{\partial t} \right) - \frac{h^2}{6} \frac{\partial^2}{\partial x^2} \left(\frac{\partial u}{\partial t} \right) \tag{5.11}$$

$$\frac{\partial \eta}{\partial t} + \frac{\partial (h + \eta)u}{\partial x} = 0 \tag{5.12}$$

For constant depth, the solitary wave and the cnoidal waves are solutions to these equations. For variable depth and a two-dimensional problem, numerical solutions by several techniques are available. Some of the more well-known include that of Abbott, Petersen, and Skovgaard (1978), which is based on finite-difference methods.

Several recent developments have extended the use of the Boussinesq equations in coastal engineering. One of these is the modification of the equations to permit their use in deeper water than theoretically justified. These extensions include those of Madsen, Murray, and Sørenson (1991), Nwogu (1993), and Wei et al. (1995). The second effort has been to include the effects of wave breaking so that the Boussinesq models can be used across the surf zone. Some examples are Schäffer, Madsen, and Deigaard (1993) and Kennedy et al. (2000).

If the initial wave field is expanded in terms of slowly varying (in x) Fourier modes, Boussinesq equations yield a set of coupled evolution equations that predict the amplitude and phase of the Fourier modes with distance (Freilich and Guza 1984; Liu, Yoon, and Kirby 1985; and Kaihatu and Kirby 1998). Field applications of the spectral Boussinesq theory show that the model predictions agree very well with normally incident ocean waves (Freilich and Guza 1984). Elgar and Guza (1985) have shown that the model is also able to predict the skewness of the shoaling wave field, which is important for sediment transport considerations.

The KdV equation (from Korteweg and deVries 1895) results from the Boussinesq theory by making the assumption that the waves can travel in one direction only. A large body of work exists on the mathematics of this equation and its derivatives such as the Kadomtsev and Petviashvili (1970) or K–P, equation, for KdV waves propagating at an angle to the horizontal coordinate system.

In the surf zone and on the beach face, the simpler nonlinear shallow water equations (also from Airy) can provide good estimates of the waveform and velocities because these equations lead to the formation of bores, which characterize the broken waves:

$$\frac{\partial u}{\partial t} + u\frac{\partial u}{\partial x} = -g\frac{\partial \eta}{\partial x} \tag{5.13}$$

$$\frac{\partial \eta}{\partial t} + \frac{\partial (h + \eta)u}{\partial x} = 0 \tag{5.14}$$

Hibberd and Peregrine (1979) and Packwood (1980) were the early developers of this approach, and Kobayashi and colleagues have produced several working models (Kobayashi, De Silva, and Watson 1989; Kobayashi and Wurjanto 1992).

5.2.2 WAVE REFRACTION AND SHOALING

As waves propagate toward shore, the wave length decreases as the depth decreases, which is a consequence of the dispersion relationship (Eq. (5.3)). The wave period is fixed; the wavelength and hence the wave speed decrease as the wave encounters shallower water. For a long crested wave traveling over irregular bottom depths, the change in wave speed along the wave crest implies that the wave changes direction locally, or it *refracts*, much in the same way that light refracts as it passes through media with different indices of refraction.* The result is that the wave direction turns toward regions of shallow water and away from regions of deep water. This can create regions of wave focusing on headlands and shoals.

The simplest representation of wave refraction is the refraction of waves propagating obliquely over straight and parallel offshore bathymetry. In this case, Snell's law, developed for optics, is valid. This law relates the wave direction, measured by an angle to the x-axis (drawn normal to the bottom contour), and the wave speed C

---

* The classic physical example is a pencil standing in a water glass. When viewed from the side, the part of the pencil above water appears to be oriented in a different direction than the part below water.

in one water depth to that in deep water:

$$\frac{\sin\theta}{C} = \frac{\sin\theta_0}{C_0} = \text{constant}, \tag{5.15}$$

where the subscript $_0$ denotes deep water.

Wave refraction diagrams for realistic bathymetry provide a picture of how waves propagate from the offshore to the shoreline of interest. These diagrams can be drawn by hand, if it is assumed that a depth contour is locally straight and that Snell's law can be applied there. Typically at the offshore end of a bathymetric chart, wave rays of a given direction are drawn (where the ray is a vector locally parallel to the wave direction; following a ray is the same as following a given section of wave crest). Then each ray is calculated, contour by contour, to the shore line, with each depth change causing a change in wave direction according to Snell's law. Now most of these calculations are done with more elaborate computer models or more sophisticated numerical wave models such as a mild-slope, parabolic, or Boussinesq wave model.

Another effect of the change in wavelength in shallow water is that the wave height increases. This is a consequence of a conservation of energy argument and the decrease in group velocity (Eq. (5.8)) in shallow water in concert with the decrease in C (note that n, however, goes from one-half in deep water to unity in shallow water but that this increase is dominated by the decrease in C). This increase in wave height is referred to as *shoaling*.

A convenient formula that expresses both the effects of wave shoaling and refraction is

$$H = H_0 K_s K_r, \tag{5.16}$$

where H_0 is the deep water wave height, K_s is the shoaling coefficient,

$$K_s = \sqrt{\frac{C_{g0}}{C_g}},$$

and K_r is the refraction coefficient, which for straight and parallel shoreline contours can be expressed in terms of the wave angles as follows:

$$K_r = \sqrt{\frac{\cos\theta_0}{\cos\theta}}$$

Given the deepwater wave height H_0, the group velocity C_{g0}, and the wave angle θ_0, the wave height at another depth can be calculated (when it is used in tandem with Snell's law above).

Wave diffraction occurs when abrupt changes in wave height occur such as when waves encounter a surface-piercing object like an offshore breakwater. Behind the structure, no waves exist and, by analogy to light, a shadow exists in the wave field. The crest-wise changes in wave height then lead to changes in wave direction, causing the waves to turn into the shadow zone. The process is illustrated in Figure 5.2, which shows the diffraction of waves from the tip of a breakwater. Note that the wave field looks as if there is a point source of waves at the end of the structure. In fact, diffraction can be explained by a superposition of point wave sources along the crest (*Huygen's principle*).

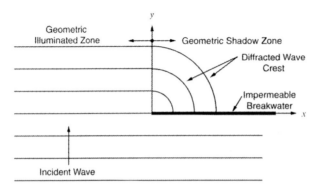

Figure 5.2 Diffraction of waves at a breakwater (from Dean and Dalrymple 1991).

5.2.3 WAVE PROPAGATION MODELS

Historically, wave models used to predict the wave height and direction over large areas were developed for a wave train with a single frequency, which is referred to as a monochromatic wave train in analogy to light. Monochromatic models for wave propagation can be classified by the phenomena that are included in the model. Refraction models can be ray-tracing models (e.g., Noda 1974), or grid models (e.g., Dalrymple 1988). Refraction–diffraction models are more elaborate, involving either finite element methods (Berkhoff 1972) or mathematical simplifications (such as in parabolic models, e.g., REF/DIF by Kirby and Dalrymple 1983).

Spectral models entail bringing the full directional and spectral description of the waves from offshore to onshore. These models have not evolved as far as monochromatic models and are the subject of intense research. Examples of such work are Brink-Kjaer (1984); Booij, Holthuijsen, and Herbers (1985); Booij and Holthuijsen (1987); and Mathiesen (1984).

Recent models often include the interactions of wave fields with currents and bathymetry, the input of wave energy by the wind, and wave breaking. For example, Holthuijsen, Booij, and Ris (1993) introduced the SWAN model, which predicts directional spectra, significant wave height, mean period, average wave direction, radiation stresses, and bottom motions over the model domain. The model includes nonlinear wave interactions, current blocking, refraction and shoaling, and white capping and depth-induced breaking.

5.2.4 WAVE BREAKING

In deep water, waves break because of excessive energy input, mostly from the wind. The limiting wave height is taken as $H_0/L_0 \approx 0.17$, where L_0 is the deep water wavelength.

In shallow water, waves continue to shoal until they become so large that they become unstable and break. Empirically, Battjes (1974) has shown that the breaking wave characteristics can be correlated to the surf similarity parameter ζ, which is

Table 5.2 Breaking Wave Characteristics and the Surf Similarity Parameter

$\zeta \rightarrow$	≈ 0.1		0.5		1.0	2.0		3.0	4.0	5.0
Breaker Type	**Spilling**				**Plunging**			**Collapsing/ Surging**		**None (Reflection)**
κ	0.8		1.0		1.1	1.2				
N	6–7	2–3		1–2	< 1		< 1			
r	10^{-3}	10^{-2}			0.1	0.4		0.8		

κ = breaking index; N = number of waves in surf zone; r = reflection from beach.
Source: Battjes (1974).

defined as the ratio of the beach slope, $\tan\beta$, to the square root of the deep water wave steepness, by the following expression:

$$\zeta = \tan\beta / \sqrt{H_0/L_0} \tag{5.17}$$

His results are shown in Table 5.2, which shows the breaker type, the breaking index, the number of waves in the surf zone, and the reflection coefficient from a beach as a function of the surf similarity parameter.

At first, simple theoretical models were proposed to predict breaking. Theoretical studies of solitary waves (a single wave of elevation caused, for example, by the displacement of a wavemaker in one direction only) in constant-depth water showed that the wave breaks when its height exceeds approximately 0.78 of the water depth. This led to the widespread use of the so-called spilling breaker assumption that the wave height within the surf zone is a linear function of the local water depth $H = \kappa h$, where κ, the *breaker index*, is on the order of 0.8. Later experiments with periodic waves pointed out that the bottom slope was important as well, leading to elaborate empirical models for breaking (e.g., Weggel 1972). The spilling breaker assumption, however, always leads to a linear dependency of wave height with water depth. In the laboratory, the wave height is often seen to decrease more rapidly at the breaking line than farther landward. In the field, on the other hand, Thornton and Guza (1982), showed that the root-mean-square wave height in the surf zone (on their mildly sloping beach) was reasonably represented by $H_{\text{rms}} = 0.42h$.

Dally, Dean, and Dalrymple (1985) developed a wave-breaking model based on the concept of a stable wave height within the surf zone for a given water depth. This model has two height thresholds, each of which depends on the water depth. As waves shoal up to the highest threshold (a breaking criteria), breaking commences. Breaking continues until the wave height decreases to the lower threshold (a stable wave height). This stable wave height concept appears in experiments by Horikawa and Kuo (1966), which involved creating a breaking wave on a slope. This breaking wave then propagates into a constant depth region. Measurements of the wave height along the wave tank showed that the waves approach a stable (broken) wave height of $H = \Gamma h$, where Γ is about 0.35–0.40 in the constant depth region.

The Dally et al. model, valid landward of the location of initial breaking, is expressed in terms of the conservation of energy equation

$$\frac{\partial EC_g}{\partial x} = -\frac{K}{h}[EC_g - (EC_g)_s],$$ (5.18)

where K is an empirical constant equal to approximately 0.17. For shallow water, the energy flux can be reduced to

$$EC_g = \frac{1}{8}\rho g H^2 \sqrt{gh},$$

and thus the preceding equation relates the wave height H to the water depth h. For planar beaches, where the depth is given by $h = mx$, an analytic solution can be obtained,

$$\frac{H}{H_b} = \sqrt{\left[\left(\frac{h}{H_b}\right)^{\left(\frac{K}{m} - \frac{1}{2}\right)}(1 + \alpha) - \alpha\left(\frac{h}{H_b}\right)^2\right]},$$ (5.19)

where m is the beach slope and

$$\alpha = \frac{K\Gamma^2}{m\left(\frac{5}{2} - \frac{K}{m}\right)}\left(\frac{H_b}{H_b}\right)^2$$

For the special case of $K/m = 5/2$, a different solution is necessary:

$$\frac{H}{H_b} = \left(\frac{h}{H_b}\right)\sqrt{\left[1 - \beta \ln\left(\frac{h}{H_b}\right)\right]}$$ (5.20)

Here,

$$\beta = \frac{5}{2}\Gamma^2\left(\frac{H_b}{H_b}\right)^2$$

The range of solutions to Eqs. (5.19) and (5.20) are shown in Figure 5.3. Note that for $K/m > 3$, the wave heights can be much less than predicted by a spilling wave assumption (which coincides approximately with the $K/m = 3$ curve), whereas for

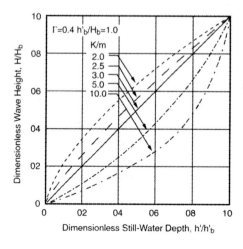

Figure 5.3   Wave height variation predicted across a planar beach (Dally et al. 1985, copyright by the American Geophysical Union).

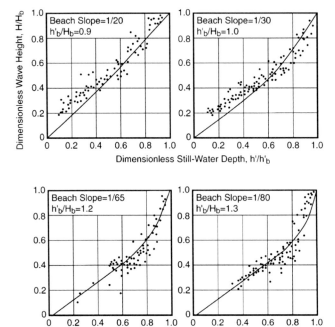

Figure 5.4 Comparisons of analytic solution of Dally et al. (1985, copyright by the American Geophysical Union) with laboratory data of Horikawa and Kuo (1966).

steeper beaches, $K/m < 3$, the wave heights are larger. Comparisons with Horikawa and Kuo data for waves breaking on a planar beach are shown in Figure 5.4.

For more complicated beach profiles, numerical solutions to Eq. (5.18) are used. Dally (1990) extended this model to include a realistic surf zone by shoaling a distribution of wave heights rather than a monochromatic wave train.

Wave breaking is one of the most difficult hydrodynamics problems. The highly nonlinear and turbulent nature of the flow field has prevented the development of a detailed model of wave breakers, which has spurred the development of macroscale models.

Peregrine and Svendsen (1978) developed a wake model for turbulent bores, arguing that, from a frame of reference moving with the wave, the turbulence spreads from the toe of the bore into the region beneath the wave like a wake develops.

A wave-breaking model for realistic wave fields was proposed by Battjes and Janssen (1978), who also utilized the conservation of energy equation, but the loss of wave energy in the surf zone was represented by the analogy of a turbulent hydraulic jump for the wave bore in the surf zone. Further, the random nature of the wave field was incorporated by breaking only the largest waves in the distribution of wave heights at a point. In the field, this model was extended by Thornton and Guza (1983), who were able to predict the root-mean-square wave height from the shoaling zone to inside the surf zone to within 9 percent.

Svendsen (1984) developed the roller model, which is based on the bore–hydraulic jump model. The roller is a recirculating body of water surfing on the front face of

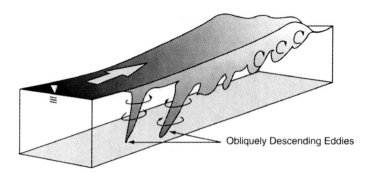

Obliquely Descending Eddies

Figure 5.5 Schematic of vertically descending eddies with the arrow showing the direction of breaker travel (from Nadaoka 1986).

the wave after breaking is initiated. This roller has mass and momentum that must be accounted for in the governing equations. He showed that agreement between theory and setup measurements was better with the roller model than a breaking-index model.

Measurements of breaking waves in the laboratory have provided valuable insights into the nature of the breaking process. Nadaoka, Hino, and Koyano (1989) have shown that spilling breakers produce what they refer to as "obliquely descending eddies," which are near-vertical vortices that remain stationary after the breaker passes. These eddies descend to the bottom, pulling bubbles down into the water column. Figure 5.5 shows a schematic of the eddies they observed. The role of the eddies in nearshore mixing processes (both of mass and momentum) is as yet unknown; further, the generation mechanism, which implies the rotation of horizontal vorticity due to the roller on the face of a spilling breaker, into near-vertical vorticity, is at yet unknown. However, Nadaoka, Ueno, and Igarashi (1988) observed these eddies in the field and showed that they are an important mechanism for the suspension of sediment and that the bubbles, drawn into the vortices, provide a buoyancy that creates an upwelling of sediment after the wave passage.

5.2.5 MEAN WAVE QUANTITIES

Associated with the passage of periodic waves are some useful quantities found by averaging in time over the wave period. For example, there is a mean transport of water toward the shoreline, the *mass transport*, which is not predicted by the linear Airy theory, which assumes that each water particle under a waveform is traveling in a closed elliptical orbit. We define the mass transport as

$$M = \frac{1}{t_2 - t_1} \int_{t_1}^{t_2} \int_{-h}^{\eta} \rho u(x, z) \, dz dt, \tag{5.21}$$

where the time interval between t_1 and t_2 is a long time (many wave periods for irregular waves; one wave period for periodic waves). If we integrate over the depth from the bottom only to the mean water surface, $z = 0$, rather than the instantaneous water surface η, we obtain M equal to zero, as predicted by linear theory. If we continue the integration up to η, the mass transport becomes

$$M = \frac{E}{C}, \tag{5.22}$$

which shows that there is a nonlinear transport of water in the wave direction due to the larger forward transport of water under the wave crest because the total depth is greater when compared with the backward transport under the trough. From this formula, the mass transport is larger for more energetic waves.

This mass transport has momentum associated with it, which means that forces will be generated whenever this momentum changes magnitude or direction by Newton's second law. To determine this momentum, we integrate the momentum flux from the bottom to the surface as follows:

$$\mathcal{M} = \frac{1}{t_2 - t_1} \int_{t_1}^{t_2} \int_{-h}^{\eta} (\rho u) u \, dz dt \tag{5.23}$$

This quantity has as a first approximation $\mathcal{M} = MC_g = En$, which indicates that the flux of momentum is described by the mass transport times the group velocity.

Offshore of the breaker line, there is a depression of the mean water level from the still water level due to the waves, which is called *setdown* and is denoted as $\bar{\eta}$. This quantity, originally elucidated by Longuet-Higgins and Stewart (1963), is (e.g., Dean and Dalrymple 1991)

$$\bar{\eta} = -\frac{H^2 k}{8 \sinh 2kh}, \tag{5.24}$$

where, again, H is the wave height. Because the wave height increases as waves shoal, the setdown increases as well, reaching a maximum at the breakerline that is approximately 5 percent of the breaking water depth.

Longuet-Higgins and Stewart (1963) introduced the concept of wave momentum flux, designating the sum of the momentum flux and the mean pressure as the *radiation stress*, based on the analog with light, which develops a radiation pressure when shining on an object. The quantities are related in the following way:

$$\mathcal{M} + \frac{1}{2}\rho g h^2 = S_{xx} + \frac{1}{2}\rho g (h + \bar{\eta})^2, \tag{5.25}$$

where S_{xx} is the radiation stress representing the flux in the x direction of the x component of momentum and $\bar{\eta}$ is the mean water level elevation. The formula for S_{xx} is

$$S_{xx} = E\left(2n - \frac{1}{2}\right) \tag{5.26}$$

There is an equivalent term for y-momentum carried in the y direction and the mean pressure, which is

$$S_{yy} = E\left(n - \frac{1}{2}\right) \tag{5.27}$$

These expressions apply for waves traveling in the x direction. If the waves were traveling in the θ direction, where θ is an angle the wave direction makes with the

x-axis, then we would have the following radiation stresses:

$$S_{xx} = E\left[n(\cos^2\theta + 1) - \frac{1}{2}\right] \tag{5.28}$$

$$S_{yy} = E\left[n(\sin^2\theta + 1) - \frac{1}{2}\right] \tag{5.29}$$

$$S_{xy} = S_{yx} = \frac{En\sin 2\theta}{2} \tag{5.30}$$

The last equation is for the flux of x momentum in the y direction, or vice versa, and it arises owing to the obliquity of the waves to the coordinate axis. This equation can be rewritten in the following form by introducing the wave celerity in numerator and denominator:

$$S_{xy} = EnC\cos\theta\left(\frac{\sin\theta}{C}\right)$$

The first part of this expression is recognized as the shoreward flux of wave energy, which, on a beach characterized by straight and parallel contours, is constant until breaking begins in the surf zone, and the second term, in parentheses, is Snell's law, which is also constant. Therefore, for this idealized beach, S_{xy} is constant from offshore to the breaker line.

5.2.6 WAVE SETUP

The wave momentum fluxes are proportional to the wave energy. If the waves break, then the momentum flux decreases. This change in momentum flux must be balanced by forces; therefore, wave breaking induces forces in the surf zone that act in the wave direction. We will now examine each of these forces by resolving them into an onshore and alongshore direction.

The onshore-directed momentum flux is S_{xx}. As the wave propagates into the surf zone, the momentum flux is equal to its value at the breaker line. At the limit of wave uprush, the value is zero. This gradient in momentum flux is balanced by a slope in water level within the surf zone $\partial\bar{\eta}/\partial x$. Balancing the forces, the following differential equation results:

$$\frac{\partial\bar{\eta}}{\partial x} = -\frac{1}{\rho g(h + \bar{\eta})}\frac{\partial S_{xx}}{\partial x} \tag{5.31}$$

Integrating the mean water level slope, we get the wave *set up* $\bar{\eta}$, as noted earlier. Using the shallow water asymptote and the spilling breaker assumption in the radiation stress terms, S_{xx} becomes

$$S_{xx} = \frac{3}{16}\rho g\kappa^2(h + \bar{\eta})^2,$$

where κ is the breaking index. As noted earlier, the value of κ is about 0.8 for spilling breakers. For a monotonic beach profile, we can solve Eq. (5.31) for $\bar{\eta}$ as follows:

$$\bar{\eta} = \bar{\eta}_b - \mathcal{K}(h - H_b), \tag{5.32}$$

where $\bar{\eta}_b$ is the mean water level at the breaker line and $\mathcal{K} = (3\kappa^2/8)/(1 + 3\kappa^2/8)$. Experiments have been carried out to verify this model in the laboratory (with good results; see Bowen, Inman, and Simmons 1968) and in the field. (This is the same wave setup that was introduced in the last chapter as a component of storm surge.)

5.3 CROSS-SHORE AND LONGSHORE CURRENTS

The wave-induced mass transport M must engender a return flow (which on a long-shore uniform beach is the *undertow*), for there can be no net onshore flow of water because of the presence of the beach. The amount of seaward mass flux is therefore equal to M. This flow is not distributed uniformly over the depth but has a distinct profile caused by the variation in wave-induced stress over the depth.

In the alongshore direction, the greatest change in radiation stress occurs owing to the S_{xy} term, which is affected, of course, by breaking. To balance this change in momentum flux, a longshore water level slope is possible where the shoreline is short; bounded at the ends by headlands, inlets, or man-made structures; or when there are rip currents. For an infinitely long uniform shoreline with uniform wave conditions, this water-level slope cannot exist, for it leads to infinite or negative water depths in the surf zone. Some other mechanism for developing a longshore balancing force is required. Mean currents flowing along the shoreline will develop bottom stresses and can balance the gradients in the radiation stress terms. The resulting *longshore current* is then directly engendered by the obliquely incident waves and the process of wave breaking. By balancing the frictional forces and the gradients in the radiation stress, Bowen (1969a), Longuet-Higgins (1970a and b), and Thornton (1970) developed equations for generating the longshore current and, in fact, since these early models, numerous studies have been made of the current field.

The steady-state equation of motion in the alongshore direction is

$$-\rho g(h+\eta)\frac{\partial\bar{\eta}}{\partial y} - \left[\frac{\partial S_{xy}}{\partial x} + \frac{\partial S_{yy}}{\partial y}\right] + [\tau_s - \tau_b] + \frac{\partial[(h+\bar{\eta})\tau_{xy}]}{\partial x} = 0, \qquad (5.33)$$

in which the last term represents lateral shear stress coupling. The y derivatives can be neglected for the case of a long straight beach because these terms would lead to infinite magnitudes of η and S_{yy} if they existed. The remaining terms (after neglecting the surface shear stress for the case of no wind and lateral shear stress coupling) can be written as

$$\frac{\partial S_{xy}}{\partial x} = -\tau_b \qquad (5.34)$$

Determining a simple expression for the bottom friction term that results from the mean current that is coexistent with the oscillatory wave field is difficult. Longuet-Higgins (1970) developed what Liu and Dalrymple (1978) call the small (incident wave) angle model defined by

$$\tau_b = \frac{\rho f}{4\pi} u_m V \qquad (5.35)$$

The f is the empirical Darcy–Weisbach friction factor, used in pipe flow calculations, which is known to be a function of the flow Reynolds number and the sand roughness, and u_m is the maximum orbital velocity of the waves, $u_m = (\eta C/h)_{max} = HC/2h \rightarrow$ $\kappa\sqrt{g(h+\bar{\eta})}/2 \approx 0.4C$ in the surf zone. This form of the bottom shear stress results from linearizing the originally nonlinear shear stress term.

Substituting for the bottom shear stress and solving Eq. (5.34) yields the first solution of Longuet-Higgins,

$$V(x) = \frac{5\pi g\kappa m^*(h+\bar{\eta})}{2f}\left(\frac{\sin\theta}{C}\right), \tag{5.36}$$

where m^* is the modified slope, $m^* = m/(1 + 3\kappa^2/8)$. This equation shows that the longshore velocity increases with water depth and with incident wave angle but decreases with bottom friction factor. The equation applies within the surf zone, resulting in an increasing velocity out to the breaker line, and then the velocity is zero offshore. Equation (5.36) is valid for any monotonic beach profile when the lateral shear stress terms are negligible.

One contribution missing in the preceding equation is the influence of the lateral shear stresses, which in this case, would tend to smooth out the velocity profile given above, particularly the breaker line discontinuity, which is not evident in laboratory experiments (Galvin and Eagleson 1965). Longuet-Higgin's second model included a lateral shear stress term τ_{xy} of the form

$$\tau_{xy} = \rho v_e \frac{\partial V}{\partial x}$$

The eddy viscosity, which has the dimensions of length^2/time, was assumed to be $v_e = Nx\sqrt{gh}$, where N is a constant and the "mixing" length scale is the distance offshore, x. His analysis yielded the following form of the longshore current, which is nondimensionalized by the values at the breaker line, $X = x/x_b$ and $V_0 = V(x_b)$, given by Eq. (5.36):

$$V/V_0 = \begin{cases} B_1 X^{p_1} + AX, & \text{for } 0 < X < 1 \\ B_2 X^{p_2}, & \text{for } X > 1, \end{cases} \tag{5.37}$$

where the coefficients and powers are

$$B_1 = \left(\frac{p_2 - 1}{p_1 - p_2}\right)A \tag{5.38}$$

$$B_2 = \left(\frac{p_1 - 1}{p_1 - p_2}\right)A \tag{5.39}$$

$$p_1 = -\frac{3}{4} + \sqrt{\left(\frac{9}{16} + \frac{1}{P}\right)} \tag{5.40}$$

$$p_2 = -\frac{3}{4} - \sqrt{\left(\frac{9}{16} + \frac{1}{P}\right)} \tag{5.41}$$

$$A = \frac{1}{\left(1 - \frac{5}{2}P\right)} \tag{5.42}$$

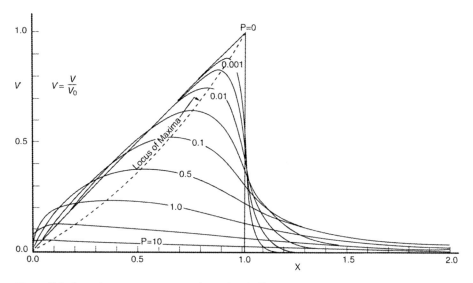

Figure 5.6 Longshore currents on a planar beach (from Longuet-Higgins 1970b, copyright by the American Geophysical Union).

The variable P is the ratio of the eddy viscosity to the bottom friction,

$$P = \frac{8\pi m N}{\kappa f}.$$

There is an additional solution for the case when P is $2/5$; the reader is referred to Longuet-Higgins (1970b) for this case. The velocity profiles for this longshore current model are shown in Figure 5.6. Comparisons with laboratory data show that values of P between 0.1 and 0.4 are reasonable. For $P = 0$, we have the same result as before (Eq. (5.36)).

The mean two-dimensional cross-shore circulation just offshore of the breaker line is also interesting. Matsunaga, Takehara, and Awaya (1988) and Matsunaga and Takehara (1992) showed in the laboratory that there is a train of horizontal vortices coupled to the undertow that are spawned at the breaker line and propagate offshore. Figure 5.7 shows a schematic of the vortex system in their wave tank. They explain the presence of these coherent eddies as an instability between the mean shoreward bottom flows and the offshore undertow. Li and Dalrymple (1998) provided a linear model for this instability based on an instability of the undertow. For very small waves, the vortices do not appear.

5.4 LOW-FREQUENCY MOTIONS AT THE SHORELINE

Wave staffs or current meters placed in the surf zone often measure extremely energetic motions at frequencies much lower than those of the incident waves. Yet,

Figure 5.7 Schematic of horizontal vortex system (from Matsunaga et al. 1988).

offshore these low-frequency motions either do not exist or are a very small portion of the total wave field. These low-frequency waves are *surf beat*, *edge waves*, and *shear waves*.

5.4.1 SURF BEAT

Surf beat was first described by Munk (1949) and Tucker (1950). Their explanation was that waves moving toward shore are often modulated into groups that force the generation of mean water level changes (Longuet-Higgins and Stewart 1963). Where the waves in a group are large, the water level is depressed (set down), whereas it is raised where the waves are smaller. As these wave groups enter the surf zone, the low-frequency forced water level variations are released, reflect from the beach, and travel offshore as free waves. This mechanism can be combined with the more likely mechanism that the large waves within the wave groups generate a larger setup on the beach, which must then decrease when the smaller waves of the group come ashore, causing an offshore radiation of low-frequency motion at the frequency associated with the wave group.

The low-frequency surf beat motion can be described by the linear shallow water wave equations, which are the equations of motion in the onshore and alongshore directions and the conservation of mass equation:

$$\frac{\partial u}{\partial t} = -g\frac{\partial \eta}{\partial x} \tag{5.43}$$

$$\frac{\partial v}{\partial t} = -g\frac{\partial \eta}{\partial y} \tag{5.44}$$

$$\frac{\partial \eta}{\partial t} + \frac{\partial uh}{\partial x} + \frac{\partial vh}{\partial y} = 0 \tag{5.45}$$

Eliminating the velocities in the continuity equation by differentiating with respect to time and then substituting for the velocities, we find the linear shallow water *wave equation* for variable depth:

$$\frac{\partial^2 \eta}{\partial t^2} = \frac{\partial}{\partial x}\left(gh\frac{\partial \eta}{\partial x}\right) + \frac{\partial}{\partial y}\left(gh\frac{\partial \eta}{\partial y}\right) \tag{5.46}$$

First, we examine the case in which the shoreline is straight and the contours are parallel to the shoreline; there is no alongshore variation in depth. Further, we assume for now that there is no variation in the long wave motion in the longshore direction and thus that the last term on the right-hand side of the equation is zero.

For a planar beach, $h = mx$, and if a wave motion that is periodic in time with angular frequency σ is assumed, Eq. (5.46) reduces to

$$\frac{\partial^2 \eta}{\partial x^2} + \frac{1}{x}\frac{\partial \eta}{\partial x} + \frac{\sigma^2}{gmx}\eta = 0 \tag{5.47}$$

A standing-wave solution to this equation is (Lamb 1945)

$$\eta(x,t) = AJ_0(2kx)\cos\sigma t, \quad \text{where } k = \frac{\sigma}{\sqrt{gh}}, \tag{5.48}$$

A is the amplitude of the motion at the shoreline, and k is a cross-shore varying wave number. The offshore dependency of the zeroth-order Bessel function, $J_0(2kx)$, determines the behavior of this wave. This solution is not directly forced by the waves but is simply a possible response to forcing. A nonlinear version of this motion is given by Carrier and Greenspan (1957).

Schäffer, Jonsson, and Svendsen (1990) included the effect of wave groups by adding forcing due to the radiation stress gradient:

$$\frac{\partial^2 \eta}{\partial t^2} = \frac{\partial}{\partial x}\left(gh\frac{\partial \eta}{\partial x}\right) + \frac{1}{\rho}\frac{\partial^2 S_{xx}}{\partial x^2} \tag{5.49}$$

This equation, with assumptions for the variation of S_{xx} within the surf zone, was solved numerically. Schäffer and Jonsson (1990) found reasonable agreement with the laboratory work of Kostense (1984).

5.4.2 EDGE WAVES

The edge wave presents a more complicated problem. These waves are motions that exist only near the shoreline and propagate *along* it. They can be described using shallow-water wave theory very simply from the work of Eckart (1951), although models for edge waves have been known since Stokes (1846).

Starting with the shallow-water wave equation, Eq. (5.46), substituting in a planar sloping beach, $h = mx$, and assuming a separable solution that is periodic in time *and* in the alongshore direction,

$$\eta(x, y, t) = Af(x)\cos\lambda y\cos\sigma t,$$

allows us to obtain

$$\frac{\partial^2 f}{\partial x^2} + \frac{1}{x}\frac{\partial f}{\partial x} + \left(\frac{\sigma^2}{gmx} - \lambda^2\right)f = 0 \tag{5.50}$$

The only solution of this equation that is bounded at the shoreline and decays offshore (remember that it is a trapped wave solution) is

$$\eta(x, y, t) = Ae^{-\lambda x}L_n(2\lambda x)\cos\lambda y\cos\sigma t \tag{5.51}$$

The function $L_n(2\lambda x)$ is a Laguerre polynomial, which has the following expansion as a function of z and n:

$$L_0(z) = 1 \tag{5.52}$$

$$L_1(z) = 1 - z \tag{5.53}$$

$$L_2(z) = 1 - 2z + \frac{1}{2}z^2 \tag{5.54}$$

$$L_n(z) = \sum_{k=0}^{n} \frac{(-1)^k n! z^k}{(k!)^2(n-k)!} \tag{5.55}$$

An instantaneous snapshot of the first four solutions, which are referred to as the zero-, first-, second-, and third-mode edge waves, is shown in Figure 5.8 and compared

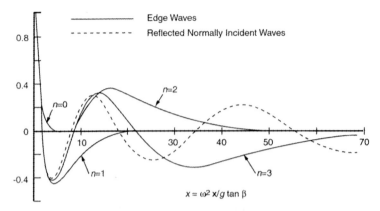

Figure 5.8 Free surface elevations corresponding to the first four edge wave modes and a fully reflected wave as a function of dimensionless distance offshore.

with the normally incident standing wave solution (Eq. (5.48)). Notice that the mode number n corresponds to the number of zero crossings of the water surface elevation.

This wave motion must satisfy the following edge wave dispersion relationship:

$$\sigma^2 = g\lambda(2n + 1)m,$$

which relates the alongshore wave number λ to the wave frequency and the beach slope. Ursell (1952) has shown that a more accurate representation of the dispersion relationship is

$$\sigma^2 = g\lambda \sin(2n + 1)m,$$

which is equivalent to the previous one for small beach slope m.

The waveform described by (Eq. (5.51)) is a standing edge wave, which will not propagate along a beach. Propagating waveforms may be found by adding together two standing waves as before. The second standing wave might be proportional to $\sin \lambda y \sin \sigma t$, in which case, we obtain

$$\eta(x, y, t) = Ae^{-\lambda x} L_n(2\lambda x) \cos(\lambda y - \sigma t), \tag{5.56}$$

a wave that propagates in the positive y direction. By subtracting, instead of adding, we can obtain a wave propagating in the opposite direction.

The edge wave solution above is valid only for planar beaches; additional means to solve Eq. (5.46) are required for other beach profiles. For those that can be described by an exponentially increasing depth, $h = h_0(1 - e^{-rx})$, where h_0 is the offshore constant depth, Ball (1967) developed analytical solutions for the wave motion. For arbitrary beach profiles, Holman and Bowen (1979) and Kirby, Dalrymple, and Liu (1981) provide numerical methods.

The effects of a longshore current on edge waves has been investigated by Howd, Bowen, and Holman (1992), who showed that the influence of the current is of the same magnitude as a variable beach profile. Furthermore, the equations governing edge waves on a uniform longshore current can be cast into the same governing equation as before through a transformation of the depth. After introducing a current-modified effective depth defined as

$$h'(x) = \frac{h(x)}{\left(1 - \frac{V(x)}{C}\right)^2},$$

Eq. (5.46) still can be used for the solution. The celerity C is the speed of the edge wave $C = \sigma/\lambda$.

The generation of edge waves in nature has been a question of great interest, particularly in conjunction with the study of nearshore circulation. Guza and Davis (1974) developed a model showing that edge waves can be generated through a nonlinear resonant mechanism with the incident wave train. An incident wave train with a frequency σ can generate two edge waves with frequencies $\sigma/2$, which are called subharmonic edge waves. This mechanism has been verified in the laboratory[*] (Guza and Inman 1975).

It should be pointed out that, from linear analyses, edge waves on a beach with straight and parallel contours cannot be caused by an incident wave train from offshore. This can be proven with Snell's law:

$$\frac{\sin \theta}{C} = \frac{\sin \theta_0}{C_0}$$

If we consider a wave train generated in shallow water but directed obliquely offshore at angle θ to the beach normal, there are two possibilities: the wave propagates offshore to deep water (so-called leaky modes), or, as the wave encounters deeper water, the wave angle becomes larger until finally the wave propagates directly alongshore. From Snell's law, the wave angle θ becomes 90°. Part of the crest will be in shallow water and part in deeper water, which will turn the wave back onshore. Then, if a significant portion of the wave is reflected back offshore, this process repeats. This means that the wave is trapped against the shoreline and it is an edge wave. In other words, if a wave generated in shallow water has a propagation angle between 0° and θ_c, the wave propagates offshore. For angles greater than θ_c, waves become trapped as edge waves.

The only way edge waves can be generated by waves incident from offshore is if the bathymetry is sufficiently irregular so that waves can approach the shallow water with large angles of incidence. For man-made structures, two examples are the trapping of waves by the ends of breakwaters (Dalrymple, Kirby, and Seli 1986) and the reflection of waves from groins and jetties such that they have the correct wave angle as they propagate onto the straight and parallel beach contours.

Gallagher (1971) proposed another mechanism for edge wave generation based on the concept that the incident wave field, which is composed of many waves and many frequencies, can have wave groups occurring at a period that satisfies the edge wave dispersion relationship. Bowen and Guza (1978) examined the limit case of two

---

[*] A very graphic laboratory experiment in a wave basin with reflecting sidewalls to illustrate this mechanism is to generate normally incident waves onto a sloping beach. The wave period should be chosen so that the width of the wave basin (length of the beach) is $n\pi/\lambda$, where $n \geq 1$. For $n = 2$, one edge wave fits along the beach. Within several minutes of starting the wave generator, the presence of the edge waves will be obvious, for the run-up on the beach is distorted strongly by their presence.

waves with slightly different frequencies in a laboratory which formed wave groups. Edge waves were observed as predicted by this mechanism.

Lippmann, Holman, and Bowen (1997) provide a review of the mechanisms for the generation of edge waves and show that edge wave generation by spatially and temporally varying radiation stresses of the incident waves can account for most of the low-frequency motion in the surf zone. They use the linear wave equation (Eq. 5.46) with radiation stress driving terms.

In the field, there are now numerous studies that show very large edge wave motions in the surf zone. In a major study, Huntley, Guza, and Thornton (1981) showed, using a longshore array of current meters, that low-frequency energy in the surf zone had wave lengths associated with the edge wave dispersion relationship given above. Figure 5.9 shows the field data and the associated dispersion relationships. This low-frequency energy in some cases can have more energy content locally within the shallow portions of the surf zone than the incident wave field, which has important ramifications (as yet not properly elucidated) for coastal processes.

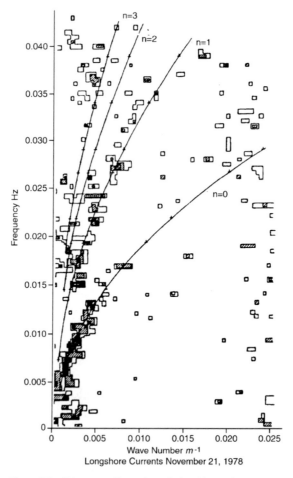

Figure 5.9 Edge wave dispersion relationships and wave numbers for Torrey Pines Beach (Huntley, Guza, and Thornton 1981, copyright by the American Geophysical Union).

A manifestation of the low-frequency wave motion is the slowly varying location of the limit of wave uprush on the beach face. An observer standing at this limit of wave uprush may soon find it necessary to move to other locations up and down the beach face as the surf zone slowly rises and falls owing to the edge wave or standing wave motions within the surf zone over the course of minutes.

5.4.3 SHEAR WAVES

In 1986, a field experiment at Duck, North Carolina, revealed a surprising behavior of the longshore current (Oltman-Shay, Howd, and Birkemeier 1989). The very strong longshore current that were caused by large waves began to oscillate with a low frequency. The alongshore wavelengths of this motion, detected by current meters distributed along the shoreline, were much less than predicted by edge wave or gravity wave theories. Bowen and Holman (1989) provided a theory for a wave motion that depended on the cross-shore shear in the longshore current velocity profile as a restoring force rather than gravity. This wave motion, occurring in the horizontal plane and moving with a speed somewhat less than the longshore current, causes the longshore current to move back and forth across the surf zone. Reniers et al. (1997) demonstrated the existence of shear waves in the laboratory. Dodd, Oltman-Shay, and Thornton (1992) showed that bottom friction retards the onset of the shear wave instability, whereas Putrevu and Svendsen (1993) demonstrated that a barred shoreline reinforces it, which explains why these waves were present at Duck and not obvious at other field sites on more planar beaches.

Allen, Newberger, and Holman (1996) showed, using a numerical model, that the incorporation of nonlinear terms allows the shear waves to evolve into large-amplitude oscillations and to migrate offshore of the surf zone as eddies. This motion looks very much like unstable rip currents. Özkan-Haller and Kirby (1999) have also modeled this nonlinear behavior, showing the interaction between shear waves, the offshore detachment of pairs of vortices, and that the shear waves cause considerable horizontal mixing in the surf zone – even greater than that due to turbulence.

This topic is still one of considerable interest to nearshore hydrodynamicists.

5.5 NEARSHORE CIRCULATION AND RIP CURRENTS

The nearshore circulation system occurring at the beach often includes nonuniform longshore currents, rip currents, and cross-shore flows.

Rip currents are jets of water issuing through the breaker line that carry sand offshore. These currents can sometimes be observed occurring (somewhat) periodically down a long straight beach; they also occur under piers, alongside jetties and groins, in the center of embayments formed by headlands, and at breaks in sandbars and offshore breakwaters. The offshore velocity in rip currents can exceed 2 m/s, and they contribute to the death toll at beaches by carrying unwary swimmers directly offshore into deep water. The general circulation pattern for a rip current system, with its feeder currents along the beach, is shown in Figure 5.10.

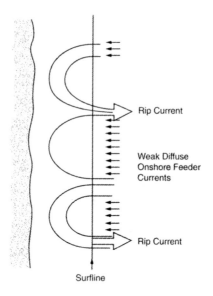

Figure 5.10 Schematic of a nearshore circulation system showing the rip current and the feeder currents.

Rip currents on a long straight beach have been observed from one horizon to the other with roughly a uniform spacing. What creates this longshore periodic phenomena? Several mechanisms could be responsible. First, the offshore bathymetry may have a periodic spacing, causing periodic refraction of the waves and leading to periodic variations in setup and the circulation system (Bowen 1969b, Sonu 1972). This simply backs us up to the question of why there is a variation in the bathymetry. Is this some edgewave-induced phenomenon (Holman and Bowen 1982)? Alternatively, the incident wave climate may have periodic modulations (Dalrymple 1974, Tang and Dalrymple 1989) that force the nearshore hydrodynamics. The Dalrymple model examines crossing wave trains of the same frequency that force a fixed longshore variation in wave height and thus wave setup, creating rip currents where the wave amplitudes cancel out. Fowler and Dalrymple (1990) showed that intersecting wave trains with nearly the same period will produce migrating rip current systems. Both of the intersecting wave models have been tested successfully in the laboratory. The Tang and Dalrymple concept is that a directional spectrum will have modulations, as discussed by Gallagher, which will force the nearshore circulation system.

Zyserman, Fredsøe, and Deigaard (1990) consider rip currents on a coast with sandbars. The rip current channels and the sediment transport are assumed to be in balance, and thus the spacing and the depth of the channels can be predicted.

Dalrymple (1978) examined the mechanism for creating rip currents in channels through offshore bars. He proposed that the wave setup created shoreward of the sandbar is higher than that created behind the rip channel. Therefore, there is a longshore gradient of mean water level driving flow toward the channel. Haller, Dalrymple, and Svendsen (1997) showed in a wave basin that the rip currents do occur via preestablished channels through longshore sandbars but that these rip currents are unstable and oscillate side-to-side because of flow instabilities triggered by the strong shear in the offshore velocity profile. Haller and Dalrymple (2001) provide a model for this instability based on the shear in the offshore velocity profile of the rip current.

Table 5.3 Mechanisms for Rip Currents

Wave–Wave Interaction	Representative References
Incident Wave–Edge Wave	
Synchronous	Bowen (1969b)
Infragravity	Sasaki (1975)
Intersecting Wave Trains	Dalrymple (1975), Fowler and Dalrymple (1990)
Wave–Current Interaction	Dalrymple and Lozano (1978)
Wave Structure Interaction	
Bottom Topography	Bowen (1969b), Zyserman et al. (1990)
Coastal Boundaries	
Breakwaters	Liu and Mei (1976)
Islands	Mei and Angelides (1977)
Barred Coastlines	Dalrymple (1978)

Primary Source: Dalrymple (1978).

Other nearshore circulation models examine instabilities and eigenvalue responses of the surf zone. For example, Dalrymple and Lozano (1978) examined the case of normally incident waves on a surf zone on a planar beach. One possible solution to this problem is no circulation with a longshore uniform setup within the surf zone. Another possibility assumes an existing rip current, which forces the incident waves, by wave–current interaction, to slow over the rip, causing the waves to refract toward the rip. This refraction inside the surf zone drives longshore currents toward the base of the rip, which then flows offshore. This self-reinforcing circulation pattern provides another stable solution to the flow in the surf zone.

Finally, the presence of structures can lead to rip currents. Liu and Mei (1976) examined the rip formed behind an offshore breakwater as a result of the wave diffraction pattern that occurs in this region. Table 5.3 lists several of the various models for rip currents.

5.5.1 NUMERICAL MODELING OF NEARSHORE CIRCULATION

Numerical modeling of the nearshore circulation system (including rip and longshore currents) permits the study of both onshore and offshore motions as well as longshore motions and can in fact include the influence of rip currents. A variety of models has been developed, for example, Noda (1974); Birkemeier and Dalrymple (1975); Vemulakonda, Houston, and Butler (1982); Kawahara and Kashiyama (1984); Wu and Liu (1985); and Van Dongeren et al. (1994). These depth-averaged models in general solve the nearshore circulation field forced by bottom variations, although Ebersole and Dalrymple (1980) examined the case of intersecting wave trains and Wind and Vreugdenhil (1986) addressed the circulation induced between two barriers, pointing out the importance of correctly modeling the lateral shear stresses. The last model by Van Dongeren et al. is quasi-three-dimensional, adding the influence of the undertow on the longshore current, as pointed out by Putrevu and Svendsen (1992).

Madsen, Sørensen, and Schaffer (1997a,b) have shown that an extended Boussinesq wave model can predict surf zone hydrodynamics quite well when a wave-breaking algorithm is included. Averaging the numerical model currents over a wave period yields the "mean" flows of the nearshore circulation system. Chen et al. (1999) compared Boussinesq model results with a physical model of the nearshore circulation on a barred shoreline with rip channels. The numerical model predicts the instabilities in the rip currents as seen in the physical model by Haller et al. (1997).

Most of these nearshore circulation models were developed by finite difference methods, although Wu and Liu use finite element techniques. Some of these models are being used for engineering work, although it should be pointed out that most of them are very computer intensive and require very small time steps (on the order of seconds) to reach steady-state solutions. This often causes problems when trying to determine the effect of several days' worth of wave conditions or to predict 1 or 50 years of coastal conditions.

Models typically have been developed only for monochromatic (single frequency) wave trains rather than for directional spectra. There is a definite need for further development of these models.

5.6 SWASH ZONE DYNAMICS

The swash zone is defined as that region on the beach face delineated at the upper limit by the maximum uprush of the waves and at its lower extremity by the maximum downrush. This portion of the beach face is intermittently affected by successive waves that traverse this zone in a zig-zag fashion, and it is an area where very interesting patterns, called beach cusps, can occur. Finally, substantial quantities of longshore sediment transport may occur within the swash zone limits. Thus, it is of interest to understand this phenomenon better and to develop a predictive capability for this region.

In this section we will examine progressively more complex and realistic cases of swash zone dynamics by using simpler (analytic) models than those numerical models previously described. As shown in Figure 5.11, friction, gravity, inertia, and pressure gradients are all potentially significant forces acting on a water element within the swash zone.

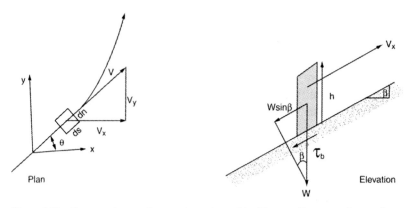

Figure 5.11 Forces acting on elemental swash particle. The coordinate x points onshore and y alongshore.

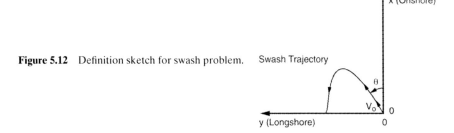

Figure 5.12 Definition sketch for swash problem.

We will consider the case of a wave approaching the beach face at some angle θ to the x-axis that points onshore. The leading edge of the wave advances up the beach face with initial velocity V_0 (see Figure 5.12). To render the problem tractable, it is assumed that it is possible to idealize the water particle as a solid particle that retains its identity as it moves up and down the beach face. The height of the particle is h, and its length and width are taken as Δs and Δn, where s is in the direction of the particle motion and n is perpendicular. The appropriate equations of motion are

$$\rho h \Delta s \Delta n \frac{dV_x}{\partial t} = -\rho g h \Delta s \Delta n \sin \beta - \rho \frac{f}{8} \Delta s \Delta n |V| V_x$$

$$\rho h \Delta s \Delta n \frac{dV_y}{\partial t} = -\rho \frac{f}{8} h \Delta s \Delta n |V| V_y, \tag{5.57}$$

or, after eliminating common terms,

$$\frac{dV_x}{dt} = -g \sin \beta - \frac{f}{8h}|V|V_x$$

$$\frac{dV_y}{dt} = -\frac{f}{8h}|V|V_y, \tag{5.58}$$

in which β is the bottom slope and f is the Darcy–Weisbach friction factor. All the terms on the right-hand side (gravity and friction) serve to decelerate the uprushing particle. Various forms of the bottom friction will be discussed in later sections.

The initial conditions (at $x = 0$) are

$$V_{x_0} = V_0 \cos \theta_0$$
$$V_{y_0} = V_0 \sin \theta_0 \tag{5.59}$$

5.6.1 OBLIQUELY INCIDENT WAVES AND NO FRICTION

Equations (5.58) are readily solved for the case of no friction. A simple integration of these equations yields the following for the two horizontal components of water particle velocity and displacement (for a particle located initially at the origin):

$$V_x = V_0 \cos \theta_0 - gt \sin \beta$$
$$V_y = V_0 \sin \theta_0$$

The position of the particle can be found by integrating the velocities with respect to time:

$$x(t) = V_0 \cos \theta_0 t - \frac{gt^2}{2} \sin \beta$$
$$y(t) = V_0 \sin \theta_0 t \tag{5.60}$$

Several results can be obtained from this simple formulation. For example, if we consider the case of normally incident waves ($\theta = 0$), the maximum uprush occurs at the time that the velocity is zero, as given by

$$t_{\max} = \frac{V_0}{g \sin \beta} \tag{5.61}$$

It is seen that the corresponding value of the maximum uprush is

$$x_{\max} = \frac{V_0^2}{2g \sin \beta}, \tag{5.62}$$

which is recognized as the familiar result for trajectory of a solid particle shot into the air vertically with speed V_0, however, in this case, the value of gravity has been reduced by the sine of the beach face slope. The time required for the particle to return to the starting location ($x = 0$) is known as the frictionless natural period of a water particle on a beach face, or the *swash period* T_{n_0}, as given by $2t_{\max}$:

$$T_{n_0} = \frac{2V_0}{g \sin \beta} \tag{5.63}$$

The natural period depends both on the initial shore-normal velocity and the slope of the beach face.

5.6.2 NONDIMENSIONAL EQUATIONS

At this stage it is convenient to express Eqs. (5.58) in nondimensional form. We will choose T_{n_0} and V_0, as the reference quantities, that is,

$$t' = \frac{t}{T_{n_0}}$$
$$V' = \frac{V}{V_0}, \tag{5.64}$$

which transforms Eq. (5.58) into

$$\frac{dV'_x}{dt'} = -2 - \gamma |V'| V'_x$$
$$\frac{dV'_y}{dt'} = -\gamma |V'| V'_y, \tag{5.65}$$

where γ is a friction parameter defined as

$$\gamma = \frac{fV_0^2}{4gh \sin \beta} \tag{5.66}$$

5.6.3 OBLIQUELY INCIDENT WAVES WITH LINEAR FRICTION

Although it is impossible to develop simple solutions to the full swash equations owing to their nonlinearity, the linearized swash equations can be solved.

The equations are

$$
\begin{aligned}
\frac{dV_x'}{dt'} &= -2 - \gamma_L V_x' \\
\frac{dV_y'}{dt'} &= -\gamma_L V_y',
\end{aligned}
\tag{5.67}
$$

where γ_L is a linearized friction term. Solving,

$$
V_x'(t') = \frac{2}{\gamma_L}\left(e^{-\gamma_L t'} - 1\right) + V_{x_0}' e^{-\gamma_L t'}
$$

$$
V_y'(t') = V_{y_0}' e^{-\gamma_L t'}
\tag{5.68}
$$

$$
x'(t') = \left(\frac{2}{\gamma_L^2} + \frac{V_{x_0}'}{\gamma_L}\right)\left(1 - e^{-\gamma_L t'}\right) - \frac{2}{\gamma_L} t'
$$

$$
y'(t') = \frac{V_{y_0}'}{\gamma_L}\left(1 - e^{-\gamma_L t'}\right)
\tag{5.69}
$$

It is of interest to compare the characteristics of the preceding swash solution with those for frictionless swash. We will consider the case of small γ_L (i.e., small friction).

The maximum value of uprush occurs when $V_x' = 0$, which, for small γ_L, yields (using $e^{-\gamma_L t'} \approx 1 - \gamma_L t' + \cdots$)

$$
t_{\max}' = \frac{V_{x_0}'}{2 + \gamma_L V_{x_0}'} \approx \frac{t_{\max_0}'}{\left(1 + \frac{\gamma_L}{2}\cos\theta_0\right)},
\tag{5.70}
$$

in which $t_{\max}'$ denotes the uprush time without friction. Thus, the effect of friction is to reduce the uprush time. The maximum uprush $x_{\max}'$ is

$$
x_{\max}' = x_{\max_0}'\left(1 - \frac{\gamma_L \cos\theta_0}{2}\right)
\tag{5.71}
$$

That is, in accordance with intuition, the effect of friction is to reduce the maximum uprush.

The maximum backrush velocity can be evaluated by first solving for t_0', where $x'(t_0') = 0$ and substituting this value of t_0' in Eq. (5.68a). Solving for t_0'

$$
t_0' = \frac{V_{x_0}'}{1 + \frac{\gamma_L \cos\theta_0}{2}},
\tag{5.72}
$$

we see that the natural period is reduced. The maximum backrush velocity $[V_B' = V'(t_0')]$ is

$$
V_x'(t_0') = -V_{x_0}'\left(\frac{1 - \frac{\gamma_L \cos\theta_0}{2}}{1 + \frac{\gamma_L \cos\theta_0}{2}}\right) \approx -V_{x_0}'(1 - \gamma_L \cos\theta_0)
\tag{5.73}
$$

It is clear that the particle has imparted a net shoreward impulse on the beach face because its upward momentum is greater than the downward momentum. The net upward impulse I per unit length of the beach face is

$$I_x = \rho \Delta x h [V_x(0) + V_x(T_n)] \tag{5.74}$$

such that for a frictionless system $V_x(T_n) = -V_x(0)$, the net impulse is zero. The average force $\bar{F}_x$ per unit length is

$$\bar{F}_x = \frac{I_x}{T} = \frac{\rho \Delta x h}{T}[V_x(0) + V_x(T_n)] \tag{5.75}$$

For the case of linearized friction just considered,

$$\bar{F} = \frac{\rho \Delta x h}{T}\left(\gamma_L V'_{x_0}\right) = \frac{\rho \Delta x h}{T}\gamma_L \cos^2 \theta_0$$

$$= \frac{\rho \Delta x h}{T} V_0[\cos^2 \theta_0 - \cos \theta_0 \cos \theta_F + \gamma_L \cos^2 \theta_0 \cos \theta_F], \tag{5.76}$$

where θ_F is the angle of the return flow.

Figure 5.13 presents a comparison of the nondimensional idealized trajectories for frictionless and frictional systems. The initial angle of obliquity is 40°, and the value of γ_L is 5.0 for the frictional system. Of particular interest is the saw-toothed shape of the trajectory for $\gamma_L = 5.0$, whereas the frictionless trajectory is symmetrical about the peak. A field method to estimate γ_L would be useful. Because it can be shown that the return angle is much less than the initial angle, this difference could be used to estimate γ_L.

Figure 5.14 presents the relationship between initial (θ_0) and final (θ_F) swash angles for varying values of non-dimensional linear friction.

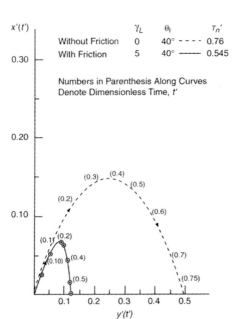

	γ_L	θ_i	T_n'
Without Friction	0	40°	0.76
With Friction	5	40°	0.545

Numbers in Parenthesis Along Curves Denote Dimensionless Time, t'

Figure 5.13 Swash trajectories, effect of linear friction.

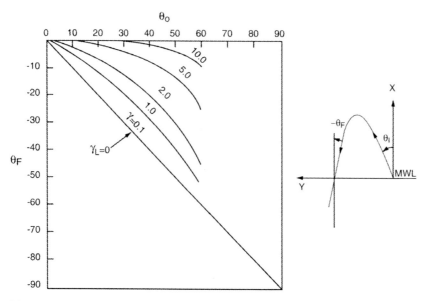

Figure 5.14 Relationship between initial and final swash angles as a function of β (for linear friction).

5.6.4 NORMALLY INCIDENT WAVES WITH NONLINEAR FRICTION

We commence with Eqs. (5.65) and note for normal incident waves, $V'_y = 0$. Because of the absolute value sign in the friction term, it is necessary to consider uprush and backrush separately.

Uprush – During uprush, $V' > 0$, and the first of Eqs. (5.65) becomes

$$\frac{dV'_x}{dt'} = -2 - \gamma V'^2, \tag{5.77}$$

a solution of which is

$$V'(t') = \sqrt{\frac{2}{\gamma}} \tan\left[\sqrt{2\gamma}(t'_{max} - t')\right] \tag{5.78}$$

in which t'_{max} is the time at which the maximum particle excursion occurs and is

$$t'_{max} = \frac{1}{\sqrt{2\gamma}} \tan^{-1}\left(\sqrt{\frac{\gamma}{2}}\right) \tag{5.79}$$

The associated uprush displacement is determined by integrating Eq. (5.78) to yield

$$x'(t') = \frac{1}{\gamma} \ell n \left\{ \frac{\cos(\sqrt{2\gamma}t'_{max})}{\cos[\sqrt{2\gamma}(t'_{max} - t')]} \right\}, \tag{5.80}$$

and the maximum displacement occurs at $t' = t'_{max}$ and is

$$x'(t'_{max}) = x'_{max} = \frac{1}{\gamma} \ell n\{\cos \sqrt{2\gamma}t'_{max}\} \tag{5.81}$$

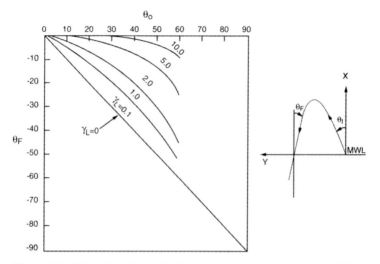

Figure 5.15   Effect of nonlinear friction parameter γ on swash characteristics.

Backrush – On the basis of a development similar to that used for the uprush, for backrush $V' < 0$ and Eq. (5.65) becomes

$$\frac{dV'}{dt'} = -2 + \gamma V'^2, \tag{5.82}$$

the solution of which can be shown to be

$$V'(t') = -\sqrt{\frac{2}{\gamma}} \, \tanh[\sqrt{2\gamma}(t' - t'_{\max})] \tag{5.83}$$

Equation (5.83) can be integrated to yield

$$x'(t') = x'_{\max} - \frac{1}{\gamma} \ell n \{\cosh \sqrt{2\gamma}(t' - t'_{\max})\} \tag{5.84}$$

The natural period is altered by nonlinear friction and can be determined from Eqs. (5.79) and (5.84) as

$$T'_{\mathrm{n}} = \frac{1}{\sqrt{2\gamma}} \left\{ \tan^{-1}\left(\sqrt{\frac{\gamma}{2}}\right) + \cosh^{-1}[e^{\gamma x'_{\max}}] \right\} \tag{5.85}$$

Figure 5.15 shows the effect of varying nonlinear friction on natural period T'_{n}, maximum downrush velocity (V'_0), and maximum uprush displacement $x'_{\max}$.

5.6.5 FIELD AND THEORETICAL STUDIES

Hughes (1992) carried out a comparison of theoretical and measured swash motions. He applied the long-wave frictionless equations to predict the swash motions on a planar beach face. It was found from the theory that the trajectory of the leading edge of the swash motion was described qualitatively by Eq. (5.60) and that the thickness of the swash decreased as the swash traversed up the beach face. The field experiments were conducted in southeast Australia on several beaches ranging in slope from 0.093 to 0.15. The beach faces were considerably steeper than the profile immediately seaward, leading to bore collapse on the beach face. The swash motions were documented by capacitance gauges and stakes driven into the beach face. Documentation from the capacitance gauges was complemented by photographing the swash motion in the vicinity of the stakes. It was found that long wave theory provided a good representation for the form of the uprush trajectories and the swash thicknesses; however, the magnitudes of both were overpredicted. The measured uprush trajectories were found to be approximately 0.65 of the theoretical, and the swash thicknesses were overpredicted by a factor of 2 to 3. It was concluded that a major reason for the differences between measurements and theory was the lack of friction and infiltration in the theoretical formulation.

REFERENCES

Abbott, M.B., H.M. Petersen, and O. Skovgaard, "On the Numerical Modeling of Short Waves in Shallow Water," *J. Hyd. Res.*, 16, 173–204, 1978.

Airy, G.B., "Tides and Waves," *Encyclopaedia Metropolitana*, London, J.J. Griffin, 1845.

Allen, J.S., P.A. Newberger, and R.A. Holman, "Nonlinear Shear Instabilities of Alongshore Currents on Plane Beaches," *J. Fluid Mech.*, 310, 181–213, 1996.

Ball, F.K., "Edge Waves in an Ocean of Finite Depth," *Deep Sea Res.*, Oceanogr. Abst., 14, 79–88, 1967.

Battjes, J.A., "Surf Similarity," *Proc. 14th Intl. Conf. Coastal Eng.*, ASCE, Copenhagen, 466–480, 1974.

Battjes, J.A., and J.P.F.M. Janssen, "Energy Loss and Set-up due to Breaking of Random Waves," *Proc. 16th Intl. Conf. Coastal Eng.*, ASCE, Hamburg, 1978.

Berkhoff, J.C.W., "Computation of Combined Refraction–Diffraction," *Proc. 13th Intl. Conf. Coastal Eng.*, ASCE, Vancouver, 471–484, 1972.

Birkemeier, W.A., and R.A. Dalrymple, "Nearshore Water Circulation Induced by Wind and Waves," *Proceedings of Modeling 75*, ASCE, San Francisco, 1975.

Booij, N., "Gravity Waves on Water with Nonuniform Depths and Current," Tech. Univ. Delft, Rpt. 81-1, Dept. Civil Eng., 1981.

Booij, N., and L.H. Holthuijsen, "Propagation of Ocean Waves in Discrete Spectral Wave Models," *J. Computational Physics*, 68, 307–326, 1987.

Booij, N., L.H. Holthuijsen, and T.H.C. Herbers, "A Numerical Model for Wave Boundary Conditions in Port Design," *Proc. Intl. Conf. Numerical and Physical Modeling of Ports and Harbours*, BHRA, Birmingham, 263–268, 1985.

Bowen, A.J., "The Generation of Longshore Currents on a Plane Beach," *J. Marine Res.*, 37, 206–215, 1969a.

Bowen, A.J., "Rip Currents, I. Theoretical Investigations," *J. Geophys. Res.*, 74, 5467–5478, 1969b.

Bowen, A.J., and R.T. Guza, "Edge Waves and Surf Beat," *J. Geophys. Res.*, 83, C4, 1913–1920, 1978.

Bowen, A.J., and R.A. Holman, "Shear Instabilities of the Longshore Current: 1. Theory," *J. Geophys. Res.*, 94, 18023–18030, 1989.

Bowen, A.J., D.L. Inman, and V.P. Simmons, "Wave 'Set-down' and Wave Set-up," *J. Geophys. Res.*, 73, 2569–2577, 1968.

Brink-Kjaer, O., "Depth-Current Refraction of Wave Spectra," *Symp. Description and Modeling of Directional Seas*, Tech. Univ. Denmark, 1984.

Carrier, G.F., and H.P. Greenspan, "Water Waves of Finite Amplitude on a Sloping Beach," *J. Fluid Mech.*, 4, 97–109, 1957.

Chen, Q., R.A. Dalrymple, J.T. Kirby, A. Kennedy, and M.C. Haller, "Boussinesq Modelling of a Rip Current System," *J. Geophys. Res.*, 104, C9, 20,617–20,638, 1999.

Dally, W.R., "Random Breaking Waves: A Closed-form Solution for Planar Beaches," *Coastal Eng.*, 14, 3, 233–265, 1990.

Dally, W.R., R.G. Dean, and R.A. Dalrymple, "Wave Height Variation Across Beaches of Arbitrary Profile," *J. Geophys. Res.*, Vol. 90, 6, 11917–11927, 1985.

Dalrymple, R.A., "A Finite Amplitude Wave on a Linear Shear Current," *J. Geophys. Res.*, 79, 30, 4498–4505, 1974.

Dalrymple, R.A., "A Mechanism for Rip Current Generation on an Open Coast," *J. Geophys. Res.*, 80, 3485–3487, 1975.

Dalrymple, R.A., "Rip Currents and Their Causes," *Proc. 16th Intl. Conf. Coastal Eng.*, ASCE, Hamburg, 1414–1427, 1978.

Dalrymple, R.A. "A Model for the Refraction of Water Waves," *J. Waterway, Port, Coastal and Ocean Eng.*, ASCE, 114, 4, 423–435, 1988.

Dalrymple, R.A., and C.J. Lozano, "Wave–Current Interaction Models for Rip Currents," *J. Geophys. Res.*, Vol. 83, No. C12, 1978.

Dalrymple, R.A., J.T. Kirby, and D.J. Seli, "Wave Trapping by Breakwaters," *Proc. 20th Intl. Conf. Coastal Eng.*, Taipei, ASCE, 1986.

Dean, R.G., "Stream Function Representation of Nonlinear Ocean Waves, *J. Geophys. Res.*, 70, 18, 4561–4572, 1965.

Dean, R.G., and R.A. Dalrymple, *Water Wave Mechanics for Engineers and Scientists*, Singapore: World Scientific Press, 353 pp., 1991.

Dingemans, M.W., *Water Wave Propagation Over Uneven Bottoms*, Vols. 1 and 2, Singapore: World Scientific Press, 967 pp., 1997.

Dodd, N., J. Oltman-Shay, and E.B. Thornton, "Shear Instabilities in the Longshore Current: A Comparison of Observations and Theory," *J. Phys Oceanogr.*, 22, 62–82, 1992.

Ebersole, B.A., and R.A. Dalrymple, "Numerical Modelling of Nearshore Circulation," *Proc. 17th Intl. Conf. Coastal Eng.*, ASCE, Sydney, 1980.

Eckart, C., "Surface Waves in Water of Variable Depth," Univ. California, Scripps Institution of Oceanography, Wave Report 100, SIO Ref. 51-12, 99 pp., 1951.

Elgar, S., and R.T. Guza, "Shoaling Gravity Waves: Comparison Between Field Observations, Linear Theory and Nonlinear Model," *J. Fluid Mechanics*, 158, 47–70, 1985.

Fenton, J.D., "A Fifth-Order Stokes Theory for Steady Waves," *J. Waterways, Port, Coastal and Ocean Eng.*, ASCE, 111, 216–234, 1985.

Fenton, J.D., and W.D. McKee, unpublished, 1989; referred to by Fenton, J.D., "Nonlinear Wave Theories," in *The Sea*, 9, A, 3–25, New York: Wiley Interscience, 1990.

Fowler, R.E., and R.A. Dalrymple, "Wave Group Forced Nearshore Circulation," *Proc. 22nd Intl. Conf. Coastal Eng.*, ASCE, 729–742, 1990.

Freilich, M.H., and R.T. Guza, "Nonlinear Effects on Shoaling Surface Gravity Waves," *Phil. Trans. Roy. Soc. London*, A, 31, 1–41, 1984.

Gallagher, B., "Generation of Surf Beat by Nonlinear Wave Interactions," *J. Fluid Mechanics*, 49, 1–20, 1971.

Galvin, C.J., and P.S. Eagleson, "Experimental Study of Longshore Currents on a Plane Beach," Tech. Memo., U.S. Army Coastal Engineering Research Center, 10, 80 pp., 1965.

Guza, R.T., and R.E. Davis, "Excitation of Edge Waves by Waves Incident on a Beach," *J. Geophys. Res.*, 79, 1285–1291, 1974.

Guza, R.T., and D.L. Inman, "Edge Waves and Beach Cusps," *J. Geophys. Res.*, 80, 2997–3012, 1975.

Haller, M.C., and R.A. Dalrymple, "Rip Current Instabilities," *J. Fluid Mechanics*, 433, 161–192, 2001

Haller, M.C., R.A. Dalrymple, and I.A. Svendsen, "Rip Channels and Nearshore Circulation," *Proc. Waves '97*, ASCE, 1997.

Hibberd, S., and D.H. Peregrine, "Surf and Run-up on a Beach: a Uniform Bore," *J. Fluid Mech.*, 95, 2, 323–345, 1979.

Holman, R.A., and A.J. Bowen, "Edge Waves on Complex Beach Profiles," *J. Geophys. Res.*, 84, C10, 6339–6346, 1979.

Holman, R.A., and A.J. Bowen, "Bars, Bumps, and Holes: Models for the Generation of Complex Beach Topography," *J. Geophys. Res.*, 87, C1, 457–468, 1982.

Holthuijsen, L.H., N. Booij, and R.C. Ris, "A Spectral Wave Model for the Coastal Zone, *Proc. 2nd Intl. Symp. Ocean Wave Measurement and Analysis*, ASCE, New Orleans, 630–641, 1993.

Horikawa, K., and C.T. Kuo, "A Study of Wave Transformation Inside the Surf Zone," *Proc. 10th Intl. Conf. Coastal Eng.*, ASCE, 217–233, 1966.

Howd, P.A., A.J. Bowen, and R.A. Holman, "Edge Waves in the Presence of Strong Longshore Currents," *J. Geophys. Res.*, 97, C7, 11,357–11,371, 1992.

Huntley, D.A., R.T. Guza, and E.B. Thornton, "Field Observations of Surf Beat: 1. Progressive Edge Waves," *J. Geophys. Res.*, 86, 6451–6466, 1981.

Hughes, M.G., "Application of a Non-Linear Shallow Water Theory to Swash Following Bore Collapse on a Sandy Beach," *J. Coastal Research*, 8, 3, 562–578, 1992.

Kadomtsev, B.B., and V.I. Petviashvili, "On the Stability of Solitary Waves in Weakly Dispersive Media," *Sov. Phys. Dokl.*, 15, 539–541, 1970.

Kaihatu, J.M., and J.T. Kirby, "Two-Dimensional Parabolic Modeling of Extended Boussinesq Equations," *J. Waterway, Port, Coastal and Ocean Eng.*, 124, 57–67, 1998.

Kawahara, M., and K. Kashiyama, "Selective Lumping Finite Element Model for Nearshore Currents," *Int. J. Numerical Methods Fluids*, 4, 71–97, 1984.

Kennedy, A.B., Q. Chen, J.T. Kirby, and R.A. Dalrymple, "Boussinesq Modeling of Wave Transformation, Breaking and Runup. I: 1D," *J. Waterway, Port, Coastal, and Ocean Eng.*, 126, 39–47, 2000.

Kirby, J.T., and R.A. Dalrymple, "A Parabolic Equation for the Combined Refraction-Diffraction of Stokes Waves by Mildly Varying Topography," *J. Fluid Mechanics*, 136, 453–466, 1983.

Kirby, J.T., R.A. Dalrymple, and P.L.-F. Liu, "Modifications of Edge Waves by Barred-Beach Topography," *Coastal Eng.*, 5, 35–49, 1981.

Kobayashi, N., G.S. De Silva, and K.D. Watson, "Wave Transformation and Swash Oscillation on Gentle and Steep Slopes," *J. Geophys. Res.*, 94, C1, 951–966, 1989.

Kobayashi, N., and A. Wurjanto, "Irregular Wave Setup and Run-Up on Beaches," *J. Waterway, Port, Coastal and Ocean Eng.*, 118, 4, 368–386, 1992.

Korteweg, D.J., and G. De Vries, "On the Change of Form of Long Waves Advancing in a Rectangular Channel, and on a New Type of Long Stationary Waves," *Philos. Mag.*, 4th Ser, 39, 422–443, 1895.

Kostense, J.K., "Measurements of Surf Beat and Set-down Beneath Wave Groups," *Proc. 19th Intl. Conf. Coastal Eng.*, ASCE, Houston, 724–740, 1984.

Lamb, Sir H., *Hydrodynamics*, 6th ed., New York: Dover Press, 1945.

Leadon, M.E., N.T. Nguyen, and R.R. Clark, "Hurricane Opal: Beach and Dune Erosion and Structural Damage Along the Panhandle Coast of Florida," Bureau of Beaches and Coastal Systems, Florida Dept. Env. Protection, State of Florida, 1997.

Li, L., and R.A. Dalrymple, "Instabilities of the Undertow," *J. Fluid Mechanics*, 369, 175–190, 1998.

Lippmann, T.C., R.A. Holman, and A.J. Bowen, "Generation of Edge Waves in Shallow Water," *J. Geophys. Res.*, 102, C4, 8663–8679, 1997.

Liu, P.L.-F., and R.A. Dalrymple, "Bottom Frictional Stresses and Longshore Currents due to Waves with Large Angles of Incidence," *J. Marine Res.*, 32, 2, 357–375, 1978.

Liu, P.L.-F., and C.C. Mei, "Water Motion on a Beach in the Presence of a Breakwater, I and II," *J. Geophys. Res.*, 81, 3079–3094, 1976.

Liu, P.L.-F., S.B. Yoon, and J.T. Kirby, "Nonlinear Refraction–Diffraction of Waves in Shallow Water," *J. Fluid Mechanics*, 153, 184–201, 1985.

Longuet-Higgins, M.S., "Longshore Currents Generated by Obliquely Incident Sea Waves, 1," *J. Geophys. Res.*, 75, 33, 6778–6789, 1970a.

Longuet-Higgins, M.S., "Longshore Currents Generated by Obliquely Incident Sea Waves, 2," *J. Geophys. Res.*, 75, 33, 6790–6801, 1970b.

Longuet-Higgins, M.S., and R.W. Stewart, "Radiation Stress in Water Waves; a Physical Discussion with Applications," *Deep Sea Res.*, 11, 4, 529–563, 1963.

Madsen, P.A., R. Murray, and O.R. Sørensen, "A New Form of the Boussinesq Equations with Improved Linear Dispersion Characteristics," *Coastal Eng.*, 15, 371–388, 1991.

Madsen, P.A., O.R. Sørensen, and H.A. Schäffer, "Surf Zone Dynamics Simulated by a Boussinesq Type Model. Part I. Model Description and Cross-shore Motion of Regular Waves," *Coastal Eng.*, 32, 255–287, 1997a.

Madsen, P.A., O.R. Sørensen, and H.A. Schäffer, "Surf Zone Dynamics Simulated by a Boussinesq Type Model. Part II. Surf Beat and Swash Oscillations for Wave Groups and Irregular Waves," *Coastal Eng.*, 32, 287–319, 1997b.

Matiessen, M., "Current-Depth Refraction of Directional Wave Spectra," *Symp. Description and Modeling of Directional Seas*, Tech. Univ. Denmark, 1984.

Matsunaga, N., K. Takehara, and Y. Awaya, "Coherent Eddies Induced by Breakers on a Sloping Bed," *Proc. 21st Intl. Conf. Coastal Eng.*, ASCE, Malaga, 234–245, 1988.

Matsunaga, N., and K. Takehara, "Vortex Train in an Offshore Zone," *Proc. 23rd Intl. Conf. Coastal Eng.*, ASCE, Venice, 3163–3177, 1992.

Mei, C.C., *The Applied Dynamics of Ocean Surface Waves*, New York: Wiley Interscience, 740 pp., 1983. Also, Singapore: World Scientific Press, 1989.

Mei, C.C., and D. Angelides, "Longshore Circulation Around a Conical Island," *Coastal Eng.*, 1, 31–42, 1977.

Miller, D.J., "Giant Waves in Lituya Bay, Alaska," Geological Survey Professional Paper 354-C, 1960.

Munk, W.H., "Surf Beats," *Trans. Amer. Geophys. Un.*, 30, 849–854, 1949.

Munk, W.H., "The Solitary Wave Theory and Its Applications to Surf Problems," *Ann. N.Y. Acad. Sci.*, 51, 376–424, 1949.

Nadaoka, K., "A Fundamental Study on Shoaling and Velocity Field Structure of Water Waves in the Nearshore Zone," Ph.D. Dissertation, Tokyo Inst. Technology, Tech. Rpt. Dept. Civil Engineering No. 36, 36–125, 1986.

Nadaoka, K., M. Hino, and Y. Koyano, "Structure of the Turbulent Flow Field Under Breaking Waves in the Surf Zone," *J. Fluid Mech.*, 204, 359–387, 1989.

Nadaoka, K., S. Ueno, and T. Igarashi, "Field Observation of Three-Dimensional Large-Scale Eddies and Sediment Suspension in the Surf Zone," *Coastal Eng. in Japan*, 31, 2, 277–287, 1988.

Newman, J.N., "Numerical Solutions of the Water–Wave Dispersion Relationship," *Appld. Ocean Res.*, 12, 1, 14–18, 1990.

Nielsen, P., "Explicit Formulae for Practical Wave Calculation," *Coastal Eng.*, 6, 4, 389–398, 1982.

Noda, E.K., "Wave-induced Nearshore Circulation," *J. Geophys. Res.*, 79, 27, 4097–4106, 1974.

Nwogu, O., "An Alternative Form of the Boussinesq Equations for Nearshore Wave Propagation," *J. Waterway, Port, Coastal and Ocean Eng.*, 119, 6, 618–638, 1993.

Oltman-Shay, J., P.A. Howd, and W.A. Birkemeier, "Shear Instabilities of the Mean Longshore Current: 2. Field Observations," *J. Geophys. Res.*, 94, 18,031–18,042, 1989.

Özkan-Haller, H.T., and J.T. Kirby, "Nonlinear Evolution of Shear Instabilities of the Longshore Current: A Comparison of Observations and Computations," *J. Geophys. Res.*, 104, C11, 25,953–25,984, 1999.

Packwood, A.R., *Surf and Run-up on Beaches*, Ph.D. Diss., School of Math., Univ. of Bristol, 1980.

Peregrine, D.H., "Long Waves on a Beach," *J. Fluid Mechanics*, 27, 815–827, 1967.

Peregrine, H.D., and I.A. Svendsen, "Spilling Breakers, Bores, and Hydraulic Jumps," *Proc. 16th Intl. Conf. Coastal Eng.*, ASCE, Hamburg, 540–549, 1978.

Phillips, O.M., *The Dynamics of the Upper Ocean*, 2nd ed., New York: Cambridge Univ. Press, 336 pp., 1980.

Putrevu, U., and I.A. Svendsen, "Shear Instability of Longshore Currents: A Numerical Study," *J. Geophys. Res.*, 97, C5, 7283–7303, 1993.

Putrevu, U., and I.A. Svendsen, "A Mixing Mechanism in the Nearshore Region," *Proc. 23rd Intl. Conf. Coastal Eng.*, ASCE, Venice, 1992.

Reniers, A.J.H., J.A. Battjes, A. Falques, and D.A. Huntley, "A Laboratory Study on the Shear Instability of Longshore Currents," *J. Geophys. Res.*, 102, 8597–8609, 1997.

Russell, J.S., "Report on Waves," *14th Mtg. Brit. Assoc. Advanc. Sci.*, 1844.

Sasaki, T., "Simulation of Shoreline and Nearshore Current," *Proceedings of Civil Eng. in the Oceans*, III, ASCE, Univ. Delaware, 179–196, 1975.

Schäffer, H.A., P.A. Madsen, and R. Deigaard, "A Boussinesq Model for Waves Breaking in Shallow Water," *Coastal Eng.*, 20, 185–202, 1993.

Schäffer, H.A., and I.G. Jonsson, "Theory Versus Experiments in Two-dimensional Surf Beats," *Proc. 22nd Intl. Conf. Coastal Eng.*, ASCE, Delft, 1131–1143, 1990.

Schäffer, H.A., I.G. Jonsson, and I.A. Svendsen, "Free and Forced Cross-shore Long Waves," in *Water Wave Kinematics*, A. Torum and O.T. Gudmestad, eds., Dordrecht: Kluwer Academic Publishers, 367–385, 1990.

Sonu, C.J., "Field Observations of Nearshore Circulation and Meandering Currents," *J. Geophys. Res.*, 77, 3232–3247, 1972.

Stokes, Sir G.G., "Report on Recent Researches in Hydrodynamics," *Brit. Ass. Rep.*, 1846. Also, in *Mathematical and Physical Papers*, Cambridge Univ. Press, 1905.

Stokes, Sir G.G., "On the Theory of Oscillatory Waves," *Trans. Camb. Philos. Soc.*, 8, 441–455, 1847. Also, in *Mathematical and Physical Papers*, Cambridge Univ. Press, 1905.

Svendsen, I.A., "Wave Heights and Set-up in a Surf Zone," *Coastal Eng.*, 8, 303–329, 1984.

Tang, E., and R.A. Dalrymple, "Rip Currents, Nearshore Circulation and Wave Groups," in *Nearshore Sediment Transport*, R.J. Seymour, ed., New York: Plenum Press, 1989.

Thornton, E.B., "Variations of Longshore Current Across the Surf Zone," *Proc. 12th Intl. Conf. Coastal Eng.*, ASCE, 291–308, 1970.

Thornton, E.B., and R.T. Guza, "Energy Saturation and Phase Speeds Measured on a Natural Beach," *J. Geophys. Res.*, 87, 9499–9508, 1982.

Thornton, E.B., and R.T. Guza, "Transformation of Wave Height Distribution," *J. Geophys. Res.*, 88, C10, 5925–5938, 1983.

Tucker, M.J., "Surf Beats: Sea Waves of 1 to 5 Minutes Period," *Proc. Roy. Soc. London*, A, 202, 565–573, 1950.

Ursell, F., "Edge Waves on a Sloping Beach," *Proc. Roy. Soc. London*, A, 214, 27–97, 1952.

Van Dongeren, A.R., F.E. Sancho, I.A. Svendsen, and U. Putrevu, "SHORECIRC: A Quasi 3-D Nearshore Model," *Proc. 24th Intl. Conf. Coastal Eng.*, ASCE, Kobe, 1994.

Vemulakonda, S.R., J.R. Houston, and H.L. Butler, "Modeling of Longshore Currents for Field Situations," *Proc. 18th Intl. Conf. Coastal Eng.*, ASCE, Cape Town, 1982.

Weggel, J.R., "Maximum Breaker Height," *J. Waterways, Harbors, Coastal Eng. Div.*, ASCE, 98, WW4, 1972.

Wei, G., J.T. Kirby, S.T. Grilli, and R. Subramanya, "A Fully Nonlinear Boussinesq Model for Surface Waves, I. Highly Nonlinear, Unsteady Waves," *J. Fluid Mech.*, 294, 71–92, 1995.

Wiegel, R.L., *Oceanographical Engineering*, Englewood Cliffs: NJ, Prentice–Hall, 532 pp., 1964.

Wind, H.G., and C.B. Vreugdenhil, "Rip-current Generation Near Structures," *J. Fluid Mechanics*, 171, 459–476, 1986.

Wu, C.-S., and P.L.-F. Liu, "Finite Element Modelling of Nonlinear Coastal Currents," *J. Waterway, Port, Coastal and Ocean Eng.*, ASCE, 111, 2, 417–432, 1985.

Zyserman, J., J. Fredsøe, and R. Deigaard, "Prediction of the Dimensions of a Rip Current System on a Coast with Bars," *Proc. 22nd Intl. Conf. Coastal Eng.*, ASCE, Delft, 959–972, 1990.

EXERCISES

5.1 For a deep-water wave with a height of 1 m, a period of 8 s, and an angle of incidence to the shoreline of $30°$, determine the wave height in 5 m of water. Assume straight and parallel contours for the offshore bathymetry.

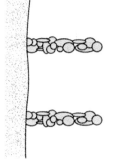

Figure 5.16 Figure for Problem 5.2.

5.2 Consider waves advancing normal to shore over a uniform fixed slope. Two groins are present, as shown in Figure 5.16, and these groins cause considerable energy losses for the waves in the immediate vicinity of the groin.

   (a) On a qualitative basis, plot isolines of setup (and setdown) in the vicinity of the groin compartment. Discuss the reasons for the shore-parallel gradients in setup and setdown. Sketch the location of the surf zone.

   (b) Sketch lines (qualitatively) of any currents in the vicinity of the groin compartment. Are forces present that drive currents outside the surf zone?

5.3 In terms of momentum and momentum flux,

   (a) Interpret shear stress and check the dimensions.

   (b) Interpret pressure and check the dimensions. Describe, in your own words, the reason for the following:

   (c) Setdown

   (d) Setup

   (e) Longshore currents as driven by oblique waves.

5.4 Consider a deep-water wave of height $H_0 = 8$ ft and period, $T = 10$ s advancing normal to shore over a bottom of uniform slope 1:50.

   (a) Calculate and plot the setdown and setup.

   (b) Discuss whether you could apply the results (from part **(a)**) to a profile of the form $h = Ax^{2/3}$ without further computations.

5.5 One feature of nonlinear waves is the asymmetry of velocity distribution, that is, the velocity is higher and of shorter duration under the crest than under the

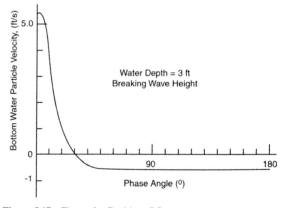

Figure 5.17 Figure for Problem 5.5.

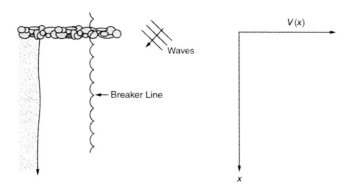

Figure 5.18 Figure for Problem 5.6.

trough. An approximate distribution is shown in Figure 5.17. An asymmetrical velocity distribution results in a nonzero average bottom shear stress in the *onshore* direction.

(a) On a qualitative basis, would the effect of the shear stress be to increase or decrease the setup owing to radiation stresses?

(b) Using a reasonable value for the shear stress coefficient ($c_f = 0.2$), the average shear stress for a water depth of 3 ft inside the surf zone is found to be $0.0070 \, \text{lb/ft}^2$. For an equilibrium beach profile and a sand size of $0.3 \, \text{mm}$, what would be the approximate percentage change in $\partial\bar{\eta}/\partial x$ at this depth?

5.6 Suppose that a shore-perpendicular structure located at $y = 0$ blocks the long-shore current such that at $y = 0$, $\bar{V} = 0$ across the entire surf zone (note that x is positive offshore in this example). Qualitatively sketch the variation of the average longshore current $\bar{V}$ with increasing distance downdrift of the structure using Figure 5.18 for reference.

5.7 In your own words, describe the significance and utility of S_{xx} and S_{xy} for the case of an offshore area characterized by straight and parallel bottom contours.

(a) An offshore area may be characterized by straight and parallel bottom contours and a steepened profile due to a reef. The profile is shown in Figure 5.19. For the wave characteristics given, is it possible to establish

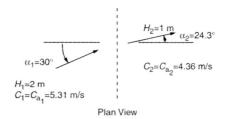

Plan View

Figure 5.19 Figure for Problem 5.7.

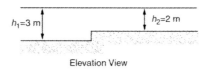

Elevation View

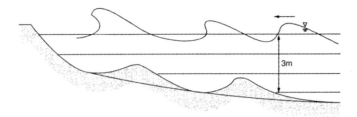

Figure 5.20   Figure for Problem 5.9.

the total shore-parallel force exerted on the bottom per unit length of shoreline? Discuss the cause and agent of this force. Assume no wave breaking on the reef.

(b) If possible, calculate the total shore-perpendicular force for part **(a)**.

(c) Sketch the form of any expected longshore current in part **(a)**.

5.8 For the barred beach profile shown in Figure 5.20, suppose that the wave height is limited by the spilling breaker assumption. For an initial breaking wave height of 3 m,

(a) Plot the setup distribution across the surf zone qualitatively. Describe principal differences between the results for the barred beach profile and those for a monotonic beach profile.

(b) Plot the longshore current distribution across the surf zone qualitatively for waves arriving at an angle. Describe principal differences between the results for the barred beach profile and those for the monotonic beach profile.

5.9 **(a)** How does the maximum wave setup depend on the beach profile if the spilling breaker assumption applies?

(b) Suppose that the initial break point occurs in accordance with $H_b = \kappa h_b$ but that for shallower depths the wave height initially is less than, but later follows, the relationship $H_b = \kappa h_b$. Will the resulting maximum setup be greater than, the same, or less than the case in which $H_b = \kappa h_b$? Discuss your answer.

(c) Suppose that the initial break point occurs in accordance with $H_b = \kappa h_b$ but that, for shallower depths, the wave height initially is greater than but later follows the relationship $H_b = \kappa h_b$. Will the resulting maximum setup be greater than, the same as, or less than the case in which $H_b = \kappa h_b$? Discuss your answer.

Waves

Figure 5.21   Figure for Problem 5.10.

5.10 For the offshore breakwater, as shown in Figure 5.21.

    (a) Sketch the current pattern that you would expect behind the structure.

    (b) Briefly describe the cause of this circulation pattern.

    (c) For the shoreline position shown, provide a qualitative sketch of the distribution of longshore sediment transport, $Q_s(y)$.

5.11 The *total* longshore thrust on the surf zone is $(S_{xy})_b$ where

$$S_{xy} = E \frac{C_G}{C} \sin \theta \cos \theta$$

and the subscript b denotes breaking conditions.

    (a) Considering a beach profile of the form $h = Ax^{2/3}$, a fixed deep-water wave direction θ_0, and wave period, T, the spilling breaker assumption, develop an expression for average shear stress $\overline{\tau}_b$ exerted on the bottom within the surf zone as a function of increasing breaking wave height. Assume all bottom shear stresses are exerted within the surf zone.

    (b) Discuss the characteristics of $\overline{\tau}_b$ versus H_b, that is, does $\overline{\tau}_b$ increase or decrease with increasing H_b? Why?

    (c) Repeat **(a)** and **(b)** for the case of a profile of uniform slope m.

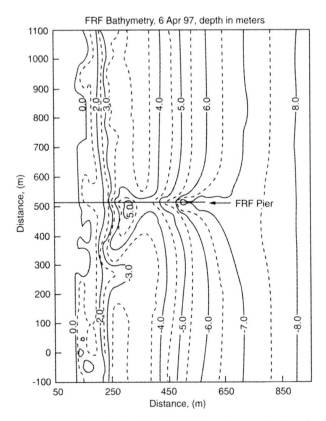

Figure 5.22 Figure for Problem 5.12. Bathymetry courtesy of U.S. Army Corps of Engineers.

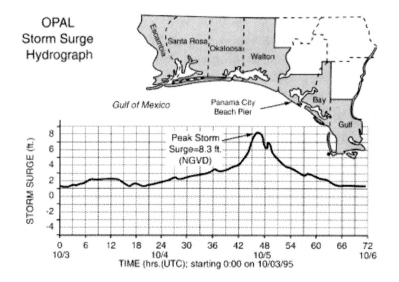

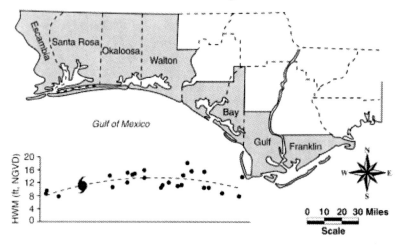

Figure 5.23　Figure for Problem 5.13 (from Leadon, Nguyen, and Clark 1997).

5.12　The bathymetry shown in Figure 5.22 is from the U.S. Army Field Research Facility at Duck, North Carolina (April 6, 1997). There is a large perturbation of the contours due to the presence of the pier, which is supported by piling. Provide plausible mechanisms for the presence of the trough under the pier.

5.13　From the sketch in Figure 5.23, the maximum storm surge measured during Hurricane Opal (October 1995) at a tide gage on a pier in Panama City Beach, Florida, was 8.3 ft. The normal (NGVD) water depth at the tide gauge location is 15 ft. Using the relationship for wave setup across the surf zone, Eq. (5.32), determine the maximum concurrent water level $\overline{\eta}_{max}$ experienced near the place at which the total water depth is zero and compare it with the visually observed high water marks.

PART THREE

COASTAL RESPONSE

CHAPTER SIX

Field Measurement Techniques and Analysis

Some of the most ambitious and daring field programs occurred during World War II with the University of California at Berkeley's field studies (directed by M.P. O'Brien*) of the beaches of northern California, Oregon, and Washington. The studies were intended to examine the nature of beaches and waves to provide information for amphibious landings on enemy beaches. The field surveys consisted of beach profiles made using military amphibious vehicles (DUKW) in surf sometimes exceeding 6 m – so high that sometimes the 10-m-long vehicles would surf the waves. The measurements were made by lead line casts along profiles separated by about 300 m. Triangulation of the DUKW by a transit operator located onshore provided the position information. Fortunately, no one was killed during these studies! (A further description of these studies can be found in Bascom 1980.)

6.1 INTRODUCTION

The elevations– and contours of a shoreline are a response to the forcing of waves, winds, and currents and depend on the supply of sediment to the beach. By measuring the dimensions of a beach repeatedly, the amounts of erosion or accretion can be determined over the period of the surveys as well as some indications of where the material may be going.

This chapter will review methods of measuring the beach and present several tools for determining beach changes such as the empirical orthogonal eigenfunction method, which is a convenient method to resolve the nature of the changes in a beach profile, and the sand budget analysis.

Field methods to measure the sediment transport directly will be discussed in Chapter 8.

---

* Dean Morrough P. O'Brien, who trained and worked as a mechanical engineer, spent a good part of his life working in the field of coastal engineering. We will encounter some of his contributions in the chapter on tidal inlets; however, his influence on the field extended beyond the University of California, Berkeley, where he retired as Dean of the College of Engineering. He also served as a civilian member of the Beach Erosion Board of the U.S. Army Corps of Engineers and its successor, the Coastal Engineering Research Board, for over three decades and charted the research direction for many coastal engineers in the United States. Wiegel (1987) gives a biographical sketch of Dean O'Brien.

6.2 BEACH PROFILE MEASUREMENTS

Beach profiles vary with time, both seasonally as the wave climate changes and over the long-term, in response to the pressures of erosion or accretion. Beach profiles measured at the same location over time can provide details about the behavior of the beach. By taking a series of profiles along a beach and then repeating the profile measurements at later times, the behavior of the entire beach can be examined in terms of shoreline recession and volumetric sand loss; moreover, an overall sand budget (sources and sinks of sand) can be determined.

Complete beach profiles extend from the dry beach (landward of expected storm damage) and extend offshore to depths beyond which the bottom does not change significantly with time. These profile measurements require a very high level of accuracy because errors over long distances along the profile can accumulate and can be interpreted erroneously as large volume changes. Fortunately, these errors are not cumulative from survey to survey.

6.2.1 WADING AND BOAT SURVEYS

Typical procedures entail two separate surveys, an offshore survey conducted by boat and the inshore survey that involves level and transit survey equipment (Inman and Rusnak 1956).

For all survey methodologies, an accurate original baseline, consisting of fixed monuments, is required, and the locations and elevations of the monuments must be known accurately. Along the baseline, the distances between profile lines is determined by the accuracy required for the volumetric calculations. Average profile spacings may range from 20 to 500 m, depending on the purpose of the survey and the length of the beach to be surveyed.

For the inshore survey, the directions of the profiles are determined and are usually oriented perpendicular to the shoreline trend. These directions are then indicated through the use of a pair of range poles or a surveyor with a theodolite to keep the surveyors on line. Using standard land surveying equipment (level, chain, and elevation rod), the survey crew determines the elevations of the dry beach along the profile line and as much of offshore as is possible to reach by wading, swimming, or both. This surveying is usually done at low tide; the deeper the waders can go offshore the better, for it is necessary to have the wading profile overlap the offshore survey, which can be done at high tide (see Figure 6.1).

The offshore portion of the profile is generally obtained using a survey vessel equipped with a fathometer and positioning system so that the position of the boat can be correlated with the depth measurements. The boat is kept on the profile line by the visual profile markers (the range poles), by radio, or more recently by modern electronic distance measuring (EDM) equipment, or lasers.

Recently, jet skis equipped with global positioning system (GPS) receivers and a fathometer, have been used for surveys. This procedure permits a single person to conduct a complete survey.

The offshore survey is usually taken out to a depth exceeding the *depth of closure*. This depth is chosen because it is the depth beyond which there is normally no change in the profile with time. Always taking profiles as far offshore as the depth of

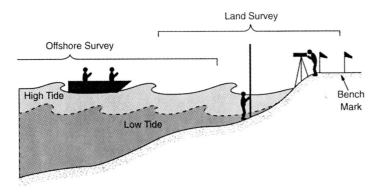

Figure 6.1 Beach profile survey technique (adapted from Nordstrom and Inman 1975).

closure will make it likely that all the profiles taken along the same line will reach the same depth at the same distance offshore. That is, the profiles through time should all converge at the offshore end of the profile line, making it possible to compute sediment gains and losses between surveys. The depth of closure can be estimated from wave and sand conditions based on empirical relationships, as discussed in Chapter 8.

Hydrographic surveys are usually done with one boat, and thus considerable time is required for the survey to be completed. For tidal bodies of water, because the sounding depths are relative to the water level at the time of sounding, a tide gauge system must be used to determine the water level associated with each profile. In all situations, tidal and nontidal, the water level with respect to a known datum is necessary to convert depths measured with the fathometer to the elevations used on the beach profiles.

An additional problem with boat surveys (beyond the required tidal corrections) is the effect of the waves, which lift and lower the boat regularly during the survey. Typically, the bathymetric traces are smoothed by eye, and the average readings are used. The disadvantage of this method is that it may result in some true irregularity of the bottom being smoothed away. More objective filtering methods to remove the waves from the bathymetric data are also used (Gable and Wanetick 1984). Waves are not a problem for more sophisticated sounding systems with vessel heave compensation.

6.2.2 OTHER PROFILING METHODS

6.2.2.1 Instrument Measurements

The principal purpose of other methods of hydrographic surveying is to remove the influence of waves and tides on the depth measurements.

The Field Research Facility of the U.S. Army Corps of Engineers at Duck, North Carolina, uses the Coastal Research Amphibious Buggy (CRAB) to obtain beach profiles (see Figure 6.2). This device, consisting of a 10.6-m-high tripod on large powered wheels, is driven along the profile lines. The CRAB carries a prism for a shore-based EDM positioning system (Birkemeier and Mason 1984). The CRAB is operational in 2-m waves, can survey to depths of 8 m, and provides high-accuracy profiles. More recently, the CRAB has been outfitted with a Global Positioning

Figure 6.2   The Corps of Engineers' CRAB surveying vehicle (photograph courtesy of W.A. Birkemeier of the Field Research Facility, Duck, North Carolina).

System (GPS), reducing substantially the number of people required to conduct measurements and the possibilities of measurement errors. The reported accuracy with GPS is ±3 cm in all three dimensions.

Another device for profiling is a so-called sea sled, which also carries a prism for the EDM. The sled is pulled over the offshore profile, and a shore-based EDM system is used for position. The advantage of the sled over the CRAB is the lower initial and operating costs. The disadvantages are the speed and the need for a boat or another device to pull the sled offshore.

The hydrostatic pressure over the beach profile can be used to measure the profile depth with a hydrostatic profiler (Seymour and Bothman 1984), which consists of a sealed fluid-filled tube with a pressure sensor at both ends. One end is kept fixed at the shoreline, and the other, mounted on a sled, is pulled offshore by a boat over the given profile and then lowered to the bottom. The armored tube is then pulled onshore 5 m at a time, and the pressures are measured. The difference in pressure between the two ends of the cable at a given time is directly related by hydrostatic pressure to the elevation difference between the fixed reference and the sled. Gable and Wanetick (1984) compared this method with the CRAB and traditional boat surveys. The boat surveys showed depth errors on the order of 20 cm, whereas the CRAB and the hydrostatic profiler reduced the errors by a factor of two. They also noted that a systematic error of 3 cm leads to an error in beach volume over a 1 km section of beach (and a profile of 1000 m of length) of the order of 30,000 m³.

Davis and Nielsen (1988) developed a unique method for measuring the small variations in the mean water level that occur across the surf zone and into the porous beach landward of the waterline. Rigid weighted tubes are laid from a common position landward of the shoreline out to various distances within the surf zone or landward of the shoreline. Filter material is placed over the outer ends of the tubes to keep sand out. The landward ends terminate at the same location on the beach and are assembled into a manometer board for easy observation of the water levels that are elevated by applying the same partial vacuum on all tubes. The elevations of the water in the tubes provide the relative elevations of the mean water levels at the ends of the tubes.

The installations by Nielsen are extremely robust, operating for years. (The method has also been applied in the laboratory for measuring wave setup and setdown.) A side benefit of the apparatus not pursued by Nielsen is its use as a wave gauge. Because the tubes are essentially rigid, the water enclosed is incompressible, and Nielsen's experience has demonstrated that any air in the tubes soon dissolves. If the landward ends of the tubes are capped, wave-induced pressure fluctuations are transmitted unattenuated over the full length of the tube. Thus, these pressure fluctuations provide a measure of the waves at the seaward locations of the tubes. Because of the wave-induced pressure attenuation in the porous medium with elevation, it would be necessary to know the elevation of the seaward end of the tube. This could be determined through long time averages and a profile measurement conducted over the tube end.

As mentioned before, global positioning systems are revolutionizing beach and hydrographic profiling. The systems utilize a set of satellites to obtain the three coordinates on the earth's surface of the survey vessel or level rod. Under ideal conditions, the GPS unit at a fixed point can provide position accuracy within centimeters. The cost and weight of these units have been decreasing rapidly, making them much more accessible and useful for coastal surveying purposes. Differential GPS (using one receiver at a fixed known location) makes it possible to improve the accuracy with which position is determined. Combining these methodologies with computer programs permits the use of computer displays and plots of the actual survey position relative to desired survey tracks. GPS units have been mounted on all types of vehicles, including boats, jet-skis, beach buggys, and even people.

Finally, new airborne laser technologies (e.g., Optec 1990; Lillycrop, Parson, and Irish 1996) provide the ability to "see" the bottom from a fast-moving helicopter or airplane. These methods allow for large areas to be surveyed nearly synoptically. For example, Irish and Lillycrop (1997) reported on helicopter surveys of New Pass, a tidal inlet in Florida, using the Corps of Engineers' Scanning Hydrographic Operational Airborne Lidar Survey (SHOALS) system. They were able to map at a rate of 5 km^2/hr with a horizontal resolution of ± 3 m and a vertical accuracy of ± 15 cm.

The SHOALS system has a limiting depth of approximately 2 Secchi depths, which, for very clear water, can be up to 30 m.* Obviously, nearshore regions

---

* The Secchi depth is the water depth to which a Secchi disk can be seen from the surface. A Secchi disk is a white, weighted horizontal disk, usually 20–30 cm in diameter, attached to a line that is lowered over the side of a ship. The Italian astronomer Pietro Angelo Secchi invented the disk around 1860 to study the Mediterranean Sea.

characterized by very turbid water or deep water will not be as amenable to this method.

6.2.2.2 Diver-Measured Elevations

It is difficult to measure bottom elevation changes (to centimeter accuracy) precisely at a particular location through the use of fathometers or even with the CRAB because it is difficult to reoccupy exactly the same horizontal position with sufficient accuracy. A second difficult measurement is the establishment of maximum scour that may occur during a storm but that may have filled in before the waves have subsided to a level where measurements can be conducted. Two labor-intensive methods are reviewed here that can be employed to determine the desired information.

As an example, it may be desired to measure small depth changes resulting from, for example, the migration of an underwater mound, which is a procedure that requires accurate horizontal and vertical control. Using divers to place vertical rods into the bottom and to monitor the changes in the distance from the tops of the rods to the bottom at various times will establish the local sand level changes quite accurately. Inman and Rusnak (1956) have used this approach to monitor offshore depth changes to determine the active depths off the southern California shoreline. To develop information relevant to a dredged channel, this method has also been applied by Harley and Dean (1982) to monitor the filling of a dredged hole for this purpose.

Dean, Chen, and Browder (1997) determined scour in the vicinity of a submerged breakwater by jetting rods into the bottom near the structure and placing a loose-fitting disk around each rod. As the sand bottom eroded, the disk rode the eroding bottom downward and was left at the maximum erosion depth. Excavating down to the disk later permitted the maximum scour between measurement periods to be determined. During the 17-month study reported by Dean et al. (1997), scour at some of the rods was up to 30 inches.

6.3 ANALYSIS OF BEACH PROFILE DATA

Once several beach profiles have been measured (for example, along a beach and for several different times), some conclusions can be made about the behavior of the beach between the profiles or between the surveys.

6.3.1 DETERMINING CHANGES IN SAND VOLUME

A convenient use of beach profiles is the determination of volumetric change of a beach, ΔV_s. Periodic beach surveys are carried out, and numerous profiles are taken along the beach of interest. An arbitrary vertical datum is then chosen; the assumed depth of closure is convenient. For each profile at a given time, the area of sand above the arbitrary datum, from the baseline to the offshore limit of the profile, is determined. For all the profiles of the same survey, the volume of sand in the beach above the datum is obtained by integrating the areas of the profiles along the beach. This is typically done by a trapezoidal rule of integration, which is to say that the area in the profile (which can be considered as volume per unit length of beach) is multiplied by the distance between profiles (with half the distance being used at each end of the beach). Surveyors refer to this method as the average end area method. (Clearly, errors in the profile result in errors in the volumetric estimates. Therefore, the accuracy of the profile – particularly the offshore portion – should be checked and the errors estimated.)

Carrying out the volumetric calculation for each of the surveys provides a time history of the volume of the beach, and by determining the volume differences between surveys the erosion or accretion of the beach can be assessed as a function of time.

The volumetric method provides advantages that are not apparent when examining, say, just the location of the mean water line. Seasonal variations (alone) in the beach profile, which can account for large swings in the location of the mean water level, will result in no change in volume. The effects of engineering works, such as beach fill, are more completely described by the volumetric analysis because adjustments to the profile during the reshaping of the beach in the cross-shore direction do not result in changes of volume.

6.3.2 EMPIRICAL ORTHOGONAL EIGENFUNCTION (EOF) METHOD

The empirical orthogonal eigenfunction (EOF) method, a widely used statistical tool, can be applied to analyze beach profiles to determine their variation through time or along a beach. It is a means of examining the variations in the profiles in a compact fashion in which the data determine the importance of the variations. The method is, however, only a descriptive tool and does not yield any information relating to the processes that govern the beach profile.

Various studies using the EOF, which is also called *principal component analysis*, for beach profiles have been conducted by Winant, Inman, and Nordstrom (1975), Aubrey (1978, 1979), and Dick and Dalrymple (1984), among others.

The data required for the EOF analysis consist of multiple beach profiles, either over time at a fixed location or over distance at a fixed time. These profiles are given in digital form; that is, a table of distances offshore and the associated depth for each profile.

6.3.2.1 Theoretical Basis of the EOF

The objective of the EOF is to describe the changes among the different beach profiles by the least number of functions, which are called *eigenfunctions*. Each of these functions consists of a contribution to the water depth as a function of the distance along the profile. The primary advantage of this method is that the first eigenfunction is selected so that it accounts for the greatest possible variance of the data (the variance is defined as the mean square of the depths). The successive eigenfunctions are each selected in turn such that they represent the greatest possible amount of the remaining variance. In this way, it is usually possible to account for a large percentage of the variance with a small number of terms.

Consider a profile starting at a stable baseline position running across the dry beach and then offshore to the depth of closure. Suppose that K surveys are conducted and that, for each survey, measurements are made at the same I locations along the profile. These measured elevations are denoted as h_{i_k}. The EOF method is based on assuming that this elevation is explained by the summation of eigenfunctions multiplied by constants.

$$h_{i_k} = \sum_{n=1}^{N} C_{n_k} e_{n_i} \quad \text{for the } i\text{th profile position and } k\text{th survey} \tag{6.1}$$

Here, e_{n_i} represents the nth spatially varying empirical eigenfunction evaluated at the ith location along the profile. The constant C_{n_k} represents a coefficient for the kth survey and the nth eigenfunction. (At this point there is a very close analogy with Fourier analysis, in which the eigenfunctions are sines and cosines.)

One property of the eigenfunctions is that they are independent of each other (orthogonal), that is,

$$\sum_{i=1}^{I} e_{n_i} e_{m_i} = \delta_{nm},$$

where $\delta_{nm} = 1$ if $n = m$ and is zero otherwise.

To obtain the value of the unknown C_{n_k}, we minimize the mean square error in the fit of h_{i_k} by the eigenfunctions; the local error ϵ_{i_k} is defined as

$$\epsilon_{i_k} = h_{i_k} - \sum_{n=1}^{N} C_{n_k} e_{n_i} \tag{6.2}$$

The minimization is carried out in the least-squares sense by minimizing the sum of the squares of the errors over the profile:

$$\text{Minimize} \ \sum_{i=1}^{I} \epsilon_{i_k}^2 \quad \text{with respect to } C_{m_k}, \text{ or} \tag{6.3}$$

$$2 \sum_{i=1}^{I} \left(h_{i_k} - \sum_{n=1}^{N} C_{n_k} e_{n_i} \right) e_{m_i} = 0 \tag{6.4}$$

Using the orthogonality relationship, we have

$$C_{m_k} = \sum_{i=1}^{I} h_{i_k} e_{m_i} \tag{6.5}$$

This equation permits us to determine the C_{m_k} for a given survey once we know the eigenfunctions. Before showing how to obtain the eigenfunctions, we examine the total mean-square variance of the profile data, σ^2, which is defined as

$$\sigma^2 = \frac{1}{IK} \sum_{k=1}^{K} \sum_{i=1}^{I} h_{i_k}^2 = \frac{1}{IK} \sum_{k=1}^{K} \sum_{i=1}^{I} \left(\sum_{n=1}^{N} C_{n_k} e_{n_i} \right) \left(\sum_{m=1}^{N} C_{m_k} e_{m_i} \right), \tag{6.6}$$

which, after repeated use of the orthogonality properties of the eigenfunctions, can be written as

$$\sigma^2 = \frac{1}{IK} \sum_{k=1}^{K} \sum_{n=1}^{N} C_{n_k}^2, \tag{6.7}$$

which is Parseval's theorem: the sum of the squares of the coefficients is equal to the square of the variance. In other words, the variance is composed of the sum of the squares of all the coefficients over all the surveys.

To find each eigenfunction, we will maximize its contribution to the variance. However, if we allow the eigenfunction to be large, the coefficient can be made arbitrarily small as well. Therefore, we will restrict the size of the eigenfunction to unity in order to find a unique solution. This in fact has been used in the orthogonality condition, which we have used, but not provided for.

Therefore, we use the Lagrange multiplier approach to introduce the constraint to our maximization problem. We maximize the function

$$\frac{1}{IK} \sum_{k=1}^{K} c_{n_k}^2 - \lambda \left(\sum_{i=1}^{I} e_{n_i}^2 - 1 \right)$$

with respect to e_{n_m}, where λ is the Lagrange multiplier. Differentiating, we obtain

$$\sum_{i=1}^{I} e_{n_i} \left(\frac{1}{IK} \sum_{k=1}^{K} h_{i_k} h_{m_k} \right) = \lambda e_{n_m} \tag{6.8}$$

Finally, if we define

$$a_{im} = \frac{1}{IK} \sum_{k=1}^{K} h_{i_k} h_{m_k}, \tag{6.9}$$

we have the symmetric matrix equation

$$\sum_{i=1}^{I} a_{im} e_{n_i} = \lambda e_{n_m} \tag{6.10}$$

This is one equation for I unknowns, but by allowing m to vary from one to I, we obtain I equations for I unknowns.

The preceding equation (6.10) is an eigenvalue matrix equation consisting of a symmetric real coefficient matrix. There are as many eigenfunctions as there are points I in the profiles; therefore, $N = I$, and each eigenfunction is associated with a distinct eigenvalue λ_n. It can be shown relatively easily that the eigenvalues are related to the total variance in the following way:

$$\sigma^2 = \sum_{n=1}^{I} \lambda_n \tag{6.11}$$

EXAMPLE 1

To give an example of the EOF method, four surveys (spaced apart in time) are taken of the same "beach" profile. The data are given in the table below for the following offshore distances: 1.0, 2.0, and 3.0 m.

		Depth h_{i_k}	
Survey, k	$i = 1$	$i = 2$	$i = 3$
1	1	2	3
2	1	2	2
3	0	1	2
4	1	3	4

These artificial surveys are shown in Figure 6.3(a) and define a matrix of depth values $\mathbf{h}$. Determining the correlation matrix a_{im} (by Eq. (6.9)), which is simply $\mathbf{h}^T \cdot \mathbf{h}/IK$, we have

$$\begin{pmatrix} 0.25 & 0.58333 & 0.75 \\ 0.58333 & 1.5 & 2.0 \\ 0.75 & 2.0 & 2.75 \end{pmatrix} \tag{6.12}$$

Note that the sum of the diagonal terms divided by IK is 4.5, the mean square of the data σ^2. Finally the eigenvalue problem is solved for each eigenvector, which gives

λ		Eigenvector	
4.4359	0.221286	0.578693	0.784951
0.0573	−0.721954	−0.443901	0.530786
0.0068	−0.655603	0.684154	−0.31956

The eigenvectors are shown also in Figure 6.3(b). From the above table of the eigenvalues and Eq. (6.11), the first eigenvector accounts for most of the variance, 4.4359 out of 4.5, or 98.58%. The second eigenvector accounts for 1.27%. The third supplies the remaining 0.15%. The variation of the eigenvectors with each survey is determined by the C_{n_k}, which are found by Eq. (6.5). They are

		C_{n_k}	
Survey, k	$n = 1$	$n = 2$	$n = 3$
1	3.73352	−0.0173975	−0.245975
2	2.94857	−0.548184	0.0735851
3	2.14859	0.617671	0.0450341
4	5.09717	0.0694875	0.118619

6.3.2.2 Applications of the EOF Method

Winant, Inman, and Nordstrom (1975) examined the variation of several beach profiles through time. A 2-year set of monthly profiles was examined with the EOF

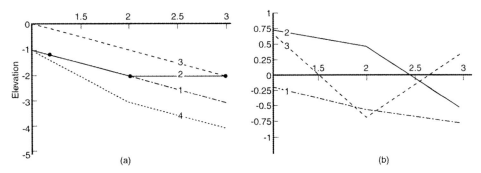

Figure 6.3 Four "beach" profiles and associated eigenvectors for the example problem. (a) Profiles; (b) eigenvectors.

method, and Figure 6.4 shows the variation for the Indian Canyon, CA Range Profile. Most of the variance of the data is explained by the mean profile eigenfunction, which accounted for 99.70 percent of the variance. This high variance percentage is generally the case for beach profiles. Note that this eigenfunction does not show the presence of a sandbar. (Note that it is this mean profile eigenfunction that should be correlated with the equilibrium profile, Eq. (7.4).)

The next eigenfunction, accounting for 0.16 percent of the variance, is shown as the solid line in Figure 6.4(b) and is referred to as the bar and berm function. In Figure 6.4(c), the weights C_{n_k} for the first three eigenfunctions are given for the profile. From an examination of the shape and the temporal variation of the bar and berm function, it is not hard to see why Winant et al. (1975) gave it the name they did. During the winter in California, when the waves generate an offshore bar and erode the beach, the bar and berm function accounts for most of the observed change of the beach, and the function's weight is large and negative. Aubrey (1979) pointed out that the bar and berm function has two zero crossings for Torrey Pines beach, implying that the formation of the bar requires sediment from both the foreshore and the offshore. In the summer, when the bar moves onshore and the beach profile is steeper, the function has a large positive weight. Over a large collection of profiles for many years, the bar and berm function exhibits a 1-year periodicity.

The third eigenfunction, the terrace function, is the dotted line in Figure 6.4(b). The terrace function explains the low tidal terrace that appears in the California profile data, and its variation is more random perhaps because the profiles were taken monthly. The variation in this function may occur in much shorter time scales, say, on the order of days, which implies that monthly sampling would alias the data. More eigenfunctions were generated in the Winant et al. (1975) study, but they concluded that the three largest eigenfunctions were sufficient to explain the variations in the profile data.

Another use of the EOF is to examine the spatial variation of a beach in the alongshore direction. Dick and Dalrymple (1984) analyzed a series of beach profiles taken along Bethany Beach, Delaware, that were all measured on the same date. The mean eigenfunction provided the beach shape, and the next eigenfunction showed the rotation of the beach about a common axis along the baseline.

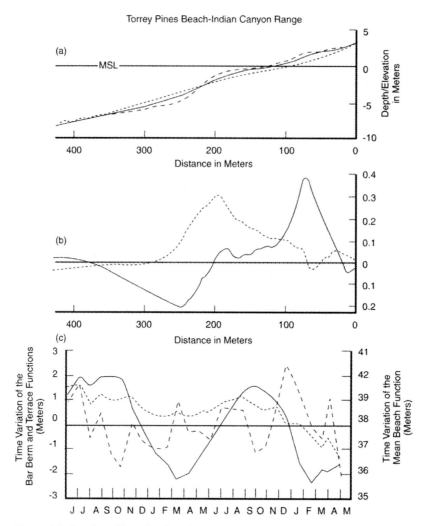

Figure 6.4 Beach profiles, eigenfunctions, and weights for Indian Canyon Range, Torrey Pines Beach, CA (from Winant et al. 1975, copyright the American Geophysical Union). The top figure (a) shows the first eigenfunction as a function of distance offshore (solid line). Two representative beach profiles showing seasonal variability are also shown. (b) Second (solid) and third (dotted) eigenfunction with distance offshore. (c) The temporal variation of the weights of three eigenfunctions by month.

Using two beach surveys (taken a season apart), determining the difference in depth along each profile, and using the EOF method on the difference data, Dick and Dalrymple also found that the first eigenfunction is similar in shape to the bar and berm function found in the temporal eigenfunction analysis.

6.3.3 COMPLEX PRINCIPAL COMPONENT ANALYSIS

An extension of EOF is the complex principal component analysis method (CPCA), which allows time-dependent forms to be gleaned from data. Few coastal engineering applications of this method have been made, and thus its full utility has yet to be explored. The first example of the method is from Liang and Seymour (1991), who examined field data at Duck, North Carolina, looking for migrating sand structures in both the alongshore and the cross-shore directions. Another is Medina et al. (1992). Bosma and Dalrymple (1996) have extended the method to three-dimensions with three-mode CPCA, analyzing the downdrift propagation of bypassed material near an inlet. The method allows surveys through time to be decomposed into pairs of eigenfunctions (with longshore and cross-shore variability).

The methodology of the CPCA is analogous to that of EOF except that the analysis first involves converting the data from a purely real series of numbers to a complex series. (The data might be from a beach profile, and thus the complex series is the variation in the offshore direction.) The complex series is created by the Hilbert transform, which is easily implemented by using fast Fourier transforms. Using this complex series, the methodology of the EOF is then followed except that everything is now done with complex numbers. The interested reader is referred to Wallace and Dickson (1972) and Horel (1984).

6.3.4 SAND BUDGET

The sand budget is simply the bookkeeping of all the sand entering, leaving, or contained within a study area; it is merely a conservation of volume argument. We examine a small control volume comprising a vertical volume element of width dx and length dy located offshore of the beach. The depth of water over the small element is $h(x, y, t)$. The change in depth of water over the element with respect to a fixed datum will be caused by the transport of sand both alongshore and in the cross-shore direction. The sand transport, with units of volume/(time-unit width), is given by the components q_x and q_y, which vary with x, y, and time t. The volume of sand entering the control volume over a time period dt is $q_x \, dy dt$ plus $q_y \, dx dt$. The amount leaving from the other side of the element can be found by a Taylor series expansion if it is assumed that dx and dy are small:

$$\left[\left(q_x + \frac{\partial q_x}{\partial x} dx \right) dy dt + \left(q_y + \frac{\partial q_y}{\partial y} dy \right) dx dt \right]$$

The net change of element volume is the amount that left the element in time dt less the volume coming in. This net change in sand volume is equal to the increase in water volume over the area (or the decrease in sand level times the area) that occurred in the time interval: $[h(t + dt) - h(t)] \, dx dy$. Using a Taylor series in time (for small dt), this volume change reduces to $(\partial h/\partial t) \, dt dx dy$. Equating the volume change to net volume transported out of the area (and dividing by $dt dx dy$) leads to

$$\frac{\partial h}{\partial t} = \frac{\partial q_x}{\partial x} + \frac{\partial q_y}{\partial y} - s, \tag{6.13}$$

where s is an additional source term indicating any material added per unit area per unit time, its dimensions are volume/(time-unit area). This exact equation is the *conservation of sand equation* and is valid for any location on a beach. Beach changes can be predicted by the conservation of sand equation provided that the volumetric sand transport rates q_x, q_y are known for all locations.

To develop a sand budget, we need to integrate this equation over the area of a beach and also with respect to time. We will do this in single steps – first, across the profile from onshore to offshore:

$$\int_{y_1}^{y_2} \frac{\partial h}{\partial t} dy = \frac{\partial}{\partial t} \int_{y_1}^{y_2} h \, dy$$

$$= \frac{\partial}{\partial x} \int_{y_1}^{y_2} q_x \, dy + q_y(x, y_2, t) - q_y(x, y_1, t) - \int_{y_1}^{y_2} s \, dy$$

The derivatives and integrations can be interchanged freely because the limits of integration are fixed locations. The term $q_y(x, y_2, t)$ is the sand transport rate out of the profile at the offshore end; $q_y(x, y_1, t)$ is the transport rate in at the inshore end of the profile – ideally, y_2 and y_1 are far enough offshore and inshore such that these quantities are usually taken as zero. This means that the profiles should be taken offshore to the depth of closure and landward to a location of no transport. This equation can be rewritten as

$$\frac{\partial A_p}{\partial t} = \frac{\partial Q_x}{\partial x} + q_y(x, y_2, t) - q_y(x, y_1, t) - \int_{y_1}^{y_2} s \, dy, \tag{6.14}$$

where A_p is the profile area above the bottom (actually the water cross-sectional area below some fixed datum) and Q_x is the total volumetric sand transport rate in the x direction. Now we integrate along the beach from one end of the study area (at $x = x_1$) to another, x_2:

$$\frac{\partial}{\partial t} \int_{x_1}^{x_2} A_p \, dx = Q_x(x_2, t) - Q_x(x_1, t)$$

$$+ \int_{x_1}^{x_2} q_y(x, y_2, t) \, dx - \int_{x_1}^{x_2} q_y(x, y_1, t) \, dx - \int_{x_1}^{x_2} \int_{y_1}^{y_2} s \, dy \, dx$$

We now define $\int_{x_1}^{x_2} A_p \, dx$ as the water volume $V(t)$. The next integration is over time:

$$V(t_2) - V(t_1) = \int_{t_1}^{t_2} Q_x(x_2, t) \, dt - \int_{t_1}^{t_2} Q_x(x_1, t) \, dt + \int_{t_1}^{t_2} \int_{x_1}^{x_2} q_y(x, y_2, t) \, dx \, dt$$

$$- \int_{t_1}^{t_2} \int_{x_1}^{x_2} q_y(x, y_1, t) \, dx \, dt - \int_{t_1}^{t_2} \int_{x_1}^{x_2} \int_{y_1}^{y_2} s \, dy \, dx \, dt \tag{6.15}$$

or, defining

V_{x_1} = the volume of sand carried into the study area alongshore,
V_{x_2} = alongshore volume of sand leaving the study area alongshore,
V_{y_1} = volume of sand transported into the area from the landward side,
V_{y_2} = volume of sand going offshore out of the area,
S = volume added artificially within the study area,

we can rewrite the preceding equation as

$$V(t_2) - V(t_1) = V_{x_2} - V_{x_1} + V_{y_2} - V_{y_1} - S. \tag{6.16}$$

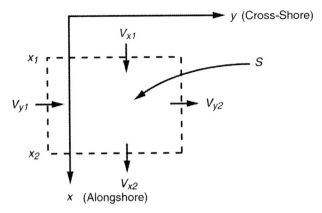

Figure 6.5 Sediment budget.

Because the change in water volume above the study area is due to sand infilling, we can compute the sand accumulated in the area ΔV_s as

$$\Delta V_s = -V(t_2) + V(t_1) = V_{x_1} - V_{x_2} + V_{y_1} - V_{y_2} + S \tag{6.17}$$

The equation now states that the change in total sand volume is related to (1) the difference in sand volume transported alongshore into and out of the study area, (2) the total amount of transport into the area from the land less the sand carried offshore, and (3) the amount S, which is the amount of sand added to the profile artificially, say through beach fill. These contributions to the sand budget are shown in Figure 6.5. Of course Eq. (6.17) could have been written based on an intuitive approach to the sand budget; however, this long derivation is intended to show that from little acorns (Eq. (6.13)) mighty oaks grow!

6.3.5 EVEN–ODD ANALYSIS OF COASTAL SHORELINE CHANGES

One method of determining the effects of a natural feature, such as an inlet or a coastal structure that impacts the longshore sediment transport, is called the *even–odd analysis* (Berek and Dean 1982). Its purpose is to separate out those shoreline changes that are symmetric about a point on the coastline (and probably not attributable to the structure) and those that are due to the presence of the coastal structure. The procedure separates shoreline or volumetric changes (here measured by ΔV_s) into an even $\Delta V_e(x)$ and an odd function $\Delta V_o(x)$ about the origin of a longshore coordinate system; this origin is placed at the feature being examined. We illustrate the method below for volumetric changes with the total change expressed as a sum of even and odd components:

$$\Delta V_s = \Delta V_e(x) + \Delta V_o(x) \tag{6.18}$$

In the most simple case, the even function (at a distance from the structure) can be interpreted as the shoreline changes that are ongoing in the absence of the structure. The odd function then depicts those changes that are due to the structure alone.

For negative values of x, we have, by definition

$$\Delta V_s(-x) = \Delta V_e(-x) + \Delta V_o(-x)$$
$$= \Delta V_e(x) - \Delta V_o(x) \tag{6.19}$$

Solving the two equations (6.18, 6.19) gives the formulas necessary to determine the even and odd functions:

$$\Delta V_e(x) = \frac{1}{2}[\Delta V_s(x) + \Delta V_s(-x)] \tag{6.20}$$

$$\Delta V_o(x) = \frac{1}{2}[\Delta V_s(x) - \Delta V_s(-x)], \tag{6.21}$$

where the function $V_e(x)$ is symmetric about the y-axis, and the odd function $V_o(x)$ is antisymmetric.

At $x = 0$, there may be two data points, say on each side of a groin (of negligible thickness). We denote the two points as $\Delta V_s(0^+)$ and $\Delta V_s(0^-)$. Then, the preceding formulas (6.20 and 6.21) can be used directly. However, if there is only one point at $X = 0$, $\Delta V(0)$, then, we take $\Delta V_e(0) = \Delta V_s(0)$ and $V_o(0) = 0$.

For example, consider a structure located at $x = 0$. Updrift and downdrift of the structure, surveys indicate that there is a shoreline recession. By determining the volume lost between surveys at different locations along the beach, $\Delta V_s(x)$, the even–odd analysis proceeds using the equations above. To determine the amount of impoundment or erosion due to the structure, it is a matter of summing or integrating the volumes $\Delta V_o(x)$ along the shoreline from the origin. This type of analysis is especially effective in identifying the interference of an inlet with the longshore sediment transport.

EXAMPLE 2

Consider a groin placed on a beach (at $x = 0$) that creates an updrift fillet and downdrift erosion. Two beach surveys taken 5 years apart (during the same season) give the following data for the location of the mean water line. Use these data to determine the long-term erosion rate and the effect of the groin.

Location →	−30 m	−20	−10	10	20	30
Survey 1 →	10 m	10	10	10	10	10
Survey 2 →	10 m	11	14	4	7	8
Net change→	0	1	4	−6	−3	−2

To determine the ongoing erosion, use the even function in the even–odd analysis based on the shoreline position: $y_e(x) = [y(x) + y(-x)]/2$. Starting with $x = -30$, we have $y_e(-30) = [0 + (-2)]/2 = -1$. Continuing for all the points, we find $y_e(x)$ is constant and equal to -1 m. This means that during the 5-year period there was 1 m of shoreline recession or $1/5$ m per year.

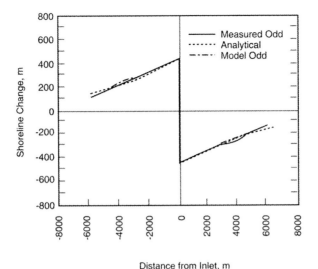

Figure 6.6 Even and odd analysis of St. Lucie Inlet, Florida (from Work and Dean 1990). Data are for the period 1928–67.

The odd analysis, based on $y_o(x) = [y(x) - y(-x)]/2$, yields the following:

Location →	−30 m	−20	−10	10	20	30
y_o →	1	2	5	−5	−2	−1

The effect of the groin on the shoreline is evident in the odd function, y_o, and it is easily separable from the ongoing erosion trend.

Figure 6.6 from Work and Dean (1990) shows the results of an even–odd analysis applied to the shoreline adjacent to the St. Lucie Inlet, Florida, area. This inlet, cut in 1892, has caused severe erosion on Jupiter Island located downdrift (south; to the right in the figure).

6.4 HISTORICAL SHORELINE CHANGES

Long-term shoreline change rates can be determined from historical data at a given site. The types of data available are charts or maps and aerial photographs.

The charts and photographs (all taken at different times) provide an historical record of shoreline position. Typically these historical positions, corrected to a common datum, are used to create an overlay of shorelines, and the shoreline changes with time are then determined.

Recently this process has become automated to a certain extent by digitizing shorelines and determining the change from long-term databases (Leatherman 1983, McBride 1989). Great care must be taken to correct the various horizontal and vertical datums to a common base.

6.4.1 CHARTS

Usually, the oldest available data are historical navigational charts. Topographic maps (e.g., U.S. Geological Survey (USGS) Quadrangles), Admiralty Charts, National Ocean Survey (NOS) bathymetric surveys, and local, state, or university surveys made at different times can be utilized. These maps have limitations, however, because, for example, the USGS maps are designed for upland use and the shorelines are usually obtained from aerial photographs.

The difficulty with these data is accuracy. The oldest maps suffer from positional inaccuracy as well as vertical elevation inaccuracy. However, these old maps provide an historical reference that may overshadow uncertainties about their accuracy. For example, the presence of an inlet on a map of a barrier island may provide information about the historical migration of the inlet. More recently, the chart accuracy is limited owing primarily to the different vertical datums that are used. For bathymetric maps, the mean lower low water (U.S. Army Corps of Engineers), the mean sea level, or the National Geodetic Vertical Datum is used. The shoreline position is different for each one. These differences in datums, however, can fortunately be corrected. Another source of error is the accuracy with which the shoreline was mapped in the survey that led to the chart. Recent charts sometimes represent a compilation of historical sounding and shoreline data rather than a complete new survey. This can lead to errors where the bathymetry has changed over time. Crowell, Leatherman, and Buckley (1991) have provided estimates of the errors present in available sources of shoreline position data.

6.4.2 AERIAL PHOTOGRAPHS

Aerial photographs often exist for a site of interest that can provide a pictorial history of the shoreline position. These photographs often present a distorted view of the coast owing to aircraft tilt and pitch as well as distortion near the edges of the photograph. A variety of tools can be used to eliminate these effects. Stereoscopic plotters and zoom transfer scopes serve to remove the tilt effects. More modern techniques use digitization and software to rectify the images.

Although the scale of aerial photographs is often based on aircraft altitude, the horizontal control is usually provided by stable coastal features such as roads, buildings, ponds, or special targets specifically placed at known locations. However, the location of the shoreline is subjective because the shoreline visible in the photograph will depend on tidal stage. Typically, the high water line is used, or the change in color between the wetted beach and the dry beach, which is assumed to represent the mean high water line.

6.4.3 SHORELINE CHANGE QUANTIFICATION

The single most important determinant in regulating coastal construction along the shoreline is the shoreline change trend. Applications of a shoreline trend include establishing setback distances for construction and determining whether and, if so, under what conditions a structure can be rebuilt following severe storm damage. Yet,

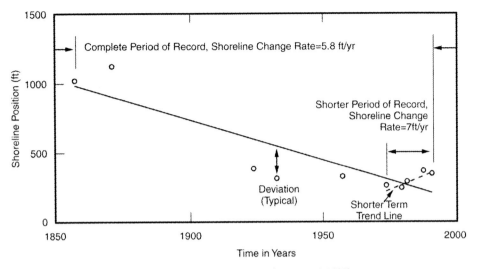

Figure 6.7 Shoreline positions at Amelia Island, Florida (Dean et al. 1998).

a casual examination of shoreline position data for several times will demonstrate that shorelines do not vary monotonically. Superposed on the trends are fluctuations with a range of magnitudes and time scales. These time scales can include seasonal, decadel, and even longer period fluctuations. Additionally, even under natural conditions, there may be reversals in the trend. The effects of human activities along the shoreline, many of which we will discuss, complicate the shoreline trends. Figure 6.7 presents an example of shoreline positions at a particular location on Amelia Island, Florida. It is seen that the long-term trend (1857–1991) is erosional at approximately −5.8 ft/yr; however, the short-term trend (1974–91) is accretional at 7.0 ft/yr. This is an area where there has been considerable beach nourishment over the past several decades, which accounts for the shoreline advancement trend.

With the preceding scenario of shoreline change, the question arises, what is the most appropriate method of extracting a shoreline trend from a set of times and shoreline positions? Two methods will be presented: the *end point method* and the *linear regression*, or least-squares, method. The shoreline trends obtained by these two methods can be substantial, differing in magnitude and even in sign. Other methods have been proposed, and the objective of some of these is to identify changes and to find and use the most appropriate portion of the data that will provide a linear trend.

The end point method simply employs the earliest and the most recent data points. This method has the advantage of simplicity, but omits all of the data other than the first and last data points. The trend m is given by

$$m = \frac{y_n - y_1}{t_n - t_1},\tag{6.22}$$

where y_1 and y_n are the first and last measured shoreline positions, in feet or meters, and t_1 and t_n are the dates of the surveys typically given in years.

Because there are errors in the data, there is an associated uncertainty in the trend m. It can be shown that for a normal distribution of uncertainty in y_1 and y_n, with standard deviations σ_1 and σ_n, the associated standard deviation of m, σ_n, is

$$\sigma_n = \frac{\sqrt{0.5\left(\sigma_1^2 + \sigma_2^2\right)}}{t_n - t_1} \tag{6.23}$$

The second method, linear regression, uses all of the shoreline position data available. Thus, if any of the data points contains an undue amount of error, the overall effect on the trend is not as large as it can be for the end point method. The issue of the most appropriate method of extracting trends from shoreline position data has been the subject of considerable interest. The reader is referred to the following references for additional detail: Crowell, Leatherman, and Buckley (1991, 1993); Dolan et al. (1980); Dolan, Fenster, and Holme (1991); Fenster, Dolan, and Elder (1993ab); Fenster and Dolan (1996); Foster and Savage (1989); and National Research Council (1990). The least-squares method is generally accepted for determining shoreline change trends.

In addition to the shoreline trend, the fluctuations about the trend are significant because these fluctuations can also threaten coastal structures. Dean, Cheng, and Malakar (1998) have analyzed the shoreline position data set for the State of Florida and, in addition to the trends, have established the standard deviations of the shoreline fluctuations about the trend line (by linear regression). These results were developed and presented on a county by county basis. Figure 6.8 presents long-term results for Indian River County along the east coast of Florida. The vertical bars about the trend value represent the standard deviations of the shoreline positions about the trend line. Note the different scale for the standard deviations. It was found that the standard deviations were much greater near the inlets than along the general coastline, as you might expect.

6.5 MAJOR FIELD CAMPAIGNS

Studying coastal processes in nature often requires a major effort. Although the measurement of beach profiles or beach cusp spacings, for example, is a straightforward surveying exercise, measuring the environmental variables (waves, winds, and currents synoptically and over a large area) requires a massive experimental campaign. One of the first large-scale series of experiments with multiple investigators was the Nearshore Sediment Transport Study in the United States supported by Sea Grant (Seymour 1989). This study carried out large-scale experiments at Torrey Pines Beach, California, (Gable, 1979) and at Santa Barbara, California, (Gable 1981). These experiments involved the first large-scale use of electromagnetic current meters set out in arrays along and across the surf zone to measure the nearshore currents. Wave measurements (staff and pressure), beach profiles, and some measures of the sediment transport were also collected. A third experiment took place at Rudee Inlet, Virginia, to examine sediment captured at an inlet. The major findings of these experiments were primarily in the area of nearshore hydrodynamics – edge wave measurements, nearshore currents, and longshore sediment transport.

A series of field experiments has been conducted subsequently at the site of the U.S. Army Corps of Engineers Field Research Facility, which has been in operation since 1977 at Duck, North Carolina. A 560-m-long research pier was installed

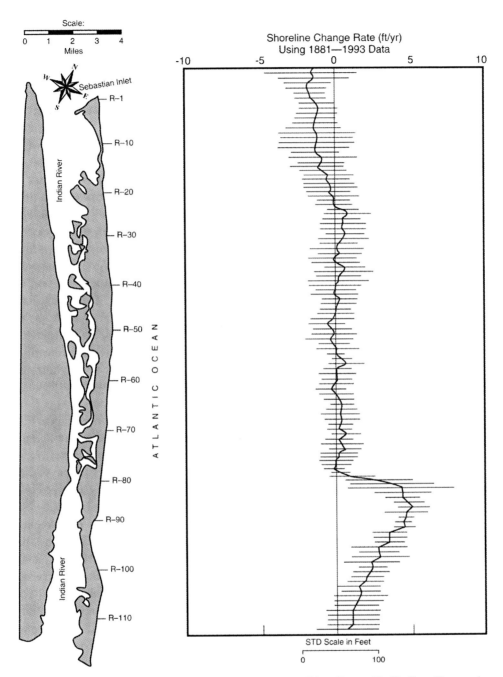

Figure 6.8 Shoreline trends and standard deviations for Indian River County, Florida (from Dean et al. 1998).

to support field measurements. The pier was constructed of concrete supported on pairs of round piling spaced every 12.2 m along the pier.* The Corps of Engineers has provided this facility and logistics to host a variety of major experiments carried out by investigators from many institutions. The first was DUCK, then SUPERDUCK,

---

* Despite this wide spacing, the presence of the pier has caused a scour trough to develop under the pier; see Figure 5.22. Experiments are carried out updrift (north) of the pier to avoid pier effects.

DELILAH, DUCK 94, and recently SandyDuck (Birkemeier, Long, and Hathaway 1996). Several important hydrodynamic and morphodynamic results have been observed during these highly instrumented experiments such as shear waves and bar migration.

In Canada, the Canadian Coastal Sediment Study (Bowen et al. 1986) conducted under the auspices of the National Research Council of Canada was a 4-year effort to study nearshore sediment transport. This $1.3 million (Canadian) effort focused on conducting two collaborative field experiments (Point Sapin, New Brunswick, during October and November 1983, and Stanhope Lane, Prince Edward Island, in the fall of 1984) and the specification (Hay and Heffler 1983) and development of an acoustic sediment profiler. The field experiments were designed to compare field measurements with coastal engineering analysis tools such as wave prediction and sediment transport calculations and to conduct basic research in coastal processes. The comparisons of field data with engineering calculations were hampered by a lack of synoptic and large-scale coverage of the waves, currents, and sediment transport; however, wave-forecasting methods, shoaling and refraction methods, and predictions of sediment transport were compared with the data available. The conclusions were that wave predictions based on good local winds were satisfactory, wave shoaling of spectra up to breaking was satisfactory (based on limited field measurements of waves in shallow water) but was considered the area of wave research needing greatest attention, and that the predictions of longshore currents and sediment transport are better made with simple formulas rather than elaborate ones given the large number of unknown variables that can exist at a beach.

The management of the Canadian study differed from that in the major U.S. studies in that the Canadian study was funded by government agencies, but had no lead agency with a mission to study coastal processes. Overall scientific and technical management was carried out by a single individual. In the U.S. studies, moneys from government funding agencies (such as the Office of Naval Research) went directly to the principal investigators, who provided the scientific leadership but were also responsible for the purchase, care, and deployment of the scientific instruments. Many of these investigators have developed an inventory of equipment over time and considerable field expertise. Furthermore, the Coastal Engineering Research Center (now subsumed into the Coastal Hydraulic Laboratory (CHL) at the Waterways Experiment Station in Vicksburg, Mississippi) has played a major logistical role for the experiments conducted at the Field Research Facility at Duck, North Carolina. Although the coordination of several principal investigators from different types of organizations (universities, government laboratories, and private companies) is difficult, the motivation, synergism, and flexibility of this approach have proven to result in successful experiments.

6.6 INNOVATIVE FIELD TECHNIQUES

6.6.1 VIDEO METHODS FOR SANDBARS

Holman and Lippmann (1987) and Lippmann and Holman (1989) have described a method of measuring sandbars and other beach features by remote sensing, which has several advantages such as the acquisition of synoptic data during storms or the ability to acquire large amounts of synoptic data without a large field experiment. The basic method used either a film camera (in the early development of this technique) or a video camera mounted in a position to view the beach. In the case of a film camera, a darkening filter is placed over the camera lens that is directed toward the beach, and the lens is left open for several minutes. In this way, the breaking zone, illuminated by the foam of the breaking waves, is highlighted. The video technique is, in principle, the same, except that numerous frames are recorded and then averaged digitally. This approach allows greater flexibility in the analysis process, including image rectification to enable the images to be viewed in their true orientation. By examining the intensity of pixels across an image, spectral information can be obtained to determine periodic beach features, for example.

Dean et al. (1997) employed a video imaging system to estimate the relative wave energy dissipation over natural and artificial reefs and in the nearshore zone.

Unmanned video stations have been set up in the United States and the Netherlands, and the video data are available over the Internet. These Argus stations consist of a personal computer connected to the Internet and a video camera located in a position to provide a good view of a beach (a tower or a tall building, for example). Some of these cameras have remote zooming and panning capabilities. The stations record an instantaneous snapshot (obtained hourly) along with a 12-min, time-averaged exposure.

6.6.2 SEDIMENTATION

In almost all cases, coastal engineering projects when first conceived will have an inadequate amount of background data to answer critical questions about the performance of the project or, in many cases, the required maintenance. Long-term wave data and sediment transport rates are crucial to the project design. One example is the dredging of deep channels, where there is inadequate data to provide reliable estimates of the amounts and frequencies of required dredging to maintain the channels to the desired depth. Other examples could include the construction of inlets or erosion mitigation structures and the associated impacts on the adjacent shorelines.

Many times the best design information, if available, includes the behavior of similar projects located within the same general area. In some cases it may be possible to allow for different exposures of the available projects from those being investigated.

Usually the available time and financial resources do not permit addressing all the design questions completely; however, in many cases, innovative field techniques can be carried out to answer, or at least to shed some light on, these matters. Three examples of unusual techniques are presented next.

6.6.2.1 Areas of Expected High Siltation

In areas where there is substantial suspended sediment, there can be great concerns about maintenance dredging if enclosed harbors are constructed or deepened in these regions. One effective and innovative approach was developed by Everts (1975) for such areas, and he evaluated the technique in Alaska. In this region the fine sediments produced by glacial outwash cause very high concentrations of suspended sediment. This, in conjunction with the high tidal ranges, can produce very high rates of siltation

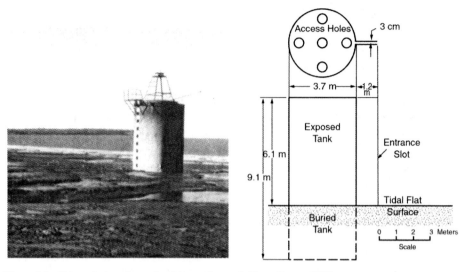

Figure 6.9   Side and plan views of sedimentation tank (from Everts 1975).

in harbors. Everts constructed a "model" harbor, which was simply a large-diameter vertical circular pipe with a small entrance running along the full height of the pipe. As shown in Figure 6.9, the upper portion of the pipe extended above the high tide. The principle of the device was that, during rising tide, the water would flow into the pipe and the suspended sediment would settle out, leaving predominantly clear water to exit the pipe during the falling tide. The field results agreed well with the sedimentation measured in a nearby harbor.

6.6.2.2 Longshore Sediment Transport

In most locations the amount of longshore sediment transport is not well quantified. In fact, within the United States, until fairly recently, even the direction of net longshore sediment transport was not well known in all areas.

During the early 1970s, there was concern about the net longshore sediment transport quantities along the Panhandle area of Florida. In order to measure this quantity, Walton (1978) constructed an experimental groin composed of large sand bags that extended well across the active surf zone. The structure was quite substantial, and it appears that it would have yielded the information desired; unfortunately, Hurricane Eloise in September 1975 completely destroyed the experimental structure before the shoreline had time to respond to its presence.

Bodge and Dean (1987) were somewhat more fortunate and relied on a much shorter time span than envisioned by Walton. Additionally, the structure was much less substantial. A sand bag groin was again constructed across most of the active surf zone in less than a day, and the impoundment was monitored through surveys during and immediately after completion of the groin construction. The distribution of the impoundment was interpreted in terms of the cross-shore distribution of the longshore sediment transport. The groin construction began early one day and was completed by the afternoon. The monitoring was continued through the night until early afternoon on the next day, at which time the experiment was considered complete and the groin was dismantled by cutting the sandbags and allowing the sand to be dispersed by the waves. A total of five such groins were constructed at different times. The disadvantage of this technique, of course, is that it provides only an instantaneous measurement of the drift rate.

Measurements of the littoral drift rate using other types of techniques are discussed in Chapter 8.

6.6.2.3 Evaluation of Anticipated Sedimentation in a Deep Navigational Channel

In the mid-1970s, the Exxon Corporation, in conjunction with Carbocol of Colombia, was searching for the location of a coal exporting facility on the north coast of Colombia. A potential site was located at Bahia Portete, as shown in Figure 6.10; however, the limiting natural depths leading into the bay were on the order of 5 m, whereas the channel depth required for the ore carriers was 22 m.

In order to develop projections of anticipated shoaling in the unnaturally deep channel, a field program was undertaken to develop oceanographic and sedimentation properties in the area. One particular study directed toward prediction of the anticipated maintenance dredging involved excavation of two test pits located in water depths of 5 and 13 m. The locations of these pits are shown in Figure 6.10. The pits were approximately 10 m on a side and 1 m below the bottom. Twenty-one brass rods were driven in the form of a plus sign across each pit, thus facilitating the monitoring of accumulated sediment by divers. In addition, the distribution of suspended sediment over depth was established by tubes suspended on taut lines below buoyant elements and attached to anchors at the bottom. The approximate geometry of the test pits and the configuration of the suspended samplers are presented in Figure 6.11. The combination of the rates at which sediment accumulated in the test pits and the deposition of suspended sediment in the tubes tethered above the bottom allowed projection of the anticipated shoaling rates in the channel. The suspended sediment was considered to deposit uniformly along the channel bottom, whereas it was recognized that the remainder of the sediment, interpreted as bed load, would only enter

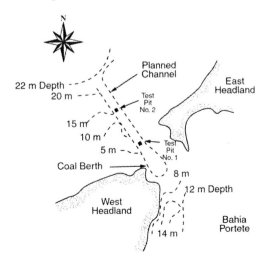

Figure 6.10 Overall layout and test pit locations for Bahia Portete sedimentation studies (from Harley and Dean 1982).

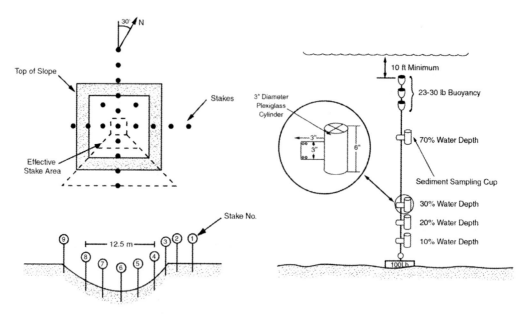

Figure 6.11 Test pit geometry and suspended sediment samplers (from Harley and Dean 1982).

the channel on the sides. Thus, the difference in width between the navigational channel and the dimensions of the test pit could be taken into account.

It was found that the deposition rates in the test pits were not uniform with time but varied over the 18-month monitoring period. An attempt was made to determine the predominant cause of the deposition. The finding was that the deposition varied more with wind speed than any other variable, and the interpretation was that the higher wind speeds in this arid region caused greater evaporation from the large surface area of Bahia Portete, thus densifying the water and causing it to sink and flow out along the bottom. Some of the heavy, suspended load that the water carried was then deposited in the test pits. Partial confirmation was provided by divers, who generally noted that a thermocline was present with the cooler, less saline surface waters overlying the more saline, warmer turbid bottom waters. Apparently there was an inflow of oceanic waters at the surface into Bahia Portete. The high evaporation rates cause this water to densify, sink, flow out along the bottom, where it entrained considerable suspended sediment.

REFERENCES

Aubrey, D.G., Statistical and Dynamical Prediction of Changes in Natural Sand Beaches, Ph.D. thesis, Scripps Institution of Oceanography, San Diego, 194 pp., 1978.

Aubrey, D.G., "Seasonal Patterns of Onshore/Offshore Sediment Movement," *J. Geophys. Res.*, 84, C10, 6347–6354, 1979.

Bascom, W., *Waves and Beaches: The Dynamics of the Ocean Surface*, Garden City, NY: Anchor Books, 267 pp., 1980.

Berek, E.P., and R.G. Dean, "Field Investigation of Longshore Transport Distribution," *Proc. 18th Intl. Conf. Coastal Eng.*, ASCE, Cape Town, 1620–1639, 1982.

Birkemeier, W.A., C.E. Long, and K.K. Hathaway, "DELILAH, DUCK94 and SandyDuck: Three Nearshore Field Experiments," *Proc. 25th Intl. Conf. Coastal Eng.*, ASCE, Orlando, 4052–4065, 1996.

Birkemeier, W.A., and C. Mason, "The CRAB: A Unique Nearshore Surveying Vehicle," *Journal of Surveying Eng.*, ASCE, 110, 1, 1984.

Bodge, K.R., and R.G. Dean, "Short-Term Impoundment of Longshore Sediment Transport," U.S. Army Corps of Engineers Coastal Engineering Research Center, MP CERC-87-7, 1987.

Bosma, K.F., and R.A. Dalrymple, "Beach Profile Analysis Around Indian River Inlet, Delaware," *Proc. 25th Intl. Conf. Coastal Eng.*, ASCE, Orlando, 1996, 2829–2842, 1996.

Bowen, A.J., D.M. Chartrand, P.E. Daniel, C.W. Glodowski, D.J.W. Piper, J.S. Readshaw, J. Thibault, and D.H. Willis, "Canadian Coastal Sediment Transport Study: Final Report of the Steering Committee," National Res. Council of Canada, TR-HY-013 (NRC No. 26603), 100 pp., 1986.

Crowell, M., S.P. Leatherman, and M.K. Buckley, "Historical Shoreline Change: Error Analysis and Mapping Accuracy," *J. Coastal Res.*, 7, 3, 839–852, 1991.

Crowell, M., S.P. Leatherman, and M.K. Buckley, "Shoreline Change Rate Analysis: Long Term versus Short Term Data," *Shore and Beach*, 61, 2, 13–20, 1993.

Davis, G.A., and P. Nielsen, "Field Measurement of Wave Set-Up," *Proc. 21st Intl. Conf. Coastal Eng.*, ASCE, Malaga, 539–552, 1988.

Dean, R.G., R. Chen, and A.E. Browder, "Full Scale Monitoring of a Submerged Breakwater, Palm Beach, Florida," *Coastal Eng.*, 3–4, 291–315, 1997.

Dean, R.G., J. Cheng, and S. Malakar, "Characteristics of Shoreline Change Along the Sandy Beaches of the State of Florida: An Atlas," Report No. UFL/COEL-98/015, Department of Coastal and Oceanographic Engineering, University of Florida, Gainesville, FL, 1998.

Dick, J.E., and R.A. Dalrymple, "Short and Long Term Beach Changes at Bethany Beach, Delaware," *Proc. 19th Intl. Conf. Coastal Eng.*, Houston, 1650–1667, 1984.

Dolan, R., M. Fenster, and S.J. Holme, "Temporal Analysis of Shoreline Recession and Accretion," *J. Coastal Res.*, 7, 3, 723–744, 1991.

Dolan, R., B.P. Hayden, P. May, and S. May, "The Reliability of Shoreline Change Measurements from Aerial Photographs," *Shore and Beach*, 48, 22–29, 1980.

Everts, C.H., "Shoaling Rate Prediction Using a Sedimentation Tank," *Proc. Civil Eng. in the Oceans, III*, ASCE, Univ. Delaware, 294–312, 1975.

Fenster, M., R. Dolan, and J.F. Elder, "A New Method for Predicting Shoreline Positions from Historical Data," *J. Coastal Res.*, 9, 1, 147–171, 1993a.

Fenster, M., R. Dolan, and J. F. Elder, "Historical Shoreline Trends Along the Outer Banks, North Carolina: Processes and Responses," *J. Coastal Res.*, 9, 1, 172–188, 1993b.

Fenster, M., and R. Dolan, "Assessing the Impact of Tidal Inlets on Adjacent Barrier Island Shorelines," *J. Coastal Res.*, 12, 1, 294–310, 1996.

Foster, E.R., and R.J. Savage, "Methods of Historical Shoreline Analysis," *Proceedings, Coastal Zone '89*, 5, 4434–4448, 1989.

Gable, C.G., and J.R. Wanetick, "Survey Techniques Used to Measure Nearshore Profiles," *Proc. 19th Intl. Conf. Coastal Eng.*, Houston, 1879–1895, 1984.

Gable, C.G., ed., Report on Data from the Nearshore Sediment Transport Study Experiment at Torrey Pines Beach, California, November–December, 1978, University of California, San Diego, Inst. of Marine Resources, IMR Ref. No. 79-8, 99 pp. with 7 Appendices, 1979.

Gable, C.G., ed. Report on Data from the Nearshore Sediment Transport Study Experiment at Santa Barbara, California, January–February, 1980, University of California, San Diego, Inst. of Marine Resources, IMR Ref. No. 80-5, 314 pp., 1981.

Harley, R., and R.G. Dean, "Channel Shoaling Prediction: A Method and Application," *Proc. 18th Intl. Conf. Coastal Eng.*, ASCE, 1199–1218, 1982.

Hay, A.E., and D. Heffler, "Design Considerations for an Acoustic Sediment Transport Monitor for the Nearshore Zone," National Res. Council of Canada, Rpt. C2S2-4, 1983.

Holman, R.A., and T.C. Lippmann, "Remote Sensing of Nearshore Bar Systems–Making Morphology Visible," *Proc. Coastal Sediments '87*, ASCE, 929–944, 1987.

Horel, J.D., "Camplex Principal Component Analysis: Theory and Examples," *J. Climate Applied Meteorology*, 23, 1660–1673, 1984.

Inman, D.L., and G.A. Rusnak, Changes in Sand Level on the Beach and Shelf at La Jolla, California, U.S. Army Corps of Engineers, Beach Erosion Board, Tech. Memo. 82, 30 pp., 1956.

Irish, J.L., and W.J. Lillycrop, "Monitoring New Pass, Florida with High Density Lidar Bathymetry," *J. Coastal Research*, 13, 4, 1997.

Leatherman, S.P., "Shoreline Mapping: A Comparison of Techniques," *Shore and Beach*, 51, 28–33, 1983.

Liang, G., and R.J. Seymour, "Complex Principal Component Analysis of Wave-like Sand Motions," *Coastal Sediments '91*, ASCE, 2175–2186, 1991.

Lillycrop, W.J., L.E. Parson, and J.L. Irish, "Development and Operation of the SHOALS Airborne Lidar Hydrographic Survey System," *SPIE Laser Remote Sensing of Natural Waters: From Theory to Practice*, 2964, 26–37, 1996.

Lippman, T.C., and R.A. Holman, "Quantification of Sand Bar Morphology: A Video Technique Based on Wave Dissipation," *J. Geophys. Res.*, 94, 995–1011, 1989.

McBride, R.A., "Accurate Computer Mapping of Coastal Change: Bayou Lafourche Shoreline, Louisiana, USA," *Proc. Coastal Zone, '89*, ASCE, 7076–719, 1989.

Medina, R., C. Vidal, M.A. Losada, and A.J. Roldan, "Three-Mode Principal Component Analysis of Bathymetric Data, Applied to Playa de Castilla," *Proc. 23rd Intl. Conf. Coastal Eng.*, ASCE, Venice, 2265–2278, 1992.

National Research Council, *Managing Coastal Erosion*, Washington, DC: National Academy Press, 182 pp., 1990.

Nordstrom, C.E., and D.L. Inman, "Sand Level Changes on Torrey Pines Beach, California," U.S. Army Coastal Engineering Research Center, M.P. 11-75, 166 pp., 1975.

Optec, Inc., Helicopter Lidar Bathymeter System, U.S. Army Coastal Engineering Research Center, Contract Report CERC-90-2, 131pp+appendixes, 1990.

Seymour, R.J., ed., *Nearshore Sediment Transport*, New York: Plenum Press, 418 pp., 1989.

Seymour, R.J., and D.P. Bothman, "A Hydrostatic Profiler for Nearshore Surveying," *Coastal Eng.*, 8, 1–14, 1984.

Wallace, J.M., and R.E. Dickson, "Empirical Orthogonal Representation of Time Series in the Frequency Domain. Part I, Theoretical Considerations," *J. Appld. Meteorology*, 11, 887–892, 1972.

Walton, T.L., "Abstract–Sediment Trap at Panama City," *Proc. Workshop on Coastal Sediment Transport, with Emphasis on the National Sediment Transport Study*, DEL-SG-15-78, Univ. Delaware, 93–94, 1978.

Wiegel, R.L., "Biographical Sketch of Morrough P. O'Brien," *Shore and Beach*, 55, 3–4, 6–14, 1987.

Winant, C.D., D.L. Inman, and C.E. Nordstrom, "Description of Seasonal Beach Changes Using Empirical Eigenfunctions," *J. Geophys. Res.*, 80, 15, 1979–1986, 1975.

Work, P.A., and R.G. Dean, "Even/Odd Analysis of Shoreline Changes Adjacent to Florida's Tidal Inlets," *Proc. 22nd Intl. Conf. Coastal Eng.*, ASCE, Delft, 2522–2535, 1990.

EXERCISES

6.1 A detached breakwater has been installed. Sand was initially placed symmetrically landward of the breakwater in an attempt to avoid trapping of sand from the littoral system. The longshore distribution of volumetric change from preplacement to approximately 18 months after placement is shown in Figure 6.12.

(a) Develop the even and odd components of the volumetric change.

(b) What direction of longshore sediment transport is reflected in the results of (a)?

(c) Comment on whether you believe that the breakwater has caused adverse effects on either of the adjacent shorelines.

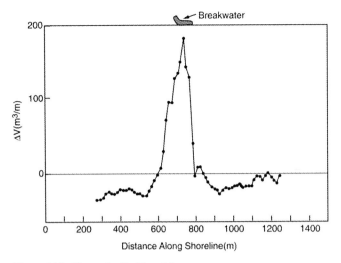

Figure 6.12 Figure for Problem 6.1.

6.2   Jetties are constructed at an inlet, and measurements of shoreline changes over a 1-year period yield the results shown in Figure 6.13.

   (a) Calculate and plot the even and odd components of shoreline change.

      (b1) From the standpoint of the entire odd function, is any volume loss associated with the odd function?

      (b2) From the standpoint of the entire even function, is any volume loss associated with the even function?

   (c) Given $B = 2$ m and $h_* = 6$ m, what is the annual loss of sediment to the system?

   (d) What is the direction of the longshore sediment transport? Discuss your response.

   (e) Assume that the updrift jetty is so short that sand is transported into the inlet. Interpret the cause of the net loss of sediment to the shoreline system.

6.3   Show that the weights of the EOF method, C_{n_k}, are also orthogonal; that is,

$$\sum_k C_{n_k} C_{m_k} = \delta_{nm}$$

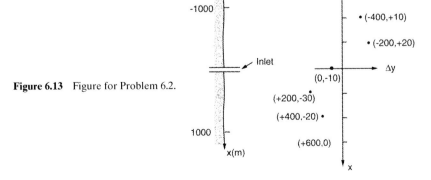

Figure 6.13 Figure for Problem 6.2.

CHAPTER SEVEN

Equilibrium Beach Profiles

During the 1970s and 1980s, the shorefront hotels and businesses at Ocean City, Maryland, an important recreational beach on the East Coast of the United States, were jeopardized by a combination of slow, pervasive shoreline erosion and by their imprudent nearness to the shoreline. Additionally, the beaches were too narrow to accommodate the numerous summer tourists.

Various measures had been attempted to maintain the beaches in the past, including beach scraping* and extensive groin construction. Neither of these approaches addressed the basic problem of an inadequate amount of sand in the system and the presence of Ocean City Inlet to the south. Thus, another method was needed, and, in 1989 and 1991, the beaches were nourished with a total of 5,000,000 m³ of sand taken from far offshore. These beach nourishments increased the beach width by 60 m. Severe storms occurred in November 1991 and January 1992, resulting in substantial reduction in this beach width. This spawned the usual criticism that the nourishment project had failed and had not been a wise use of public funds. However, subsequent surveys reported by the U. S. Army Corps of Engineers established that 90 percent of the nourishment sand was still in the project area but had moved offshore by the natural profile response to storms. In fact, the storms were variously referred to as 100- and 200-year events. The local assessment was that the fill had prevented more than $93 million in storm damages and had salvaged the $500 million in tourist revenues in the subsequent summer season. This benefit of the beach nourishment was far more than the $60 million cost of the two nourishment projects. (More details are found in Stauble et al. 1993.)

7.1 INTRODUCTION

The beach profile is the variation of water depth with distance offshore from the shoreline. The *equilibrium profile* is conceptually the result of the balance of destructive versus constructive forces. In the laboratory, it is relatively easy to construct an equilibrium profile by running a steady wave train onto a sand beach for a long time.

---

* Using bulldozers to push sand from the low-tide regions of the profile up onto the beach face.

After the remolding of the initial profile, a "final" profile results that changes little with time. This is the equilibrium profile for that beach material and those wave conditions. In nature, the equilibrium profile is considered to be a dynamic concept, for the incident wave field and water level change continuously in nature; therefore, the profile responds continuously. By averaging these profiles over a long period, a mean equilibrium can be defined.

Although the examination of the two-dimensional equilibrium beach profile neglects many of the alongshore processes, an understanding of these profiles is important for interpreting natural beach processes and for many coastal engineering applications. Examples of natural processes are the effect of relative sea level rise and beach erosion that results from storms. Engineering activities will be to determine the efficacy of beach fill projects, as discussed in Chapter 11.

Much of the material in this chapter has been summarized in Dean (1991).

7.2 METHODS OF DERIVING EQUILIBRIUM BEACH PROFILES

There are at least three possible courses of investigation for developing a theory for the equilibrium profile. These include the following:

Kinematic Approach. The motions of an individual sand grain (whether suspended or bedload) are predicted based on the applied forces, and the form of the beach profile is calculated, given that there is no net sand motion along the entire profile (Eagleson, Glenne, and Dracup 1963). Although this is an attractive approach from the standpoint of completely comprehending sediment transport processes, it appears to be beyond our present state of knowledge – particularly when you consider how many sand grains make up a beach!

Dynamic Approach. Here we postulate, in general terms, that an equilibrium profile occurs when the constructive and destructive forces acting on the bottom are balanced. Although this approach is less satisfying from the standpoint of a complete understanding of the processes, it is certainly more within our ability to carry out the results to a point at which they are applicable to actual engineering problems.

Empirical Approach. This approach is purely descriptive and attempts to describe the beach profiles in forms that are most characteristically found in nature. Experiments can relate the characteristics of these profile forms through empirical coefficients to sediment size, wave characteristics, or both. (This method was illustrated through the Empirical Orthogonal Eigenfunction approach in the previous chapter in which the first eigenfunction corresponds to the equilibrium profile.)

We will discuss all these approaches but concentrate on the dynamic approach because it presently offers the greatest likelihood of success in explaining some significant coastal processes, and it provides tools to solve some coastal engineering problems.

7.3 CONSTRUCTIVE AND DESTRUCTIVE FORCES ACTING ON BEACH PROFILES

An equilibrium beach profile represents a balance of destructive and constructive forces acting on the beach. If either of these two competing types of forces is altered as the result of a change in wave- or water-level characteristics, there is an imbalance: the larger force dominates until the evolution of the beach profile brings the forces back into balance.

Many different destructive and constructive forces affect beach profiles. At present, neither the complete identification nor the quantification of these individual forces is well understood; however, it is possible to recognize their presence and, through empirical means, to quantify the role that they play in the establishment of the equilibrium beach profile and also in profiles that are out of equilibrium.

7.3.1 DESTRUCTIVE FORCES

Of all the destructive forces, gravity is the most important, as it tries to make the equilibrium profile horizontal. An additional destructive force, which appears to be very significant, is the high turbulence level that exists within the surf zone. Breaking waves transform organized wave energy into highly chaotic turbulent fluctuations. These turbulent fluctuations act to dislodge sediment particles that are marginally stable and, in concert with the gravitational forces acting on these sediment particles, transport them in an offshore direction. The significance of the turbulence is evident to one swimming in the surf zone with different beach profiles. A profile with a mild beach slope will be characterized by relatively low turbulence levels because the breaking process is distributed over a wide surf zone, and, in some cases, it is doubtful whether the turbulence actually penetrates with sufficient strength to the bottom to cause any disturbance of the sediment. In a surf zone with a steep beach profile, a wave will dissipate its energy within a very limited volume, and thus the magnitude of the turbulent fluctuations is necessarily much higher and extends much deeper into the water column. We will find that, because of their abilities to remain stable under various levels of turbulent energy, beaches composed of fine and coarse sediments will be associated with mild and steeper slopes, respectively.

Undertow, the seaward return of the wave-induced mass transport discussed in Chapter 5, transports suspended sediment offshore and causes seaward-directed shear stresses that contribute to an offshore bedload transport.

7.3.2 CONSTRUCTIVE FORCES

At least three individual constructive forces can be identified as agents that form beach profiles. The first is due to the net onshore shear stresses at the bottom that result from the nonlinear (asymmetric) form of a shallow-water wave. Both the wave profile and the water particle velocities beneath a periodic nonlinear wave are symmetric about the crest, and the higher velocities occur under the crest but extend over a shorter time period than the negative velocities that occur under the trough. Even though the theoretical velocity predictions result in a zero mean velocity at the bottom (the velocity is a series of sines or cosine functions that each average to zero

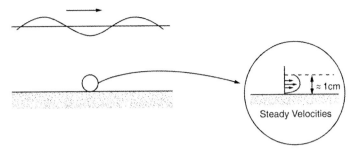

Figure 7.1 Streaming velocities in the bottom boundary layer.

over a wave period), the mean shear stress $\overline{\tau_b}$ because of its quadratic relationship to the near-bottom velocity U_b

$$\overline{\tau_b} = \frac{\rho f}{8}\overline{U_b|U_b|}$$

results in a mean onshore shear stress. Dean (1987) has presented nondimensional mean shear stresses as a function of the wave parameters H/L_0 and h/L_0.

A second constructive force is that due to the streaming velocities at the bottom. These mean velocities are due to the energy dissipation within the bottom boundary layer and resulting local momentum transfer. This phenomenon, which occurs both within and seaward of the surf zone, has been investigated in the laboratory by Bagnold (1947), and the results are sketched in Figure 7.1. Longuet–Higgins (1953) has also evaluated this phenomenon theoretically for a Newtonian fluid and determined that the (onshore) bottom streaming velocity is given as

$$\overline{U_b} = \frac{3\sigma k H^2}{16 \sinh^2 kh},$$

which is the mean velocity immediately above the boundary layer; σ is the angular frequency of the wave, k is the wave number, H is the wave height, and h is the water depth. Although the streaming velocity is a result of viscous effects, the magnitude of $\overline{U_b}$ is surprisingly independent of viscosity.

The final constructive force to be mentioned here is due to the intermittent suspensions and selective transport of the particles by the wave-induced crest (shoreward) velocities. The greatest turbulence generated during the breaking of a wave can be observed to occur under the wave crest. If this turbulence, in conjunction with water particle velocities, is sufficiently strong, sediment will be suspended and advected by the prevalent velocities until deposition. Thus, a particle suspended at the wave crest will be acted upon first by onshore velocities and, very qualitatively, if the fall time is less than one-half of a wave period, a net *onshore* sediment transport will result. If the fall time is greater than one-half a wave period but less than one wave period, a net *offshore* sediment transport can occur. This argument by Dean (1973) will be discussed further in Chapter 8 in Section 8.4.1.

Owing to our present inability to quantify each of the constructive and destructive forces, the procedure presented in the following sections will concentrate on the effect due to one identifiable destructive force; through comparison with actual profiles, we will attempt to quantify its effect.

7.4 DEVELOPMENT OF EQUILIBRIUM BEACH PROFILE THEORIES

We will now examine several models to predict the manner in which the depth varies across the surf zone. Each of the models will be based upon linear wave theory, and we will need to use some results from Chapter 5 as expressed for shallow water. The necessary relationships are

Wave energy per unit surface area: $E = \frac{1}{8}\rho g H^2$
Energy flux: $\mathcal{F} = E C_g$
Group velocity (in shallow water): $C_g = \sqrt{gh}$
Spilling breaker assumption (in the surf zone): $H = \kappa h$

Here ρ is the density of water, g the acceleration of gravity, H the local wave height, h the local water depth, and κ the breaking index, which is about 0.8.

Commencing in this section and continuing into the later sections, a right-handed coordinate system will be adopted with the x-axis directed *alongshore* with the positive direction to the right of an observer looking seaward and the y-axis pointing *offshore*. This will prove particularly useful in our later discussions of beach planform evolution.

7.4.1 UNIFORM WAVE ENERGY DISSIPATION PER UNIT VOLUME

The first equilibrium profile theory will be based on the assumption that the turbulence in the surf zone, created by the breaking process, is the dominant destructive force. The amount of turbulence is represented by the amount of energy dissipated per unit water volume by the breaking waves. We will not attempt to identify or quantify all the constructive forces that will coexist with the destructive forces, but rather the concept is, simply that, if a sediment of a given size is considered to be able to withstand a given level of wave energy dissipation per unit water volume, then the energy dissipation per unit volume may be considered to be representative of the magnitude of turbulent fluctuations (destructive forces) per unit volume.

This uniform energy dissipation per unit volume for a given grain size is $\mathcal{D}_*(d)$; written in terms of the energy conservation, it is

$$\frac{1}{h}\frac{d\mathcal{F}}{dy'} = -\mathcal{D}_*(d) \tag{7.1}$$

in which y' is now the shore-normal coordinate directed onshore. This equation states that any change in wave energy flux $\mathcal{F}$ over a certain distance divided by the water depth must be equal to the average wave energy dissipation per unit volume for which the sediment is stable. As a first approximation, if we consider the allowable wave energy dissipation per unit volume for an equilibrium beach profile to be a function only of sediment size d and not a function of distance offshore, then from Eq. (7.1) and the definitions above, we have

$$\frac{d\left(\frac{1}{8}\rho g \kappa^2 h^2 \sqrt{gh}\right)}{dy'} = -h\mathcal{D}_*(d) \tag{7.2}$$

Taking the derivative and simplifying, the dissipation per unit volume is

$$\mathcal{D}_*(d) = \frac{5}{16}\rho g^{3/2}\kappa^2 h^{1/2}\frac{dh}{dy}, \tag{7.3}$$

which is directly dependent on the beach slope and the square root of the water depth. (We have removed a minus sign by reverting to the offshore direction y instead of the onshore coordinate y'.) In Eq. (7.2), note that the depth h is the only variable that varies with y, and thus we can integrate for h,

$$h(y) = \left(\frac{24\mathcal{D}_*(d)}{5\rho g\sqrt{g}\kappa^2}\right)^{2/3}y^{2/3} = A(d)y^{2/3}, \tag{7.4}$$

where again y is oriented in an offshore direction with the origin at the mean water line. This equation for the equilibrium beach profile is very useful (hence, the box around it), and it will be applied in the remaining chapters to describe the beach profile. The dimensional parameter A is the *profile scale factor* and is a function of the energy dissipation and indirectly the grain size of the beach. This profile formula (Eq. (7.4)) indicates that, for a given sediment size, the water depth is proportional to the distance offshore to the two-thirds power. If we regard the wave energy dissipation per unit volume that a sediment particle can sustain to increase with the size of that sediment particle, then Eq. (7.4) indicates that sediments of larger sizes would have steeper profiles, which is an observation in accordance with nature.

The power law profile we have developed has an encouraging feature: the profile is concave upwards and thus similar to natural profiles. It has drawbacks as well: the beach slope goes to infinity at the shoreline, the A scale parameter is dimensional, and the profile deepens monotonically in the offshore direction; thus, it is unable to describe sandbars. Some of these drawbacks will be rectified later in this chapter, but the others do not preclude the use of this very simple conceptual description of the beach profile for some design purposes.

7.4.2 UNIFORM WAVE ENERGY DISSIPATION PER UNIT AREA

There are other possible dynamic arguments for the derivation of an equilibrium profile; for example, it is possible that it is the energy dissipation per *unit surface area* rather than unit volume that leads to the equilibrium profile. If the preceding derivation is repeated starting with Eq. (7.1) without the $1/h$ term, which would now be the dissipation per unit surface area, the resulting beach profile will be

$$h = A_2 y^{2/5}, \tag{7.5}$$

where A_2 is a dimensional constant.

7.4.3 UNIFORM BOTTOM SHEAR STRESS

Alternatively, one could argue that an equilibrium profile exists when the bottom shear stress is a constant across the surf zone. In Chapter 5, the force balance in the longshore direction due to the breaking waves (neglecting lateral shear stresses) was

$$\tau_b = -\frac{dS_{yx}}{dy}, \tag{7.6}$$

where τ_b is the average bottom shear stress in the longshore direction developed by the breaking wave and the resulting longshore current, and S_{yx} is the radiation stress for the longshore momentum carried in the y direction. Using shallow water representations for the radiation stress term and Snell's law, we have

$$\tau_b = -\frac{1}{8} \frac{d(\rho g \kappa^2 h^2 \sqrt{gh} \, (\sin \theta / C))}{dy}$$

Finally, after integrating for h, we obtain

$$h = \left(\frac{8\tau_b}{\rho g \kappa^2 \sqrt{g}} \frac{C}{\sin \theta} \right)^{2/5} y^{2/5} = A_3 y^{2/5}, \tag{7.7}$$

where the scale factor A_3 is a dimensional constant. This model can only exist where the waves approach the beach at an angle. Once again, we obtain a two-fifths power law for the equilibrium profile. As will be shown below, neither of these two relationships provide as good agreement with the field data as the two-thirds power law.

7.4.4 A SEDIMENT TRANSPORT ARGUMENT

Bowen (1980) argued that an equilibrium profile would occur when the cross-shore sediment transport at each location on the profile was zero. Using Bagnold's sediment transport models (discussed in Chapter 8), he developed two models. The first was based on a zero net suspended sediment transport at each location, giving (surprisingly) $h = Ay^{2/3}$, where

$$A = \left(\frac{(7.5w)^2}{g} \right)^{1/3},$$

showing a dependency on the fall velocity.

His more complete profile model, including both suspended and bedload transport, relates the bottom slope to the Dean number (Eq. (3.1)) and the steepness of the waves. The greater the Dean number, the smaller the slope.

7.4.5 VERIFICATION OF THE $Ay^{2/3}$ PROFILE

Bruun (1954), in a field study of beach profiles at Monterey Bay, California, and along the coast of Denmark, was the first to empirically identify the two-thirds power law (Eq. (7.4)) as an appropriate representation of natural profiles.

Dean (1977) examined some 502 beach profiles in the United States, extending from the eastern end of Long Island around the Florida Peninsula to the Texas–Mexico border. In this evaluation, a least-squares fit was done to each of the profiles

Table 7.1 Characteristics of Beach Profile Data Groups

Data Group	Profiles	Location From	Location To	ϵ^* (ft) (All Profiles)	Average Profile $A(\text{ft}^{1/3})$	Average Profile m	Average Profile ϵ (ft)
I	1–35	Montauk Point NY	Rockaway Beach NY	2.39	0.398	0.533	1.42
II	36–78	Sandy Hook NJ	Cape May NJ	2.74	0.0793	0.822	1.18
III	79–116	Fenwick Light DE	Ocean City Inlet MD	2.13	0.0945	0.762	1.54
IV	117–145	Virginia Beach VA	Ocracoke NC	2.18	0.128	0.709	1.54
V	145–159	Folly Beach SC	Tybee Island GA	1.19	0.243	0.523	0.73
VI	160–394	Nassau Sound FL	Golden Beach FL	1.89	0.255	0.594	0.36
VII	395–404	Key West FL	Key West FL	1.33	0.155	0.520	0.76
VIII	405–439	Caxambas Pass FL	Clearwater Beach FL	2.90	0.277	0.554	1.05
IX	440–477	St. Andrew Pt. FL	Rollover Fish Pass TX	1.02	0.113	0.644	0.43
X	478–504	Galveston TX	Brazos Santiago TX	0.73	0.138	0.620	0.17

* Note that ϵ represents the standard deviation of the measured from the fitted beach profiles.

with the following generalized power law profile:

$$h(y) = Ay^m$$

The average value of the exponent was found to be approximately 0.66, which is in very good accord with the result derived in Eq. (7.4). These profiles were then formed into 10 groups based on geographic regions as presented in Table 7.1 and averaged. Then, a best least-squares fit was found for each of the profile averages. The results of these averages are presented in Figures 7.2 and 7.3, where the value of the exponent m can be seen to vary from 0.52 to 0.82 but again with a mean value about 0.66. Unfortunately, little definitive sediment data were available to correlate with the various A parameters associated with the individual profiles; however, Dean noted that 99 percent of all the computed values of A were between 0.0 and 0.3 ft$^{1/3}$ with the majority of values $0.1 < A < 0.2$ ft$^{1/3}$.

Hughes and Chiu (1978) studied beach profiles and associated sediment characteristics at different locations in the state of Florida and at Lake Michigan. They found that Eq. (7.4) described the beach profiles reasonably well and that, for

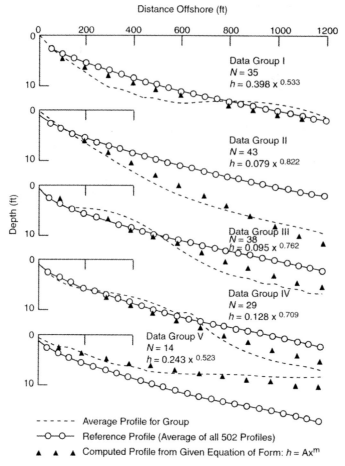

Figure 7.2 Comparison of beach profiles for Data Groups I–V.

profiles analyzed on the east coast of Florida, there was good qualitative correlation between the scale parameter A and sediment size. Figure 7.4 presents plots of variations of A and sand size (in ϕ units) at various locations on the Florida east coast. A strong correlation is seen with larger A values and larger diameters. Profiles from Lake Michigan and Gulf county on the west coast of Florida were analyzed together with a lesser (if any) correlation between the scale parameter and sediment size (see Figure 7.5).

Moore (1982) examined numerous profiles for which sediment sizes were available to quantify the relationship between the scale parameter A and the effective diameter of the sediments across the surf zone. The results from several laboratory and field profiles representing a wide range of sediment sizes were assembled; the maximum sediment size of an individual profile ranged from 0.1 mm to 30 cm in diameter. The resulting relationship between the profile scale parameter and the sand diameter is presented as the bold line in Figure 7.6. Later, Dean (1987) simply transformed Moore's relationship of A versus d to A versus w, the fall velocity, and found the surprisingly simple linear (on a log–log plot) relationship

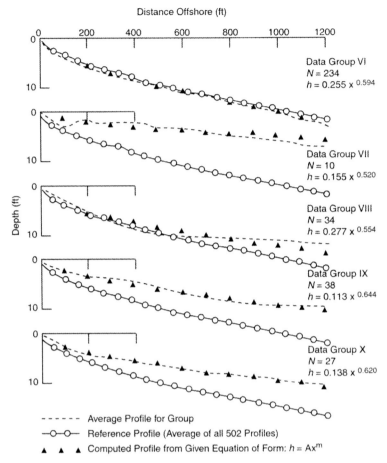

Figure 7.3 Comparisons of beach profiles for Data Groups VI–X.

in Figure 7.6.

$$A = 0.067w^{0.44}, \tag{7.8}$$

where A now has the dimensions of $m^{1/3}$ and w is in cm/s.*

Of special interest in Moore's analysis are the unusually steep profiles composed of large cobbles and profiles with predominantly shell material. Figure 7.7 presents measured and analytical fits for a profile with sediment diameter ranging from 15 to 30 cm (approximately the size of bowling balls), yielding an A parameter of $0.82\,m^{1/3}$. In the figure, the water depth is 10 m at a distance of only 42 m offshore. Figure 7.8 presents an equilibrium profile fit for a beach composed of whole and broken shell.

It has been found that the A values vary fairly smoothly over long distances and are reasonably constant with time. Figure 7.9 presents results developed by Balsillie (1987) for a 67-km segment of shoreline in Brevard County, Florida, for profiles surveyed in 1972 and 1986. Although no sediment size information is available, the increase in A values toward the south is presumably due to a corresponding increase in sediment size.

* Note that A $(ft^{1/3}) \simeq 1.5A$ $(m^{1/3})$.

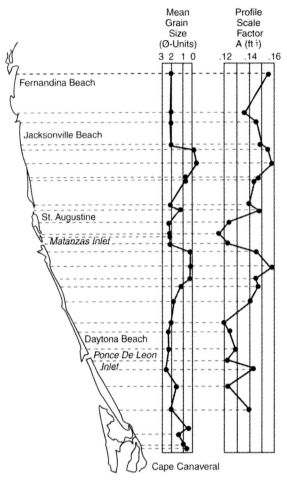

Figure 7.4 Variation of profile scale factor A (ft$^{1/3}$) and mean grain size, ϕ units, for the northeast coast of Florida (from Hughes and Chiu 1978).

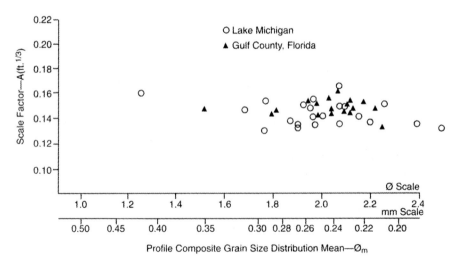

Figure 7.5 Profile scale factor A versus grain size for Lake Michigan and Gulf County, Florida, profiles (from Hughes and Chiu 1978).

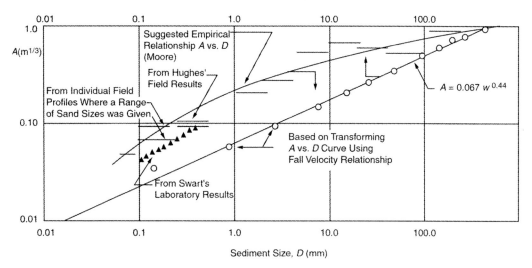

Figure 7.6 Profile scale factor *A* versus sediment diameter *d* and fall velocity *w* (from Dean 1987; adapted in part from Moore 1982).

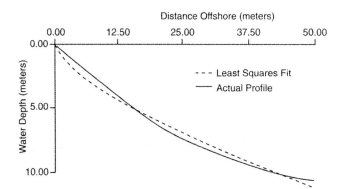

Figure 7.7 Profile P4 from Zenkovich (1967) from a boulder coast in eastern Kamchatka; sand diameter from 150 to 300 mm; least-squares value of $A = 0.82$ m$^{1/3}$.

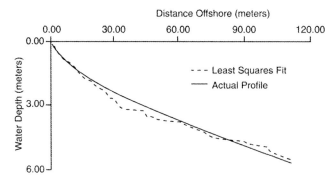

Figure 7.8 Profile P10 from Zenkovich (1967) from near the end of spit in western Black Sea; whole and broken shell; least-squares value of $A = 0.25$ m$^{1/3}$.

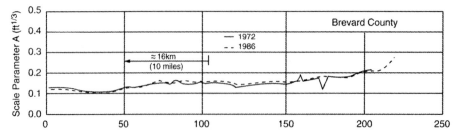

Figure 7.9 Variation of equilibrium beach profile scale parameter A versus distance (horizontal axis represents Florida Department of Natural Resources Monument numbers) along the Brevard County shoreline (data from Balsillie 1987).

In summary, field comparisons indicate that the 2/3 power law relationship is the most representative of the power law expressions for equilibrium beach profiles. For convenience, Table 7.2 presents A values for sand in 0.01-mm increments (Dean, Walton, and Kriebel 1994). The values in the table are given in $m^{1/3}$ and encompass sand sizes ranging from 0.10 to 1.09 mm. To convert to $ft^{1/3}$, multiply by 1.5. For example, for a sediment size of 0.22 mm, $A = 0.106\ m^{1/3} = 0.159\ ft^{1/3}$.

7.4.6 INDEPENDENT FIELD EVALUATION OF EQUILIBRIUM BEACH PROFILES

Pruzak (1993) examined beach profiles at Lubiatowo on the Baltic Sea and Gold Beach on the Black Sea. The profiles on the Baltic Sea were measured over a period of 28 years and analyzed in terms of the variation of the sediment scale parameter A with time for profiles of the form

$$h = Ay^{2/3}.$$

The mean profile was characterized by several bars, and the mean sediment diameter was 0.22 mm. He found that there was a long-term cyclic variation of A with an average equal to $0.075\ m^{1/3}$ and variations from 0.053 to $0.097\ m^{1/3}$. At Gold

Table 7.2 Summary of Recommended A Values ($m^{1/3}$) for Diameters from 0.10 to 1.09 mm

d (mm)	0.00	0.01	0.02	0.03	0.04	0.05	0.06	0.07	0.08	0.09
0.1	0.063	0.0672	0.0714	0.0756	0.0798	0.084	0.0872	0.0904	0.0936	0.0968
0.2	0.100	0.103	0.106	0.109	0.112	0.115	0.117	0.119	0.121	0.123
0.3	0.125	0.127	0.129	0.131	0.133	0.135	0.137	0.139	0.141	0.143
0.4	0.145	0.1466	0.1482	0.1498	0.1514	0.153	0.1546	0.1562	0.1578	0.1594
0.5	0.161	0.1622	0.1634	0.1646	0.1658	0.167	0.1682	0.1694	0.1706	0.1718
0.6	0.173	0.1742	0.1754	0.1766	0.1778	0.179	0.1802	0.1814	0.1826	0.1838
0.7	0.185	0.1859	0.1868	0.1877	0.1886	0.1895	0.1904	0.1913	0.1922	0.1931
0.8	0.194	0.1948	0.1956	0.1964	0.1972	0.198	0.1988	0.1996	0.2004	0.2012
0.9	0.202	0.2028	0.2036	0.2044	0.2052	0.206	0.2068	0.2076	0.2084	0.2092
1.0	0.210	0.2108	0.2116	0.2124	0.2132	0.2140	0.2148	0.2156	0.2164	0.2172

Beach, the profile was characterized by a single bar, the mean sediment size was 0.4 mm, and measurements were available only over a period of 6 years. During this time, there was an increasing trend of the sediment scale parameter, which ranged from 0.15 to 0.24 $m^{1/3}$. For the time periods when profile data were available for the Black Sea site and the A values were increasing, they were also increasing in the longer term data set for the Baltic Sea site. These trends may be due to long-term changes in wave climate and emphasize the shortcoming of an A representation in terms of only one parameter (the sand size). It is noted that the ranges of A values found by Pruzak are in general agreement with those presented in Figure 7.6. The profile data were also analyzed by empirical orthogonal function (EOF) methods.

7.4.7 A REEXAMINATION OF THE DESTRUCTIVE FORCES

The development leading to the form of the equilibrium profile considered only the wave energy dissipation per unit volume as a destructive force, yet it is clear that gravity should also be considered as a destructive force. One consequence of this limited consideration of destructive forces is that the equilibrium beach profile, given by $h = Ay^{2/3}$, has an unrealistic vertical slope at the mean water level intersection $y = 0$,

$$\frac{dh}{dy} = \frac{2}{3}Ay^{-1/3}$$

In principle, this infinite slope is, of course, incorrect, and here we will incorporate the effect of gravity into a modified equilibrium profile. But, in practice, the unrealistic slope is confined to a very small region near the shoreline. For example, to find a reasonable nearshore slope of $1/20$ (with a profile scale factor of $A = 0.1\ m^{1/3}$), we only need to go offshore 2.4 m from the still water line, reaching a water depth of 0.18 m. Thus, if we neglect this very nearshore portion of the profile, the $2/3$ power law continues to be of value and will be utilized throughout the text. Nevertheless, it is possible to develop more elaborate profile forms that do not have this drawback.

We hypothesize that gravity acts through the beach slope as a destructive force. The stable dissipation per unit volume within the surf zone is expressed then as

$$\mathcal{D}_* - Bg\frac{dh}{dy} = \frac{1}{h}\frac{dEC_g}{dy} \tag{7.9}$$

The factor B is an unknown constant. Expressing EC_g in terms of h, as in Eq. (7.2), and differentiating, we obtain

$$\frac{5}{24}\rho g\sqrt{g}\kappa^2\frac{dh^{3/2}}{dy} + Bg\frac{dh}{dy} = \mathcal{D}_* \tag{7.10}$$

Integrating and substituting for A from Eq. (7.4), we obtain

$$h^{3/2} + \frac{Bg}{\mathcal{D}_*}A^{3/2}h = A^{3/2}y \tag{7.11}$$

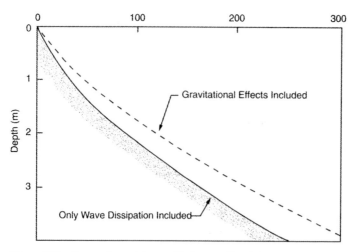

Figure 7.10 Comparison of the equilibrium profile with and without gravitational effects (corresponding to a sand size of 0.2 mm).

In shallow water, the second term on the left-hand side dominates, and the water depth varies linearly with distance offshore, just as it occurs (at least approximately) in nature:

$$h(y) = \frac{\mathcal{D}_*}{Bg} y$$

for small y. Although we do not know the value of $\mathcal{D}_*/Bg$, this can be determined from measured beach profiles, for it is the beachface slope. For larger depths, the first term on the right-hand side dominates, and the same relationship is obtained for the equilibrium profile as before. Figure 7.10 shows the equilibrium profile, including the gravitational term. Note that the A term, used here to represent the same terms as in Eq. (7.4) has a different magnitude than those discussed earlier because the profile now has a linear contribution from the term involving B.

It is not straightforward to use Eq. (7.11) to calculate the depth h for a given value of y because of the various powers of h that appear. It is often easier to find y for a given value of h.

7.4.8 LARSON'S MODEL

Larson (1988) modified Dean's approach by replacing the spilling breaker assumption with the breaking model described in Chapter 5 derived from Dally, Dean, and Dalrymple (1985), which assumes that the dissipation due to breaking in the surf zone is given by

$$\frac{d\mathcal{F}}{dy} = -\frac{K}{h}(\mathcal{F} - \mathcal{F}_s) \tag{7.12}$$

where the energy flux $\mathcal{F} = EC_g$, K is an empirical constant taken as 0.17, and $\mathcal{F}_s$ is the stable energy flux based on a stable breaking wave height $H_s = \gamma h$, where γ is about 0.4. Equating this energy flux relationship to Dean's constant energy dissipation per unit volume yields

$$h\mathcal{D}_* = \frac{K}{h}\left(\frac{1}{8}\rho g\sqrt{gh}(H^2 - \gamma^2 h^2)\right) \tag{7.13}$$

Solving for H, Larson obtained

$$H = \sqrt{\frac{8h^2\mathcal{D}_*}{\rho g\sqrt{gh}K} + \gamma^2 h^2}, \tag{7.14}$$

which provides the wave height across the surf zone as a function of the profile depth h. To proceed, this expression for wave height is substituted into the energy flux equation (Eq. (7.1)), which leads to

$$2\frac{h}{K} + \frac{5}{24}\rho g^{3/2}\left(\frac{\gamma^2 h^{3/2}}{\mathcal{D}_*}\right) = y \tag{7.15}$$

This equation, of the same form as Eq. (7.11), relates the equilibrium water depth to the distance offshore. Thus, it too is more easily solved for the distance y in terms of h. This relationship also includes a linear term that removes the infinite slope at the shoreline that occurs in Eq. (7.4). For very small y values, we have

$$h = \frac{K}{2}y,$$

giving the beachface slope of $K/2 = .085 \approx 1/12$, which is not an unreasonable value (see Chapter 2).

The relationship between $\mathcal{D}_*$ and the wave height can be determined: we first need to determine the breaking depth h_b. At this depth, we assume that the breaking index is the correct predictor of breaking wave height

$$H_b = \kappa h_b \tag{7.16}$$

$$= \sqrt{\frac{8h_b^2\mathcal{D}_*}{\rho g\sqrt{gh_b}K} + \gamma^2 h_b^2} \tag{7.17}$$

from Eq. (7.14). Solving for h_b, we have

$$h_b = \left(\frac{8\mathcal{D}_*}{\rho g^{3/2}K(\kappa^2 - \gamma^2)}\right)^2 = \left(\frac{\mathcal{D}_*}{\beta}\right)^2, \tag{7.18}$$

which defines the parameter β for later use. Rewriting, we obtain

$$\mathcal{D}_* = \frac{K}{8}\rho g\sqrt{gh_b}(\kappa^2 - \gamma^2) = \frac{K}{8}\rho g\sqrt{\frac{gH_b}{\kappa}}(\kappa^2 - \gamma^2), \tag{7.19}$$

which indicates that $\mathcal{D}_*$ is dependent on the incident wave height, which is a relationship not given by the power law.

From these expressions we can determine the width of an equilibrium surf zone. Denoting W_* as this width, we have, from Eq. (7.15),

$$W_* = 2\frac{h_b}{K} + \frac{5}{24}\rho g^{3/2}\left(\frac{\gamma^2 h_b^{3/2}}{\mathcal{D}_*}\right) \tag{7.20}$$

$$= 2\frac{h_b}{K}\left[1 + \frac{5}{6}\left(\frac{\gamma^2}{\kappa^2 - \gamma^2}\right)\right] \tag{7.21}$$

By putting in the values of κ and γ, we have $W = 15.0h_b = 19.3H_b$ for the equilibrium profile width for this model.

The Larson model shows that the wave height distribution across the surf zone and the beach profile depends on $\mathcal{D}_*$, the volumetric dissipation, which in Eq. (7.19) is shown to be related to the incident wave height (through the breaking depth). Also, the model predicts a profile that is independent of grain size.

Although this model provides wave heights across the surf zone in addition to the water depths, the slope is in general milder than that given by the power law, similar to the previous model that included the gravity term.

7.4.8.1 Other Profile Shapes

Other forms of the equilibrium profile have been proposed. Bodge (1992) and Komar and McDougal (1994) have recommended the form:

$$h(y) = h_0(1 - e^{-Ky}), \tag{7.22}$$

which asymptotically approaches a uniform depth h_0 in the seaward direction. The beach slope is

$$\frac{dh}{dy} = h_0 K e^{-Ky}, \tag{7.23}$$

which shows that the slope decreases exponentially with distance offshore. An advantage of this profile is that the beachface slope at the shoreline is nonzero, $h_0 K$. Bodge compared Eq. (7.22) to the averages of the 10 groups of profiles used by Dean (1977; here, in Section 7.4.5) and found an improved prediction over the power law profiles in 6 of the 10 groups if both A and m were allowed to vary in the power law relationship (Eq. (7.4)). If m was fixed at 2/3, he found an improved fit in 8 of the 10 groups.

Komar and McDougal (1994) emphasized the importance of a finite slope at the shoreline and replaced h_0 by m/K such that, with a measured value of the slope m, the preceding profile has only one parameter to be quantified (K). The limiting depth for offshore is given by m/K. The procedure was demonstrated with an application to one profile of the Nile River Delta.

Using several southern California beach profiles, Inman, Elwany, and Jenkins (1993) recommended an equilibrium profile consisting of two segments termed the outer shore segment and the inner bar–berm segment. The two segments are joined at the breakpoint bar. Each of the segments is of the form $h = Ay^m$, and they found $m \simeq 0.4$ for both segments. In all, they used seven parameters to describe the profile, making a good tool for analyzing profiles but one of restricted use for engineering applications.

An equilibrium model for strongly barred shorelines was proposed by Wang and Davis (1998). Although the two-power law model of Inman et al. (1993) was good for California beaches, those with larger pronounced sandbars required an additional segment on the landward face of the sandbar. This model has nine fitting parameters.

There have been other attempts to find the equilibrium beach profiles; for example, models for irregular sea states and numerical models such as Creed et al. (1992) following the work of Roelvink and Stive (1989).

7.4.9 NONUNIFORM SAND SIZES

A realistic beach is composed of a variety of sand sizes, with the sand usually becoming finer in the offshore direction, owing to the hydrodynamic sorting that takes place across the beach. All of the preceding models have assumed the sand size to be uniform across the surf zone. Here we will develop equilibrium profiles for the case of varying sand size across the profile.

It is possible to obtain analytical solutions for the beach profile if the variation of A across the profile is simple. Two analytical forms of the A parameter that will be considered here are a linear and an exponential variation, as follows:

$$A_1 = A_{0_1} - my \tag{7.24}$$

$$A_2 = A_{0_2} e^{-Ky} \tag{7.25}$$

Generally, m and $K > 0$. These equations, when substituted in Eq. (7.29) below and integrated, yield

$$h_1(y) = \left[\left(\frac{2}{5m} \right) \left(A_{0_1}^{5/2} - \left(A_{0_1} - my \right)^{5/2} \right) \right]^{2/3} \tag{7.26}$$

$$h_2(y) = \left[A_{0_2}^{3/2} \left(\frac{2}{3K} \right) \left(1 - e^{-3Ky/2} \right) \right]^{2/3} \tag{7.27}$$

For small values of Ky or my/A_{0_1}, these equations reduce, as you might expect, to the standard profile: $h_i = A_{0_i} y^{2/3}$, where i is 1 or 2.

For arbitrary distributions of sand size across the profile, numerical approaches are used by discretizing the beach profile into different sections. The equation used for the derivation of the equilibrium beach profile (Eq. (7.2)) was

$$\frac{d \left(\frac{1}{8} \rho g \kappa^2 h^2 \sqrt{gh} \right)}{dy} = h \mathcal{D}_* \tag{7.28}$$

Taking the derivative, we find

$$\frac{dh^{3/2}}{dy} = \frac{24}{5} \frac{\mathcal{D}_*}{\rho g \sqrt{g} \kappa^2} = A^{3/2}, \tag{7.29}$$

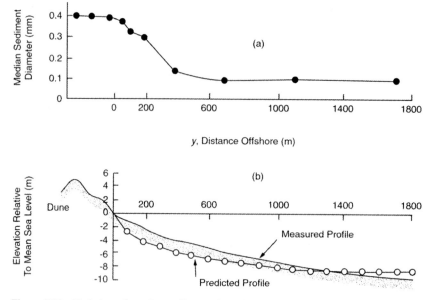

Figure 7.11 Variation of median sediment size with location across beach profile, North Jupiter Island, Florida, and predicted equilibrium profile. (a) Cross-shore sediment size distribution; (b) measured and predicted profile.

where A is a function of y. Integrating, we find

$$h(y) = \left[\int A^{3/2}\, dy + c \right]^{2/3} \tag{7.30}$$

Suppose that the distribution of grain size across the surf zone is as shown in Figure 7.11. We can approximate this distribution with piecewise constant values of grain size and hence piecewise constant profile scale parameters A. In this case, for each interval, a separate equilibrium equation for depth will apply, each with a different origin. To obtain the form of the profile over the segment, y_n to y_{n+1}, we integrate the Eq. (7.30) for a constant value of A:

$$h^{3/2} = A^{3/2}y + c,$$

where c is a constant of integration. Evaluating c, using the condition that at y_n, $h = h_n$, gives us the final form

$$h(y) = \left(h_n^{3/2} + A_n^{3/2}(y - y_n) \right)^{2/3}, \quad \text{for } y_n < y < y_{n+1}, \tag{7.31}$$

where A_n is the profile scale parameter valid over the distance between $y_n < y < y_{n+1}$. Given A_n for each section of the profile, this equation can be solved to determine the entire beach profile.

7.4.9.1 Comparisons to Field Profiles

Dean and Charles (1994) reported on an extensive field test of the equilibrium beach profile methods based on field data from the east coast of Florida. More than 1000 sand samples were collected across a total of 165 profiles. The sand samples were collected at nominal depths of 0, 1, 2, 3, 5, and 8 m, and the distance y at each offshore sampling location was measured. The calculated water depths at the locations of interest were based on the assumption of a linearly varying sediment scale parameter between the two adjacent sampling locations

$$A(y) = A_n + \left(\frac{A_{n+1} - A_n}{y_{n+1} - y_n} \right) (y - y_n), \tag{7.32}$$

which is applicable for $y_n < y < y_{n+1}$. The A values at the locations of the sediment samples were determined from Table 7.2. Given the calculated depth at one sampling location, y_n, the calculated depth at the next sampling location, y_{n+1}, is

$$h_{n+1} = \left\{ h(y_n)^{3/2} + \frac{2}{5m_n} \left[A_{n+1}^{5/2} - A_n^{5/2} \right] \right\}^{2/3} \tag{7.33}$$

in which m_n is the slope of A between y_n and y_{n+1}, that is,

$$m_n = \left(\frac{A_{n+1} - A_n}{y_{n+1} - y_n} \right)$$

In general it was found that considerable scatter existed between the measured and calculated individual profiles; however, as shown in the lower panel of Figure 7.12, there was good general agreement between the average measured and calculated profiles out to depths on the order of 4 to 4.5 m, which were interpreted as the limiting depths to which breaking waves shape the profile. The upper panel in Figure 7.12 presents the average sand sizes at the various distances offshore, and the lower panel presents the averages of the measured and calculated profiles. With respect to constructive and destructive forces active across the profile, in regions where breaking waves are active across the profile, the destructive forces are greater than in regions where breaking waves are less dominant and the slopes are reduced. However, seaward of the normally active breaking zone, the constructive forces are relatively dominant and the profile slopes are steeper. The standard deviations associated with the measured and calculated profiles are presented in Table 7.3 for various measured depths.

In addition to testing the averages of the measured and calculated profiles, Dean and Charles attempted to determine if improved A versus D relationships could be developed over the sand size range included in the data. The average A value, $\overline{A}$, was defined to correspond to the approximate average sand diameter ($\overline{D} = 0.2$ mm), and the existing A value was allowed to vary in accordance with

$$A'(D) = \overline{A}(\overline{D}) + \delta(A(D) - \overline{A}(\overline{D})), \tag{7.34}$$

in which δ ranges from 0 to unity; for a δ value of unity, the A value is unchanged, and, for a δ of 0, the A value is equal to the constant value of $\overline{A}(\overline{D})$, which in this case is 0.1 m$^{1/3}$. From the data, an improved fit resulted with $0.3 < \delta < 0.5$; that is, a substantially reduced dependency of A on grain size. However, no changes are

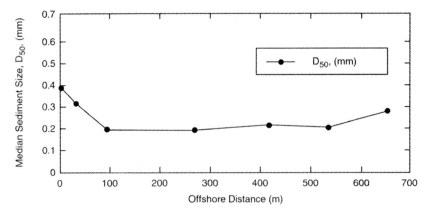

(a) Averages for 165 Florida East Coast Profiles

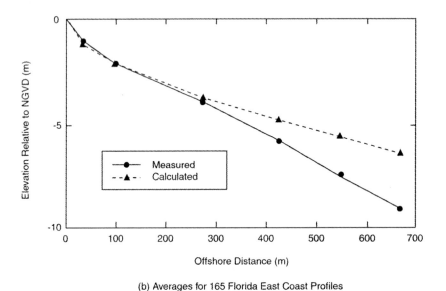

(b) Averages for 165 Florida East Coast Profiles

Figure 7.12 Comparison of average measured and predicted beach profiles for 165 locations along Florida's east coast (from Dean and Charles 1994). (a) Measured sediment sizes, D_{50} (mm); (b) comparison of measured and predicted profiles.

recommended yet for the A values in Table 7.2 until more data are available for confirmation.

Figures 7.11 and 7.13 provide additional tests of the methods discussed here. Figure 7.11 presents as the upper panel, the grain size distribution across a profile on Jupiter Island, Florida. The lower panel compares the measured profile with the calculated profile based on Eq. (7.31), the local A parameter determined from Figure 7.6, and the measured grain size distribution. The second comparison is for a profile that was surveyed repeatedly (five times) on the Egyptian coast to the west of the Rosetta Outlet of the Nile River. The five profiles are presented in Figure 7.13(a),

Table 7.3 Standard Deviations for Various Depths and Sediment Scale Parameter, A, Relationships

Factor	Standard Deviations Between Measured and Calculated Depths (m) for These Measured Water Depths (m)					
δ	0.94	2.03	3.99	5.85	7.60	9.14
0.0	0.39	0.60	1.06	1.23	1.47	1.90
0.1	0.38	0.57	1.04	1.24	1.53	2.00
0.2	0.38	0.56	1.08	1.26	1.61	2.11
0.3	0.37	0.55	1.03	1.30	1.71	2.24
0.4	0.37	0.54	1.05	1.36	1.83	2.37
0.5	0.37	0.55	1.07	1.44	1.95	2.52
0.6	0.37	0.56	1.11	1.53	2.08	2.67
0.7	0.38	0.58	1.16	1.63	2.22	2.82
0.8	0.39	0.60	1.21	1.74	2.37	2.98
0.9	0.40	0.63	1.27	1.86	2.51	3.14
1.0	0.41	0.66	1.34	1.98	2.67	3.30

and their average is the solid line in Figure 7.13(b). The computed profile is the dashed line in Figure 7.13(b). This computed profile is based on the limited sediment size information that was available; that is, the sand sizes at the waterline and at 600 m offshore were reported to be 0.2 and 0.1 mm, respectively. The calculated profile in Figure 7.13(b) actually represents both Eqs. (7.26) and (7.27); however, the two are basically indistinguishable. The agreements for these examples are quite encouraging.

Dean, Healy, and Dommerholt (1993) have conducted additional blindfold tests of the equilibrium profile, including the gravity term (Eq. (7.11)), by comparing measured and calculated profiles based on the measured cross-shore distributions of the sediment sizes. The calculated profiles were based on the measured beach face slope m for $\mathcal{D}_*/Bg$ (see Eq. (7.9)) and by integrating across the profile in accordance with

$$h(y_{i+1}) = h(y_i) + \frac{dh}{dy}\,\Delta y, \quad \text{where} \quad \frac{dh}{dy} = \left(\frac{1}{m} + \frac{3}{2}\frac{h^{1/2}}{A^{3/2}}\right)^{-1} \tag{7.35}$$

The calculated and predicted profiles were compared, with varying agreement, in their article.

7.4.10 EQUILIBRIUM PROFILE, INCLUDING THE EFFECT OF WAVE SETUP

So far, we have not included the effect of wave setup on the water level inside the surf zone. Further, the equilibrium profile derived earlier has a vertical slope at the intersection of the still water level with the beach. By including the setup, we provide a more realistic (but more complicated) form of the equilibrium profile.

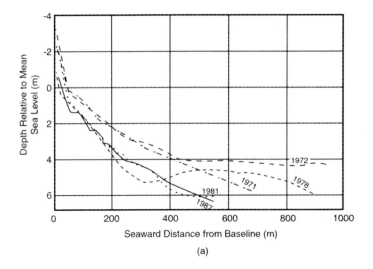

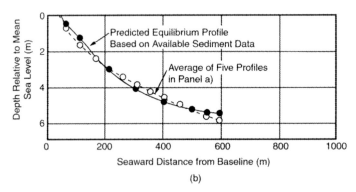

Figure 7.13 Blindfold test of equilibrium profile methods for cross-shore variation of sediment size west of the Nile's Rosetta Outlet, Egypt. (a) Profiles for five different years; (b) comparison of average measured and predicted equilibrium profiles.

The setup in the surf zone, as discussed in Chapter 5, is due to the transfer of momentum from the organized wave motion to the surf zone and, according to linear shallow water wave theory, this setup is described by the relationship

$$\bar{\eta}(y) = \bar{\eta}_b + \mathcal{K}(h_b - h(y)),\tag{7.36}$$

in which $\mathcal{K}$ is a constant involving the breaking index κ as follows:

$$\mathcal{K} = \frac{3\kappa^2/8}{1 + 3\kappa^2/8}$$

A plot of the set-up is shown in Figure 7.14. In the shallower water portions of the surf zone, the wave setup acts as an increase in water level and may, therefore, much as a "local" tide, contribute directly to the beach erosion. We can include the setup $\bar{\eta}$ and a surge S in the previous derivation of the equilibrium profile by expressing the local water depth, not by $h(y)$, but by $h(y) + S + \bar{\eta}(y)$. The energy dissipation per

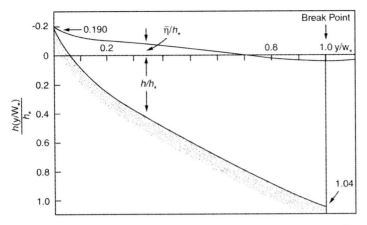

Figure 7.14 Nondimensional wave set-up and associated equilibrium profile.

unit volume in the surf zone can now be expressed as

$$\frac{dEC_g}{dy'} = -(h + S + \bar{\eta})\, \mathcal{D}_*, \tag{7.37}$$

where $\mathcal{D}_*$ is the uniform dissipation rate. Substituting the definitions of the energy and group velocity as before,

$$\frac{\rho g \kappa^2 \sqrt{g}}{8(h + S + \bar{\eta})} \frac{d(h + S + \bar{\eta})^{\frac{5}{2}}}{dy'} = -\mathcal{D}_* \tag{7.38}$$

Integrating, we find the not-too-surprising result

$$(h + S + \bar{\eta}) = Ay^{2/3}, \tag{7.39}$$

which is similar to our previous formula (Eq. (7.4)) except that the depth is now the total local depth, including the set-up. It is important to note that the origin for y is now at the location where $(h + S + \bar{\eta}) = 0$.

With consideration confined to the case for $S = 0$, this equation for the equilibrium profile may be combined with the set-up equation, Eq. (7.36), to determine both the still water depth, $h(y)$, and $\eta(y)$. Substituting for $\bar{\eta}$ in Eq. (7.39), we have

$$h(y) = \frac{Ay^{2/3} - \bar{\eta}_b - \mathcal{K}h_b}{(1 - \mathcal{K})} \tag{7.40}$$

and

$$\bar{\eta} = \frac{\bar{\eta}_b + \mathcal{K}h_b - \mathcal{K}Ay^{2/3}}{(1 - \mathcal{K})}, \tag{7.41}$$

which can be easily checked for $\bar{\eta} = \bar{\eta}_b$, $h = h_b$, and $h_b + \bar{\eta}_b = Ay^{2/3}$.

In nondimensional form, we have

$$\bar{\eta}' = \frac{\bar{\eta}}{h_b}; \quad h' = \frac{h}{h_b}; \quad \text{and} \quad y' = \frac{y}{y_b}$$

in which $h_b = H_b/\kappa$, and $y_b = (H_b/A\kappa)^{3/2}$. The resulting nondimensional expressions for $\bar{\eta}'$ and h' are

$$\bar{\eta}' = \frac{\bar{\eta}_b}{h_b} + \frac{\mathcal{K}}{(1-\mathcal{K})}\left(1 - y'^{2/3}\right) \tag{7.42}$$

$$h' = \frac{\left(y'^{2/3} - \mathcal{K}\right) + \frac{(1-\mathcal{K})}{20}}{(1-\mathcal{K})} \tag{7.43}$$

From Chapter 5,

$$\bar{\eta}_b \approx -\frac{H_b}{20}$$

These equations are plotted in Figure 7.14. They are also used in the next section to explore the recession contributions due to waves, including the associated wave setup and storm surges.

7.5 APPLICATIONS OF THE EQUILIBRIUM PROFILE

The concept of an equilibrium profile, which is the average beach response to the natural forcing makes it possible to determine several beach responses to changes in forcing. In this section, we will often be concerned with changes in the shoreline position Δy, which can represent either a shoreline advancement ($\Delta y > 0$) or a recession ($\Delta y < 0$).

7.5.1 SEA LEVEL RISE

An increase in water level, caused by either a gradual rise in relative sea level or a more rapidly rising storm surge, will result in a greater depth of water for any given location within the surf zone. The dissipation per unit volume is related to the product of the local depth and the bottom slope from Eq. (7.3)

$$\mathcal{D}_* \propto \sqrt{h}\,\frac{\partial h}{\partial y}.$$

A greater depth requires a milder beach slope for equilibrium.

7.5.1.1 The Bruun Rule

The earliest relationship between increased water level and profile response was presented by Bruun (1962) and has since become known as the Bruun rule. Bruun's method does not require any specific form of the equilibrium profile but only that the form be known. The response is considered in terms of the horizontal recession R of the profile and the sea level rise S. Two requirements must be satisfied by the new profile. First, the new equilibrium profile relative to the new water level intersection with the profile must be the same as the previous profile, that is, the profile shape does not change with respect to the water line. Second, the sand volume in the profile must be conserved.

With respect to the first requirement, the profile is considered to translate landward and upward without change in form; the two *required* components of this translation (R, S) are considered separately and are required to conserve the sand volume. The required sand volume from a sea level rise is the product of the sea level rise S and the width W_* of the active profile. This follows from the requirement that the profile relative to the new water level remain unchanged,

$$\text{Volume Required: } \Delta V_- = W_* S$$

The volume *generated* by a horizontal profile recession R is the product of R and the vertical dimension of the profile out to a distance W_*. We can express this vertical dimension as $(h_* + B)$ (see Figure 7.15).

$$\text{Volume Generated: } \Delta V_+ = R(h_* + B)$$

This expression may not be intuitively obvious, but if you consider that the volume we are discussing is obtained by moving the profile laterally, it is relatively easy to see that ΔV_+ is as given above.

We now equate the two volumes generated by recession that are required for the profile to be maintained against sea level rise, giving us

$$\Delta y = -R = -S \frac{W_*}{(h_* + B)} = -\frac{S}{\tan \theta}, \tag{7.44}$$

in which $\tan \theta$ is the average slope over the active profile, as shown in Figure 7.15. This is the Bruun rule. It is somewhat surprising that, although Eq. (7.44) was derived for the case of an arbitrary profile shape, it is precisely the amount of horizontal encroachment (inundation) that would result owing to a water level rise of S on a *fixed* inclined surface of slope $\tan \theta$.

EXAMPLE 1

A beach has a grain size of 0.2 mm and a berm height of 2 m. If the depth of the active profile is 6 m, what is $\tan \theta$ and what is the shoreline response to a sea level rise S?

If the equilibrium profile is assumed to be $h = Ay^{2/3}$, the width of the profile can be determined, for A is known (Table 7.2) to be 0.1 m$^{1/3}$ and $W_* = (h_*/A)^{3/2} = 465$ m. Therefore, $\tan \beta = 0.0172$.

From Eq. (7.44), $\Delta y = -58.1S$, or 1 m of sea level rise results in almost 60 m of shoreline retreat. The denominator, $\tan \beta$, is generally a small number, and thus at many locations around the world, according to the Bruun rule, retreat rates are 50 to 100 times the sea level rise rate.

7.5.1.2 Barrier Islands

During storms accompanied by low storm tides, the erosive forces on barrier islands are concentrated below the island crest level, and the eroded sand is carried seaward

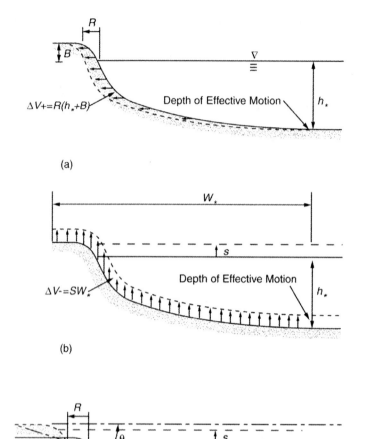

Figure 7.15 Equilibrium profile response to sea level rise: Bruun's rule. (a) Volume of sand generated by horizontal retreat R of equilibrium profile over vertical distance $(h_* + B)$; (b) volume of sand required to maintain an equilibrium profile of active width W_* owing to a rise S in mean water level; (c) landward (R) and upward (S) components of profile translation to achieve equilibrium relative to increased sea level.

to form a bar. With further relative rise in sea level, the island becomes more and more susceptible to overtopping by a storm of given intensity. Some severe storms will have such large storm surges that the barrier island will be overtopped, which is an event that causes changes in the erosion–deposition response of the island. With water flow over the barrier island, sand is transported there, where a portion or all of it will be deposited. This process of sand deposition on barrier islands during major storm events was first identified by Godfrey and Godfrey (1973) as an important element in the evolution of barrier islands. It is the process that, on a long-term average, can maintain barrier island elevation relative to the rising sea level. The

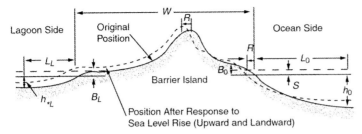

Figure 7.16 Barrier island equilibrium response (from Dean and Maurmeyer 1983).

process of landward transport of sand is referred to as *overwash*, whereas the deposit is called *washover*. If the major overwash events occur relatively infrequently, the associated magnitudes of individual washover deposits can be large. These deposits can extend inland on the order of 100 m or more. During the March 1962 storm, washover deposits up to nearly 2 m in thickness occurred along the Delaware coast. Hurricane Opal in 1995 caused washover deposits of 1.5-m thickness along portions of the Pensacola Beach shoreline.*

As a generalization to the Bruun rule to account for this effect of overtopping and migration, consider Figure 7.16 in which the island width W and relevant berm heights and active coastal zones are defined. With sea level rise, the entire barrier island unit will migrate landward with a corresponding increase in vertical elevation, requiring the volumes eroded and deposited to be the same. On the assumption that the barrier translates landward and upward with a constant form, the volume of sand generated by a landward displacement R is $R(h_{*_0} - h_{*_L})$. The volume required as a result of a sea level rise S is $S(L_O + W + L_L)$. Equating the two volumes yields an extended Bruun rule for barrier islands:

$$R = S\frac{(L_O + W + L_L)}{(h_{*_0} - h_{*_L})} \tag{7.45}$$

This equation, introduced by Dean and Maurmeyer (1983), predicts larger recession rates than the original Bruun rule because all of the sediment eroded on the ocean side no longer is used to increase the level of the ocean bottom as before; some of this material is deposited on the barrier island, thereby increasing its elevation, and the remaining portion is deposited on the bay side to maintain the equilibrium profile there. If the effective depths of the ocean and bay approach each other, the rate of recession of the barrier island increases for a given rise in sea level.

In cases where the bay depth equals or exceeds the ocean depth, it is impossible for the barrier island to migrate and still maintain its original form. In these situations,

---

* Current practice in developed areas after an overwash event is to remove the deposit and place the sand back in the active beach system or to remove the sand completely from the coast. On the premise that washover deposits are nature's way of maintaining the elevation of barrier islands in the presence of sea level rise, the problem of dealing with poststorm deposits presents a complex coastal engineering and social question. If the deposits are removed, with increasing time and sea level the area will become increasingly vulnerable to storms. Yet, no practical way has been devised for developed areas to leave the washover deposits, thus maintaining the long-term relationship between land elevations and the rising sea level. Moreover, placing the washover deposits back on the beach reduces erosion.

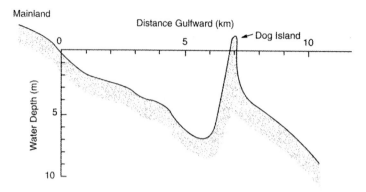

Figure 7.17 Dog Island Reef, Florida.

the evolutionary phase would probably result in a barrier island's migrating landward and, because the island cannot maintain its original form, it would gradually narrow and eventually be completely submerged. Many authors (Shepard 1960, Sanders and Kumar 1975, for example) have concluded, based on geological evidence, that several barrier islands may have drowned in place; for example, some features off the present shoreline of Long Island, New York, and northern New Jersey are presumed to have originated in this fashion. Figure 7.17 presents an example off western Florida where the evidence suggests the drowning of a barrier island, leaving a sand shoal named Dog Island Reef. The cross section shown in the figure is adjacent to the shoal.

7.5.1.3 The Edelman Method

One restriction of the Bruun rule is that the sea level rise should be much less than the berm height $(S \ll B)$ because the berm height relative to the rising sea level is considered fixed. Edelman (1972) removed this limitation of relative small changes in sea level by accounting for the progressive decrease in *relative* dune elevation. To follow his development, we consider the beach profile to be always in equilibrium and the horizontal rate of shoreline recession to be given by U and the vertical rate of rise of the sea bottom to be V. During an increment of time Δt, a typical element on the profile, moves from point a to point b, as shown in Figure 7.18. The dune height B is considered to vary with time.

If the profile maintains equilibrium at each instant of time, then by the same consideration used in the Bruun rule, we have

$$(h_* + B)U \, \Delta t = W_* V \, \Delta t \tag{7.46}$$

or

$$U = \frac{W_* V}{(h_* + B)} \tag{7.47}$$

The total recession R is

$$R = -\Delta y = \int_{t_0}^{t} U \, dt \tag{7.48}$$

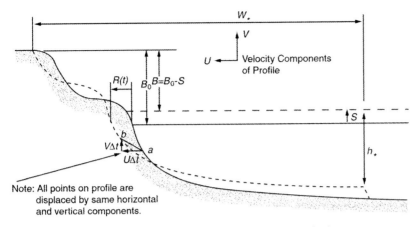

Figure 7.18   Erosional response of the profile by the Edelman method.

and

$$V = \frac{dS}{dt} \quad \text{and} \quad h_* + B = h_* + B_0 - S,$$ (7.49)

for the dune height decreases with the sea level rise.

Substituting Eqs. (7.46), (7.47), and (7.49) into Eq. (7.48), we have

$$R = \int_{t_0}^{t} \frac{W_*}{(h_* + B_0 - S)} \frac{dS}{dt} dt,$$

which is then integrated to yield

$$R = -\Delta y = W_* \ln \left(\frac{h_* + B_0}{h_* + B_0 - S} \right),$$ (7.50)

This recession equation (Eq. (7.50)) will always yield a somewhat larger value of R than Bruun's rule (Eq. (7.44)) because the latter assumes a constant dune height above sea level. However, for small values of $S/(h_* + B_0)$, say less than 0.3, Edelman's method reduces to the Bruun rule. Also, the more nearly complete the dune erosion is, the greater the relative rate of erosion. This is simply because, as sea level rises, less and less material is present in the dune to contribute to the volume eroded.

7.5.2 STORM RESPONSE OF THE EQUILIBRIUM PROFILE

A storm surge represents a rise in sea level, which, coupled with large storm waves, begins the change of the equilibrium profile to a storm profile. As in the Bruun rule argument, the sea level rise S and a breaking wave height H_b will occur during the storm, resulting in a new equilibrium profile should the storm last long enough. If the volume eroded is the same as caused by the rise in water level by the surge, then, by referring to Figure 7.19, we see that eroded volumes must equal the deposited volumes,

$$V_{E_1} + V_{E_2} = V_D$$

or, adding the common volume V_C to both sides, we have

$$V_{E_1} + V_{E_2} + V_C = V_D + V_C$$

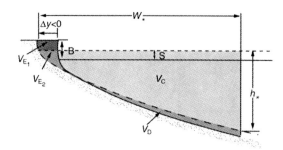

Figure 7.19 Shoreline change Δy due to sea level rise S and breaking depth h_*.

The sum of the volumes on the left-hand side of the equation can be expressed as

$$V_{E_1} + V_{E_2} + V_C = -\Delta y(B - S) + \int_0^{W_*} Ay^{2/3}\,dy, \tag{7.51}$$

where the integral includes $V_{E_2} + V_C$. (The use of a common volume to introduce the simple integration of the area above the equilibrium profile is a helpful trick, and we will use it frequently.) The sum of volumes on the right-hand side consists of the area of the rectangle above the original profile and the area contained above the original profile up to the mean sea level.

$$V_D + V_C = (W_* + \Delta y)S + \int_0^{W_* + \Delta y} Ay^{2/3}\,dy \tag{7.52}$$

Equating Eqs. (7.51) and (7.52) and carrying out the integration, we obtain

$$-\Delta y B + \frac{3}{5}AW_*^{5/3} = W_*S + \frac{3}{5}A(W_* + \Delta y)^{5/3},$$

which can be simplified to

$$\frac{\Delta y}{W_*} = -\frac{S}{B} + \frac{3h_*}{5B}\left[1 - \left(1 + \frac{\Delta y}{W_*}\right)^{5/3}\right] \tag{7.53}$$

in which $h_* = AW_*^{2/3}$ has been used. The results can be condensed even further by denoting the following dimensionless quantities:

$$\Delta y' = \Delta y/W_*$$
$$S' = S/B$$
$$B' = B/h_*,$$

which gives us the following dimensionless form:

$$\Delta y' = -S' + \frac{3}{5B'}\left[1 - (1 + \Delta y')^{5/3}\right] \tag{7.54}$$

This equation is plotted in Figure 7.20 with the abscissa representing the dimensionless storm-breaking depth and the ordinate, the dimensionless storm tide. The isolines signify nondimensional horizontal shoreline recession.

In general, the horizontal shoreline recession is small compared with the width of the surf zone; that is, for $\Delta y' \ll 1$, Eq. (7.54) can be approximated, using a binomial

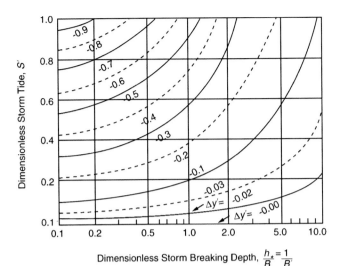

Figure 7.20 Isolines of dimensionless shoreline changes $\Delta y'$ versus dimensionless breaking depth h_*/B and dimensionless storm tide S', where $m = 2/3$ (from Dean 1991).

expansion, as

$$\Delta y' = -\frac{S'B'}{(1 + B')}$$

or in dimensional form,

$$\frac{\Delta y}{W_*} \simeq -S' - \frac{\Delta y'}{B'} = -\frac{S}{(h_* + B)}, \tag{7.55}$$

which is again the Bruun rule. As discussed for the Edelman method, Eq. (7.54) would always yield more recession than the Bruun rule because the reduced berm height with increased water level is taken into consideration.

7.5.3 EQUILIBRIUM PROFILES ON PLANAR BEACHES

A planar beach may at first appear to be an unrealistic shape, yet such beaches do occur, both artificially and naturally. For example, many wave tank studies of equilibrium profiles are begun with an (artificial) initially planar slope. Beach nourishment is often placed on a beach with a nearly planar offshore slope, and revetments and causeways exposed to waves are frequently constructed with planar slopes. Finally, shorelines of emergence, or shorelines after major storms, often have nearly planar foreshores. We will determine the amount of shoreline change Δy that would occur if these planar shorelines were shaped by nature into an equilibrium profile.

As presented in Figure 7.21, there are five types of equilibrium profiles that can form, depending on the initial slope m_i and the sediment and wave characteristics. Wave tank tests by Sunamura and Horikawa (1974) identified profile types 1, 2, and 5.

For the Type 1 profile occurring in laboratory wave tanks (see Figure 7.22), the initial slope is much steeper than that for the equilibrium profile and only seaward sediment transport occurs. An additional characteristic is that a scarp is formed at

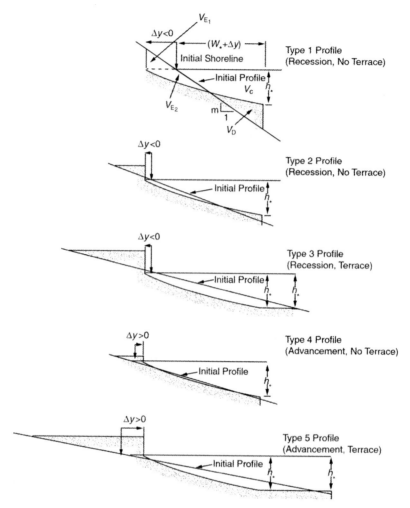

Figure 7.21 Illustration of five equilibrium profile types commencing from an initially planar slope (from Dean 1991).

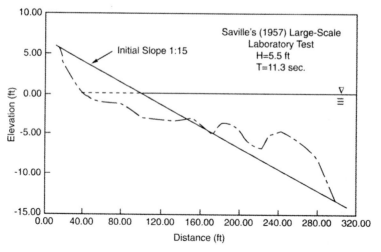

Figure 7.22 Type 1 profile from large wave tank data of Saville (1957).

the shoreline and no berm is deposited. The Type 2 profile, also a case of shoreline recession, occurs for a somewhat milder relative slope (initial slope compared with equilibrium), sediment transport occurs in both the onshore and offshore directions and a berm is formed at the shoreline. With still milder relative slopes, sediment transport occurs only shoreward, resulting in a Type 3 profile characterized by shoreline recession in which all of the sediment transported is deposited as a berm feature. A terrace or "beach" (here assumed horizontal) is formed at the seaward end of the equilibrium profile. This type of profile is probably the least likely to occur owing to the unrealistically high berm elevation required. The Type 4 profile is one of shoreline advancement occurring for still milder initial slopes and is characterized by sediment transport in both the landward and seaward directions. Finally, the Type 5 profile is one of shoreline advancement with only landward sediment transport and leaves a horizontal terrace or bench at the seaward end of the equilibrium profile.

The following paragraphs quantify the profile characteristics and shoreline changes for each of these five types. As described previously, shoreline recession and advancement will be denoted by negative and positive Δy, respectively. It can be shown that the nondimensional shoreline change $\Delta y' (\equiv \Delta y / W_*)$ is a function of the nondimensional depth of limiting profile change, $h'_* (\equiv h_* / W_* m_i)$ and nondimensional berm height $B' (\equiv B / W_* m_i)$. The developments associated with these profile types will not be presented in detail. Methods are similar to those applied earlier, for example, for the case of shoreline recession due to an elevated water level. Figure 7.23 presents isolines of $\Delta y'(h'_*, B')$ and the associated regions of occurrence for the five profile types.

Suh and Dalrymple (1988) investigated equilibrium profiles commencing from an initially planar profile and identified Type 1 (erosional) and Type 4 (accretional) profiles. For Type 1 profiles, as shown in Figure 7.21, two areas are eroded and one

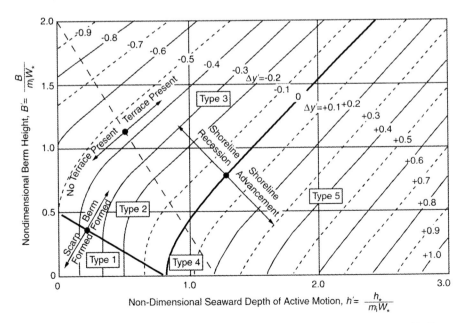

Figure 7.23 Regimes of equilibrium profile types commencing from an initially planar profile (from Dean 1991).

is deposited; the volume balance equation is then

$$V_{E_1} + V_{E_2} + V_c = V_D + V_c \tag{7.56}$$

with the common volume V_c added to both sides. These volumes are computed as

$$\frac{1}{2}m_i (\Delta y)^2 + \int_0^{W_*} Ay^{2/3} \, dy = \frac{1}{2}m_i(W_* + \Delta y)^2 \tag{7.57}$$

Carrying out the integration, Suh and Dalrymple showed that

$$\frac{\Delta y}{W_*} = \frac{3h_*}{5m_i W_*} - \frac{1}{2}, \tag{7.58}$$

which in nondimensional form is

$$\Delta y' = \frac{3h'}{5} - \frac{1}{2}, \tag{7.59}$$

where $\Delta y < 0$ as the shoreline retreats in this case. Because $B = -m_i \, \Delta y$ for this case, we also have a similar relationship for B':

$$B' = \frac{3h'}{5} - \frac{1}{2}$$

Types 2 and 3 (erosional) and Type 5 (accretional) profiles were not identified; however, it was found that for small ratios of $B/\Delta y$ and h_*/W_*, the Type 4 (accretional) profile could be represented by Eq. (7.58). Suh and Dalrymple defined the equilibrium planar slope m_e as

$$m_e = \frac{h_*}{W_*}, \tag{7.60}$$

and, based on an inspection of Eq. (7.58), written as

$$\Delta y' = \frac{3}{5}\left(h' - \frac{5}{6}\right),$$

established the condition separating eroding (Type 1) and accreting (Type 4) profiles,

$$h' = \frac{m_e}{m_i} = \begin{cases} >5/6, & \text{advancing profiles} \\ <5/6, & \text{recessionary profile} \end{cases} \tag{7.61}$$

Equations (7.58), (7.60), and (7.61) were tested by comparison with laboratory data of Rector (1954); Saville (1957); Eagleson, Glenne, and Dracup (1963); Paul, Kamphuis, and Brebner (1972); and Sunamura and Horikawa (1974). The results, presented in Figure 7.24, demonstrate good agreement.

7.5.4 EQUILIBRIUM PROFILES IN FRONT OF VERTICAL BARRIERS

Sandy beaches occur at the foot of noneroding, or very slowly eroding, cliffs. A manmade analog would be a beach in front of a seawall. The difference between the analysis here and those we have examined above is that no erosion will be permitted landward of the barrier. All changes in the profile near the barrier must be balanced by other changes offshore.

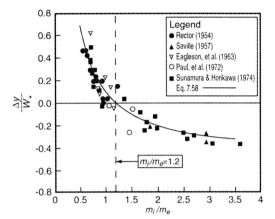

Figure 7.24 Laboratory results for shoreline recession and accretion; $\Delta y / W_*$ versus m_i / m_e. Recession occurs to the right of the vertical dashed line (from Suh and Dalrymple 1988).

It is well-known that during storms a scour trough often will occur adjacent to a seawall. For our purposes here it is appropriate to consider this scour as a profile lowering owing to two components: (1) the localized and probably dominant effect due to the interaction of the seawall with the waves and tides, and (2) the effect due to sediment transport offshore to form a profile in equilibrium with the elevated water level. By applying equilibrium profile concepts, it is possible to calculate only the second component.

The system of interest is presented in Figure 7.25. The profile is considered to be in equilibrium with virtual origin $y_1 = 0$. For a water level elevated by an amount S, the equilibrium profile will now be different and will have a virtual origin at $y_2 = 0$. We denote the distances from these virtual origins to the wall as y_{w_1}, and y_{w_2} for the original and elevated water levels, respectively. As in previous cases, the approach is to establish the origin y_{w_2}, (now virtual) such that the sand volumes seaward of the seawall and associated with the equilibrium profiles are equal before and after the increase in water level. In the following, all depths (h values) are referenced to the original water level except h_*, which, as described previously, is a reference depth related to the breaking wave heights.

Equating volumes as before, we obtain

$$\int_{y_{w_1}}^{W_* - y_{w_2} + y_{w_1}} h_1(y_1) \, dy_1 = \int_{y_{w_2}}^{W_*} h_2(y_2) \, dy_2, \qquad (7.62)$$

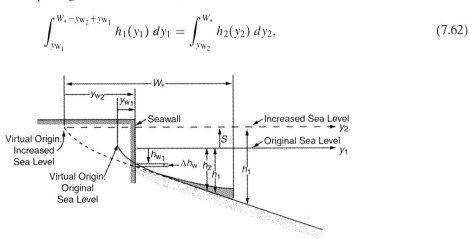

Figure 7.25 Definition sketch of profile erosion due to sea level increase and the influence of the seawall.

which can be integrated using $h_1(y_1) = Ay_1^{2/3}$ and $h_2(y_2) + S = Ay_2^{2/3}$ and simplified to yield

$$1 - \left[\frac{h'_{w_2} + S'}{h'_*}\right]^{5/2} - \left[1 - \left(\frac{h'_{w_2} + S'}{h'_*}\right)^{3/2} + \left(\frac{1}{h'_*}\right)^{3/2}\right]^{5/3}$$

$$+ \left(\frac{1}{h'_*}\right)^{5/2} - \frac{5}{3}\left(\frac{S'}{h'_*}\right)\left[1 - \left(\frac{h'_{w_2} + S'}{h'_*}\right)^{3/2}\right] = 0 \qquad (7.63)$$

in which the primes represent nondimensional quantities defined as

$$h'_{w_2} = \frac{h_{w_2}}{h_{w_1}}$$

$$h'_* = \frac{h_*}{h_{w_1}}$$

$$S' = \frac{S}{h_{w_1}}$$

Equation (7.63) is implicit in h'_{w_2} and must be solved by iteration, defining the change in depth at the wall Δh_w as

$$\Delta h_w = h_{w_2} - h_{w_1}$$

and in nondimensional form

$$\Delta h'_w = \frac{\Delta h_w}{h_{w_1}}$$

The quantity $\Delta h'_w$ is now a function of the following two nondimensional variables: S' and h'_*. The relationship $\Delta h'_w(h'_*, S')$ is presented in Figure 7.26, where it is seen that for a fixed h'_* and increasing S', the nondimensional scour $\Delta h'_w$ first increases and then decreases to zero. Figure 7.27 presents a specific example for $h'_* = 6$. The interpretation of this form is that, as S' increases, the profile is no longer in equilibrium and sand is transported seaward to develop the equilibrium profile and the water depth adjacent to the seawall increases. However, as sea level rises further, with the same total breaking depth, the *active* surf zone width decreases such that less sand

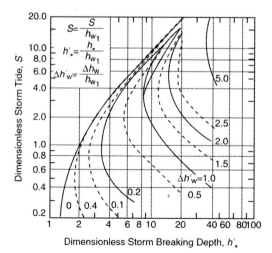

Figure 7.26 Isolines of dimensionless seawall scour, $\Delta h'_w$ versus dimensionless storm tide S' and dimensionless storm breaking depth h'_* (from Dean 1991).

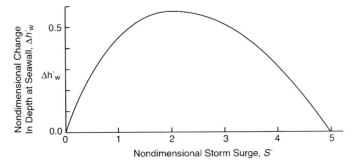

Figure 7.27 Dimensionless change in water depth at wall $\Delta h'_w$ as a function of dimensionless storm surge S', example for $h'_* = 6$ (from Dean 1991).

must be transported seaward to satisfy the equilibrium profile. With increasing storm tide, the surf zone width approaches zero at the limit

$$S + h_{w_1} = h_*$$

or

$$S' = h'_* - 1,$$

which corresponds to the upper line in Figure 7.26. Note that, as discussed previously, the increased depth at the seawall predicted here does not include the scour interaction effect of the seawall and waves.

7.5.5 RECESSION OF A NATURAL PROFILE DUE TO STORM SURGES AND WAVES

Beach recession is known to be related to augmented water levels and wave heights, yet the role played by each factor is poorly understood. On the basis of equilibrium beach profiles and with the wave height effect considered to be solely due to the wave setup, it is possible to identify the separate effects of these two factors if the profile response progresses to completion.

Consider the situation presented in Figure 7.28 in which both a storm tide S and a wave of breaking height H_b are shown. The resulting equilibrium beach profile and associated wave setup were developed in earlier sections and are given by Eqs. (7.42) and (7.43).

The development of the equilibrium shoreline change Δy is determined such that the volume of sand eroded from the foreshore is equal to that deposited offshore to form an equilibrium profile relative to the storm surge and wave setup. This requires that the volume of water and air below berm level landward of the break point for both the reference and storm profiles be equal. Thus,

$$\int_{\Delta y}^{0} [B - S - \bar{\eta}(y)] \, dy + \int_{\Delta y}^{W_* + \Delta y} A(y - \Delta y)^{2/3} \, dy$$

$$= \int_{0}^{W_* + \Delta y} A y^{2/3} \, dy + \int_{0}^{W_* + \Delta y} [S + \bar{\eta}(y)] \, dy \tag{7.64}$$

By definition, $W_1 = W_2 - R$, and integrating the above, we obtain

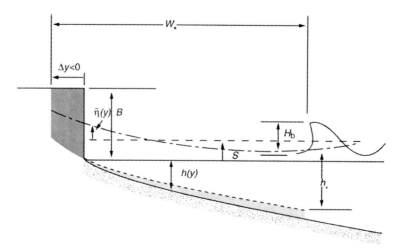

Figure 7.28 Beach recession due to waves and tides, including wave set-up (from Dean 1991).

$$\Delta y' + \frac{3}{5B'}[1 + \Delta y']^{5/3} = \frac{1}{B'}\frac{(3/5 - \mathcal{K})}{(1 - \mathcal{K})} - S' - \bar{\eta}'_b, \tag{7.65}$$

which relates the dimensionless recession $\Delta y / W_*$ to the dimensionless berm height B/h_* and the dimensionless storm tide S/B. The term $\mathcal{K}$ is defined on page 184.

The question of the relative roles of breaking waves and storm surges can be addressed by simplifying Eq. (7.65) for the case of small relative shoreline change $\Delta y'$. Employing this approximation and simplifying further for $\kappa = 0.78$, we obtain

$$\frac{\Delta y}{W_*} \cong -\frac{0.068 H_b + S}{B + 1.28 H_b} \tag{7.66}$$

Care must be taken in interpreting this equation because the surf zone width W_* includes the effect of the breaking wave height. In particular, $W_* = (H_b / \kappa A)^{3/2}$. The dimensionless shoreline change $\Delta y'$ is much more strongly related to the storm surge than wave height, as shown in Eq. (7.66). A storm surge is approximately 15 times as effective. However, during storms, the breaking wave height may be two or three times as great as the storm tide, and the larger breaking waves may persist much longer than the peak storm tides. The apparent reason that the storm tide plays a much greater role than the breaking wave set-up is evident from Figures 7.14 and 7.28, where it is seen that wave setup and setdown act to reduce the mean water level over a substantial portion of the surf zone.

When there is no storm surge and the ratio of the breaking wave height to berm height is large, Eq. (7.66) can be simplified to $\Delta y / W_* = -0.079$, or

$$\Delta y = -0.079 \left(\frac{H_b}{\kappa A} \right)^{3/2}$$

Further inspection of this equation in conjunction with Figure 7.6 will show that the potential beach recession increases for larger wave heights and finer grained beaches.

It is of interest that Eq. (7.66) predicts that it is possible for the beach to accrete for negative storm surges.

7.5.6 COMPARISON WITH EMPIRICAL ORTHOGONAL FUNCTIONS

It is instructive to compare results of profiles obtained from a simple application of the equilibrium beach profile methodology with those developed by various researchers (e.g., Winant, Inman, and Nordstrom 1975 and Weishar and Wood 1983) in their application of Empirical Orthogonal Function (EOF) methods to time series of natural beach profiles. The EOF method was described in Chapter 6. For our purposes here, we note that the first EOF is analogous to the equilibrium beach profile, and the second EOF is termed the berm–bar function.

We will consider the change in profile elevation resulting from a *single* elevated water level and wave and sediment conditions that would mobilize sediment out to a depth h_*. On the basis of Figure 7.19 and Eq. (7.54), the first EOF is the average profile, and the second EOF can be shown to be approximately

$$\frac{\Delta h}{h_*} = \Delta h' = \begin{cases} B'(1 - S') + (y' - \Delta y')^{2/3}, & \Delta y' < y' < 0 \\ (y' - \Delta y')^{2/3} - y'^{2/3} - S'B' & 0 < y' < 1 + \Delta y', \end{cases} \qquad (7.67)$$

where the primed (nondimensional) quantities are as defined in Eq. (7.54). Figure 7.29 presents a comparison between Eq. (7.67) and the second EOF, as determined by Winant et al. (1975) based on field measurements at Torrey Pines, California. The similarities between the EOF obtained by these investigators and those developed by equilibrium profile considerations are quite evident.

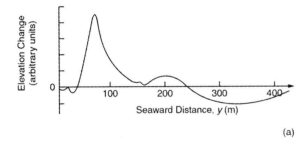

(a)

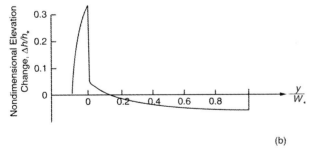

(b)

Figure 7.29 Comparison of beach profile elevation changes by equilibrium profile concepts with results from EOF analysis of field measurements example data: $h_*/B = 4$ and $S/B = 0.5$ (from Dean 1991). (a) Most significant eigenfunction of profile change (from Winant et al. 1975), (b) nondimensional elevation changes based on equilibrium beach profiles.

REFERENCES

Bagnold, R.A., "Sand Movement by Waves: Some Small Scale Experiments with Sand of Very Low Density," *J. Institution of Civil Engineers*, 27, 4, Paper 5554, 447–469, 1947.

Balsillie, J.H., "Offshore Profile Description Using the Power Curve Fit, Part II: Standard Florida Offshore Profile Tables," Division of Beaches and Shores, Florida Dept. of Natural Resources, Tech. and Design Memo. No. 82-1-IIa, 1987.

Bodge, K.R., "Representing Equilibrium Beach Profiles with an Exponential Expression," *J. Coastal Res.*, 8, 1, 47–55, 1992.

Bowen, A.J., "Simple Models of Nearshore Sedimentation; Beach Profiles and Longshore Bars," in *The Coastline of Canada*, S.B. McCann, ed., Geol. Survey of Canada, Paper 80-11, 1–11, 1980.

Bruun, P., "Coast Erosion and the Development of Beach Profiles," U.S. Army Corps of Engineers, Beach Erosion Board, Tech. Memo. No. 44, 1954.

Bruun, P., "Sea Level Rise as a Cause of Shore Erosion," *J. Waterway, Port, Coastal and Ocean Eng.*, ASCE, 88, 117, 1962.

Creed, C.G., R.A. Dalrymple, D.L. Kriebel, and J.M. Kaihatu, "Equilibrium Beach Profiles with Random Seas," *Proc. 23rd Intl. Conf. Coastal Eng.*, ASCE, Venice, 1992.

Dally, W.R., R.G. Dean, and R.A. Dalrymple, "Wave Height Variation Across Beaches of Arbitrary Profile," *J. Geophys. Res.*, 90, 6, 1985.

Dean, R.G., "Heuristic Models of Sand Transport in the Surf Zone," *Proc. Conf. Eng. Dynamics in the Surf Zone*, Sydney, 208–214, 1973.

Dean, R.G., "Equilibrium Beach Profiles: U.S. Atlantic and Gulf Coasts," Department of Civil Engineering, Ocean Engineering Report No. 12, University of Delaware, January, 1977.

Dean, R.G., "Coastal Sediment Processes: Toward Engineering Solutions," *Proc. Coastal Sediments*, ASCE, 1–24, 1987.

Dean, R.G., "Equilibrium Beach Profiles: Principles and Applications," *J. Coastal Res.*, 7, 1, 53–84, 1991.

Dean, R.G., and L. Charles, "Equilibrium Beach Profiles: Concepts and Evaluation," Rept. UFL/COEL-94/013, Department of Coastal and Oceanographic Engineering, University of Florida, 1994.

Dean, R.G., and E.M. Maurmeyer, "Models for Beach Profile Response," in *CRC Handbook of Coastal Processes and Erosion*, P.D. Komar, ed., Boca Raton: CRC Press, 151–166, 1983.

Dean, R.G., T.R. Healy, and A. Dommerholt, "A 'Blindfolded' Test of Equilibrium Beach Profile Concepts with New Zealand Data," *Marine Geology*, 109, 253–266, 1993.

Dean, R.G., T.L. Walton, and D.L. Kriebel, "Cross-shore Sediment Transport," in *Coastal Engineering Manual*, U.S. Army Coastal & Hydraulics Laboratory, 2001.

Eagleson, P.S., B. Glenne, and J.A. Dracup, "Equilibrium Characteristics of Sand Beaches," *J. Hyd. Div.*, ASCE, 89, 1, 35–57, 1963.

Edelman, T., "Dune Erosion during Storm Conditions," *Proc. 13th Intl. Conf. Coastal Eng.*, ASCE, 1305–1312, 1972.

Godfrey, P.J., and M.M. Godfrey, "The Role of Overwash and Inlet Dynamics in the Formation of Salt Marshes on North Carolina Barrier Islands," in *Ecology of Halophytes*, R. Reimold, ed., New York: Academic Press, 407–427, 1973.

Hughes, S.A., and T.Y. Chiu, "The Variations in Beach Profiles When Approximated by a Theoretical Curve," UFL/COEL/TR-039, Coastal and Oceanographic Engineering Department, Univ. Florida, 136 pp., 1978.

Inman, D.L., M.H.S. Elwany, and S.A. Jenkins, "Shorerise and Bar–Berm Profiles on Ocean Beaches," *J. Geophys. Res.*, 98, C10, 18,181–18,199, 1993.

Komar, P.D., and W.G. McDougal, "The Analysis of Exponential Beach Profiles," *J. Coastal Research*, 10, 1, 59–69, 1994.

Larson, M., "Quantification of Beach Profile Change," Rept. 1008, Dept. Water Resources Eng., Univ. Lund, 1988. This material is also presented in Larson, M., and N.C. Kraus,

"SBEACH: Numerical Model for Simulating Storm Induced Beach Change," U.S. Army Corps of Engineers, CERC, Tech. Rpt. CERC-89-9, 1989.

Longuet-Higgins, M.S., "Mass Transport in Water Waves," *Philos. Trans. Royal Soc. of London*, Series A, 245, 535–581, 1953.

Moore, B.D., "Beach Profile Evolution in Response to Changes in Water Level and Wave Height," MCE Thesis, Department of Civil Engineering, University of Delaware, 164 pp., 1982.

Paul, M.J., J.W. Kamphuis, and A. Brebner, "Similarity of Equilibrium Beach Profiles," *Proc. 13th Intl. Coastal Eng. Conf.*, ASCE, Vancouver, 1217–1256, 1972.

Pruzak, Z., "The Analysis of Beach Profile Changes Using Dean's Method and Empirical Orthogonal Functions," *Coastal Eng.*, 19, 245–261, 1993.

Rector, R.L., "Laboratory Study of Equilibrium Profiles of Beaches," Tech. Memo. 41, Beach Erosion Board, Corps of Engineers, Washington, 1954.

Roelvink, J.A., and M.J.F. Stive, "Bar-generating Cross-shore Flow Mechanisms on a Beach," *J. Geophys. Res.*, 94, C4, 4785–4800, 1989.

Sanders, J.E., and N. Kumar, "Evidence of Shoreface Retreat and In-place 'Drowning' During Holocene Submergence of Barriers, Shelf off Fire Island, New York," *Bull. Geolog. Soc. America*, 86, 65, 1975.

Saville, T., Jr., "Scale Effects in Two Dimensional Beach Studies," *Proc. 7th General Meeting*, Intl. Assoc. Hyd. Res., A3-1–A3-10, 1957.

Shepard, F.P., "Gulf Coast Barriers" in *Recent Sediments, Northwest Gulf of Mexico.* F.P. Shepard, F.B. Phleger, and T.H. Van Andel, eds., 197–220; Am. Assoc. of Petrol. Geologists, 394 pp., 1960.

Stauble, D.K., A.W. Garcia, N.C. Kraus, W.G. Grosskopf, and G.P. Bass, "Beach Nourishment Project Response and Design Evaluation: Ocean City, Maryland," U.S. Army Corps of Engineers, CERC, Tech. Rpt. CERC-93-13, 1993.

Suh, K., and R.A. Dalrymple, "Expression for Shoreline Advancement of Initially Plane Beach," *J. Waterway, Port, Coastal and Ocean Eng.*, ASCE, 114, 6, 770–777, 1988.

Sunamura, T., and K. Horikawa, "Two-Dimensional Beach Transformation Due to Waves," *Proc. 14th Intl. Conf. Coastal Eng.*, ASCE, Copenhagen, 920–938, 1974.

Wang, P., and R.A. Davis, Jr., "A Beach Profile Model for a Barred Coast – Case Study from Sand Key, West-Central Florida," *J. Coastal Res.*, 14, 3, 981–991, 1998.

Weishar, L.L., and W.L. Wood, "An Evaluation of Offshore and Beach Processes on a Tideless Coast," *J. Sedimentary Petrology*, 53, 3, 1983.

Winant, C.D., D.L. Inman, and C.E. Nordstrom, "Description of Seasonal Beach Changes Using Empirical Eigenfunctions," *J. Geophys. Res.*, 80, 15, 1979–1986, 1975.

Zenkovich, V.P., *Processes of Coastal Development*, London: Oliver and Boyd, 1967.

EXERCISES

7.1 Determine an expression for the parameter A_2 in Eq. (7.5).

7.2 Calculate and draw the idealized equilibrium beach profiles for sand sizes of 0.2 and 0.4 mm diameter and a berm height of 2.5 m.

7.3 Calculate, and superimpose on the original profiles determined in the previous problem, the profiles resulting from a storm tide of 2 m and a wave height of 2.5 m. Compare the recessions for the two profiles.

7.4 Why, generally, are coarser sediments found on the landward portions of a beach profile and finer sediments farther seaward?

7.5 For a sand size of 0.2 mm, a storm tide of 2 m, and a wave height of 2.5 m, calculate and compare the equilibrium profiles (for storm and normal conditions) for the case in which a seawall is located in a water depth $h_{11} = 0.3$ m for the normal condition.

7.6 For an initial planar slope of 1:15, calculate and compare the equilibrium slope for a sand size of 0.2 mm, a breaking wave height of 1.8 m, and a 3-s period. The equilibrium slope is based on the breaking depth.

7.7 Consider an equilibrium beach profile and a sand diameter, $d = 0.2$ mm, for which $A = 0.1$ m$^{1/3}$. The breaking wave height and direction are $H_b = 2$ m, and $\theta_b = 10°$. Assume that lateral mixing is negligible and that the contours are straight and parallel. Determine the following:

(a) The average shear stress $\langle \tau_b \rangle$ across the surf zone (of width y_b)?

$$\langle \tau_b \rangle = \frac{1}{y_b} \int_0^{y_b} \tau_b(y)\,dy$$

(b) The maximum shear stress in the surf zone.

(c) The wave energy dissipation per unit water volume in the surf zone. Comment on the cross-shore distribution of this quantity within the surf zone.

7.8 Given an equilibrium beach with a sand particle of 0.3 mm diameter, for a breaking wave height of 2 m and a breaking wave direction of 10°, plot the component of bottom shear stress parallel to the beach for the case of no shear coupling of adjacent water columns. Plot the expected modifications resulting from lateral shear coupling of adjacent water columns qualitatively.

7.9 For the same storm tide and wave height as in Problem 7.3, interpret (from a geometric point of view) why beaches composed of finer sand experience a greater recession than beaches composed of coarser sand.

7.10 Compare the equilibrium model with a gravity term and the Larson model. Express both in terms of A. If both models were identical, what is B?

7.11 Consider an idealized beach profile that is initially in equilibrium.

(a) For a change (increase or decrease) in water level, describe in general terms what causes the profile to adjust to a new equilibrium form.

(b) Suppose that a quantity of sand coarser than that originally present is placed in the shallow water portions of the profile. Sketch, qualitatively, the resulting equilibrium beach profile. See Figure 7.30.

(c) Suppose that a quantity of sand finer than that originally present is placed in the shallow water portions of the profile. Sketch, qualitatively, the resulting equilibrium beach profile.

(d) For a decrease in sea level, with the breaking conditions remaining the same, sketch the initial and resulting equilibrium beach profiles.

(e) For an increase in water level, what portion of the profile would you expect to adjust most rapidly initially? State the basis for your answer. Illustrate with one or more sketches, if helpful.

(f) For a decrease in water level, what portion of the profile would you expect to adjust most rapidly initially? State the basis for your answer. Illustrate with one or more sketches if helpful.

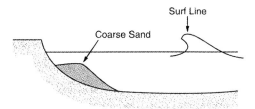

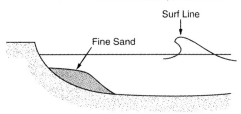

Figure 7.30   Profiles for Problem 7.11.

7.12   For an initially planar slope of 1:50, calculate the equilibrium slope for a sand size of 0.2 mm, a breaking wave height of 2 m, and a berm height B of $K H_b$, where $K = 0.8$. What is the shoreline change Δy? What is the profile type? For what value of K would there be no shoreline change? What is this profile type?

7.13   Hydrocarbon extraction is causing ground subsidence (sinking) at a rate of 2 mm/yr. The sand size is 0.2 mm, the berm height B is 2 m, and the closure depth h_* is 6 m. If no nourishment is carried out, what would be the effect on the shoreline position? Quantify any shoreline change rate. What would be the annual nourishment requirements per unit of beach length in order to obtain a stable shoreline?

7.14   Consider a beach fill composed of sand coarser than the native profile such that the profiles intersect yet the A values are unknown. The profiles intersect at h', the shoreline displacement in Δy, and the berm height is B. Develop an expression for the volume of sand per unit length of beach needed.

7.15   A beach nourishment project is being planned in which sand will be placed in the surf zone, as shown in Figure 7.31. Consider only onshore–offshore motion and discuss with sketches what the possible results are if the following conditions obtain:

   (a) The sand is the same as the native;
   (b) The sand is finer than the native.

7.16   Compare the beach recessions to be expected for sand sizes of 0.1 and 0.3 mm for the following storm conditions: $H_b = 2$ m, $S = 1$ m, $B = 2$ m. Use the simplest method (Bruun's) of the three available.

Figure 7.31   Nourishment profile for Problem 7.14.

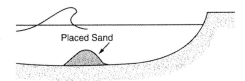

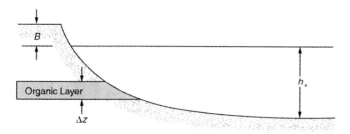

Figure 7.32   Profile with organic layer for Problem 7.19.

7.17   Demonstrate, by integration over the active profile, that the volume change V due to a uniform horizontal recession R acting over a total vertical dimension $h_* + B$ is simply

$$V = R(h_* + B)$$

7.18   Show that Edelman's result (Eq. (7.50)) yields the same equation for horizontal recession as Bruun's if the total vertical rise S_{max} is considered small. (Note: it is helpful to employ the approximation

$$\ln(1 + x) = x - \frac{x^2}{2} + \cdots \quad \text{(where x is small.)}$$

7.19   Suppose that you introduce a fairly small amount of a poorly sorted sand into the surf zone.

(a) At what location would you expect the coarse fraction to be located eventually? Why?

(b) At what location would you expect the fine fraction to be located eventually? Why?

7.20   Bruun's rule equates the "volume required" by sea level rise to the volume "yielded" by shoreline retreat. Consider the situation shown in Figure 7.32 with a berm height B, a depth of closure h_*, and a layer of organic material (i.e., no sand) of thickness Δz.

(a) Develop a relationship for the ratio R/S (shoreline retreat/sea level rise), including the effect of the organic layer thickness Δz.

(b) Calculate the shoreline retreat for the following conditions:

$$B = 2 \text{ m}$$
$$h_* = 8 \text{ m}$$
$$S = 0.1 \text{ m}$$
$$A = 0.1 \text{ m}^{1/3}$$

Carry out these calculations for $\Delta z = 0$ and $\Delta z = 2$ m. Discuss the differences.

7.21   A groin is constructed in an area of longshore sediment transport with the expected shoreline advancement and recession on the updrift and downdrift sides, respectively, as shown in Figure 7.33. The sand is varied in size. Note

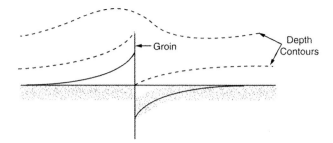

Figure 7.33 Schematic diagram of groin for Problem 7.20.

that coarse sand tends to collect preferentially on the updrift side of the groin. Provide an explanation of why this coarse sand accumulates in this location.

7.22 It is planned to nourish a shoreline with sand of 0.15 mm diameter considerably finer than the natural size of 0.3 mm. Under natural conditions, an offshore bar forms from November through March. The seasonal variation in wave height is presented in Figure 7.34. Assume the wave period to be nearly constant on an annual basis. (Assume the criterion for bar formation is $H_0/wT > 2.75$; $w =$ sediment fall velocity).

(a) Would a bar be present for a longer or shorter duration with the nourished beach compared with the natural beach?

(b) During what months would you expect a bar to be present on the nourished beach?

7.23 **(a)** In general, would you expect the equilibrium slope of the beach face to increase or decrease with increasing permeability if other factors remained the same?

(b) Discuss any mechanisms that could cause either a tendency for increased or decreased stability of particles on the beach face due to increased permeability.

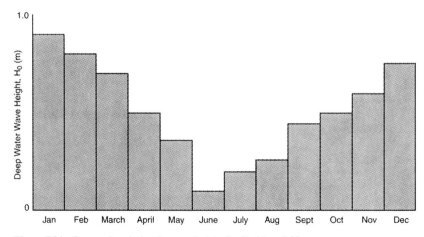

Figure 7.34 Seasonal variation in wave height for Problem 7.22.

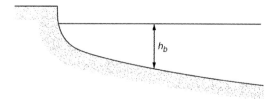

Figure 7.35 Profile for Problem 7.24.

7.24 Consider the two-dimensional beach profile in Figure 7.35, which is initially in equilibrium. The breaking depth is h_b.

(a) In Case A, sand of the same size as the original is added at a constant rate at the mean water line over the time interval $0 < t < t_1$.

(a-1) Sketch the profile at some time (say $t_1/2$) during which sand is being added to the profile. Discuss differences in form between this profile and the equilibrium profile. Show, by arrows, the direction of sediment transport across the profile.

(a-2) Sketch and describe the characteristics of the profile after equilibrium has again been attained (i.e., for $t \gg t_1$). Compare with the original.

(b) In Case B, sand is removed at a constant rate from the mean waterline over the time interval, $0 < t < t_1$.

(b-1) Sketch the profile at some time (say $t_1/2$) during which sand is being removed. Discuss differences in form between this profile and the equilibrium profile. Show, by arrows, the direction of sediment transport across the profile.

(b-1) Same as (a-2) (but for Case B).

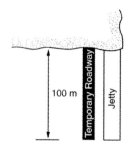

Plan View

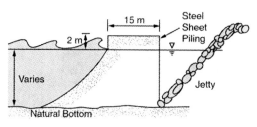

Cross Section

Figure 7.36 Sketch of temporary roadway: Problem 7.28.

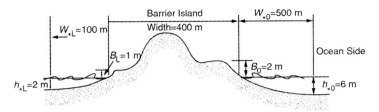

Figure 7.37 Sketch for Problem 7.28.

7.25 In Section 7.4.7, the distance and depth at which a given slope m occurs on an equilibrium profile was discussed. Show that the equations for these calculations are

$$y = \left(\frac{2}{3} \frac{A}{m} \right)^3 \tag{E7.1}$$

$$h = \frac{4}{9} \frac{A^3}{m^2} \tag{E7.2}$$

7.26 For a beach profile, the sediment size versus distance offshore is as follows: What is the water depth at a distance of 250 m from the shoreline?

y	d
0–100 m	0.3 mm
>100 m	0.15 mm

7.27 The average seasonal shoreline change can range from 10 m for the Florida coastline to 100 m for the south shore of Long Island and a few other coastal locations. Examine the role of waves and mean water levels on this seasonal change using Eq. (7.66) to calculate the shoreline position change in Florida from summer to winter. Consider a berm height $B = 2$ m and a median sediment size $d_{50} = 0.2$ mm. The summer and winter breaking wave heights are 1 and 4 m, respectively. Use the monthly mean sea levels (in feet) for Pensacola, Florida, given in Figure 3.4, for the summer and winter sea levels. (Note that the sea level heights in the figure are given on an arbitrary scale and it is probably best to subtract an approximate mean value from those given.)

7.28 A jetty is to be constructed from a temporary construction roadway with steel sheet piling used to support one side. The other side of the temporary roadway will be a beach, as shown in Figure 7.36. The roadway will be 100 m in length. The sand size of the natural beach is 0.22 mm, and the size of the roadway fill sand is 0.3 mm. What is the total volume of sand required to construct the temporary roadway?

7.29 Consider the barrier island system shown in Figure 7.37. The sea level is rising at a rate of 20 cm/100 yr. The berm heights and depths of closure on the two sides of the island are $B_0 = 2$ m, $h_{*_0} = 6$ m, and $B_L = 1$ m, $h_{*_L} = 2$ m for the seaward and landward sides, respectively. The island is 5-km long. What annual rate of beach nourishment would be required to maintain the island system in a stable state in the presence of the sea level rise?

CHAPTER EIGHT

Sediment Transport

*In 1927, a detached offshore breakwater was built 1000 ft off the coast of Santa
Barbara, California, to provide protection for a harbor. Within 1 year, a large
amount of sand accumulated behind the structure, threatening to shoal the
deep-water end of the harbor. A 600-ft extension of the breakwater, attaching it
to the western shoreline, was constructed in 1930 to prevent the encroachment of
sand from the west into the harbor. Subsequent updrift impoundment of sand by
this extension formed what is now called Leadbetter Beach, which is shown in
Figure 8.1. This wide recreation beach was caused by the approximately
270,000 yd³ (200,000 m³) of sand transported annually by the waves around the
rocky headlands to the west (and updrift) of the harbor. (A volume of
270,000 yd³ of sand is a large load to be moved per year by the waves. To grasp
this figure, one would need to imagine filling a football field with a pile of sand
150 ft (46 m) high each year!*) By 1931, sand was being transported along the
offshore portion of the breakwater and deposited in the navigational channel,
threatening to seal off the harbor.*

*The almost total trapping of sand by the Santa Barbara breakwater caused
serious erosion to the sandy beaches downdrift of the harbor, for these beaches
no longer received their annual supply of material. (This led to the first court
case on "sand rights" in the United States – the Miramar case.) Since 1935, the
harbor has been dredged. At first the sand was placed offshore in a large
submerged mound with the thought that waves would move the material onto the
beaches. This approach was unsuccessful, and subsequent dredged material was
placed directly onto downdrift beaches.*

*At present, a floating dredge is maintained at the harbor to bypass sand as
required from the spit at the end of the breakwater to the downdrift beaches.
(Most of this material appears in Wiegel 1963.)*

8.1 INTRODUCTION

The theoretical basis for sediment transport along shorelines and offshore is not fully
developed. This is largely because at the present time we do not have a complete

---

* An American football field is 100 yd long by 53.3 yd wide. A good trivia question!

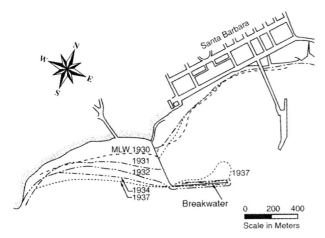

Figure 8.1 Historical shorelines at Santa Barbara, California (from Johnson 1957). Leadbetter Beach is on the left.

understanding of sediment transport even under simpler conditions, such as unidirectional sediment transport in rivers and channels, despite the efforts of many investigators over the last century to put forward theories to understand it. For example, estimates of the sediment discharge of a river using different theories can differ by several orders of magnitude, as shown by Figure 8.2.

Further, several effects known to be important in fluvial hydraulics have yet to be incorporated into coastal engineering practice owing to the paucity of data with which to verify these effects. Temperature, for example, is known to play an important role in the suspended sediment transport of a river. Lane, Carlson, and Manson (1949) showed that a 10°F decrease in water temperature results in a 33-percent increase in suspended sediment transport. This effect on coastal sediment transport has not

Figure 8.2 Sediment discharge calculations and data for the Colorado River (from Raudkivi 1967. © A.A. Balkema Publishers; used with permission).

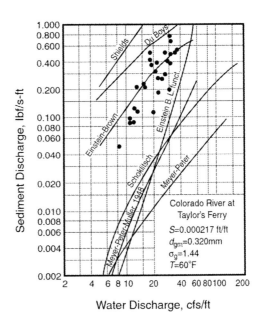

been investigated despite annual water temperature changes several times greater than this at many beaches.

The effects of breaking waves and turbulence within the surf zone add a level of complexity to coastal processes not encountered in hydraulics. The waves, with their oscillatory velocities, the nearshore circulation, and winds at the beach drive beach materials alongshore and offshore in a complicated fashion by providing a time-dependent source of momentum and energy.

This coastal sediment transport will be referred to as *littoral transport*, which can be decomposed into longshore transport and cross-shore transport. This chapter will discuss each of these components separately, although this division should be viewed purely as artificial rather than natural. But before this dichotomy takes place, we will discuss the incipient motion of the sediment under wave action and the so-called depth of closure, which figures prominently in the equilibrium profile calculations of Chapter 7.

8.2 INCIPIENT SAND MOTION AND DEPTH OF CLOSURE

The motion of a particle of sand is caused by forces acting on the particle. If the forces are not strong enough to dislodge the particle from its resting place, it does not move. We will first examine the problem of incipient motion for steady flow, as discussed by hydraulic engineers, and then we will address the unsteady flow problem due to the oscillatory wave field. The influence of wave–breaking-induced turbulence will be shown to be an important mobilizing agent within the surf zone.

8.2.1 STEADY FLOW

A sand grain sitting on a sandy bottom experiences several forces in a steady flow, as is illustrated in Figure 8.3. The *drag* force F_D acts in the flow direction, the *lift* force F_L acts perpendicular to and away from the sand bed, and the *weight* W_s, acts downwards. These forces, for an idealized spherical particle of diameter d are expressed in terms of the flow velocity U as

$$F_D = \frac{1}{2}\rho C_D U^2 A_p \tag{8.1}$$

$$F_L = \frac{1}{2}\rho C_L U^2 A_p \tag{8.2}$$

$$W_s = (\rho_s - \rho)g V_p, \tag{8.3}$$

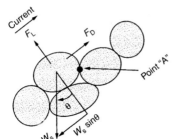

Figure 8.3 Forces on a sand particle in an inclined bed. Point A denotes the point of contact between two particles.

where C_D, C_L are the drag and lift coefficients, which depend on the flow Reynolds number, and $A_p (= \pi d^2/4)$ and $V_p (= \pi d^3/6)$ are the particle's projected area and volume. The term $(\rho_s - \rho)g$ is the submerged specific weight of the sediment.

Now a naive view will tell us that when the lift force exceeds the weight of the particle, it will lift out of its resting place. But this is a far more stringent requirement for motion than necessary because the particle can begin to roll before it is lifted into the flow. The initiation of rolling is resisted by the interparticle contacts. In Figure 8.3, point A shows where the particle of interest touches its downstream neighbor. For the particle to roll out of position, it must pivot about A. Therefore, we can compute the force moments about the point A, and when all of the moments due to the flow exceed the restoring moment of gravity, the particle will move. The moment arm for each of the forces will be somewhat different, but they all will be proportional to d, the grain size. The moment balance then is

$$F_D \alpha_1 d + F_L \alpha_2 d - W_s \alpha_3 d = 0, \tag{8.4}$$

where the α_i $(i = 1, 2, 3)$ are factors to multiply times the grain size d to find the appropriate moment arms. The bed shear stress, τ_b, can be introduced by the following definition:

$$\tau_b = \frac{1}{8} \rho f U^2,$$

where f is the Darcy–Weisbach friction coefficient, and $f/4 = C_D$. Therefore, $F_D = \tau_b d^2$, and $F_L = C_L F_D / C_D$. Substituting these expressions and the definitions for A_p and V_p into the preceding equation, we have

$$\tau_c \left(\alpha_1 + \frac{C_L}{C_D} \alpha_2 \right) = (\rho_s - \rho) g \alpha_3 d,$$

or, expressed in dimensionless form,

$$\boxed{\frac{\tau_c}{(\rho_s - \rho)gd} = f(\mathbf{Re}),} \tag{8.5}$$

where the τ_c indicates *critical* bed shear stress for incipient motion, and the function on the right-hand side depends on the Reynolds number $\mathbf{Re}$, for the drag and lift coefficients are functions of $\mathbf{Re}$, and the moment arm factors α_i may be as well. For a uniform depth, the left-hand side of Eq. (8.5) is known as the critical *Shields parameter,** denoted as $\mathbf{\Psi}_c$, which is used as an indicator of incipient motion. The dimensionless Shields parameter can be viewed as the ratio of the shear forces on the sediment particle acting to mobilize it to the submerged weight of the particle, which acts to keep it stable. Figure 8.4 shows the empirical relationship between the Shields parameter and the Reynolds number based on u_*, the shear velocity, which

---

* Kennedy (1995) has presented an interesting account of Shields' contributions, which were not recognized until well after his death.

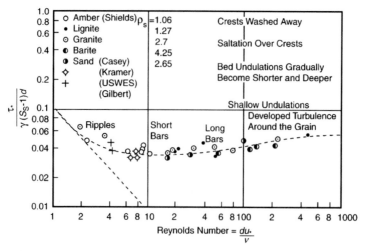

Figure 8.4 Shields curve for the initiation of motion for steady flow (from Raudkivi 1967. © A.A. Balkema Publishers; used with permission).

is a measure of the bed shear stress. This Reynolds number is defined as

$$\mathbf{Re} = \frac{u_* d}{\nu},$$

in which ν is the kinematic viscosity and the shear velocity u_* is defined as

$$u_* = \left(\frac{\tau_b}{\rho}\right)^{1/2}.$$

The data in Figure 8.4 were taken by a number of laboratory investigators who determined when the bed began to move for various sediments and steady-flow magnitudes.

From the figure, it is clear that, if the Shields parameter Ψ, which is now defined with τ_b replacing τ_c,

$$\Psi = \frac{\tau_b}{(\rho_s - \rho)gd} = \frac{\tau_b}{\gamma(S_s - 1)d}, \tag{8.6}$$

is greater than 0.03, there is a likelihood that the sand grains will move. Here, $\gamma = \rho g$ and $S_s = \rho_s/\rho$. If Ψ exceeds the value of 0.1, then it is almost certain that the bed is moving. For values of Ψ exceeding the incipient motion criterion, different bed forms may result. For example, for $\mathbf{Re} < 10$, ripples form on the bed, whereas for the approximate range, $10 < \mathbf{Re} < 100$, dunes exist. Finally, for larger values of $\mathbf{Re}$, the bed is flat but moving in *sheetflow*, which is a flat blanket of moving sand. Nielsen (1979) pointed out that if $\Psi > 0.83$, then sheetflow occurs.

As mentioned before, in the surf zone, wave breaking injects a considerable amount of turbulence into the water column, which provides an additional mobilizing effect that may allow sediment motion at considerably lower velocities than predicted from the Shields curve.

8.2.2 UNSTEADY FLOW

For oscillatory flows, Madsen and Grant (1975) and Sleath (1984) have suggested the following representations of the critical Shields parameter:

$$
\frac{\tau_c}{(\rho_s - \rho)gd} = f_1\left(\frac{(\rho_s - \rho)gd^3}{\rho \nu^2}\right) = f_2\left(\frac{d}{4\nu}\sqrt{\frac{(\rho_s - \rho)}{\rho}gd}\right)
$$

$$
= f_3\left(\sqrt[3]{\frac{(\rho_s - \rho)g}{\rho \nu^2}}\,d\right),
\tag{8.7}
$$

where f_1, f_2, f_3 are new functions of the given arguments and where the second and third arguments are basically the square root and cube root of the first.[*] These new forms of the critical Shields relationship have an advantage over Eq. (8.5) because they do not explicitly involve the shear velocity u_*, which is a function of time in unsteady flows and involves the friction factor. Each of the preceding forms is equally valid; some investigators prefer one over another.

The Shields relationship, modified for oscillatory flows, has been examined by several investigators: theoretically, such as Madsen and Grant (1975) and Komar and Miller (1975), and experimentally by others. The importance of their work is that the Shields parameter can be used for both steady flow and unsteady flow under waves. The redrawn curve is shown in Figure 8.5.

From the Shields curve, it is possible to determine the wave parameters to initiate sediment motion outside the surf zone far from the effects of wave breaking. It is a complicated procedure, however, in that we have to calculate the bottom shear stress using empirical friction coefficient data of Jonsson (1966) or Kamphuis (1975) to determine the bottom shear stress and then determine the wave that will provide the appropriate velocity. There are of course many combinations of wave height and period that will create a given maximum velocity. Such calculations have been carried out, for example, by Komar and Miller (1975) to determine the water depths for which a 15-s period wave will initiate sediment motion. Their results are shown in Figure 8.6, where it is easy to see that this long-period wave can create bottom motions in well over 100 m of water.

---

[*] The various forms can be derived from the original Shields representation (Eq. (8.5)). The first relationship is derived by introducing $(\rho_s - \rho)gd$ into the definition of u_*:

$$
u_* = \left(\frac{\tau_c}{\rho}\right)^{1/2} = \left(\frac{\tau_c}{(\rho_s - \rho)gd}\frac{(\rho_s - \rho)gd}{\rho}\right)^{1/2}
$$

The Reynolds number can now be written as

$$
\frac{u_* d}{\nu} = \left(\frac{\tau_c}{(\rho_s - \rho)gd}\frac{(\rho_s - \rho)gd^3}{\rho \nu^2}\right)^{1/2}
$$

The Reynolds number now involves the Shields parameter. Returning to the equation for the Shields relationship, Eq. (8.5), we have

$$
\frac{\tau_c}{(\rho_s - \rho)gd} = f(\mathbf{Re}) = f\left(\left(\frac{\tau_c}{(\rho_s - \rho)gd}\frac{(\rho_s - \rho)gd^3}{\rho \nu^2}\right)^{1/2}\right)
$$

Now because the unknown function f on the right-hand side involves the Shields parameter, which is what we are solving for, we factor it out, leaving a new function f_1, as given in Eq. (8.7).

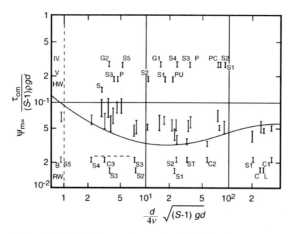

Figure 8.5 Shields curve for motion initiation for unsteady flows (modified from Madsen and Grant 1975). Vertical bars show the ranges from different experiments (from Komar and Miller 1975).

8.2.3 DEPTH OF CLOSURE

The offshore depth beyond which beach profiles taken over time at a given site coincide is known as the *depth of closure*. Seaward of this depth, although the waves can move sediment, the net sediment transport does not result in significant changes in mean water depth.

Hallermeier (1978) examined the depth of closure using wave tank tests of the equilibrium beach profile established on originally planar beaches. The cut depth h_c is defined as in Figure 8.7. Hallermeier used the Shields parameter to determine the cut depth:

$$\frac{\rho u_b^2}{(\rho_s - \rho)gh_c} = 0.03 \tag{8.8}$$

(or, by introducing the sediment diameter d, we have $\Psi = 0.03fh_c/8d$, where f

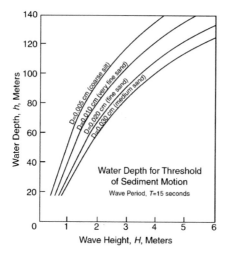

Figure 8.6 Wave heights and water depths for incipient motion for a wave with a 15-s period (from Komar and Miller 1975).

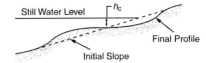

Figure 8.7 Definition of the cut depth (from Hallermeier 1978).

is the bottom friction coefficient, which varies with Reynolds number and bottom roughness). By substituting $u_b = a\sigma/\sinh kh$ from linear wave theory, Eq. (8.8) can be written as

$$kh_c \sinh^2 kh_c \tanh kh_c = \frac{\rho a^2 \sigma^4}{0.03 g^2 (\rho_s - \rho)} \qquad (8.9)$$

This expression can be solved iteratively for kh_c. However, it is more convenient to use deep water values for the wave characteristics. Therefore, introducing shoaling, we can convert Eq. (8.9) to

$$kh_c \sinh^2 kh_c \tanh^2 kh_c \left(1 + \frac{2kh_c}{\sinh 2kh_c}\right) = \frac{329 \rho H_0^2}{(\rho_s - \rho) L_0^2} \qquad (8.10)$$

EXAMPLE 1

Determine the depth of closure for waves with an offshore wave height of 0.5 m and a period of 8 s on a sandy beach of 0.3 mm.

The right-hand side of Eq. (8.10) is

$$[329 \, (0.5)^2/(1.65 * 99.8^2)] = 0.0050$$

with $L_0 = gT^2/(2\pi)$. As a first approximation, we will assume that kh_c will be small; therefore, we can approximate the expression in Eq. (8.10) as $2(kh_c)^5 = 0.005$, or our first approximation is $kh_c = 0.3018$. By iteration or other numerical methods with Eq. (8.10), we finally obtain $kh_c = 0.305$. To determine h_c, we use the dispersion relationship

$$h_c = \frac{gkh_c \tanh kh_c}{\sigma^2},$$

which yields $h_c = 1.44 \, \text{m}$.

Hallermeier converted Eq. (8.10) to conditions valid for use on the open coast that involve the annual wave statistics. He proposed that the depth of closure on the open coast is the result of erosion of the beach by the largest waves, and therefore the wave height to be used in Eq. (8.9) would be the significant wave height that is exceeded only 12 hours per year, H_e. This "effective" wave height, which occurs only 0.137 percent of the time can be related to the mean wave height by $H_e = \bar{H} + 5.6\sigma_H$, where σ_H is the standard deviation in annual wave heights. The expression for h_c is then

$$h_c = 2.28 H_e - 68.5 \left(\frac{H_e^2}{g T_e^2}\right) \qquad (8.11)$$

based on the assumption that $(\rho_s - \rho)/\rho = 1.65$.

Birkemeier (1985), utilizing numerous beach profiles taken at the U.S. Army Field Research Facility with the CRAB (discussed in Chapter 6), evaluated Hallermeier's relationship and found a more appropriate relationship for his field data is

$$h_c = 1.75 H_e - 57.9 \left(\frac{H_e^2}{g T_e^2} \right), \tag{8.12}$$

and a good approximation to the data is given simply by

$$h_c = 1.57 H_e$$

with an average error of 0.5 m.

Nicholls, Birkemeier, and Hallermeier (1996) have generalized the closure depth concept to include the consideration of a time frame other than 1 year. Equation (8.11) becomes

$$h_c(t) = 2.28 H(t)_e - 68.5 \left[H(t)_e^2 / g T(t)_e^2 \right], \tag{8.13}$$

where all the variables are now functions of time measured in years, $H(t)_e$ is the significant wave height that is exceeded during only 12 h in the time t, and $T(t)_e$ is the associated period. Because t is always increasing, the relevant significant wave height and period continue to increase as well and, as a result, the depth of closure increases monotonically. When this equation was evaluated with the beach profile data set for Duck, North Carolina, Nicholls et al. found that the predictions provided an upper bound to the measured closure depths for individual events and annual wave climates. This degree of conservatism in the predictions relative to the measurements increased with the length of time considered. An unexpected reason for this conservatism is that constructive forces (onshore) must be operative at the observed depths of closure.

Hallermeier (1978) pointed out several applications of the depth of closure formulas. In planning beach surveys, knowledge of the wave climate permits the depth of closure to be calculated and thus dispenses with the need to do several profiles a year to determine the depth of closure empirically. This knowledge also makes it possible to ascertain whether the field survey lines have been extended far enough offshore. For offshore disposal of material: (1) if it is sand, which is expected to be moved by the waves onto the beach, the material should be placed in depths shallower than h_c, (2) but if the material is *not* wanted on the beaches, such as would be the case for most dredged material from harbors, it should be placed offshore of the depth of closure. Finally, Hallermeir noted that the depth of closure is about half that depth predicted for incipient motion, as discussed in the last section.

8.3 LONGSHORE SEDIMENT TRANSPORT

The sediment moved along a coastline under the action of the waves and the longshore currents is transported in several modes: *bedload transport*, which is either in sheet flow or rolled along the bottom; *suspended load*, which is carried up within the fluid column and moved by currents; and *swash load*, which is moved on the beach face by the swash. It is not entirely clear which of these motions predominates for various wave conditions, sediment types, and locations on the profile or even whether it is important to distinguish the different mechanisms.

Littoral transport can occur in two alongshore directions, depending on the wave direction. By convention, an observer, looking out to sea, will denote longshore transport as positive when the sediment is transported to the observer's right. Typically the longshore transport at a site will consist of positive drift for one or more seasons, and negative drift for the remainder of the year. The *net* drift is the sum of the positive and negative components, and the *gross* drift is the sum of the drift magnitudes. All of these parameters can be important for a coastline. For example, for inlet and channel shoaling at unjettied entrances, the direction of the drift is likely not to be so important; it is the gross drift that represents the amount of dredging likely to be needed for the site. The net drift for this site could, in fact, be zero, yet the dredging requirements for the inlet could be quite high if the gross transport is high.

Unfortunately, as discussed in Chapter 6, there is no meter or gauge that can measure the longshore sediment transport. The total transport is often estimated by such measures as the impoundment of sand at a jetty or breakwater or the deposition of sediment in an inlet or harbor. These indirect measures are more or less correct, depending on how efficient the coastal structure is at trapping the material.* Moreover, directional wave data necessary for correlation with measured impoundments are usually lacking. Dredging records at inlets, for example, can underestimate or overestimate the net longshore transport, for a considerable amount of material may be bypassed by tidal currents or waves, or deposition may occur through bidirectional transport. Further, the possible seasonal variation of transport direction can lead to a very small net transport, which can be the difference between two large numbers, leading to large errors in estimation. Finally, the interannual variation in net transport can be large at a site, causing even correct measurements to give misleading predictions for future transport quantities, because the net drift direction and magnitude could change significantly from year to year.

8.3.1 ENERGY FLUX MODEL

Historically, the total amount of material moved along the shoreline has been related to the amount of energy available in the waves arriving at the shoreline. In the simplest model of longshore sand transport, the flux of energy in the wave direction is determined to be $\mathcal{F}\Delta l$, where $\mathcal{F}$ is the energy flux of the waves per unit crest width (recall from Chapter 5 that $\mathcal{F} = EC_g$), and Δl is a length along the wave crest. To determine the amount of this energy flux per unit length of coastline, Δx, the following geometric relationship is used: $\Delta x = \Delta l / \cos\theta$, where θ is the angle the wave ray makes with the onshore (y) direction, as in Figure 8.8. Now it is supposed that the energy flux in the alongshore direction is responsible for the longshore sediment transport; therefore, we multiply the energy flux per unit length ($\Delta x = \text{unity}$) of beach by $\sin\theta$ to obtain

$$\mathcal{F}\cos\theta\sin\theta \equiv P_\ell = EC_g\sin\theta\cos\theta = \frac{1}{16}\rho g H^2 C_g \sin 2\theta, \qquad (8.14)$$

---

* The construction of a jetty can result in large quantities of sand being moved ashore from offshore shoals that were stable under the preexisting hydraulic regime. Good preconstruction surveys are important to quantify the sand impoundments and shoals that exist, and studies should be undertaken to predict the behavior of these sand bodies during and after construction.

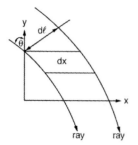

Figure 8.8   Schematic diagram for the alongshore energy flux.

after using the trigonometric identity, $2 \sin \theta \cos \theta = \sin 2\theta$. For many years, this alongshore energy flux per unit length of beach, P_ℓ, was correlated empirically with the volume of sand moved by the waves. This correlation was not easy to carry out, for there is a problem in measuring the volume of sand in transport, as was mentioned earlier. In the past, the impoundment of sand at jetties and groins has been used, along with wave measurements (height, period, and direction) to determine long-shore transport rates. Tracer experiments have also been used (Ingle 1966, Komar and Inman 1970). The established relationship is

$$Q = CP_\ell^n, \tag{8.15}$$

where C is a dimensional constant of proportionality. The power n has been found to be close to unity and has most often been taken as such. Watts (1953), Caldwell (1956), and Inman and Bagnold (1963) have all examined field data with this equation. Inman and Bagnold suggested that $n = 1$ and the dimensional coefficient C is 125, when P_ℓ has the units of millions of foot-pounds per day per foot of beach and Q has the units of cubic yards per day. This gives us $Q = 125 P_\ell$.

The volumetric transport is dependent on the wave angle through $\sin 2\theta$, which increases from no transport for normally incident waves to a maximum at 45° and then decreases for even larger angles of wave incidence.

EXAMPLE 2

Consider the amount of material moved with laboratory scale waves. If the breaking wave height is 0.5 ft, the wave period equal to 1 s, and the angle of wave incidence at breaking is 20°, then $P_\ell = \frac{1}{8} \rho g H_b^2 \sqrt{g h_b} \sin 20° \cos 20° = 0.246$ million ft-lb/day. (It is necessary to use $h_b = H_b / \kappa$ and to convert from ft-lb/s to ft-lb/day.) Therefore $Q = 125(.246) = 30.8$ yd³/day, or 11,200 yd³/yr. This becomes almost 17,500 yd³/yr with 45° of incidence. This little laboratory wave would fill a football field to a depth of more than 10 feet during the course of the year!

Inman and Bagnold (1963) criticized the use of Eq. (8.15) as being dimensionally incorrect. The left-hand side of the equation is volume per unit of time, whereas the right-hand side, which is power per unit length of beach per unit of time, has the dimensions of weight per unit of time (for $n = 1$). They introduced the immersed weight transport rate, I_ℓ, which will be discussed shortly, with the implication that the preceding transport relationship should be written in the following dimensionally correct form:

$$Q = \frac{KP_\ell}{(\rho_s - \rho)g(1 - p)},$$ (8.16)

where p is the porosity of the sediment, which is typically about 0.3 to 0.4. The porosity is introduced to convert sand weight to sand volume, taking into account the voids that occur within the sand. The dimensionless parameter K found by Komar and Inman (1970) was 0.77. Numerous other investigators have found different values for K, ranging from 0.2 (Kraus et al. 1982) to 2.2 in Caldwell (1956). Dean et al. (1982), using the Santa Barbara Harbor as a sand trap, found K to be 1.23. The determination of K will be discussed more fully in Section 8.3.4.

The group velocity can be eliminated in Eq. (8.14) for the alongshore energy flux by utilizing the shallow water approximation that $C_g = \sqrt{gh}$ and $h = H_b/\kappa$, where H_b is the breaking wave height and κ is the breaking index. This reduces Eq. (8.16) to

$$Q = C' H_b^{\frac{5}{2}} \sin 2\theta_b$$ (8.17)

Here,

$$C' = \frac{\rho K \sqrt{g/\kappa}}{16(\rho_s - \rho)(1 - p)},$$

which gives us $C' \approx 0.325$ in the same system of units used in Eq. (8.15). The longshore transport will have a positive or minus sign, depending on the breaking wave angle θ_b.

8.3.2 ENERGETICS MODEL

Bagnold (1963) developed a general model for sediment transport based on the amount of power utilized by the flow to transport sediment. Inman and Bagnold (1963) then adapted the concept for oscillatory flow within the surf zone.

Bagnold's bedload model begins by considering the interface between a fluid and a granular bottom. This interface is inclined at an angle β with respect to the horizontal. If there is no motion, the sand grains are nested against one another with many points of contact. Allow the fluid to move; it now exerts a shear stress on the boundary that causes the sand to move. We will envision this motion as the top layer of sand grains moving over a lower layer. The motion results in a *normal* stress as well as a tangential stress. This normal stress develops when the upper layer of moving grains moves up and slides over the lower layer as the grains become unnested just as in the initiation of motion discussed earlier. That is, the sand layers must dilate to permit the motion of the layer of sand grains; otherwise, the grains would not be able to move. This normal stress must also be equal to the normal component of the weight of the moving layer of sand, for this component of weight must be supported by the bed. The downstream component of sediment weight is also balanced by a shear stress.

Bagnold carried out an experiment to determine if a tangential shear stress can in fact create a normal stress. On the basis of his results, he hypothesized that the tangential stress T could be related to the normal stress P

$$T = P \tan \phi, \tag{8.18}$$

where ϕ is the angle of repose of the sediment.

To determine the immersed weight sediment transport, a certain mass m of sand (per unit area) is being moved by the waves and currents. The submerged weight of the sand can be found by dividing m by the density of the sediment ρ_s to get the volume of sand per unit area and then multiplying by the submerged specific weight. This gives us the following expression:

$$\text{Immersed weight} = \frac{m(\rho_s - \rho)g}{\rho_s}$$

Therefore,

$$P = \frac{m(\rho_s - \rho)g}{\rho_s} \cos \beta \tag{8.19}$$

The tangential shear stress necessary to lift the layer of moving sand grains is approximately related to the normal force by Eq. (8.18). Therefore, the total stress exerted on the bed to move the layer of sand is composed of the intergranular stress less the downstream component of weight.

$$T = \frac{m(\rho_s - \rho)g}{\rho_s} \cos \beta (\tan \phi - \tan \beta) \tag{8.20}$$

If the bed slope β approaches the angle of repose ϕ, then very little shear stress is necessary to move the grains because they are approaching an avalanche state. The power expended by the work of the shear stress at the bottom is

$$TU = \left(\frac{m(\rho_s - \rho)g}{\rho_s} \cos \beta U \right) (\tan \phi - \tan \beta) = I(\tan \phi - \tan \beta),$$

where the dynamic transport rate I is introduced, which involves the nearbed velocity U of the fluid. Now not all the power available in the flow field (defined as ω) is available for transporting sediment; therefore, a parameter ϵ_b is introduced, which is less than unity and represents the portion of the power available to move the sediment. Rewriting the previous equation, we have

$$\epsilon_b \omega = I(\tan \phi - \tan \beta)$$

or

$$I = \frac{\epsilon_b \omega}{(\tan \phi - \tan \beta)} \tag{8.21}$$

So far, the influence of waves and currents has been neglected. Inman and Bagnold (1963) argued that the influence of the oscillatory motion of the waves is to suspend the moving sand layer, and the mean currents then transport the sediment. By dividing the dynamic transport rate I by the oscillatory velocity U to obtain the immersed weight of the sand lifted from the bottom and then multiplying by the mean current, say, U_c, we have the transport rate for the wave-induced sediment transport:

$$I_c = \frac{m(\rho_s - \rho)g}{\rho_s} U_c = K\omega\frac{U_c}{U},$$ (8.22)

where U is the maximum orbital velocity under the waves and K is a constant. For water waves, the power, ω, is known,

$$\omega = EC_g;$$

therefore, the final expression for dynamic transport is

$$I_c = KEC_g\frac{U_c}{U}$$ (8.23)

in the direction of the current U_c, which could be due to waves, tides, or wind-induced currents. For the surf zone, however, the wave-induced longshore current is the obvious candidate for the current, and Komar (1971) replaced U_c with the average longshore current (such as given by Longuet-Higgins 1970) and showed that the dynamic transport rate can be expressed as

$$I_c = K'EC_g\sin\theta_b\cos\theta_b,$$ (8.24)

which is of the same form that the energy flux approach gave us previously, except that K' is now dimensionless; it includes the factor K and another term $\tan\beta\cos\theta_b/C_f$, which is a ratio of beach slope to Chezy bottom friction factor ($C_f = f/4$). Because K and K' are considered constants, Komar assumed that $\tan\beta\cos\theta_b/C_f$ must be constant as well.

The energetics approach can be used to derive the *suspended load* as well, as shown by Inman and Bagnold (1963) and Komar (1971). The sand that is suspended in the fluid falls with the fall velocity w, and the power lost in suspending the grains is Pw, where P is the normal stress as before. Substituting for P, we obtain

$$Pw = \frac{m(\rho_s - \rho)g}{\rho_s}\cos\beta w = I_s\frac{w}{U_s}$$ (8.25)

after introducing the dynamic transport rate I_s and the sediment velocity U_s. Now a portion of the available power is used to maintain the suspension; therefore, we can introduce an efficiency factor ϵ_s, which is the portion of the remaining power $(1 - \epsilon_b)$ used to suspend the sediment.

$$I_s = \epsilon_s(1 - \epsilon_b)\omega\left(\frac{1}{w/U_s - \tan\beta}\right)$$ (8.26)

Now if we introduce the Longuet-Higgins formula for U_s and the energy flux for ω and neglect the bottom slope term, we obtain

$$I_s = K''EC_g\sin\theta_b\cos\theta_b,$$ (8.27)

where K'' depends on ϵ_b, the beach slope, and inversely on the fall velocity of the sediment.

Bagnold's energetics model has provided bedload and suspended transport formulas, which can be reduced to the same form and combined so that the total dynamic transport rate is

$$I_\ell = K P_\ell \tag{8.28}$$

Converting the dynamic transport to a volumetric transport,

$$I_\ell = Q(\rho_s - \rho)g(1 - p),$$

yields the same form as the alongshore energy flux model, which accounts for the continued use of Eq. (8.16). Komar (1971) also provided a swash transport model.

Bailard (1981) extended the Bagnold models for sediment transport by integrating the instantaneous bedload and sediment transport equations in time to obtain equations involving the higher moments of the orbital wave velocities. Bailard provided the local mixed bedload and suspended load model in terms of the transport vector i, which includes magnitude and direction:

$$\overline{i(y)} = \rho C_f \frac{\epsilon_b}{\cos \beta \tan \phi} \left(\overline{|u|^2 u} - \frac{\tan \beta}{\tan \phi} \overline{|u|^3 j} \right)$$

$$+ \rho C_f \frac{\epsilon_s}{w} \left(\overline{|u|^2 u} - \frac{\epsilon_s}{w} \tan \beta \overline{|u|^5 j} \right) \tag{8.29}$$

The first term on the right-hand side is the local bedload transport, and the second is the suspended load. The coefficient C_f is a friction factor, and j is a unit vector pointing shorewards. Making some reasonable assumptions and integrating across the surf zone, Bailard (1984) found that his relationship could be expressed in the usual form, Eq. (8.16), with the following representation for K:

$$K = 0.050 + 2.6 \sin^2 2\theta_b + 0.0096 \tan \beta + 0.0073 \frac{u_m}{w}, \tag{8.30}$$

which shows that K varies with the wave breaking angle θ_b, the beach slope $\tan \beta$, and inversely with grain size (through the fall velocity w, but only for the suspended portion of the total transport).

8.3.3 SUSPENDED TRANSPORT MODEL

Dean's (1973) model for sediment transport is based on the suspended load within the surf zone. As in the Inman and Bagnold (1963) suspended sediment model, a portion of the available energy flux into the surf zone is dissipated by the falling sand grains. This dissipation $\mathcal{D}$ for a single sand grain is due to the loss of potential energy by the particle and is given by the product of the submerged weight of the particle and the fall velocity w:

$$\mathcal{D} = (\rho_s - \rho)g \frac{\pi d^3}{6} w, \tag{8.31}$$

where it is assumed that the grain is nearly spherical with diameter d. The number of grains of sand suspended in the surf zone N_s per unit length of beach front can be found by dividing the available energy flux per unit length of beach by the dissipation caused by a single grain:

$$N_s = \frac{\epsilon E C_g \cos \theta}{\mathcal{D}} = \epsilon \frac{\rho H_b^2}{8} \frac{C_g \cos \theta_b}{(\rho_s - \rho)(\pi d^3/6)w},$$

where ϵ is the fraction of the wave energy flux dissipated by falling sand grains. The volumetric suspended concentration of sand is then defined as the product of the number of sand grains times the volume of each grain divided by the cross-sectional area of the surf zone:

$$C_s = \frac{N_s(\pi d^3/6)}{A_C} \tag{8.32}$$

Now the amount of material in transport will be the product of the volumetric concentration times the mean flow in the surf zone times the cross-sectional area, or,

$$Q = C_s A_c V, \tag{8.33}$$

where V is the average longshore current in the surf zone. Utilizing an averaged form of Longuet-Higgins longshore current formula, Dean also obtained

$$Q = C P_\ell, \tag{8.34}$$

where again Q has the dimensions of cubic yards per day and P_ℓ is in millions of foot-pounds per foot of beach and

$$C = 34.3 \times 10^3 \epsilon \frac{\sqrt{H_b/\kappa} \tan \beta \cos \theta_b}{C_f \sqrt{g} \, (\rho_s - \rho)(1 - p)w} \tag{8.35}$$

By comparing C to the value of 125 in the *Shore Protection Manual* (c.f., U.S. Army, 1984), Dean concluded that $\epsilon = 0.002$, or that about 0.2 percent of the available wave power suspends sand within the surf zone.

8.3.4 LABORATORY AND FIELD STUDIES OF K

The value of the constant K in the longshore transport equation

$$Q = \frac{K P_\ell}{(\rho_s - \rho)g(1 - p)} \tag{8.36}$$

is important in engineering applications because, for example, as we will show later, in a beach nourishment project, if K is doubled, the project life will only be one-half that for the smaller K. In the design of a sand-bypassing facility at an inlet for a given wave climate, the sand transfer requirements of the facility vary linearly with K. There is still no consensus as to whether K is constant or possibly varies with sand size, density, fall velocity, and shape characteristics, the beach profile, water temperature, and the wave angle of incidence, as might be expected from Eqs. (8.35) and (8.30).

Note that the value of the sediment transport coefficient K often varies significantly because the standard value of 0.77 is based on the energy of the waves, whereas the significant wave height is sometimes used in the formula, resulting in a smaller value of K. The value based on significant wave heights is sometimes stated to be one-half the energy-based value because, for a Rayleigh distribution of wave heights $H_s \approx \sqrt{2} \, H_{rms}$; however, because the breaking group velocity also depends on wave height, the value of K based on significant wave height should be approximately 0.35 of the energy-based value. Beware of this possible confusion.

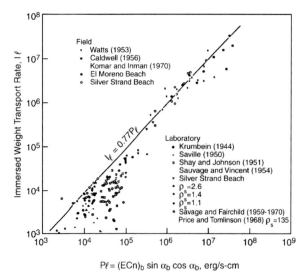

Figure 8.9 Laboratory and field correlations of I_ℓ and P_ℓ (from Komar and Inman 1970). Data for large values of P_ℓ are field values; the remaining data are from laboratory studies.

Several laboratory and field studies have been carried out to determine values of K. Figure 8.9 presents results of I_ℓ versus P_ℓ assembled by Komar and Inman (1970). Since then, many additional studies have been conducted.

If a single value of K represented the laboratory and field data in Figure 8.9, all the data would lie along a single straight line. Of particular note is that a best-fit line through the laboratory data passes below a similarly fitted line for the field data. The reason for this is not known precisely but is likely due to "scale effects." Perhaps with a better understanding of sediment transport mechanics, we could derive useful information from the laboratory studies. For example, it may be possible to scale up these results, most of which were conducted with beach sand, which is abnormally large when scaled up to prototype size. (Chapter 10 discusses scale effects in more detail.)

Because of the scale effects, recent efforts have focused on field experiments for the determination of K values. Several field approaches have been utilized. One relatively short-term approach is sand tracing. This technique overcomes one of the disadvantages of field experiments – the unpredictable nature of the waves over the course of the experiment.

In a tracer experiment, sand of similar size characteristics* is "tagged" and introduced across the active surf zone. The tagging is most often done with fluorescent dyes that are glued to the sand (care must be taken not to glue the grains together). Several colors of dye can be used on sands to be injected at different locations. Another tagging method has been to irradiate the sand, but this can lead to legal problems in some locales.

After tracer injection, the wave conditions are recorded, and synoptic sand samples are taken over the active surf zone. Two types of information are obtained from

_____

* Perhaps taken directly from the beach under study.

the sand samples that, when combined, yield an estimate of the longshore sediment transport rate. The first is the field of tracer concentration determined from sampling at numerous cross-shore and longshore locations. The concentration field of tracer is analyzed to determine the longshore center of gravity. Over the life of the experiment, the movement of this center of gravity is representative of the transport direction and rate. The remaining information required is the thickness of the moving layer. This can be estimated from putting vertical "plugs" of tracer into the bottom. The penetration depth of the plugs must be greater than the actual depth of sediment transport. At the conclusion of the experiment, the thickness of the moving layer is determined by the layer of "new" sand overlying the remaining part of the tracer plug.

The net transport rate Q is calculated as

$$Q = \frac{1}{\Delta t} \int \Delta z(y) \Delta x(y) \, dy, \tag{8.37}$$

where the integration is across the surf zone, $\Delta z(y)$ represents the depth of motion within the beach, and $\Delta x(y)$ is the alongshore displacement of the center of gravity of the tagged sand. In general, the depth of motion, $\Delta z(y)$, is on the order of centimeters.

The tracer method has been applied in field studies of longshore sediment transport by Ingle (1966), Duane (1970), Komar and Inman (1970), and Kraus, Farinato, and Horikawa (1981), among many. The method has the advantage of being applicable over relatively short periods of time, thereby providing results for (approximately) constant wave conditions and negating the need for averaging.

However, there can be problems with this type of experiment: tracer burial, if the profile is in a recovery mode; tracer loss, if the beach is eroding rapidly; and tracer dispersal, if rip currents are present that transport the tracer outside of the surf zone. It is vitally important that a large fraction of the original amount of tracer be accounted for in the sampling; otherwise, a significant transport mechanism is being overlooked, increasing the experiment's uncertainty. Of course, one tracer experiment only provides information of sediment transport for the conditions extant during the experiment. It does not give any information about annual conditions, for example.

A second method of providing short-term estimates of longshore sediment transport is through measurements of the suspended sediment concentration and the longshore component of the water velocity. (The bed-load transport is not easily measured.) The longshore transport is given by

$$Q = \frac{1}{t'} \int_0^{y_{max}} \int_0^{t'} \int_{-h}^{\eta} u(y, z, t) c(y, z, t) \, dz \, dt \, dy, \tag{8.38}$$

where $c(y, z, t)$ is the volumetric concentration of sand at a given point. Expressing both u and c as the sum of a steady term and a fluctuation about the average $(u = \bar{u} + u'; c = \bar{c} + c')$, this equation can be rewritten as

$$Q = \int_0^{y_{max}} \int_{-h}^{\bar{\eta}} \overline{u} \, \overline{c} \, dz \, dy + \frac{1}{t'} \int_0^{y_{max}} \int_0^{t'} \int_{-h}^{\eta} \overline{u' c'} \, dz \, dt \, dy \tag{8.39}$$

If existing transport models were able to predict the complete velocity and concentration fields, the local and total suspended sediment transport could be determined. Many modelers have argued that the first term is adequate to represent the full

228 SEDIMENT TRANSPORT

transport and have neglected the second term. However, Hanes (1990) has shown, based on field measurements, that neglect of the second term can result in an incorrect cross-shore transport direction. Jaffe, Sternberg, and Sallenger (1985) and Hanes and Vincent (1987) have also recognized the significance of the second term in Eq. (8.39). It is thus likely that this term is also important for longshore sediment transport.

Two different ways of measuring the suspended samples have been used. Downing, Sternberg, and Lister (1981) utilized optical backscatter (OBS) devices to sense the concentrations of sand in the vicinity of the device. The studies were conducted at Leadbetter Beach, California (near Santa Barbara), during the Nearshore Sediment Transport Study (NSTS, Seymour 1989). The associated water particle velocities were measured with electromagnetic (EM) current meters. Surprisingly, it was found that the total longshore suspended sediment transport could account for all of the transport predicted by Eq. (8.36), with a K value of 0.77. Kraus (1987) and Kraus and Dean (1987) also measured suspended sediment using fine-mesh streamer traps supported by wire frames. Again the velocities were measured by EM current meters. This study, carried out at the Corps of Engineers Field Research Facility at Duck, North Carolina, resulted in a suspended sediment transport from 7 to 29 percent of that calculated by Eq. (8.36) with $K = 0.77$ (Bodge and Kraus 1991).

An approach that uses nature to integrate the total volume over a reasonably long period of time is the total trap method, which involves using a special coastal site that includes a total barrier to the littoral transport, thus trapping the entire sediment transport. The sediment trap is surveyed periodically to quantify the transport that occurred between surveys. This method was employed by Berek and Dean (1982) at Leadbetter Beach, Santa Barbara, California, during the NSTS experiment there and by Dean (1989) at Rudee Inlet, Virginia. Figure 8.10 shows the survey plan and the wave gauge locations at Leadbetter Beach. The 74 profiles documented the profile changes, and the "trap" consisted of a portion of Leadbetter Beach, the spit at the tip of the breakwater, and the channel in the lee of the Santa Barbara Harbor breakwater. As mentioned previously, the best-fit K value of 1.23 was determined based on seven

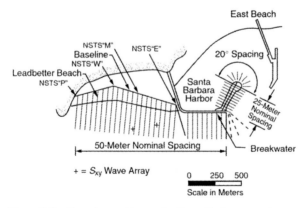

Figure 8.10 Experimental layout for Santa Barbara experiment (from Dean 1989).

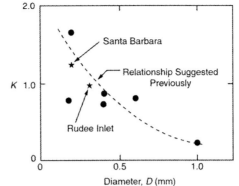

Figure 8.11 Variation of K with sediment size (modified from Dean 1989).

intersurvey periods, during which 288,000 m³ accumulated. In addition, it was found that the relationship

$$I_\ell = K' S_{xy} \tag{8.40}$$

provided a slightly better fit to the data than Eq. (8.16). In this equation, S_{xy} is the radiation stress component, or the total longshore thrust per unit length of beach, discussed in Chapter 5. The Santa Barbara field studies yielded a K' value of 2.6 m/s.

Dean (1989) assembled K values from various field experiments and interpreted the results in terms of a variation in sediment size, as shown in Figure 8.11. He suggested that the variation in K may be due to the sand size of the drift materials, with larger grain sizes associated with smaller values of K. Komar (1977) had earlier argued that K was independent of grain size. His argument was based on plotting I_ℓ/P_ℓ versus fall velocity and also $\sqrt{gH_b} \tan\beta \cos\theta_b/(C_f W)$. In both cases, no obvious dependency was found. Dean's conclusion, however, is also supported by laboratory results of Kamphuis et al. (1986) and the field studies of del Valle, Medina, and Losada (1993), who also determined that the coefficient decreases inversely with grain size. Kamphuis and Readshaw (1978) proposed that K varies with the surf similarity parameter ζ_b defined in Chapter 9 as

$$\zeta_b = \frac{m}{\sqrt{H_b/L_0}}, \tag{8.41}$$

where H_b is taken as the breaking significant wave height for field data. There is a logical basis for this dependency because the breaker type is related to ζ_b, as discussed in Chapter 5. Bodge and Dean (1987) showed K versus ζ_b for laboratory and field data, demonstrating that K apparently increases with ζ_b with a range of $0.5 < K < 1.1$.

Wang and Kraus (1999) have recently presented results from the rapid installation of a littoral barrier under low wave-energy conditions. Both the updrift and downdrift volumes were measured and found to be approximately the same (as indeed they should be). Four values of the longshore sediment transport coefficient K were determined and found to range from 0.044 to 0.541.

8.3.5 TRACTION MODELS

Another class of sediment transport models has been developed for outside the surf zone based on the body of knowledge about traction models developed for open channel flow. A key variable is the Shields parameter, $\Psi = \tau_b/(\rho_s - \rho)gd$, which was defined earlier in this chapter. If the bottom shear stress τ_b exceeds its critical value, then the sediment moves. A dimensionless transport can be defined as

$$\phi = \frac{q_s}{wd}, \tag{8.42}$$

where q_s is the transport rate per unit width normalized by the product of the fall velocity of the sediment w and the grain size d. Utilizing this form and experience that workers in steady flow have developed, Madsen and Grant (1976) have argued that the Einstein–Brown formula is valid for oscillatory flows in addition to steady flow. The concession to oscillatory flow is that everything is time dependent,

$$\phi(t) = 40\Psi(t)^3, \tag{8.43}$$

where in fact $\phi(t)$ is the transport vector. For purely oscillatory flows, ϕ is zero; only when there is an asymmetry in the flow field (and hence the shear stress term in $\Psi(t)$) is there a net transport.

To obtain a mean transport rate, Madsen and Grant integrated Eq. (8.43) over half a wave period to account for the time that the oscillatory flow in one direction begins to the time when it stops to reverse direction.

$$\overline{\phi} = \frac{2}{T} \int_{t_1}^{T/2-t_2} 40\Psi(t)^3 dt, \tag{8.44}$$

where t_1 is the time at which the incipient motion criterion is exceeded and $(T/2 - t_2)$ is the time when the transport ceases. The result of this integration is

$$\overline{\phi} = 12.5\Psi_m^3, \tag{8.45}$$

where Ψ_m is the magnitude of $\Psi(t)$. This relationship shows that the transport is proportional to the maximum velocity to the sixth power. To obtain the transport during the next half of the wave period, during which the transport is in the opposite direction, the same procedure is followed. The difference in the two transport rates is the net transport over a wave period. For sinusoidal waves and no mean current, the net would be zero.

8.3.6 OTHER TRANSPORT RELATIONSHIPS AND COMPARISON WITH FIELD DATA

Equation (8.16) is often referred to as the CERC (Coastal Engineering Research Center) equation for longshore sediment transport and is the relationship recommended in the *Shore Protection Manual* (U. S. Army Corps of Engineers 1984). Several other equations have been developed to represent longshore sediment transport. Schoones and Theron (1996) have assembled an extensive field data set of 123 cases for which the variables required to evaluate the predictive skill of such equations were available. Fifty-two relationships were evaluated, and it was determined that the Kamphuis (1991) equation provides the best fit to the data. The next best two relationships are those by Van Hijum, Pilarczyk, and Chadwick (see Chadwick 1989) and Van der Meer (1990). The Kamphuis relationship is

$$Q = 2.27 H_{sb}^2 T_p^{1.5} \tan \alpha_b^{0.75} d^{-0.25} \sin^{0.6}(2\theta_b), \tag{8.46}$$

in which Q is the total longshore sediment in kg/s, α_b is the beach slope out to the break point, and all other variables are in metric units. Both the beach slope and sediment size d are represented in this equation. Use of the equilibrium beach profile relationships from Chapter 7 will show for normal beach sand sizes, that the transport increases slightly with sand size! Also of interest is the exponent of 0.6 on the $\sin(2\theta_b)$ term which, for small values of breaking wave angles, results in a greater sensitivity to wave direction.

8.3.7 DISTRIBUTION OF THE LONGSHORE TRANSPORT ACROSS THE SURF ZONE

The littoral transport formulas developed in the preceding subsections were for the total transport, I_ℓ. It is also useful to determine the distribution of the transport across the surf zone, particularly for coastal models of shoreline erosion and for construction activities within the surf zone, such as laying a pipeline. The relationship between the two is

$$I_\ell = \int_0^\infty i_\ell(y)\, dy \tag{8.47}$$

From laboratory measurements and field studies, it is apparent that the maximum longshore transport is between the breaker line and the midpoint of the surf zone. This is also supported by theoretical studies. Komar (1971), using Eq. (8.23), which relates the wave power and the local current to the transport rate, used the Longuet–Higgins transport formula to calculate the transport distribution on a long, straight beach and showed that the maximum transport on a planar beach is located at 80 percent of the distance to the breaker line from shore. Thornton (1972), using field data as well as theory, found a similar result.

Bodge (1989) reviewed fifteen existing theories for the cross-shore distribution of the longshore sediment transport, including the two above. Many of the proposed equations consider the sediment to be mobilized locally and advected by the longshore current or to be transported by the peak shear stress or bottom velocities. Most of the models, when applied to a planar beach, predict a transport maximum between the mid-surf zone and the breaker line. Field and laboratory data demonstrate that the distribution depends greatly on the detailed profile and the breaker type (i.e., plunging, spilling, or collapsing) and may be characterized by multiple maxima. Figure 8.12 presents laboratory studies for collapsing waves and field data for spilling or plunging breakers. The field data suggest that approximately 30 percent of the transport occurs seaward of the breaker line and that, for collapsing waves, a substantial portion of the transport may occur in the swash zone. On the basis of his laboratory and field studies, Bodge recommended the following relationship for the local immersed weight transport rate:

$$i_\ell(y) = K_\alpha \left\{ \frac{1}{h} \frac{\partial E C_g}{\partial y} \right\} V_\ell(y) \left(\frac{\partial h}{\partial y} \right)^r, \tag{8.48}$$

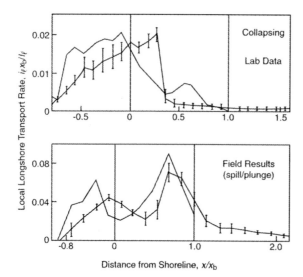

Figure 8.12 Field and laboratory data for cross-shore distribution of littoral transport (from Bodge 1989). The line with error bars is the experimental result; the other line is the prediction based on Eq. (8.48).

in which K_α is a dimensional constant with units of time, the quantity in braces is the wave energy dissipation per unit volume, V_ℓ is the local mean longshore current velocity, and r is an empirical exponent, $(0 < r < 0.5)$. Figure 8.12 provides a comparison of Eq. (8.48) with these laboratory and field data.

8.4 CROSS-SHORE SEDIMENT TRANSPORT

The longshore sediment transport is mostly due to the wave-induced longshore current, whereas cross-shore transport is a result of the water motions due to the waves and the undertow. However, the coupling between the hydrodynamics (discussed in Chapter 5) and the sediment transport is not all that well understood.

Seasonal shoreline changes are usually considered to be in response to the greater incidence of storms during winter and the associated seaward sand transport and storage in nearshore bar features. There is no consensus on the precise cause of offshore bars; however, it is generally believed that the undertow associated with breaking waves is responsible. With the greater wave heights associated with storms, the bar forms farther offshore where the water is deeper and the entire scale of the bars is likewise increased, requiring a greater volume of sand, which is provided in part by erosion of the subaerial portion of the beach profile.

Of interest is the substantially different magnitudes of the seasonal shoreline changes associated with different geographic regions. For example, Dewall and Richter (1977) found the seasonal shoreline fluctuations at Jupiter Island, Florida, to be on the order of 15 m, whereas at Duck, North Carolina, at the site of the CERC pier, Lee and Birkemeier (1993) found that the shoreline fluctuations were less, although the offshore bars migrated more than 150 m in some years. Dewall (1979) determined the seasonal fluctuations at Westhampton Beach, Long Island,

New York, to be approximately 20 to 40 m, and, at the nearby Jones Beach, the seasonal variations were established by Tanski, Bokuniewicz, and Schubert (1990) to be approximately 30 to 40 m. Katoh and Yanagishima (1988) measured 7 months of shoreline variation, wave energy flux, and beach slope with time at the Hazaki Research Pier, near Kashima, Japan, as shown in Figure 8.13. For this experiment, a 30-m variation in shoreline position occurred that was strongly correlated with the

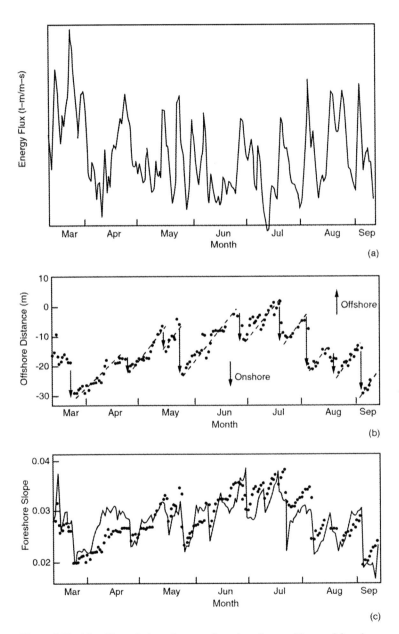

Figure 8.13 Monthly variation of energy flux, shoreline position, and foreshore slope (from Katoh and Yanagishima 1988). (a) Change of daily mean wave energy flux; (b) on–offshore changes of shoreline position; (c) changes of foreshore beach slope.

wave energy flux: large fluxes associated with storms caused a shoreline retreat. After the storms, the shoreline recovery was rapid.

In the next sections, various models for the formation of offshore bars are discussed.

8.4.1 FALL TIME MODEL

One of the original tenets of coastal engineering was that an offshore bar and a flatter foreshore would occur when the offshore wave steepness H_0/L_0 was greater than 0.025. This result was based on laboratory experiments. However, as more data were examined, it appeared that perhaps the distinction between storm profile and normal profile was not simply the offshore wave steepness but rather a combination of parameters.

Dean (1973) developed a heuristic model for cross-shore transport in the surf zone based on the suspension of sand grains by wave breaking and the eventual settling of sand to the bottom. If we consider that a wave breaking in the surf zone lifts sand from the bottom up into the water column an average distance S, then the time that it will take the sand to fall back to the bottom is

$$t = \frac{S}{w},$$

where w is the fall velocity of the sand, which is dependent on the size of the sand grain (Chapter 2). The distance S is dependent on the wave height, for it seems reasonable that larger waves would lift sand to greater distances from the bottom than smaller waves. Therefore, Dean assumed that $S = \beta H_b$, where H_b is the breaking wave height and β is a constant. Now, under a breaking wave crest, the wave particle velocities are directed onshore, and the resulting net displacement is onshore for half of a wave period. Therefore, if the fall time of a sand particle is less than $T/2$, where T is the wave period, the sand grain should move onshore. Alternatively, if the fall time is greater the $T/2$, the sand particle would be carried offshore. Therefore, for onshore motion,

$$\frac{\beta H_b}{w} < \frac{T}{2}, \tag{8.49}$$

or, rearranging,

$$\frac{H_b}{wT} = D < \frac{1}{2\beta}, \tag{8.50}$$

where D is the Dean number. Dean's original formulation introduced the deep water wave steepness by dividing Eq. (8.49) by the deepwater wave length $L_0 = gT^2/(2\pi)$. Then, he compared the resulting expression with data from wave tank tests that had resulted in either storm bars or normal beach profiles. The results of the comparison led him to the conclusion that

$$\frac{H_0}{L_0} = 1.7\frac{\pi w}{gT} \tag{8.51}$$

is the dividing line between storm and normal beach profiles. This expression can be rewritten in terms of $\mathbf{D}$ as $\mathbf{D}_0 = 0.85$. For wave steepness greater than that in Eq. (8.51), storm beach profiles result.

From Dean's results, it is clear that the sand size (through the fall velocity) plays a major role in the formation of the different seasonal profiles as does the wave period. Beaches with finer sand (and smaller fall velocities) require a smaller value of wave steepness for the formation of a storm profile. Also, from the comparison with the data, β was found to be about 0.3.

Kriebel, Dally, and Dean (1986) carried out several laboratory tests of beach profiles and reevaluated Dean's 1973 criterion. They found, using more full-scale tests, that there is a difference between laboratory tests (used by Dean) and full-scale tests (those of Saville 1957 and Kajima et al. 1982) owing to scale effects in the models. From the new study, the following relationship was found:

$$\frac{H_0}{L_0} = c_1 \frac{\pi w}{gT} \tag{8.52}$$

and $\mathbf{D}_0 = c_2$, where the constants fall in the following ranges: $4 < c_1 < 5$, and $2 < c_2 < 2.5$. The low values appeared to be valid for the small-scale laboratory tests and the large values for the full-scale tests (and for natural beaches). The lines A and B in Figure 8.14 show the relationships.

Kraus and Larson (1988) examined large wave tank data and found that the separation between barred and nonbarred profiles was distinguished by the curve

$$\frac{H_0}{L_0} = 0.00070 \left(\frac{H_0}{wT} \right)^3 \tag{8.53}$$

For a given deep water wave steepness, if the Dean number $\mathbf{D}$ exceeds that given by this equation, then the beach is in a storm profile.

Dalrymple (1992), utilizing Kraus and Larson's results, showed that the Froude number, $\mathbf{F_r} = w/\sqrt{gH_0}$, is also an important parameter for determining the presence of sandbars. He found that a single parameter, the so-called profile parameter $\mathbf{P}$

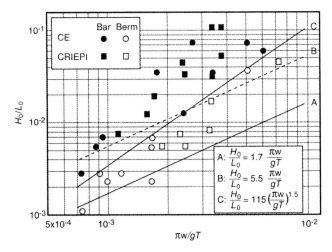

Figure 8.14 Criteria for the generation of storm or normal profiles (from Kraus and Larson 1988).

could be used to tell when bars would be present. The empirical relationship is

$$\mathbf{P} = \frac{\mathbf{D}}{\mathbf{F}_r^2} = \frac{g H_0^2}{w^3 T} = 10,400 \tag{8.54}$$

If $\mathbf{P}$ is significantly greater than this value, then bars are present in the laboratory. Kraus and Mason (1993) showed that this expression works for the field as well when the monochromatic (mean) wave height is converted to the deep-water significant wave height:

$$\mathbf{P}_s = \frac{g H_{s,0}^2}{w^3 T} = 26,500. \tag{8.55}$$

8.4.2 SIMPLE CROSS-SHORE TRANSPORT MODEL

This model was first proposed by Moore (1982) and later modified by Kriebel (1982) and Kriebel and Dean (1985). The basic concept is that, for a uniform sand size across the profile and an equilibrium beach, there is a constant energy dissipation rate per unit volume just as we assumed to obtain the $A y^{2/3}$ profile. If the beach profile differs from this equilibrium form, then the dissipation rate across the surf zone must differ from the constant value. Therefore, it is assumed that the amount of sediment moved will be dependent on the difference between the actual energy dissipation rate and that for an equilibrium profile $\mathcal{D}_*$:

$$q_s = K(\mathcal{D} - \mathcal{D}_*), \tag{8.56}$$

where q_s is the volumetric cross-shore sediment transport rate per unit width in the offshore direction and K is a new dimensional constant. The equilibrium energy dissipation rate per unit volume depends on the sand grain size and the equilibrium profile scale factor A, for

$$A = \left(\frac{24 \mathcal{D}_*}{5 \rho g \kappa^2 \sqrt{g}} \right)^{\frac{2}{3}},$$

as given in Eq. (7.4).

   If $\mathcal{D}$ is greater than the equilibrium value $\mathcal{D}_*$, there is a greater turbulence level in the surf zone than that for the equilibrium profile. Destructive forces then are able to destabilize the sediment and the quantity in Eq. (8.56) is positive, where we will define positive sediment transport in the offshore direction. On the other hand, for values of $\mathcal{D}$ less than the equilibrium value, onshore transport will occur. The value of $\mathcal{D}$ can be obtained from Eq. (7.2), that is,

$$\mathcal{D}_* = \frac{5}{16} \rho g \kappa^2 \sqrt{gh} \frac{dh}{dy},$$

which is dependent on the water depth (but only weakly because of the square root) and the local bottom slope, which has a stronger effect. If the bottom slope is greater than the equilibrium value, then $\mathcal{D}$ will be greater than the equilibrium value $\mathcal{D}_*$ and there will be offshore transport. Alternatively, if the bottom slope is milder than

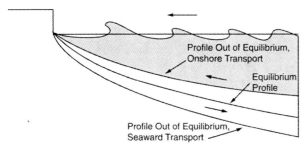

Figure 8.15 Cross-shore transport based on equilibrium dissipation model.

the equilibrium value, then there will be onshore transport. This is illustrated in Figure 8.15.

Next, a conservation of sand argument is needed to be able to examine profile change. For a given unit area of beach, the volume of sand moving in the offshore direction per unit time into the area is $q_s(y)$. The volumetric rate at which sand flows out of the area, which is dy long in the offshore direction, is $q_s(y + dy)$, which may be different owing to a different wave regime or beach conditions. The amounts of sand entering the area and leaving the area must then be balanced by the accumulation or loss of volume per unit time in the area. For a beach profile, a loss of sand volume per unit time is an increase of depth h with time multiplied by the cross-sectional area. Proceeding, we use a Taylor series expansion on the transport rate and set it equal to the volume change:

$$q_s(y + dy) - g_s(y) = \left[q_s(y) + \frac{\partial q_s}{\partial y} dy + \cdots \right] - q_s(y) = \frac{\partial h}{\partial t} dy$$

Therefore the one-dimensional conservation of sand argument leads to

$$\frac{\partial h}{\partial t} = \frac{\partial q_s}{\partial y} \tag{8.57}$$

Coupling these two equations (8.56 and 8.57) provides a means to predict profile evolution, which Moore (1982) did, yielding a reasonably constant value for K ($K = 2.2 \times 10^{-6}$ m^4/N) when comparing with laboratory data.

Dean and Zheng (1994) and Zheng and Dean (1997) have noted that various laboratory tests of cross-shore sediment transport demonstrate apparently widely different time constants. For example, Dette and Uliczka (1987) suggested an equilibration time scale of tens of hours, whereas the results of Swart (1974) suggested an equilibration time of thousands of hours. The problem of cross-shore sediment transport relationships was approached by considering Froude scaling relationships (discussed in the next chapter) from which it may be shown readily that

$$q_r = L_r^{\frac{3}{2}}, \tag{8.58}$$

in which the subscript r indicates a ratio of model to prototype values, L is a generic length scale, and the time ratio scales as the square root of the length ratio from Froude scaling. If K is considered to be a constant, the scaling ratio associated

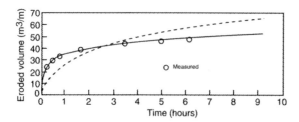

Figure 8.16 Comparison of linear ($n = 1$; dashed line) and nonlinear ($n = 3$; solid line; Eq. (8.60) transport rates for dune without foreshore data of Dette and Uliczka (1987) (from Zheng and Dean 1997).

with Eq. (8.56) can be shown to be

$$q_r = L_r^{\frac{1}{2}} \tag{8.59}$$

One possibility of an improved transport relationship is then

$$q = K(\mathcal{D} - \mathcal{D}_*)^3, \tag{8.60}$$

which scales properly. This nonlinear transport relationship can explain the different time scales discussed earlier. In general, this transport relationship provided much better agreement, an example of which is shown in Figure 8.16, which presents a comparison with the data from Dette and Uliczka (1987).

The disequilibrium model provides a macroscale look at the erosion of a beach profile but does not examine the details of the transport within the profile. Next we will look more at the details of the cross-shore modes of transport.

8.4.3 A TRACTION MODEL FOR CROSS-SHORE TRANSPORT

Here we will examine models that are more process based. The Madsen and Grant model discussed previously is also applicable to cross-shore sediment transport owing to the vector nature of the dimensionless transport ϕ. Watanabe (1982) has proposed a different form of the transport than Eq. (8.45) for onshore transport based on laboratory data obtained in a wave tank,

$$\phi = 7(\Psi_m - \Psi_c)\Psi_m^{\frac{1}{2}}, \tag{8.61}$$

where Ψ_c is the critical Shields parameter. This equation is very similar to the Grant and Madsen result for most of the available data; however, for very small and large values of Ψ_m, the equations differ, as shown in Figure 8.17.

Trowbridge and Young (1989) also used a traction model for the cross-shore sediment transport offshore of the breaker line, but related ϕ linearly to the Shields parameter, rather than the cube as in Madsen and Grant, based on comparison to the laboratory data of Horikawa, Watanabe, and Katori (1982).

$$\overline{\phi(t)} = \overline{\Psi(t)}$$

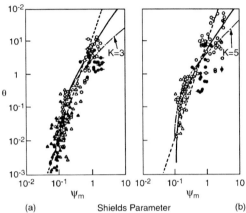

Figure 8.17 Comparison of sediment transport equations with data (from Shimizu et al. 1985). (a) Onshore transport; (b) offshore transport.

The immediate drawback of this scheme is that the averaged bottom shear stress, using first-order wave theory, yields a zero mean bottom shear stress. Therefore, they developed a scheme based on boundary layer theory that shows that the bottom shear stress can be related to

$$\bar{\tau}_b = -\rho \frac{f}{2} \frac{\overline{U^2|U|}}{\sqrt{gh}}$$

Utilizing long wave representations for the wave-induced velocity, they obtained

$$\bar{\phi} = -\sqrt{\frac{2}{\pi}} K \rho f \frac{w}{(\rho_s - \rho)} g^{-\frac{3}{4}} \mathcal{F}^{\frac{3}{2}} h^{\frac{11}{4}} \tag{8.62}$$

The energy flux of the waves appears in the $\mathcal{F}$ parameter defined in Chapter 9. The sediment transport is always onshore with this model, which is concerned only with the bedload (sheet flow) transport. Field comparisons with the migration of a sandbar observed at the Field Research Facility of the Corps of Engineers were very encouraging.

8.4.4 ENERGETICS MODELS

Stive (1986) and Roelvink and Stive (1989) examined the use of Bailard's sediment transport relationship (Eq. (8.29)) for cross-shore flows. Stive successfully examined the evolution of offshore shoals that were a result of wave action. Roelvink and Stive examined Bailard's equation coupled with a conservation of sand equation (Eq. (8.57)) to model the behavior of beach profiles. They used a random wave-breaking model (Battjes and Janssen) and included wave nonlinearity, wave grouping, undertow, and wave-induced turbulence. They concluded that their model predicted the wave hydrodynamics well but that the Bailard formulation was inadequate for regions outside the surf zone probably owing to the strong vertical variation of the flow. They were able to create sandbars with their model but not in the same locations as the laboratory results.

8.4.5 RIPPLE MODELS

Offshore of the breaker line the bed is often rippled, and the ripples affect the sediment transport in this region. As the wave-induced motion increases over a ripple, the flow separates from the crest of the ripple and forms a vortex in the trough before the next ripple. This vortex captures and carries sediment. As the flow decelerates and then reverses, this vortex rises and then is entrained in the flow. It then begins to decay owing to viscous effects. The reduction in vortex strength permits sand to fall from the vortex, and thus the sand is primarily carried in the reverse flow direction. Only when the flow is symmetric is the transport the same in both directions. Nielsen (1986), discussing simple models for suspended transport, indicates that the transport primarily moves in a direction opposite to the principal wave direction. In fact, with waves and currents in the same direction, it is still possible for sediment to move upstream against the current.

Ripples occur over a range of length scales, depending on water particle motions and sand sizes. Typical heights and lengths are on the order of centimeters and decimeters, respectively. Ripples can also have symmetric or asymmetric profiles. It is commonly believed that asymmetric ripples migrate in the direction of the steeper face, which is also in the direction of wave propagation (Evans 1941).

Field studies by Dingler and Inman (1976) and Jette (1997) found that the mobility number is significant to the dynamics of ripples. The *mobility number* Ψ_m is the Shields parameter (Eq. (8.6)) with the near-bottom orbital velocity replacing the shear velocity,

$$\Psi_m = \frac{(A\sigma)^2}{gd(s-1)},$$

where A^* is one-half the near-bottom water particle excursion, and σ is the angular frequency of the wave. (The advantage of this formulation of the Shields parameter is that the bottom shear stress, with its associated bottom friction factor, does not need to be calculated.) When the mobility number exceeds 150, ripples tend to be obliterated, whereas for $\Psi_m < 50$, ripples form very slowly. For mobility numbers between 50 and 100, ripples can form very rapidly within several wave periods.

Bagnold (1936) classified ripples into two categories. Those steep enough to cause flow separation and vortex shedding were termed "vortex ripples," whereas milder steepness ripples were "rolling grain ripples." Clifton (1976) provided another scheme, separating ripples into "orbital ripples," which have a length proportional to the near-bottom water particle excursion, and shorter ripples are described as being "anorbital." An intermediate class is referred to as "suborbital."

---

* Defined in Section 5.2.

Several empirical models for ripple geometry have been developed based on field and laboratory observations. Nielsen (1981) defined ripple characteristics in terms of the mobility number. Nielson's equation for the ripple height η in nondimensional form is

$$\frac{\eta}{A} = \begin{cases} 21\Psi_m^{-1.85}, & \Psi_m > 10 \\ 0.275 - 0.22\Psi_m^{0.5}, & \Psi_m < 10, \end{cases} \tag{8.63}$$

and the equation for the ripple wave length λ is

$$\frac{\lambda}{A} = \exp\left(\frac{693 - 0.37\ln^8\Psi_m}{1000 + 0.75\ln^8\Psi_m}\right) \tag{8.64}$$

Wiberg and Harris (1994) have another classification of ripples based on the orbital diameter $2A$.

Considerable advances in field instrumentation have resulted in a rapid increase in the availability of data about ripples. Downward-aimed acoustic transducers (Dingler and Inman 1976; Greenwood, Richards, and Brander 1993) were the first sophisticated bottom measurement devices introduced into the nearshore zone. More recently, rotating side-scan sonars (Hay and Wilson 1994) have shown the time-dependent evolution of ripples with changing wave energy. Further Hay, Craig, and Wilson (1996) have shown the three-dimensional nature of ripples with lunate ripples migrating in the direction of their horns at speeds between 0.006 and 0.045 cm/s. Jette (1997) has introduced the multiple transducer array, which is a rack of ultrasonic transducers mounted in a linear array located approximately 40 cm above the bottom.

8.5 LITTORAL DRIFT APPLICATIONS

Estimates for net annual littoral transport exist for many coastlines. These are often obtained from dredging records and sediment impoundment at coastal structures and may be roughly correct, or catastrophically wrong, depending on how the data were obtained. Often we do not even have coastal wave data to provide us with estimates of the annual wave climate. Thus, we often have to take refuge in offshore wave data, either measured or hindcast based on weather data, which are more abundant and more detailed. Estimates of shallow water directional wave climates have been developed by the U.S. Army Corps of Engineers through their Wave Information Study (WIS) based on hindcasts from meteorological information. Their estimates, based on more than 20 years of data, are available in 10 m of water (based on linear shoaling and refraction) for the Atlantic, Pacific, and Gulf of Mexico coastlines as well as those of the Great Lakes; see, for example, Jensen (1983). This database is continuing to expand.

The transport formulas developed in this chapter, specifically Eq. (8.16), can be coupled with an annual compilation of wave data for a particular site to provide annual littoral transport estimates. Ideally, wave data, height, period, and direction are available in, say, 3-hour intervals for the entire year over a span of many years. As shown in the next section, if the data are from offshore, simple shoaling and refraction schemes are used to bring the data to the shoreline.

8.5.1 LITTORAL DRIFT COMPUTATIONS BASED ON DEEP WATER DATA

The longshore sediment transport Q was expressed in Eq. (8.16) as

$$Q = \frac{K(EC_g \cos\theta \sin\theta)_b}{\rho g(s-1)(1-p)}, \tag{8.65}$$

where s is the specific gravity of the sediment, $s = \rho_s/\rho$. Note that the terms in the numerator are referenced to breaking conditions denoted by the subscript $_b$. If the bathymetry is regarded as straight and parallel and the energy losses from deep water are neglected, it is possible to express the transport solely (well, almost) in terms of deep water wave conditions. By rewriting Eq. (8.65) as

$$Q = \frac{K(EC_g \cos\theta)_b C_b}{\rho g(s-1)(1-p)} \left(\frac{\sin\theta_b}{C_b}\right), \tag{8.66}$$

it is apparent that the conservation of energy and Snell's law can be used,

$$(EC_g \cos\theta)_b = (EC_g \cos\theta)_0 \tag{8.67}$$

$$\left(\frac{\sin\theta}{C}\right)_b = \left(\frac{\sin\theta}{C}\right)_0, \tag{8.68}$$

where the subscript $_0$ denotes deep water conditions. To express the remaining C_b in the numerator in terms of deep water conditions, we use Eq. (8.67) with the shallow water asymptotes $C_b = \sqrt{gh_b}$, $H_b = \kappa h_b$ and with the deep water asymptote $L_0 = gT^2/2\pi$ to obtain h_b. Then, putting this breaking water depth into the definition of C_b gives us,

$$C_b = \left(\frac{H_0^2 T \cos\theta_0 g^3}{4\pi \cos\theta_b \kappa^2}\right)^{0.2}$$

Now, using this expression, Eq. (8.67), and Eq. (8.68) in Eq. (8.66) gives the desired result,

$$Q = \frac{K H_0^{2.4} g^{0.6} T^{0.2} \cos^{1.2}\theta_0 \sin\theta_0}{8(s-1)(1-p)2^{1.4}\pi^{0.2}\kappa^{0.4} \cos^{0.2}\theta_b} \tag{8.69}$$

This equation has the advantage that deep water parameters only (except for θ_b) are used to determine Q. Because wave refraction reduces θ_b to a small value at the breakerline and because of the small exponent on $\cos\theta_b$, it is possible to approximate $\cos^{0.2}\theta_b$ by unity, thereby leaving Q entirely in terms of deep water conditions. However, recall that energy losses have been neglected by assumption.

8.5.1.1 Littoral Drift Variability

Ideally, many years of wave data are available to provide statistical reliability to the estimates of the drift. Mann and Dalrymple (1986) made littoral drift calculations for the Atlantic shoreline of Delaware using 20 years of hindcast wave data from the CERC Wave Information Study. Their estimates showed that the 20-year mean of the calculations corresponded to previous estimates of the mean drift rate; however, the annual variations were significant. The standard deviation of the annual drift rate was as large as the mean. This is likely to be true almost everywhere owing to the annual variability in wave climate. This large annual variability implies that coastal designs must be extremely resilient to variations in littoral transport. Designs based on a fixed transport rate may fail if they do not have the capacity to deal with excessive (or minimal) transport.

8.5.2 LITTORAL DRIFT ROSE

An interesting tool for studying a section of shoreline that has a variety of shoreline planform orientations is the littoral drift rose (Walton and Dean 1973), for it shows how the littoral drift changes with the shoreline orientation (because the angle of incidence of the waves relative to the shoreline will change).

The drift rose is constructed using Eq. (8.16) and the available wave data for the site of interest. Let us assume that only deep water wave data are available. These data are totally independent of the shoreline orientation; therefore, refraction of the waves is needed to determine the local breaking angle and height. (For a realistic coastline, a computer model will be necessary for this calculation.) The principle behind the littoral drift rose is that a range of shoreline orientations is considered. These orientations correspond to the range that exists at the site of interest. For each possible shoreline orientation, the annual (or monthly) littoral positive and negative drifts along the shoreline are calculated. These then are plotted in a polar plot, as shown in Figure 8.18 for the St. Augustine, Florida, area. By convention, a solid line

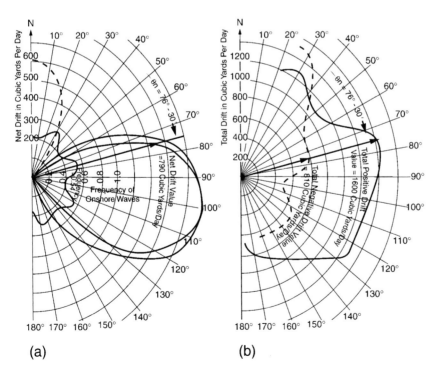

(a)   (b)

Figure 8.18 Littoral drift rose for St. John's River to St. Augustine Inlet (Walton and Dean, 1973). (a) Net transport; (b) transport components.

denotes positive transport (to the right looking offshore), and the negative drift is shown with a dashed line. From the plots it is convenient to determine the littoral transport rate for any given shoreline orientation by drawing the shoreline normal on the littoral drift rose and determining the positive and negative drift for the site. In the figure, the shoreline orientation for Ponte Vedra Beach, Florida, is shown. The $(+)$ southerly drift is 1600 yd^3/day, and the northerly drift is 810 yd^3/day, which yields a net drift of 790 yd^3/day. The net annual drift is then 288,000 yd^3/yr at this site.

While constructing the littoral drift roses as described above, Walton and Dean (1973) noted that the net littoral drift roses were similar in shape with symmetric lobes, as shown by the solid line in Figure 8.18(a). With reference to Eq. (8.69), it can be shown for a single wave train that the sediment transport can be expressed as

$$Q = \frac{K H_0^{2.4} g^{0.6} T^{0.2} \sin 2 \cos \theta_0}{16(s-1)(1-p)2^{1.4}\pi^{0.2}\kappa^{0.4}} \left(\frac{\cos \theta_0}{\cos \theta_b} \right)^{0.2} = \hat{Q} \sin 2\theta_0 \left(\frac{\cos \theta_0}{\cos \theta_b} \right)^{0.2} \tag{8.70}$$

Now, if we approximate the fraction involving the ratio of the cosines by unity, we see that Q varies with $\sin 2\theta_0$, which, in polar form, looks like that for the net littoral drift rose. Moreover, if we consider the more realistic case of the superposition of many waves of varying height originating from many directions, when expressing the wave angle of incidence in terms of a given shoreline orientation β, $\theta_0 = \beta - \alpha_0$, we can sum the littoral transport contributions from each wave train.

$$Q_N(\beta) = \sum_{n=1}^{N} \hat{Q}_n \sin 2(\beta - \alpha_n), \tag{8.71}$$

which also can be rewritten

$$Q_N = \sin 2\beta \sum_{n=1}^{N} \hat{Q}_n \cos 2\alpha_n - \cos 2\beta \sum_{n=1}^{N} \hat{Q}_n \sin 2\alpha_n = Q_T \sin 2(\beta - \alpha_T),$$

where

$$Q_T = \sqrt{ \left(\sum_{n=1}^{N} \hat{Q}_n \cos 2\alpha_n \right)^2 + \left(\sum_{n=1}^{N} \hat{Q}_n \sin 2\alpha_n \right)^2 } \tag{8.72}$$

$$\alpha_T = \frac{1}{2} \tan^{-1} \left(\frac{\sum_{n=1}^{N} \hat{Q}_n \sin 2\alpha_n}{\sum_{n=1}^{N} \hat{Q}_n \cos 2\alpha_n} \right) \tag{8.73}$$

We can therefore expect realistic net littoral drift roses to vary with shoreline orientation as $\sin 2\theta_0$. In the above equations, the effective deep water wave direction is α_T, that is, a wave with this direction would cause no transport for a beach aligned perpendicular to it. The reader is invited to determine an expression for the wave height associated with α_T.

Several simple applications of the littoral drift roses follow.

8.5.2.1 Change in Shoreline Orientation at a Source or Sink

Considering Figure 8.18, which applies for the St. Augustine, Florida, area, we see that the net transport depends on the shoreline orientation. Conversely, in locations where the transport is limited, the shoreline orientation will adjust over many decades to the sediment supply. If a localized source or sink of sand occurs, it follows that there will be an abrupt discontinuity in the shoreline orientation.

River deltas are the most obvious example of a discontinuity in sediment supply causing an abrupt change in shoreline orientation. For waves approaching normal to the ambient shoreline, one-half the sediment supply delivered by the river to the coast is carried to the left and one-half to the right. Thus, the shoreline orientation on either side would be reflective of this transport, and the difference in orientation would represent the total transport of sand to the shoreline.

Submarine canyons or inlets can act as sinks, thereby reducing the sediment supply to the downdrift shorelines, which would also have different orientations. Figure 8.19 shows the long-term orientation of the shoreline in the vicinity of Port

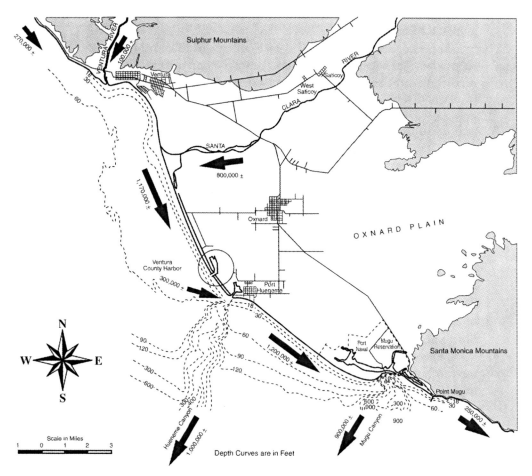

Figure 8.19 Sediment transport rates (yd^3/yr) in the vicinity of Port Hueneme, California (from Herron and Harris 1966).

Hueneme, California, where the net longshore sediment transport at Oxnard Shores is known to be approximately 750,000 m³/yr to the southeast. A submarine canyon forms an offshore extension of Port Hueneme and is known to reduce the sediment transport to the downdrift shoreline.* Problem 8.6 at the end of the chapter provides a hypothetical example for channel entrances.

8.5.2.2 Null Point and Equilibrium Planforms

For a long shoreline with variations of shoreline orientation, it is possible that there is a null point (or a nodal point) in the littoral transport. This location occurs at the shoreline angle for which the positive and negative littoral transports have the same magnitude, yielding no net transport. Along the U.S. East Coast, there are several of these null points such as just north of the Delaware–Maryland state boundaries and in New Jersey near Barnegat Inlet. In both instances, the shoreline orientation changes significantly near these locations, and thus the net drift is to the south of these sites and to the north, north of the site of the null point. Mann and Dalrymple (1986) examined the location of the Delaware null point and concluded that there was a significant variation in its annual location corresponding to the variation in the annual littoral drift rates.

For pocket beaches, bounded at each end by a headland or other type of structure, such as beaches between large groins, the waves tend to orient the beach such that the waves attack the beach normally. For straight sections of beach, this means that it will be oriented so that the beach normal corresponds to the null point angle. This can be predicted in advance (Chapter 11), providing design information in the construction of these beaches.

8.5.2.3 Stable and Unstable Shorelines

Walton and Dean (1973) examined the stability of shorelines using the littoral drift rose. Figure 8.20 shows an unstable littoral drift rose and the island orientation. Applying this rose to the island in the figure, you can see that if a negative perturbation (recession) is caused, the induced transport is away from the perturbation, resulting in its growth (hence the term "unstable"). Likewise, inspection of Figure 8.20 will demonstrate that the induced transport response to a positive perturbation is toward the perturbation, causing it to grow. Therefore, a shoreline with a littoral drift rose of the type in Figure 8.20 is unstable with respect to positive and negative perturbations. The unstable littoral drift rose *may* be responsible for several features observed along the shoreline. First, many barrier islands along the southwest coast of Florida tend to have a "dogbone" shape with the central portion exhibiting an erosional tendency and the ends tending to grow. The associated littoral drift roses along the southwest coast of Florida tend to be unstable. Also, the cuspate features, discussed at the end of Chapter 3, may be the result of an unstable littoral drift rose. In summary, these features most often occur along the shorelines of elongate water bodies, where

---

* From knowledge of the littoral drift at Port Hueneme and the effect of the submarine canyon acting as a sand sink, try sketching an approximate net littoral drift rose for this area.

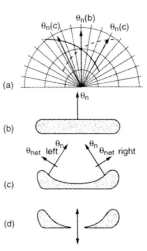

Figure 8.20 Hypothetical littoral drift rose for an unstable barrier island (Walton and Dean 1973). (a) Type I null point; (b) island oriented toward null point with zero net drift; (c) perturbation in system causes concave orientation of the island with associated drift pattern as shown; (d) instability leads to eventual breakthrough.

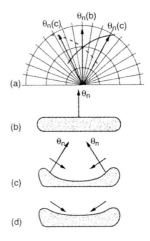

Figure 8.21 Hypothetical littoral drift rose for a stable barrier island (Walton and Dean 1973). (a) Type II null point; (b) island oriented toward null point with zero net drift; (c) perturbation in system causes concave orientation with associated drift pattern as shown; (d) self-stabilizing drift pattern restores island to original configuration.

it can be shown that unstable littoral drift roses would be likely to occur. Figure 8.21 shows a stable system.

What causes an unstable littoral drift rose? Figure 8.22 presents the littoral transport as a function of wave approach direction with a maximum occurring at $45°$ because of the $\sin 2\theta$ factor. Considering waves approaching at angles to the shoreline at less than $45°$, we see that as the angle of incidence increases, the littoral drift increases. For angles exceeding $45°$, as the angle increases, the transport decreases.[*]

Another possibility is that the littoral drift rose is as shown in Figure 8.21 . In this case the null point is stable because perturbations now result in transport that tends to reduce rather than accentuate the perturbation. In this case, there is no similar mechanism for island breaching. The east coast of Florida is characterized by this type of littoral drift rose, and in fact there are relatively few natural inlets along this coast.

---

[*] For elongated lagoons, with their associated fetch limitations, wave climates of $|\theta| > 45°$ would tend to occur. This could explain the growth of cuspate features on their (long) shorelines.

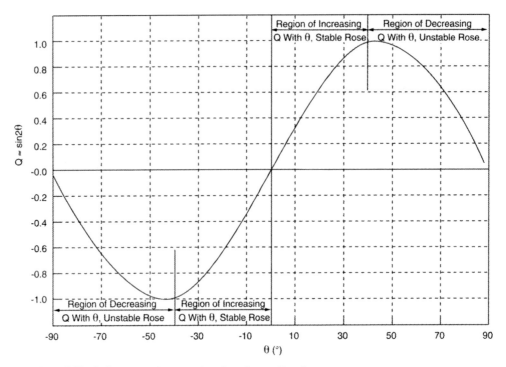

Figure 8.22 Sediment transport as a function of wave direction.

8.6 OVERWASH AND WASHOVER

Overwash is a process by which sand is transported landward from the nearshore and beach area and deposited in a layer or "plaque" of sediments on the prestorm surface. Overwash occurs during storms that include conditions of elevated water levels and energetic wave conditions. The minimum combination of storm surge and wave run-up must exceed the land level for overwash to occur. The deposit of sand that occurs as a result of overwash is called "washover." Overwash is the process (like overtopping), and washover is the associated deposit (Schwartz 1975). An excellent annotated bibliography on barrier island migration, which includes the contributions of washover deposits, appears in Leatherman (1981(a)). A second valuable reference is the collection of thirty-two papers, most of which address overwash, in Leatherman (1981(b)).

Washover deposits can be identified immediately after an overwash event as a sheet of sand completely devoid of vegetation and with a reasonably abrupt and distinct landward edge of the deposit. The deposit appears to develop progressively, as schematized in Figure 8.23, with the deposit extended by pulses of sediment transport associated with wave groups, water level fluctuations, and surges that are somewhat higher than the average. In many cases, the washover deposits can extend completely

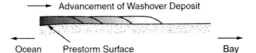

Figure 8.23 Illustration of the advancement of washover deposits as pulses.

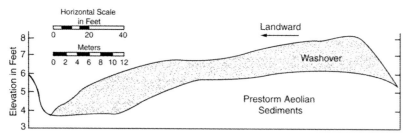

Figure 8.24   Cross section through overwash deposit on Pea Island, North Carolina (from Schwartz 1975).

to, and past, the bay shoreline, thus representing a nourishment there. Figure 8.24 presents a cross section through a washover deposit on Pea Island, North Carolina (Schwartz 1975). With time, the visual evidence of overwash becomes less and less apparent. Vegetation can rapidly colonize an overwash deposit and, during the vegetation process, aeolian processes can reshape the flat surface into undulating dunes, removing the fine fraction and leaving a "lag" layer of the more coarse material. The evidence of an overwash deposit that remains can be found by excavating or augering down through the deposit to an interface that may be evident as a lag layer or a decayed organic layer left by the buried vegetation. Washover deposits can occur as sheet deposits that are relatively uniform along significant segments of shorelines or as fan-shaped deposits through gaps in dunes.

Under natural conditions, overwash deposits would be left where formed, soon become vegetated, and, in conjunction with aeolian deposits, contribute to the upward migration of the barrier island. Also, the barrier island would retreat in response to the loss of the sand from the nearshore active system until, as retreat progresses, the overwash deposits would eventually be reincorporated back into the system. The three mechanisms of landward sediment transport are aeolian processes, inlet transport to flood tidal deltas, and overwash processes. Some investigators have found that inlet processes are more significant quantitatively (Fischer 1961, Pierce 1969, Armon 1975, Fisher and Simpson 1979), whereas Dillon (1970) has found that washover dominates in Rhode Island. In some areas, sand deposited as overwash may be transported back to the beach through aeolian processes (Leatherman 1976, Fisher and Stauble 1977). Consider two scenarios under completely opposite ends of the overwash management spectrum. In the first scenario, the overwash deposits are removed and placed back in the beach system, which is a common practice. The beach does not experience a loss of sand; however, the barrier island is not elevated to keep pace with the relative rise of sea level, and future storm events will overwash the island with increasing frequency. In the second scenario, the sand is removed from the system because it may contain a substantial amount of debris that is unsuitable for placement on the beaches. This debris is from damaged buildings and their contents, which can be strewn over extensive areas by the winds, tides, and waves. In this scenario, the beach system has lost sand and the land does not advance upward by the sand contained in the overwash deposit. This issue of the most appropriate long-term management of overwash deposits in developed areas has not received adequate attention (Godfrey and Godfrey 1973, Dolan 1972, Leatherman 1976). From short-term considerations, overwash volumes represent a loss of sand from the nearshore system, and this loss, unless compensated by beach nourishment, will result in shoreline recession.

Washover thicknesses can vary from less than 10 cm (McGowen and Scott 1975) to almost 2 m (Leatherman 1979) and can penetrate up to 1.6 km inland. Up to 100 m^3/m of dune breach can be deposited in a single storm event (Leatherman 1979), although values of 20 m^3/m to 40 m^3/m are much more common. It is clear that overwash processes can result in large quantities of landward transport, especially during the more severe storms. Laboratory efforts to examine the mechanics of overwash processes have been carried out by Williams (1978).

The surf zone hydrodynamics associated with overwash extending into bay systems differ from those of nonoverwash events. Overwash into bays reduces or negates the requirement for return flows that usually occur as undertow or as rip currents (Chapter 5). The lack of, or reduction of, undertow reduces the seaward forces, resulting in a tendency for greater landward sediment transport.

8.7 AEOLIAN SEDIMENT TRANSPORT

The process of moving sand by wind is termed *aeolian transport* and can be a significant element in coastal sediment budgets. Aeolian transport can remove sand from beach systems and move it into large dunes that can either remain stationary or migrate a considerable distance inland. In some cases, migrating sand dunes can pose a hazard by threatening burial of buildings, highways, roads, and even villages. In the more normal case, aeolian transport is the mode by which sand is supplied to dunes subsequent to their erosion after a storm. As depicted in Figure 8.25, during

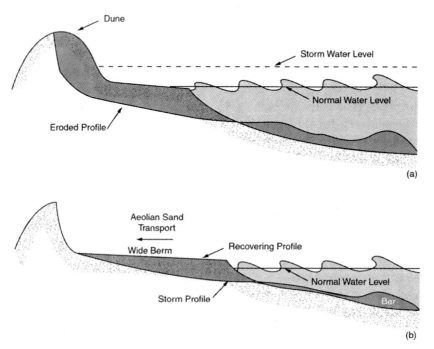

Figure 8.25 Dune erosion and profile recovery with wide berm leading to enhanced aeolian transport. (a) Dune erosion during elevated water levels and high waves; (b) profile recovery with wide berm.

a storm, the elevated water levels and energetic waves erode significant volumes of sand from dunes and transport this sand seaward, where it is stored as a bar. (This process is somewhat self-limiting, for the bar acts to cause wave breaking and energy dissipation, and thus the waves will have less erosive power when they reach the dune.) After the passage of the storm and the onset of normal wave conditions, the sand in the bar is gradually moved landward and stored temporarily as a wide berm. This berm will tend to be less moist near its landward extremity, and the sand there is most readily transported by wind. The average winds on most beaches are landward owing to sea breezes caused by the differential heating of the land during the day and the associated expansion of the air over the land. This warm air then rises, causing a convection cell in which air flows landward in the lower layers and seaward in the upper layers of the cell. Additionally, during storms there may be strong onshore winds that transport sand landward across the berm. This sand will be trapped by dune vegetation, resulting in the eventual recovery of the dune. If the sand is not trapped by vegetation, it may be blown so far inland that no recycling with the beach system will occur within human lifetime scales. Dune recovery is enhanced by strong onshore winds, a wide berm that provides the necessary sand, a low moisture content within the berm that makes the sand more transportable, and fine-grained sands that are also more transportable by wind. Aeolian transport will also selectively remove the finer fraction of the material residing in the berm, leaving a so-called lag layer – a surface layer of coarser particles.

Here, we present some of the quantitative relationships that have been developed for prediction of aeolian transport. The relationships have been developed and tested under laboratory and field conditions. Some of the field arrangements have included trenches in the beaches serving as traps (Kubota, Horikawa, and Hotta 1982) or sand fences (Sherman et al. 1996). Most of the laboratory studies have been conducted in wind tunnels (Belly 1964, Hotta et al. 1984). Vanoni (1975) has provided a valuable review of aeolian transport.

8.7.1 WIND CHARACTERISTICS

Most aeolian sand transport relationships are expressed in terms of the shear velocity u_*, which is defined as $u_* = \sqrt{\tau/\rho_a}$, where τ is the shear stress acting on the sand and ρ_a is the density of air. The wind velocity profile over a stationary sand surface is described by the logarithmic relationship

$$u(z) = \frac{u_*}{\kappa} \ln\left(\frac{z}{z_0}\right), \tag{8.74}$$

in which κ is the von Kármán constant (≈ 0.4) and z_0 is the reference height at which the velocity is zero and is recommended by Zingg (1953) as

$$z_0 = 0.035 \ln\left(\frac{d}{0.18}\right),$$

where d is the representative sediment diameter in millimeters. Thus, if the reference wind velocity is known (usually at a standard elevation of 10 m above the surface), the shear velocity can be calculated by knowing the sediment size. If the sand is moving (sand transport is occurring), the wind profile is modified and is given by

$$u = \frac{u_*}{\kappa} \ln\left(\frac{z}{z'}\right) + u',$$ (8.75)

where Zingg (1953) recommends a modified value of the von Kármán constant κ of 0.375, $z' = 10d$, and $u' = 894d$, where d is in millimeters and u' is in centimeters per second.

8.7.2 CRITICAL SHEAR VELOCITY, u_{*c}

Some of the equations proposed for sediment transport include a critical shear velocity or threshold shear velocity, whereas others do not. The rationale is that a minimum (threshold) shear velocity u_{*c} is required to cause any sediment movement.

Bagnold (1943) recommended the critical shear velocity u_{*c}

$$u_{*c} = A \sqrt{\frac{\rho_s - \rho_a}{\rho_a} g d}$$

in which ρ_s and ρ_a are the mass densities of sand and air, respectively, and A was taken as 0.085. Other investigators have recognized the effects of sand moisture on the critical shear velocity. Belly (1964) conducted a series of laboratory tests and recommended the following factor be applied to the Bagnold equation for critical shear velocity to account for the sediment moisture content c:

$$F_B = (1.8 + 0.6 \log_{10} c)$$ (8.76)

where c is the percent moisture content in the sediment by weight. Equation (8.76) shows that, if the moisture content is 0.046, the u_{*c} is unaffected, and that, for greater moisture content values, the threshold shear velocities are increased. On the basis of wind tunnel tests, Hotta et al. (1984) have recommended that the following factor be applied to the Bagnold equation:

$$F_H = \left(1.0 + 7.5 \frac{c}{u_{*c}}\right),$$

where $0.02 < c < 0.08$ and $0.2 \, \text{mm} < d < 0.8 \, \text{mm}$. Kawata and Tsuchiya (1976) considered the surface tension associated with moist sediment and developed the following factor:

$$F_K = \sqrt{1 + B},$$

where

$$B = \frac{2\sqrt{6}}{5} \sqrt{\alpha_1 \alpha_2} \sqrt{n_0} \sqrt{\frac{\rho_s}{\rho_w} \frac{T \sqrt{c} \cos \zeta}{(\rho_s - \rho_a) g d}}$$

and α_1 and α_2 are constants, n_0 is the number of grain contact points, ρ_s and ρ_w are the densities of sand and water, respectively, T is the surface tension, and ζ is the water grain contact angle.

8.7.3 TRANSPORT RELATIONSHIPS

A relationship for aeolian transport was proposed by Bagnold (1936, 1943) based on his studies of sand transport in wind tunnels:

$$q = C\sqrt{\frac{d}{d_{\text{ref}}}}\,\rho_a u_*^3$$

in which q is the sediment flux per unit width taken as mass per unit width per unit time, C is a constant ranging from 1.5 to 3.5, d is the mean grain diameter in millimeters, and d_{ref} is a reference sediment diameter and is taken to be 0.25 mm. It is somewhat surprising that in this relationship and in one other to be reviewed, the transport is greater for the larger sediment sizes. Later, Bagnold (1943) proposed a different equation based on his studies of desert sand storms in Egypt

$$q = 5.2 \times 10^{-4}(u - u_c)^3,$$

in which u is the wind velocity in centimeters at 1 m per second above the sand surface; u_c, the threshold velocity, is 400 cm/s; and q is in metric tons per hour per meter of width.

Kawamura (1951) proposed the following relationship:

$$q = C\frac{\rho_a}{g}\,(u_* - u_{*c})\,(u_* + u_{*c})^2,$$

with $C = 2.78$. Note that sediment size enters in this relationship through the critical shear velocity and that there would be greater transport for the smaller sediment sizes.

Lettau and Lettau (1977) proposed the slightly different form

$$q = C\frac{\rho_a}{g}\sqrt{\frac{d}{d_{\text{ref}}}}\,(u_* - u_{*c})u_*^2$$

with a recommended value of $C = 4.2$. This equation contains a mixed dependency with the terms under the square root resulting in greater transport for the coarser particles and the effect of u_{*c} causing greater transport for the finer particles.

O'Brien and Rindlaub (1936) conducted field studies and proposed the following aeolian transport formula that does not require the determination of the shear velocity:

$$G = 0.036u_5^3,$$

where G is the transport rate in pounds per day of sand per foot of beach width and u_5 is the wind velocity at 5 ft above the sand surface. It is clear that this equation must be applied in the units for which it was developed.

With two exceptions (the equation proposed by O'Brien and Rindlaub and the second Bagnold equation), all the transport equations are dimensionally homogeneous and can be applied in any consistent set of units.

8.8 ILLUSTRATIONS OF SAND TRANSPORT AND DUNE ACCUMULATION

After a major storm, a barrier island may be left with a barren, relatively flat surface. With a wide berm, aeolian sand transport can be substantial. Unless this sand is trapped to form a dune, it can be blown so far inland as to be lost to the active coastal system. To stabilize the blowing sand as dunes, either vegetation can be planted or sand fences can be constructed. The vegetation or sand fence elements transfer some of the shear stress directly to the ground, thus decreasing the shear stress available to act on the sand surface and causing the sand in transport to deposit. Sand fences perform much like snow fences, resulting in an approximately triangular cross-sectional deposit of sand with the peak at the sand fence. An advantage of vegetation is that, as it is buried, the stalks extend themselves upwards, whereas the sand fences need to be raised or more fencing added upon burial.

The rates at which dunes can increase in volume due to trapping of aeolian transported sand are impressive. Figure 8.26 presents a cross section showing dune growth as a result of grass planting on the Outer Banks of North Carolina. The average volumetric rate of dune growth from January 1965 to January 1967 was 3.6 yd³/ft/yr. Considerable field experimentation has been conducted on the most appropriate use of sand fencing in areas where survival of vegetation is uncertain or where vegetation may not be economical. Figure 8.27 from Savage (1962) shows similar growth of a dune as a result of sand fencing on the Core Banks.

8.9 COHESIVE SEDIMENTS

Cohesive sediments differ from the granular sediments we have discussed so far in two important respects. First, the strength characteristics of cohesive sediments change markedly with time after deposition, and secondly, their properties depend on their ion exchange potential. Cohesive sediments are of interest to the coastal

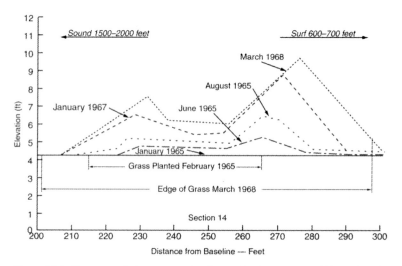

Figure 8.26 Dune growth due to grass planting on the Outer Banks of North Carolina (from Savage and Woodhouse 1965).

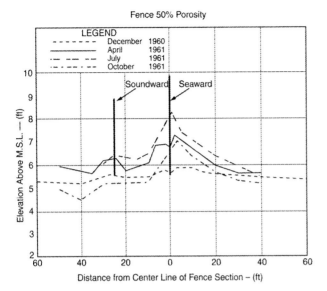

Figure 8.27 Dune growth due to sand fencing on Core Banks (from Savage 1962).

engineer because much of the required dredging in deepened channels is due to cohesive sediment deposition. In the following, our discussion of cohesive sediments will be relatively brief and will focus on the properties of such sediments and their critical shear stress and erosion potential. The cohesive nature of sediments usually becomes apparent at sizes less than approximately 0.074 mm (74 μm; $\phi > 3.76$). The critical shear stress for cohesive sediments is greater than would be anticipated from estimates based on the sediment size alone. The transport of cohesive sediments occurs largely in suspension, although aggregates on the bottom referred to as a fluid mud may move along the bottom as a viscous flow. For a particular shear stress, the erosion will occur down to the depth at which the critical shear strength corresponds to the applied shear stress.

Cohesive sediments are deposited out of suspension through flocculation and settlement to the bed. At this stage the sediment is very weak and the bulk density is very low. However, if the sediment is allowed to consolidate, its strength increases with time and depth into the sediment bed. Figure 8.28 presents the distribution with depth into the sediment bed of the critical shear stress of sediments that have been allow to consolidate out of a slurry for 12, 24, and 48 hours. Note that the total vertical elevation shown in Figure 8.28 is less than 1 cm.

Much of the available information regarding cohesive sediments has been determined through laboratory measurements.

8.9.1 CRITICAL SHEAR STRESSES

Techniques for, and limitations of, determining critical shear stresses under laboratory and field conditions are reviewed in this subsection.

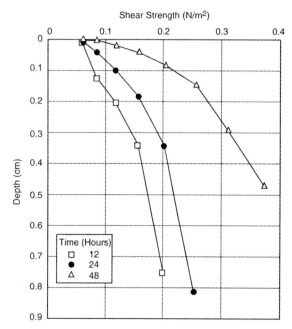

Figure 8.28   Distribution of critical shear stress with depth and consolidation time (from Johansen and Meldgaard 1993).

8.9.1.1 Laboratory Measurements

Laboratory measurements of the critical shear stress of cohesive sediments can be conducted in an annular flume or in a regular hydraulic flume. Annular flumes were developed by Etter and Hoyer (1965), Partheniades et al. (1966), and Partheniades and Kennedy (1966). These flumes consist of two concentric circular cylinders fixed to the bottom with the observations and measurements conducted on the sediment sample and overlying water placed in the annulus between the two concentric cylinders. The motion of the water is a result of rotation of the upper "lid," which is an annular disk placed at the water surface, and possibly a counterrotation of the lower channel apparatus. The moment exerted by the "lid" on the fluid is measured by a torque apparatus, and the results are interpreted in terms of the shear stress acting on the bottom sediments. If only the "lid" is rotated, secondary currents are generated that cause radial shear stresses in addition to the desired tangential shear stresses. However, computations indicate that these radial shear stresses are approximately 20 percent of the tangential shear stresses. Through a counterrotation of the lower channel apparatus, the secondary flows and radial shear stresses can be reduced further.

A second type of apparatus is shown in Figure 8.29 in which the sediment sample is remolded into a cylinder that is held stationary within a concentric rotating outer cylinder. Water is present between the inner cylinder formed by the sediment sample and the outer cylinder. The rotating outer cylinder induces a shear stress on the inner cylinder formed by the sediment sample; this shear stress is determined by measuring the torque experienced by the inner cylinder. The sediment concentration is monitored, and the critical shear stress documented.

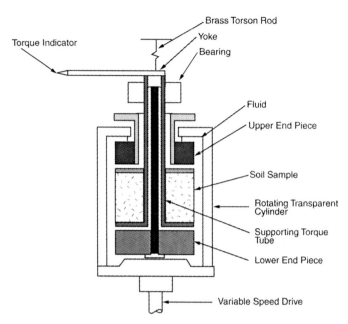

Figure 8.29 Remolded sediment sample as inner stationary cylinder; the outer cylinder rotates (from Moore and Masch 1962).

The final method of measuring critical shear stress in the laboratory is through the use of a hydraulic flume. The water in the hydraulic flume is recirculated, and the shear stress caused by the flow of the water over the bottom sediments is established. Measurements of this type have been reported by several investigators.

8.9.1.2 Field Measurements

Several methods have been developed for field measurement of the critical shear stresses of sediments. One of these involves the placement on the ocean floor of an inverted water tunnel, that is, one that is open on the bottom and generates a flow and shear stresses on the ocean floor. Observations, measurements, or both are carried out in the tunnel to determine the shear stress at which erosion of the surface sediments occurs. These inverted channels can be straight or circular in planform. The water inside the channel can be "once through," as in a straight channel with the water propelled by a pump, or "recirculating" in an annular flume with the propulsion provided by a rotating lid in much the same way as the laboratory experiments discussed previously. Young and Southard (1978) were among the first to deploy inverted flumes (the "Seaflume"). Other investigators using this type of equipment have included Nowell, McCave, and Hollister (1985) (the "Seaduct") and Maa and Lee (1991) (the "Carousel"). The maximum shear stresses some of these devices can impart to the sediments are limited. For example, the maximum shear stress of which the Carousel is capable is 8 dyn/cm^2 (Maa and Lee 1997).

A second field approach is the placement on the ocean floor of instrumented tripods with the capability of documenting the conditions under which erosion occurs and making the associated measurement of the water velocities that cause the erosion. These techniques were pioneered by Sternberg in the mid-1960s in the Puget Sound area (e.g., Sternberg 1965, 1967, 1968). Both tripods and water tunnels may cause some disturbance to the bottom sediments, resulting in a downward bias to the critical shear stresses determined. A second more serious limitation for cohesive sediments is that both methods only sample the shear stresses of the upper weaker layer of sediments (Figure 8.28), again resulting in an underestimate of the general strength characteristics. Interpretation of such results can be of special concern in a depositional environment owing to the presence of a weak, recently deposited surface mud. Finally, tripods can only sample those hydrodynamic conditions that occur during their deployment; rare environmental conditions of primary interest may not be encountered during the deployment period.

8.9.1.3 Calculation Procedures

The prediction of whether cohesive sediments will erode requires knowledge of the critical shear stresses, which are relatively difficult to measure directly. Therefore, correlations have been established between critical shear stresses and other more readily measured properties.

One of the most widely employed correlations is between critical shear stress and the vane shear strength of a cohesive sediment. The so-called vane shear apparatus provides a simple approach for determining the strength of sediments. This apparatus is a shaft with a cross-vane at one end of the shaft (see Figure 8.30). The size of the vane can vary; however, typical vane sizes are on the order of 5.5 to 6.5 cm in diameter,

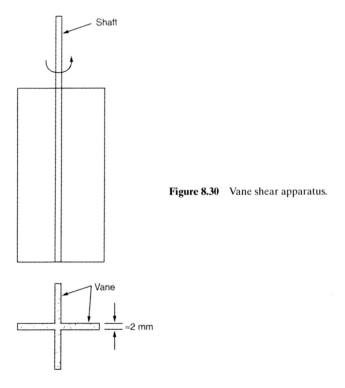

Figure 8.30 Vane shear apparatus.

the height is usually twice the diameter, and the vanes are approximately 2 mm in thickness (Terzaghi, Peck, and Mesri 1996 or Whitlow 1995). The vane is inserted into a sediment and a torque applied to the shaft until the surface area of the sediment cylinder circumscribing the vanes fails. With allowances for corrections for sediment resistance at the end(s) of the sediment cylinder, the shear strength is calculated. The application of this apparatus is simple, and it can be used in the laboratory or field to determine vane shear strengths in a matter of minutes once the sample is available.

Dunn (1959) recommended the following simple linear expression relating critical shear stress τ_c to vane shear strength, τ_y:

$$\tau_c = 0.01\tau_y \tag{8.77}$$

Migniot (1968) identified two ranges of vane shear strength in which different relationships apply:

$$\tau_c = 0.25\tau_y, \quad \tau_y > 1.6\,\text{Pa}$$

$$\tau_c = \sqrt{0.1\tau_y}, \quad \tau_y < 1.6\,\text{Pa} \tag{8.78}$$

Dade and Nowell (1991) have developed a relationship for the critical shear stress that includes, in addition to other parameters, the effects of density and the vane shear strength:

$$\tau_c = \frac{40\rho v^2}{\tan\phi\, d^2}\left\{\sqrt{\frac{g\rho_s' d^3 \tan^2\phi}{240\rho v^2}\left[1 + 3(1-\cos\phi)\frac{\tau_y}{g\rho_s' d}\right] + 1} - 1\right\}, \tag{8.79}$$

in which d is the particle diameter, ρ is the fluid density, ρ_s' is the excess density of the sediment, v is the fluid kinematic viscosity, and ϕ is the bed packing angle with a recommended value of 65°. Dade and Nowell (1992) explain that Eq. (8.79) provides a basis for bridging the range between cohesive and noncohesive sediments. Figure 8.31 illustrates the relationship (Eq. (8.79)) between critical shear stress τ_c

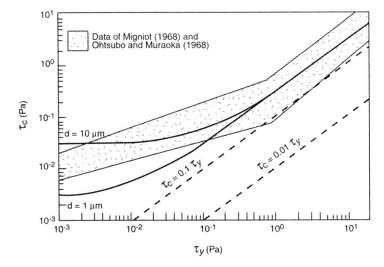

Figure 8.31 Variation of critical shear stress τ_c with vane shear strength τ_y (modified from Dade and Nowell 1991).

versus the vane shear strength, τ_y. The linear relationships

$$\tau_c = C\tau_y$$

have been added to Figure 8.31 for values of $C = 0.01$ and 0.1. It can be shown from Eq. (8.79) that for reasonably large values of τ_y, Eq. (8.79) reduces to

$$\tau_c = 0.31\tau_y, \tag{8.80}$$

which is very close to the proportionality constant recommended by Migniot (1968) for the larger vane shear strengths and approximately 30 times the proportionality constant proposed by Dunn. Equation (8.80) is consistent with Figure 8.31 for vane shear strengths τ_y greater than approximately 1 Pa (10 dyn/cm^2). For smaller values of τ_y, the value of τ_c always exceeds that given by Eq. (8.80).

Several numerical models have been developed in Europe to simulate the erosion and transport of muds. These include MUDFLOW-2D, a model by Pathirana (1993), and a model by Praagman (1986). The critical shear stress in these models is represented in terms of the bed dry density T_s. The relationships are as follows:

$$\tau_c = KT_s^m,$$

in which τ_c is in N/m^2, T_s is in kg/m^3, K is a constant, and m is an exponent. The two pairs of constants (K, m) are $(1.26 \times 10^{-9}, 3.56)$ and $(7.95 \times 10^{-10}, 3.64)$, respectively, as recommended by Praagman (1986) and Pathirana (1993). Studies by Parchure and Mehta (1985), Van Rijn (1989), and Mehta (1986) have all concluded that the critical shear stress is primarily related to the bed dry density T_s.

8.9.2 RATE OF EROSION

Because shear stresses greater than the critical are required for the onset of erosion, in a bed that has increasing shear resistance with depth into the bed, erosion will cease at a depth at which the applied shear stress is equal to the critical shear stress (see Figure 8.28).

Several models are available for quantification of the erosion rate as follows. Ariathurai and Arulanandan (1978) proposed the relationship

$$E = M_1 \frac{(\tau - \tau_c)}{\tau_c}, \tag{8.81}$$

which applies only for $\tau > \tau_c$, M_1 is a constant, and τ is the bottom shear stress. Thorn and Parsons (1980) have recommended the following modification of Eq. (8.81):

$$E = M_2(\tau - \tau_c)$$

In the two preceding equations, the value recommended for M_1 is 0.009, $M_2 = 9.73 \times 10^{-8}$, and the erosion rates E are expressed in kilograms per square meter per second and the shear stresses are in Pascals. Parchure and Mehta (1985) have proposed

$$E = E_0 \, e^{\alpha(\tau - \tau_c)^{1/2}}$$

in which E_0 ranged from 0.5 to 3.2 gm/cm^2/min and α ranged from 4.2 to 25.6 m/N$^{1/2}$.

REFERENCES

Ariathurai, R., and K. Arulanandan, "Erosion Rate of Cohesive Soils," *J. Hydraul. Div.*, ASCE, 104, HY2, 279–283, 1978.

Armon, J.W., The Dynamics of a Barrier Island Chain, Prince Edward Island, Canada, Ph.D. Dissertation, McMaster Univerity, 546 pp., 1975.

Bagnold, R.A., "The Movement of Desert Sand," *Proc. Roy. Soc. London*, A157, 594–620, 1936.

Bagnold, R.A., *The Physics of Blown Sand and Desert Dunes*, New York: William Morrow and Co., 1943.

Bagnold, R.A., "Mechanics of Marine Sedimentation," in *The Sea*, M.N. Hill, ed., 3, 507–532, New York: Interscience, 1963.

Bailard, J.A., "An Energetics Total Load Sediment Transport Model for a Plane Sloping Beach," *J. Geophys. Res.*, 86, C11, 10,938–10,954, 1981.

Bailard, J.A., "A Simplified Model for Longshore Transport," *Proc. 19th Intl. Conf. Coastal Eng.*, ASCE, Houston, 1454–1470, 1984.

Belly, P.-Y., "Sand Movement by Wind," U.S. Army Corps of Engineers, Coastal Engineering Research Center, Tech. Memo No. 1, 1964.

Berek, E.P., and R.G. Dean, "Field Investigation of Longshore Transport Distribution," *Proc. 18th Intl. Conf. Coastal Engg.*, ASCE, Cape Town, 1620–1639, 1982.

Birkemeier, W.A., "Field Data on Seaward Limit of Profile Change," *J. Waterway, Port, Coastal and Ocean Eng.*, ASCE, 111, 3, 598–602, 1985.

Bodge, K.R., "A Literature Review of the Distribution of Longshore Sediment Transport Across the Surf Zone," *J. Coastal Res.*, 5, 2, 307–328, 1989.

Bodge, K.R., and R.G. Dean, "Short-Term Impoundment of Longshore Sediment Transport," U.S. Army Corps of Engineers, Coastal Engineering Research Center, MP CERC-87-7, 1987.

Bodge, K.R., and N.C. Kraus, "Critical Examination of Longshore Transport Rate Magnitudes," *Proc. Coastal Sediments '91*, ASCE, 139–155, 1991.

Burt, N., R. Parker, and J. Watts, *Cohesive Sediments*, Chichester: John Wiley and Sons, 1997.

Caldwell, J.M., "Wave Action and Sand Movement Near Anaheim Bay, California, U.S. Army Corps of Engineers, Beach Erosion Board Tech. Memo. 68, 21 pp., 1956.

Chadwick, A.J., "Field Measurements and Numerical Model Verification of Coastal Shingle Transport," in *Advances in Water Modelling and Measurement*, M.H. Palmer, ed., British Hydromechanics Research Association, Cranfield, 1989.

Clifton, H.E., "Wave-Formed Sedimentation Structures – A Conceptual Model," *Soc. Econ. Paleontologists and Mineralogists*, Spec. Publ. 24, 126–148, 1976.

Dade, W.B., and A.R.M. Nowell, "Moving Muds in the Marine Environment," *Proc. Coastal Sediments '91*, ASCE, 54–71, 1991.

Dade, W.B., A.R.M. Nowell, and P.A. Jumar, "Predicting Erosion Resistance of Muds," *Mar. Geol.*, 105, 285–297, 1992.

Dalrymple, R.A., "Prediction of Storm/Normal Beach Profiles," *J. Waterway, Port, Coastal, and Ocean Eng.*, ASCE, 118, 2, 193–200, 1992.

Dean, R.G., "Heuristic Models of Sand Transport in the Surf Zone," *Proc. Conf. Eng. Dynamics in the Surf Zone*, Sydney, 208–214, 1973.

Dean, R.G., "Measuring Longshore Transport with Traps," in *Nearshore Sediment Transport*, R.J. Seymour, ed., New York: Plenum Press, 313–336, 1989.

Dean, R.G., E.P. Berek, C.G. Gable, and R.J. Seymour, "Longshore Transport Determined by an Efficient Trap," *Proc. 18th Intl. Conf. Coastal Eng.*, ASCE, Cape Town, 954–968, 1982.

Dean, R.G., and J. Zheng, "Cross-Shore Sediment Transport Relationships," Report UFL/COEL-94-018, Department of Coastal and Oceanographic Engineering, University of Florida, 1994.

del Valle, R., R. Medina, and M.A. Losada, "Dependence of the Coefficient K on the Grain Size," *J. Waterway, Port, Coastal and Ocean Eng.*, ASCE, 118, 6, 568–574, 1993.

Dette, H., and K. Uliczka, "Prototype Investigation on Time-Dependent Dune Erosion and Beach Erosion," *Proc. Coastal Sediments '87*, ASCE, New Orleans, 1430–1443, 1987.

Dewall, A.E., "Beach Changes at Westhampton Beach, New York," CERC MR 79-5, U.S. Army Coastal Engineering Res. Center, 1979.

Dewall, A.E., and J.J. Richter, "Beach and Nearshore Processes in Southeastern Florida," *Proc. Coastal Sediments '77*, 425–443, 1977.

Dillon, W.P. "Submergence Effects on Rhode Island Barrier Island and Lagoon and Influence on Migration of Barriers," *J. Geology*, 78, 94–106, 1970.

Dingler, J.R., and D.L. Inman, "Wave-Formed Ripples in Near-Shore Sands," *Proc. 15th Intl. Conf. Coastal Eng.*, ASCE, Hawaii, 1976.

Dolan, R., "The Barrier Dune System Along the Outer Banks of North Carolina, a Reappraisal," *Science*, 1972–1974, 1972.

Downing, J.P., R.W. Sternberg, and C.R.B. Lister, "New Instrumentation for the Investigation of Sediment Suspension Processes in the Shallow Marine Environments," *Mar. Geology*, 42, 19–34, 1981.

Duane, D.B., "Tracing Sand Movement in the Littoral Zone: Progress in the Radio-isotopic Sand Tracer (RIST) Study, July 1968–February 1969," MP 4-70, U.S. Army Coastal Engineering Research Center, 1970.

Dunn, I.S., "Tractive Resistance of Cohesive Sediment," *J. Soil Mech. and Foundations*, ASCE, 85, No. SM3, Proc. Paper 2062, 1–24, 1959.

Etter, R.J., and R.P. Hoyer, "A Laboratory Apparatus for the Study of Transport of Cohesive Sediments," M.S. Thesis, Civil Eng. Dept., Massachusetts Institute of Technology, Cambridge, MA, 1965.

Evans, O.F., "The Classification of Wave-Formed Ripple Marks," *J. Sed. Petrology*, 11, 1, 37–41, 1941.

Fischer, A.G., "Stratigraphic Record of Transgressing Seas in Light of Sedimentation on Atlantic Coast of New Jersey," *Am. Assoc. Petroleum Geologists Bulletin* 45, 1656–1666, 1961.

Fisher, J.J., and E.J. Simpson, "Washover and Tidal Sedimentation Factors in Development of a Transgressive Barrier Shoreline," in *Barrier Islands*, (S.P. Leatherman, ed.), New York: Academic Press, 127–149, 1979.

Fisher, J.S., and D.K. Stauble, "Impact of Hurricane Belle on Assateague Island," *Geology*, 5, 765–768, 1977.

Godfrey, P.J., and M.M. Godfrey, "Comparison of Ecological and Geomorphic Interactions Between Altered and Unaltered Barrier Systems in North Carolina," in *Coastal Geomorphology*, D.R. Coates, ed., State University of New York, Binghampton, NY, 329–358, 1973.

Greenwood, B., R.G. Richards, and R.W. Brander, "Acoustic Imaging of Sea Bed Geometry: A High Resolution Remote Tracking Sonar (HRRTS II)," *Mar. Geology*, 112, 207–218, 1993.

Hallermeier, R.J., "Uses for a Calculated Limit Depth to Beach Erosion," *Proc. 16th Intl. Conf. Coastal Eng.*, ASCE, Hamburg, 1493–1512, 1978.

Hanes, D.M., "The Structure of Events of Intermittent Suspension of Sand Due to Shoaling Waves," in *The Sea*, B. Le Mehaute and D.M. Hanes, eds., Chapter 28, Vol. 9 A, New York: Wiley Interscience, 941–952, 1990.

Hanes, D.M., and C.E. Vincent, "Detailed Dynamics of Nearshore Suspended Sediment," *Proc. Coastal Sediments '87*, ASCE, 285–299, 1987.

Hay, A.E., and D.J. Wilson, "Rotary Sidescan Images of Nearshore Bedform Evolution During a Storm," *Mar. Geology*, 119, 57–65, 1994.

Herron, W.J., and R.L. Harris, "Littoral Bypassing and Beach Restoration in the Vicinity of Port Hueneme, California," *Proc. Intl. Conf. Coastal Eng.*, ASCE, Tokyo, 651–675, 1966.

Horikawa, K., A. Watanabe, and S. Katori, "Sediment Transport under Sheet Flow Condition," *Proc. 18th Intl. Conf. Coastal Eng.*, ASCE, Cape Town, 1335–1352, 1982.

Hotta, S., S. Kubota, S. Katori, and K. Horikawa, "Sand Transport on a Wet Sand Surface," *Proc. 19th Intl. Conf. Coastal Eng.*, ASCE, Houston, 1265–1281, 1984.

Ingle, J.C., *The Movement of Beach Sand*, New York: Elsevier, 221 pp., 1966.

Inman, D.L., and R.A. Bagnold, "Littoral Processes," in *The Sea*, M.N. Hill, ed., 3, 529–533, New York: Interscience, 1963.

Jaffe, B.E., R.W. Sternberg, and A.H. Sallenger, "The Role of Suspended Sediment in Shore-Normal Beach Profile Changes," *Proc. 19th Intl. Conf. Coastal Eng.*, ASCE, 1983–1996, 1985.

Jensen, R.E., "Atlantic Coast Hindcast, Shallow Water, Significant Wave Information," U.S. Army Corps of Engineers, Waterways Experiment Station, WES Wave Information Studies No. 9, 1983.

Jette, C.D., "Wave Generated Bedforms in the Nearshore Sand Environment," Ph.D. Dissertation, University of Florida, 1997.

Johansen, C., T. Larsen, and O. Petersen, "Experiments on Erosion of Mud from the Danish Wadden Sea," Chap. 21, *Cohesive Sediments*, Eds. Burt, N., R. Parker, and J. Watts, Chichester, John Wiley and Sons, 1997.

Johansen, C., and H. Meldegaard, "Cohesive Sediment Transport in the Gradyb Tidal Area," (In Danish) M. Sc. Thesis, Dep. Civil Eng., University of Aalborg, Aalborg, Denmark, 1993.

Johnson, J.W., "The Littoral Drift Problem at Shoreline Harbors," *J. Waterways and Harbors Div.*, ASCE, 83, WW1, 1957.

Jonsson, I.G., "Wave Boundary Layers and Friction Factors," *Proc. 10th Intl. Conf. Coastal Eng.*, ASCE, Tokyo, 1966.

Kajima, R., T. Shimizu, K. Maruyama, and S. Saito, "Experiments on Beach Profile Change with a Large Wave Flume," *Proc. 18th Intl. Conf. Coastal Eng.*, ASCE, Cape Town, 1982.

Kamphuis, J. W., "Alongshore Sediment Transport Rate," *J. Waterway, Port, Coastal and Ocean Eng.*, ASCE, 117, 6, 624–640, 1991.

Kamphuis, J.W., "Friction Factor under Oscillatory Waves," *J. Waterways, Harbors, Coastal Eng.*, ASCE, 101, 1975.

Kamphuis, J.W., M.H. Davies, R.B. Nairn, and O.J. Sayao, "Calculation of Littoral Sand Transport Rate," *Coastal Eng.*, 10, 1–21, 1986.

Kamphuis, J.W., and J.S. Readshaw, "A Model Study of Alongshore Sediment Transport," *Proc. 16th Intl. Conf. Coastal Eng.*, ASCE, Hamburg, 1978.

Katoh, K. and S. Yanagishima, "Predictive Model for Daily Changes of Shoreline," *Proc. 21st Intl. Conf. Coastal Eng.*, ASCE, Malaga, 1253–1264, 1988.

Kawamura, R., "Study of Sand Movement by Wind," 1951, Translated as University of California Hydraulics Engineering Laboratory Rpt. HEL 2-8, Berkeley, 1965.

Kawata, Y., and Y. Tsuchiya, "Influence of Water Content on the Threshold of Sand Movement and the Rate of Sand Transport in Blown Sand," *Proc. Japanese Soc. Civil Eng.*, 249, 95–100, 1976.

Kennedy, J.F., "The Albert Shields Story," *J. Hydraulic Eng.*, ASCE, 121, 11, 766–772, 1995.

Komar, P.D., "The Mechanics of Sand Transport on Beaches," *J. Geophys. Res.*, 76, 3, 713–721, 1971.

Komar, P.D., "Beach Sand Transport: Distribution and Total Drift," *J. Waterway, Port, Coastal, and Ocean Eng.*, ASCE, 103, WW2, 225–240, 1977.

Komar, P.D., and D.L. Inman, "Longshore Sand Transport on Beaches," *J. Geophys. Res.*, 75, 30, 5914–5927, 1970.

Komar, P.D., and M.C. Miller, "Sediment Threshold under Oscillatory Waves," *Proc. 14th Intl. Conf. Coastal Eng.*, ASCE, 756–775, 1975.

Kraus, N.C., "Application of Portable Traps for Obtaining Point Measurements of Sediment Transport Rates in the Surf Zone," *J. Coastal Res.*, 2,2, 139–152, 1987.

Kraus, N.C., and J.L. Dean, "Longshore Sediment Transport Rate Distributions Measured by Trap," *Proc. Coastal Sediments '87*, ASCE, 881–896, 1987.

Kraus, N.C., and J.M. Mason, Discussion of 'Prediction of Storm/Normal Beach Profiles,' *J. Waterway, Port, Coastal, and Ocean Eng.*, ASCE, 119, 4, 466–473, 1993.

Kraus, N.C., R.S. Farinato, and K. Horikawa, "Field Experiments on Longshore Sand Transport in the Surf Zone: Time-Dependent Motion, On-Offshore Distribution and Total Transport Rate," *Coastal Eng. in Japan*, 24, 171–194, 1981.

Kraus, N.C., M. Isobe, H. Igarashi, T.O. Sasaki, and K. Horikawa, "Field Experiments on Longshore Sand Transport in the Surf Zone," *Proc. 18th Intl. Conf. Coastal Eng.*, ASCE, Cape Town, 969–988, 1982.

Kraus, N.C., and M. Larson, "Beach Profile Change Measured in the Tank for Large Waves, 1956–1957 and 1962," Tech. Rpt. CERC-88-6, U.S. Army Coastal Engineering Research Center, Vicksburg, 1988.

Kriebel, D.L., "Beach and Dune Response to Hurricanes," M.Sc. Thesis, University of Delaware, 349 pp., 1982.

Kriebel, D.L., and R.G. Dean, "Numerical Simulation of Time-Dependent Beach and Dune Erosion," *Coastal Eng.*, 9, 3, 221–245, 1985.

Kriebel, D.L., W.R. Dally, and R.G. Dean, "Undistorted Froude Model for Surf Zone Sediment Transport," *Proc. 20th Intl. Conf. Coastal Eng.*, ASCE, Taipei, 1296–1310, 1986.

Kubota, S., K. Horikawa, and S. Hotta, "Blown Sand on Beaches," *Proc. 18th Intl. Conf. Coastal Eng.*, ASCE, pp. 1181–1198, 1982.

Lane, E.W., E.J. Carlson, and O.S. Manson, "Low Temperature Increases Sediment Transportation in Colorado River," *Civil Eng.*, Sept., 1949.

Leatherman, S.P., "Quantification of Overwash Processes," Ph.D. Dissertation, Department of Environmental Sciences, University of Virginia, 1976.

Leatherman, S.P., "Beach and Dune Interactions During Storm Conditions," *Quart. J. Eng. Geology*, 12, 281–290, 1979.

Leatherman, S.P., "Barrier Island Migration: An Annotated Bibliography," Department of Geography, University of Maryland, College Park, MD, 49 pp., 1981a.

Leatherman, S.P., ed., *Overwash Processes*, Benchmark Papers in Geology, Vol. 58, Stroudsburg, PA: Hutchinson Ross Publishing Company, 1981b.

Lee, G.-H., and W.A. Birkemeier, "Beach and Nearshore Survey Data: 1985–1991 CERC Field Research Facility," Tech. Rpt. CERC-93-3, U.S. Army Coastal Eng. Res. Center, 1993.

Lettau, K., and H. Lettau, "Experimental and Micrometeorological Field Studies of Dune Migration," In *Exploring the World's Driest Climate*, Eds. K. Lettau and H. Lettau. University of Wisconsin–Madison, IES Report 101, pp. 110–147, 1977.

Longuet-Higgins, M.S., "Longshore Currents Generated by Obliquely Incident Sea Waves, 1," *J. Geophys, Res.*, 75, 33, 6778–6789, 1970.

Maa, P.-Y., and C.-H. Lee, "Variation of the Resuspension Coefficients in the Lower Chesapeake Bay," Special Issue No. 25, *J. Coastal Res.*, 63–74, 1997.

Maa, P.-Y., W.T. Shannon, C. Li, and C.-H. Lee, "In-Situ Measurements of the Critical Bed Shear Stress for Erosion," in *Environmental Hydraulics*, Eds. J.H.W. Lee and Y.K. Chung, 627–632, Rotterdam: Balkema, 1991.

Madsen, O.S., and W.D. Grant, "The Threshold of Sediment Movement Under Oscillatory Waves: A Discussion," *J. Sed. Petrology*, 45, 360–361, 1975.

Madsen, O.S., and W.D. Grant, "Quantitative Description of Sediment Transport by Waves," *Proc. 15th Intl. Conf. Coastal Eng.*, ASCE, Honolulu, 1093–1112, 1976.

Mann, D.W., and R.A. Dalrymple, "A Quantitative Approach to Delaware's Nodal Point," *Shore and Beach*, 54, 2, 13–16, 1986.

McGowen, J. H., and A. J. Scott, "Hurricanes as Geologic Agents," in *Estuarine Research*, Ed. L.E. Cronin, New York: Academic Press, 23–46, 1975.

Mehta, A.J., "On Estuarine Cohesive Sediment Suspension Behavior," *J. Geophys. Res.*, 94, C10, 14,303–14,314, 1986.

Migniot, C., "Etude des Properietes Physiques de Differents Sediments Tres Fins et de Leur Comportement Sous des Actions Hydrodynamiques," *La Houille Blanche*, 12, 169–187, 1968.

Moore, B.D., Beach Profile Evolution in Response to Changes in Water Level and Wave Height, MCE Thesis, Department of Civil Engineering, University of Delaware, 164 pp., 1982.

Moore, W. L., and F. D. Masch, "Experiments on the Scour Resistance of Cohesive Sediments," *J. Geophys. Res.*, 67, 4, 1437–1449, 1962.

Nicholls, R.J., W.A. Birkemeier, and R.J. Hallermeier, "Application of the Depth of Closure Concept," *Proc. 25th Intl. Conf. Coastal Eng.*, ASCE, Orlando, 3874–3887, 1996.

Nielsen, P., Some Basic Concepts of Wave Sediment Transport, Instit. Hydrodyn. Hydraul. Engrg. (ISVA), Series Paper 20, Tech. University of Denmark, 1979.

Nielsen, P., "Dynamics and Geometry of Wave-Generated Ripples," *J. Geophys. Res.*, 6647–6472, 1981.

Nielsen, P., "Suspended Sediment Concentration Under Waves," *Coastal Eng.*, 10, 1, 23–32, 1986.

Nowell, A.R.M., I.N. McCave, and C.D. Hollister, "Contributions of HEBBLE to Understanding Marine Sedimentation," *Mar. Geol.*, 66, 397–409, 1985.

O'Brien, M.P., and B.D. Rindlaub, "The Transportation of Sand by Wind," *Civil Eng.*, 6, 5, 325–327, 1936.

Ohtusbo, K., and K. Muraoka, "Resuspension of Cohesive Sediments by Currents," in *River Sedimentation, Proceedings, Third Symp. on River Sedimentation*, Eds. S.Y. Wang, H.W. Shen, and L.Z. Ding. 3, 1680–1689, 1986.

Parchure, T.M., and A.J. Mehta, "Erosion of Soft Cohesive Sediment Deposits," *J. Hyd. Eng.*, ASCE, 111, 10, 1308–1326, 1985.

Partheniades, E., and J.F. Kennedy, "Depositional Behavior of Fine Sediment in a Turbulent Fluid Motion," *Proc. 10th Intl. Conf. Coastal Eng.*, ASCE, 707–729, 1966.

Partheniades, E., J.F. Kennedy, R.J. Etter, and R.P. Hoyer, "Investigations of the Depositional Behavior of Fine Cohesive Sediments in an Annular Rotating Channel," Report No. 96, Massachusetts Institute of Technology, Cambridge, MA, 1966.

Pathirana, K.P.P., "Cohesive Sediment Transport in Estuaries and Coastal Environments," Ph.D. Thesis, Civil Eng. Dept., K. U. Leuven, Leuven Belguim, 1993.

Pierce, J.W., "Sediment Budget Along a Barrier Island Chain," *Sedimentary Geology*, 3, 5–16, 1969.

Praagman, N., "TRASIL: A Numerical Simulation Model for the Transport of Silt," Internal Report, Svasek, B. V. Rotterdam, 1986.

Raudkivi, A.J., *Loose Boundary Hydraulics*, Oxford, UK: Pergamon Press, 331 pp., 1967.

Roelvink, J.A., and M.J.F. Stive, "Bar-Generating Cross-shore Flow Mechanisms on a Beach," *J. Geophys. Res.*, 94, No. C4, pp. 4785–4800, 1989.

Savage, R.P., "Experimental Dune Building on the Outer Banks of North Carolina," *Shore and Beach*, 30, 2, 23–28, 1962.

Savage, R.P., and W.W. Woodhouse, "Creation and Stabilization of Coastal Barrier Dunes," *Proc. 11th Intl. Conf. Coastal Eng.*, ASCE, 671–700, 1965.

Saville, T., "Scale Effects in Two-Dimensional Beach Studies," *Trans. 7th Gen. Mtg.*, Intl. Assoc. Hyd. Res., 1, A3-1–A3-10, 1957.

Schwartz, R.K., "Nature and Genesis of Some Washover Deposits," U.S. Army Corps of Engineers, Coastal Engineering Research Center, Technical Memorandum No. 61, Fort Belvoir, VA, 1975.

Schoones, J.S., and A.K. Theron, "Improvement of the Most Accurate Longshore Transport Formula," *Proc. 25th Intl. Conf. Coastal Eng.*, ASCE, 3652–3665, 1996.

Seymour, R.J., ed., *Nearshore Sediment Transport*, New York: Plenum Press, 418 pp., 1989.

Sherman, D.J., B.O. Bauer, P.A. Gares, and D.W.T. Jackson, "Wind Blown Sand at Castroville, California," *Proc. 25th Intl. Conf. Coastal Eng.*, 4214–4226, 1996.

Shimizu, T., S. Saito, K. Maruyama, H. Hasegawa, and K. Kajima, "Modelling of Onshore/Offshore Sand Transport Rate Distributions," Rpt. 384028, Civil Eng. Lab, Central Research Institute of Electric Power Industry, (in Japanese), 1985.

Sleath, J.F.A., *Sea Bed Mechanics*, New York: John Wiley & Sons, 335 pp., 1984.

Sternberg, R.W., "An Instrument System to Measure Boundary-Layer Conditions at the Sea Floor," *Mar. Geol.*, 3, 475–482, 1965.

Sternberg, R.W., "Measurements of Sediment Movement and Ripple Migration in a Shallow Marine Environment," *Mar. Geol.*, 5, 195–205, 1967.

Sternberg, R.W., "Friction Factors in Tidal Channels with Varying Bed Roughness," *Mar. Geol.*, 6, 243–260, 1968.

Stive, M.J.F., A Model for Cross-shore Sediment Transport, in *Proc. 20th Intl. Conf. Coastal Eng.*, pp. 1550–1564, ASCE, Taipei, 1986.

Swart, D.H., "Offshore Sediment Transport and Equilibrium Profiles," Delft Hydraulics Laboratory, Public. 131, 1974.

Tanski, J., H.J. Bokuniewicz, and C.E. Schubert, "An Overview and Assessment of the Coastal Processes Data Base for the South Shore of Long Island," Spec. Rpt. 104, New York Sea Grant Program, 77 pp., 1990.

Terzaghi, K., R.B. Peck, and G. Mesri, *Soil Mechanics in Engineering Practice*, 3rd ed., New York: John Wiley and Sons, 1996.

Thorn, M.C.F., and J.G. Parsons, "Erosion of Cohesive Sediments in Estuaries: An Engineering Guide," in *Proc. Third Intl. Conf. on Dredging Tech.*, Ed. H.S. Stephens BHRA, Bordeaux, France, 349–358, 1980.

Thornton, E.B., "Distribution of Sediment Transport Across the Surf Zone," *Proc. 13th Intl. Conf. Coastal Eng.*, ASCE, Vancouver, 1049–1068, 1972.

Trowbridge, J., and D. Young, "Sand Transport by Unbroken Water Waves Under Sheet Flow Conditions," *J. Geophys. Res.*, 94, C8, 10,971–10,991, 1989.

U.S. Army Corps of Engineers, Coastal Engineering Research Center, *Shore Protection Manual*, 1984.

Van der Meer, J.A.W., "Static and Dynamic Stability of Loose Materials," in *Coastal Protection*, Ed. K.W. Pilarczyk, Balkema, Rotterdam, 1990.

Vanoni, V.A., ed., *Sedimentation Engineering*, New York: ASCE, 745 pp., 1975.

Van Rijn, L.C., "Handbook Sediment Transport by Currents and Waves," Delft Hydraulics, Rept. H. 461, 12.1–12.24.

Walton, T.L., and R.G. Dean, "Application of Littoral Drift Roses to Coastal Engineering Problems," *Proc. Conf. Eng. Dynamics in the Surf Zone*, Sydney, 22–28, 1973.

Wang, P., and N.C. Kraus, "Longshore Sediment Transport Rate Measured by Short-Term Impoundment," *J. Waterway, Port, Coastal and Ocean Engineering*, ASCE, 125, 3, 118–126, 1999.

Watanabe, A., "Numerical Models of Nearshore Currents and Beach Deformation," *Coastal Eng. in Japan*, 25, 147–161, 1982.

Watts, G.M., "A Study of Sand Movement at South Lake Worth Inlet, Florida," U.S. Army Corps of Engineers, Beach Erosion Board Tech. Memo. 42, 24 pp., 1953.

Whitlow, R., *Basic Soil Mechanics*, 3rd ed., Essex, UK: Longman, Scientific and Technical, 1995.

Wiberg, P.L., and C.K. Harris, "Ripple Geometry in Wave Dominated Environments," *J. Geophys. Res.*, 99, 775–789, 1994.

Wiegel, R.L., *Oceanographical Engineering*, Englewood Cliffs, NJ: Prentice–Hall, Inc., 532 pp., 1963.

Williams, P.J., "Laboratory Development of a Predictive Relationship for Washover Volume on Barrier Island Complexes," Master of Civil Engineering Thesis, Dept. Civil Eng., University of Delaware, 141 pp., 1978.

Young, R.N., and J.B. Southard, "Erosion of Fine-Grained Marine Sediment: Sea-floor and Laboratory Experiments," *Geol. Soc. Amer. Bull.*, 89, 663–672, 1978.

Zheng, J., and R.G. Dean, "Numerical Models and Intercomparision of Beach Profile Evolution," *Coastal Eng.*, 30, 169–201, 1997.

Zingg, A.W., "Wind Tunnel Studies of the Movement of Sedimentary Material," *Proc. 5th Hydraulics Conf.*, Bull. 34, 111–135, 1953.

EXERCISES

8.1 Consider a long, straight, uninterrupted coastline. A single groin is installed as shown in Figure 8.32.

    (a) How will the volume of material deposited updrift of the groin compare with that eroded downdrift of the groin?

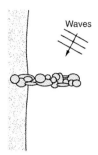

Figure 8.32 Single groin for Problem 8.1.

(b) Develop a convincing argument proving your answer in **(a)**.

(c) Does your answer depend on whether bypassing of the groin occurs? Discuss.

8.2 The longshore sediment transport $Q_s(x)$ shown in Figure 8.33 occurs on the initially straight beach. Qualitatively sketch on the lower diagram, the beach planform a short time later.

8.3 A beach nourishment project is being carried out in an area where the waves advance *normal* to the original shoreline. Two retaining structures have been built to keep the sand in the area of interest. See Figure 8.34.

(a) If waves always propagate as shown, could this method of retaining the fill cause significant adverse effects to adjacent beach segments? Any beneficial effects?

(b) The sand is added at a constant rate of 300 m³/day in the corner location shown in the sketch in Figure 8.34. For the littoral drift rose shown in this figure, sketch the approximate qualitative planform at 5, 15, and 20 days, on the assumption that 20 days are required to add the volume required to fill the compartment.

(c) Comment on the shoreline orientations at Groins A and B during and after the filling process. In particular, what would be the limiting shoreline slope (in plan view) at Groin A?

8.4 Consider unidirectional longshore sediment transport along a long, uninterrupted shoreline. Three groins are installed, as shown in Figure 8.35.

(a) Sketch the shoreline after the groins are installed but before bypassing of Groin 1 (G 1) occurs. Neglect any effect of diffraction between the groins and show the detail of the shoreline adjacent to the structures and between the groins.

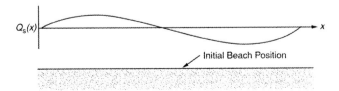

Figure 8.33 Transport rate for Problem 8.2.

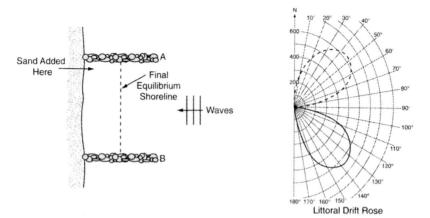

Figure 8.34 Groin compartment for Problem 8.3.

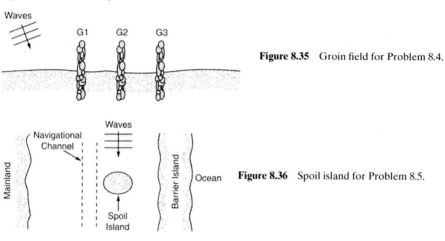

Figure 8.35 Groin field for Problem 8.4.

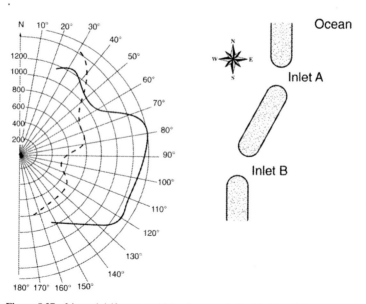

Figure 8.36 Spoil island for Problem 8.5.

Figure 8.37 Littoral drift rose and island segments for Problem 8.6.

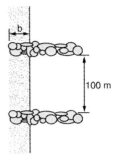

Figure 8.38 Two groins for Problem 8.6**(c)**.

100 m

(b) Sketch the shoreline after bypassing occurs around Groin 1 but before it occurs around Groin 2.

(c) Sketch the shoreline after bypassing occurs around Groin 2 but before it occurs around Groin 3.

(d) Compare the volume of material eroded downdrift of the groins with that deposited updrift of, and within, the groins.

8.5 Frequently, when navigational channels are dredged along inland waterways, nearly circular spoil islands are formed of sand-sized material. Consider such an island on which the waves are only from the north along the axis of the waterway, as shown in Figure 8.36. Neglect any effect of the dredged channel on the island.

(a) Sketch and discuss the littoral drift rose for this island (qualitatively).

(b) Sketch and discuss the evolution of this island with time.

8.6 The littoral drift rose for a particular area is given in Figure 8.37. Three shoreline segments are separated by Inlets A and B.

(a) Quantitatively describe the sand budget at Inlet A.

(b) Quantitatively describe the sand budget at Inlet B.

(c) Two groins are constructed north of Inlet A, as shown in Figure 8.38. With diffraction excluded from consideration, how far should the groins be extended landward ($b = ?$) beyond the shoreline to avoid flanking before bypassing occurs.

8.7 Groins are installed on two different beaches and, after several years, the shoreline planforms come to equilibrium, as shown in Figure 8.39. The wave heights and periods at the two locations are approximately the same. On the basis of the two shoreline planforms in the vicinity of the groins, what more can you say about the wave climate?

8.8 A lake is initially circular in planform, as shown in Figure 8.40. At time $t = 0$, the lake is subjected to the longshore sediment transport shown. Qualitatively sketch the lake shoreline sometime later.

8.9 The littoral drift rose shown in Figure 8.41 is appropriate for an area of interest. The net longshore sediment transport along the updrift shoreline is 200,000 m³/yr, and the flood tidal shoals of an inlet store 50,000 m³/yr. The

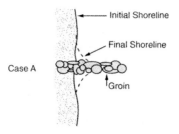

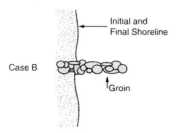

Figure 8.39   Case A and B for Problem 8.7.

updrift and downdrift shorelines on the opposite sides of the inlet are initially in equilibrium with the waves.

(a) What are the initial orientations of the updrift and downdrift shorelines?

(b) The equilibrium is altered for navigation by the removal of an additional 100,000 m^3/yr of sand from the inlet. What will be the long-term orientation of the downdrift shoreline?

(c) Is the littoral drift rose stable or unstable? Demonstrate your answer.

(d) Groin compartments are constructed as shown in Figure 8.41 along the updrift and downdrift shorelines. With wave diffraction neglected, what would be the shoreline orientation within the groin compartments before bypassing occurs? Sketch the ultimate shoreline position within the groin compartments.

8.10 Consider two different updrift jetties (shown as Cases (a) and (b) in Figure 8.42) with the longshore sediment transport directed toward the jetties.

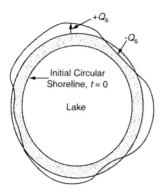

Figure 8.40   Circular lake for Problem 8.8.

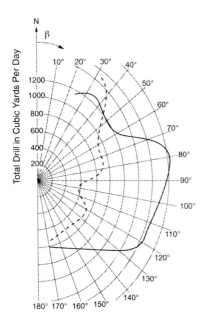

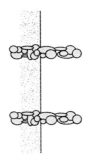

Figure 8.41 Littoral drift rose for Problem 8.9

The longshore sediment transport distribution across the surf zone is shown at a sufficient distance updrift from the jetties that they do not influence the distribution. The jetties are placed at $t = 0$. The elevation of the portion of the weir jetty (case (b)) is at elevation MSL.

(a) For each jetty, qualitatively sketch the shoreline positions and the sediment transport vectors at some time such that the shorelines have responded (but have not necessarily equilibrated) to the presence of the jetties. Discuss the reasons for any differences for the two shorelines, that is, cases (a) and (b).

(b) For each jetty, present qualitative plots of contours of the bathymetry immediately updrift of the two jetties.

(c) Discuss the physical reasons causing the sediment transport vectors immediately updrift of the jetties to differ from the vectors far updrift.

8.11 A shore-perpendicular structure is placed in a location where the unaffected longshore transport is Q_0 (see Figure 8.43). Diffraction effects may be ignored.

(a) For the case where bypassing has not yet occurred, as shown in Figure. 8.43 **(a-1)**, qualitatively sketch in the figure the longshore sediment transport rate updrift and downdrift of the structure.

(b) For the case in which bypassing has commenced, as shown in Figure 8.43 **(b-1)**, qualitatively sketch in the figure the longshore sediment transport rate updrift and downdrift of the structure.

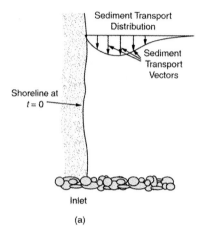

(a)

Figure 8.42   Solid and weir jetty for Problem 8.10. (a) Solid impermeable jetty; (b) weir jetty.

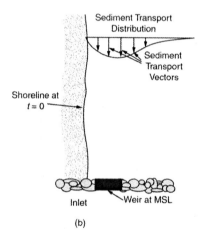

(b)

8.12   Consider 5 years of wave data for a given beach:

Year	H_e(m)	T_e(s)
1	1.7	11
2	1.5	10
3	2.1	14
4	2.4	16
5	1.8	12

What is the predicted depth of closure for the entire 5-year period? How does this 5-year depth of closure compare with the depth of closure for each year? What is the likelihood that the maximum annual depth of closure will occur in the first year?

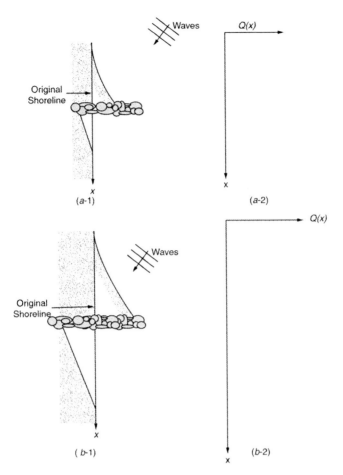

Figure 8.43 Sketch for Problem 8.11.

8.13 Consider the system in Figure 8.44 in which all of the sediment that passes a breakwater is stored temporarily in a spit at the end of the breakwater (Point B) until it is removed by dredge and transferred to the downdrift beach. The time variation of the net and gross longshore sediment transport components at Point A are desired.

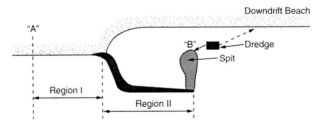

Figure 8.44 Sketch for Problem 8.12.

(a) If sufficient repeated profiles are measured between Points A and B and accurate dredging records are maintained, is it possible to determine the net and gross longshore sediment transport characteristics? Discuss your answer.

(b) For the following data collected over a 9-month period during which the dredge removed 100,000 m³, determine the net annual longshore sediment transport rate past Point A.

Region	Volume Change (m³)
I	+160,000
II	+120,000
Spit	−20,000

CHAPTER NINE

Miscellaneous Coastal Features

Nature leaves subtle evidence of antecedent conditions in the landforms and vegetation patterns that are present in the coastal areas. A significant challenge to coastal engineers is to attempt to interpret these clues in order to obtain a better understanding of the overall dynamics. An example is provided by the beach ridge patterns shown in Figure 9.1. The ridges in the center and left of the photograph are more or less shore parallel, as might be expected in the case of sand being transported shoreward and causing a shoreward advancement. However, the beach ridges to the right of the photograph are puzzling owing to their decidedly different orientation. Several questions are posed by this photograph. It is clear that at some time in the past the forces and sediment supply in the system were such that the beach was accreting. Is the beach still accreting and, if not, what were the causes that tipped the force balance to erosion? If erosion is presently occurring, will it continue or increase in the future? What were the forces that caused the different orientations of the beach ridges and, if they can be discerned, how can this information be applied to future shoreline trends? Usually, questions of this type cannot be answered without additional information. This problem can be likened to that of attempting to discern a picture with only a small number of pieces of a jigsaw puzzle. No matter how daunting and difficult the interpretation of landforms, the coastal engineer and geologist should utilize all information and evidence available to obtain more pieces of the puzzle, thereby increasing the understanding of the natural system and gaining confidence for decisions and designs.

9.1 INTRODUCTION

A variety of surprisingly regular and geometric coastal landforms are created by the daily interaction of the waves and tides with the beach. Some of these features, such as crenulate bays, take many years to form, whereas others, such as beach cusps, may form in a matter of minutes.

An important test of our understanding of coastal processes is how well we can explain these features. It is also problematic that there may be more than one

Figure 9.1 Beach ridges on the west coast of Florida (1970).

explanation for a single type of feature, thus complicating our efforts. This multiplicity of possible mechanisms for the same phenomenon distinguishes this field of science from many others.

In this chapter, morphological stages, crenulate bays, multiple off-shore sandbars, and beach cusps are examined to determine the mechanisms that may cause such curious features.

9.2 NEARSHORE MORPHOLOGY

The planforms of shorelines vary rapidly in time. During a major storm, an off-shore sandbar is usually created. After this storm, the sandbar generally migrates slowly back onshore as the ridge–runnel system mentioned in Chapter 3, recreating the straight beach. In a study of 26 beaches in Australia, Wright and Short (1984) noted that a series of steps unfold between these two end stages. Figure 9.2 shows the six morphological stages of the shoreline transitions. These states were called (1) dissipative, (2) long-shore bar trough, (3) rhythmic bar and beach, (4) transverse bar and rip, (5) ridge-and-runnel and low-tide terrace, and (6) reflective. The intermediate stages (2–5) are far more three-dimensional than the traditional ridge–runnel system implies. Only at the end stages (dissipative: a straight linear bar, and reflective: a steep beach) are the planforms straight. The transition from a long, linear storm bar to a rhythmic bar that is periodic along the beach occurs rapidly (Lippmann and Holman 1989, 1990), and more often than not the beach has some longshore periodicity that is associated with rip currents.

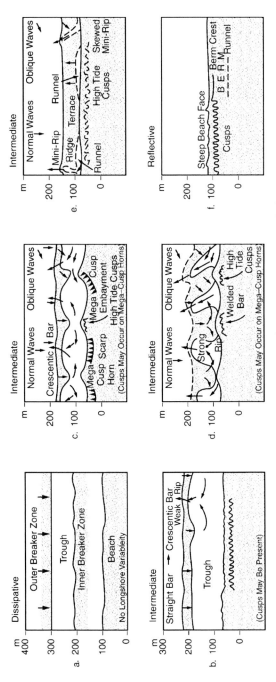

Figure 9.2 Morphological beach states with progression from (a) to (f) (from Wright and Short 1984).

277

Wright and Short have attempted to correlate various beach states with different parameters: the "surf-scaling parameter," the Dean number, the tide, and offshore wave groups (Wright and Short 1984, Wright et al. 1987). The surf-scaling parameter (Guza and Inman 1975) is

$$\epsilon_s = \frac{\sigma^2 H_b}{2g \tan^2 \beta},$$

where σ is the wave angular frequency, H_b is the breaking wave height, and β is the beach slope. This parameter can easily be shown to be related to the surf similarity parameter as $\epsilon_s = \pi/\zeta_b^2$.

Wright and Short show that the Dean number, $\mathbf{D} = H_b/(wT)$ (which includes a sediment parameter, the fall velocity w) is a reasonable predictor of beach stage, although there is generally a time lag between a change in Dean number, owing to changing wave conditions, and the formation of a new beach state. As a rough guide, because their data overlap, the dissipative state occurs for $3 < \mathbf{D}$. The second beach stage for $1.5 > \mathbf{D} > 7$; the third, $1 < \mathbf{D} < 5$; the fourth, $0.6 < \mathbf{D} < 3$; the fifth, $0.4 < \mathbf{D} < 2.5$; and, the sixth, $\mathbf{D} < 1$. Wright et al. (1987) add that tide range and wave groupiness can be important discriminators to help choose the appropriate state for a given Dean number. For example, higher tide ranges tend to move the beach to the stage with the less bathymetric relief.

9.3 CRENULATE BAYS

9.3.1 FORMATION

Crenulate bays are formed by the erosion between two headlands or nonerodible "controls," and they are one class of a more general feature termed a pocket beach. Pocket beaches can have any degree of asymmetry, whereas crenulate bays are asymmetric relative to the control points owing to the predominant wave direction, refraction, and diffraction processes. Figure 9.3 shows Pt. Reyes, California, where two

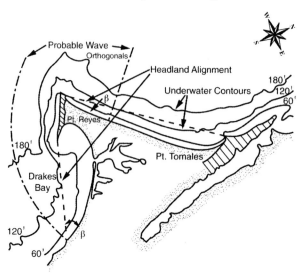

Figure 9.3 Crenulate bays at Point Reyes, California (from Silvester 1970).

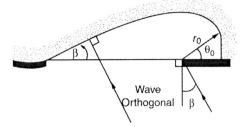

Figure 9.4 Schematic of crenulate bays (modified, with permission, from Yasso 1965. Copyright © University of Chicago Press).

different crenulate bays can be seen. Crenulate bays are also referred to as hooked bays, half-heart bays, and spiral bays. They were first studied by Yasso (1965), who identified two embayments near Sandy Hook, New Jersey, and Point Reyes and Drakes Bay, California, as examples of this type. Yasso noted that these embayments could be fit by a log spiral of the form

$$r(\theta) = r_0 e^{\alpha(\theta - \theta_0)},$$

in which, as shown in Figure 9.4, r is the distance from the center, θ is the angular coordinate with θ_0 the reference value, $r = r_0$, and α is a factor controlling the rate at which the radius increases with θ, $0.4 < \alpha < 1.1$.

Crenulate bays can be quite ubiquitous in areas where rocky promontories extend beyond the beach line. On the east coast of the Malaysian peninsula, for example, in Kuantan Province, there are a long series of these bays, all with similar orientation.

Silvester (1970) and Silvester and Ho (1972) have studied natural crenulate bays, primarily in Australia, and have conducted laboratory studies and developed a relationship for the indentation ratio as a function of wave direction β, as presented in Figure 9.5.* In general, it is clear that, in addition to the wave direction, the indentation ratio depends on the ratio of longshore sediment supply Q to potential sediment transport Q_*, which is the amount that the waves are capable of transporting. If these two transports are equal, the indentation is zero.

Terpstra and Chrzastowski (1992) have presented interesting documentation of log spiral bay development following the construction of a marina on Lake Michigan. The initial shoreline downdrift of an artificial headland was reasonably straight, and the evolution was monitored for a 9-month period, comprising a total of 11 surveys. For each survey subsequent to the initial one, the parameters associated with the log spiral were determined graphically. Well-defined decreases in the log spiral parameters α and the radius r_0 were documented.

Crenulate bays, created by "artificial headlands," have been proposed as a method of shoreline protection in areas with a sediment supply deficit. These systems serve to align the waves with the shoreline and hence reduce the sediment transport. The difficult design feature is the connection of the artificial headland with the land that is susceptible to degradation and breaching, especially during storms. One approach to

---

* An approximation to the curve in Figure 9.5 can be developed by considering that the shoreline will line up with the diffracted and refracted wave crest, that the wave crest will be formed by a circular segment within the geometric shadow zone, and that the wave direction will be unaffected in the geometric illuminated zone. Under these assumptions, you may wish to show that $a/b = \sin \beta$, which overestimates the correct value by about 50 percent for large angles.

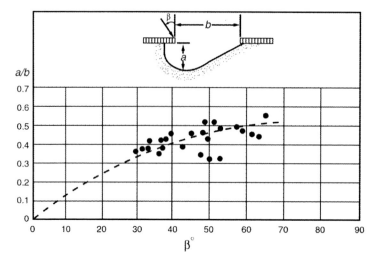

Figure 9.5 Indentation ratio versus wave angle (from Silvester 1970).

ensuring the integrity of these connections is the use of a structural "stem" to attach the structure to the shoreline. Silvester and Ho (1972) designed a shore protection scheme for reclaimed land in Singapore based on the crenulate bay concept. A second application is inside the throat of inlets, where the sediment supply is very small to nonexistent owing to sand-tight jetty structures. If left unprotected, these inlet banks will erode most seriously at the landward terminus of the jetty. Figure 9.6 presents a schematic diagram of before and after the installation of such structures inside Ocean City Inlet, Maryland, at the north end of Assateague Island.

Wind (1994) has developed an analytical model for the time-dependent evolution of a crenulate bay that has only one hard boundary condition at the updrift side. The numerical model is process based and represents the growth of the crenulate bay in terms of the product of a spatial and a temporal function in the form

$$r(\theta, t) = r_0 f(\theta)e(t),\qquad\qquad(9.1)$$

where r is the radial displacement of the shoreline in polar coordinates, r_0 is a constant, $f(\theta)$ is a shape function that varies with angle θ, and $e(t)$ is a time function. It has been found that the growth occurs with $t^{1/3}$ in the diffraction region and $t^{1/2}$ in the refraction region. The model is calibrated and verified with the laboratory data of Vichetpan (1971).

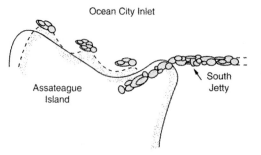

Figure 9.6 Artificial headlands at Ocean City Inlet, Maryland.

9.3.2 DESIGN IN CRENULATE BAYS

The planform of a crenulate bay is in near equilibrium with the incident waves, whether it varies with the seasonal waves for small bays or responds to the most energetic of the annual waves such as the winter wave climate. As discussed in Chapter 1 in relation to Port Orford, Oregon, design of structures or harbors in these bays can be fraught with difficulties, for the additions create perturbations in the system.

Durban, South Africa, lies within a crenulate bay with a headland to the south that serves to protect a harbor entrance just inside the updrift end of the bay, and then the bathing beaches are on the downdrift end of the embayment. As the city grew, so did the harbor's need for channel deepening and protective jetties, resulting in erosion of the cities' beaches as the littoral transport within the bay system was disrupted by the harbor. In 1982, a sand bypassing scheme designed for 280,000 m^3/yr was put into place involving a floating dredge in the harbor entrance, a fixed shore-based storage and pumping facility, and a pipeline with four booster stations spaced 700 m apart along the beaches. The dredge takes sand from an excavated sand trap updrift of the harbor jetties and discharges it into the sand-pumping facility, where it is stored and then redredged and placed into the pipeline. From the pipeline, the sand can be diverted to numerous points along the shoreline.

Consider the problem of designing a harbor site in or near a crenulate bay on an otherwise straight coastline. There are several possibilities for harbor siting. One is to put the harbor updrift of the bay. This implies that (1) the harbor will have the most wave exposure and (2) will intercept all the littoral drift headed to the embayment, resulting in erosion of the bay's beaches. The second option is similar to that used at Durban and Port Orford and entails placing the harbor just inside the updrift end of the bay, where it is sheltered from the waves. This will result in a change in planform of the bay owing to the subsequent erosion resulting from the trapping of sand by the harbor and the influence of the jetties. Finally, the harbor can be sited in the downdrift end of the bay. This will have less impact on the overall bay, and there will be some wave sheltering within the embayment from the presence of the headland.

9.4 SAND WAVES

Sand waves, which are large (kilometer scale) periodic features in the shoreline planform, are often described as propagating along the shoreline. They also persist for long periods (decades).

Bruun (1954) examined offshore sandbars in the Netherlands, Denmark, and the United States. He distinguished between sandbars moving alongshore and those with on–offshore motions. In Denmark, he noticed alongshore–migrating waves of sand with lengths of 0.2–2 km and amplitudes of 60–80 m (in the onshore–offshore direction) by plotting the location of a given contour as a function of time. These sand waves propagated in the same direction as the net longshore sand transport.

Bakker (1968) reported sand waves on the (former) island of Vlieland in the Wadden Sea. Plotting the high and low water lines at six sites using a hundred years of data showed that two sand waves propagated northeasterly along the island with a migration speed of 1/6 to 1/3 km/yr. The amplitude of these waves was 300 m, decreasing to 76 m after 5 km of travel. Bakker speculated that the cause of these sand waves was sand bypassing events at the inlet between Vlieland and the neighboring island of Texel when the outer shoal melded to Vlieland. Bakker carried out an analysis using a one-line shoreline model to be discussed in Chapter 10 and his results corresponded very well with the measured data. A recent analysis of the Dutch coastline by Verhagen (1989) showed sand waves with amplitudes of 30–500 m and alongshore propagation speeds of 50–200 m/yr.

Grove, Sonu, and Dykstra (1987) reported on the behavior of a "beach fill" at San Onofre, California. They monitored the disposition of 200,000 yd^3 of sand over the course of 2 years as it migrated along the shore at a speed of about 2 m/day and decayed rapidly in amplitude. In front of this migrating sand hump was a migrating region of erosion, and thus the original sand wave developed a negative (landward) leading edge.

Thevenot and Kraus (1995) measured sand waves at Southhampton Beach, N.Y. These westward propagation waves, with lengths less than a kilometer, were attributed to an inlet that periodically opened and closed.

On the basis of these observations, we can classify sand waves as three different types. The first is the *spatially* periodic sand wave migrating along the shore; the second class, the *temporally* periodic sand wave, is due to the periodic dumping of sand at a point (say, bypassing at an inlet, or beach fills). This class would include the Vlieland data of Bakker. The final class is the *solitary wave* class, which would include the behavior of beach fills and San Onofre results of Grove et al. In Chapter 10, we will discuss a simple analytical model for the behavior of these sand waves.

9.5 MULTIPLE OFFSHORE SANDBARS

One of the advantages of the simple cross-shore sediment transport models is that they allow us to examine some of the models for the formation of off-shore sandbars. Dean's heuristic model for the presence of bars (Chapter 8) provides us with one mechanism, but, as with many of the phenomena that occur in coastal processes, it is probably not the only mechanism.

First, let us examine the characteristics of offshore bars. They may occur as a single bar, located just inshore of the breaker line (as is often the case in laboratory flume tests), which is the *breakpoint bar*. The jet of water in the plunging breaker has sufficient momentum to penetrate the water column and provide bottom turbulence sufficient to scour the bottom onshore of the breaker point. Bagnold (1947) carried out the first laboratory studies showing that bars are formed by the breaking waves. The higher the waves, the farther offshore the bar is located in keeping with the breakpoint model, for larger waves break farther offshore. These breakpoint bars are mostly associated with beaches steep enough for plunging breakers, which implies a surf similarity parameter (Eq. (5.17)) of about unity. Further, bars will migrate with the wave climate, with larger waves moving the bar offshore. In some areas, bars are seasonal, occurring generally during winter when there is a harsher wave climate.

Sometimes bars occur in numerous rows parallel to the shoreline. There are locations in the Chesapeake Bay, Figure 9.7, where up to 17 rows of bars occur. These bars are characterized by long crests and amplitudes up to 0.5 m. Many large lakes

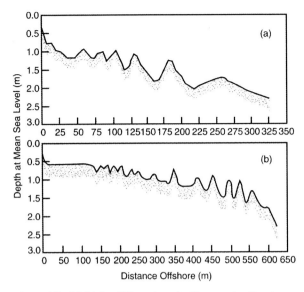

Figure 9.7 Multiple offshore bars in Chesapeake Bay (reprinted from Dolan and Dean 1985 by permission of the publisher Academic Press Limited, London).

and bays have multiple offshore bars; it appears that the occurrence of multiple bars is more likely for smaller tidal ranges (King and Williams 1949). In the Chesapeake Bay, the bar fields are persistent and clearly visible in aerial photographs spanning 16 years (Dolan and Dean 1985).

Dolan and Dean (1985) have developed a field data set of 18 profiles from the Chesapeake Bay, augmented this set with published profiles from Alaska and Lake Michigan, and tested the measured bar spacings against those predicted by the following four candidate mechanisms: (1) normally reflecting long waves, (2) edge waves, (3) breaking waves, and (4) "overtake distance," which is the distance required for a fundamental wave to overtake a free harmonic (one-half the period) of that fundamental. The tests were based on the characteristic spacings of the mechanisms. For example, for the first mechanism, normally reflecting shallow water waves, the spacing should be proportional to the square root of the depth. Other mechanisms have similar characteristics, thus allowing discrimination. The study concluded that, at least for the data set examined, the bars appeared to be formed by multiple breaking of the waves and that multiple bars are most likely to occur in locations where the general bottom slope is less than 1:140. In the next sections, we examine four multiple bar-generation mechanisms.

9.5.1 VARIABLE WAVE CLIMATE MODEL

One possibility for forming multiple bars is that the wave climate is sufficiently variable that small, locally generated waves create a breakpoint bar on the inshore portion of the profile. Storm waves break farther offshore, creating an additional, deeper break point bar (Evans 1940). An energetic wave climate with a spectrum of waves would have components such that breaking would occur on both bars.

Alternatively, there could be large swings in mean water level, which would permit the inshore bars to be acted upon by the waves at high water level and the offshore bars at low water level. This mechanism would not be applicable to the Chesapeake Bay, however, because of the small tide range.

9.5.2 PARTIAL STANDING WAVE MODEL

A second possibility is that partial standing waves are formed in the nearshore region by reflection from the shoreline or coastal structures. This would permit the formation of multiple bars, for a bar could grow at each of the antinodal locations of the partial standing wave system. At the antinodal locations, the wave-induced particle velocities are primarily vertical and do not scour the bottom, whereas under the nodal positions, large horizontal velocities occur. Lettau (1932) examined the possibility that standing waves could cause bars, and Carter, Liu, and Mei (1973) examined this hypothesis in theoretical and laboratory studies, looking in detail at the streaming velocities in the boundary layers at the bottom. Lau and Travis (1973) included the effect of a bottom slope, which showed that the bar spacing should increase with water depth. A test of this hypothesis is that the spacing between the offshore bars would increase with water depth owing to the change in wave length of the waves.

Long-period waves, which do not break on a sloping beach, can be described using shallow water wave theory and, for a planar beach, $h = mx$, the solution (Chapter 5) is

$$\eta(x, t) = a J_0(2kx) \cos \sigma t,$$

where J_0 is the zeroth order Bessel function, a is the shoreline amplitude (as $J_0(0) = 1.0$), and $k = \sigma/\sqrt{gh(x)}$ and σ are the wave number and angular frequency of the wave, respectively. Noda (1968) demonstrated that sediment would be moved to the antinodal positions of the wave profile, leading to the formations of bars. Short (1975) examined multiple bars on the shore of the Beaufort Sea and concluded that the bars were generated by standing infra-gravity waves that had a period coinciding with the peak wave period of a storm (75–150 s).

9.5.3 EDGE WAVE MODEL

Edge waves (Chapter 5) are standing or progressive waves that travel along a shoreline. The edge wave water surface displacement decays offshore; for high-mode edge waves, there are many zero crossings (or nodal points) in the offshore direction. Each of the nodes is associated with a maximum in the cross-shore velocities. Bowen and Inman (1971) argued that edge waves could be responsible for large crescentic bars that occur on many shorelines and have produced these longshore periodic bar features in a laboratory experiment.

Holman and Bowen (1982) explained that the superposition of several edge waves creates currents and bottom shear stresses that cause longshore varying bar patterns. For example, the combination of edge waves with equal and opposite wave numbers leads to a periodically varying (in the longshore direction) steady flow field that can give rise to periodic sediment transport patterns, which Holman and Bowen calculated. All of these edge wave models lead to alongshore varying bar patterns.

Kirby, Dalrymple and Liu (1981), solving the mild-slope equation for varying water depth, show that the presence of a bar can modify the attendant long wave field in such a way that it maintains a given antinodal position even if the frequency of oscillation changes. The bars tend to trap the antinodal wave positions over the bar crest, implying that once a bar field is established, the wave-trapping can lead to bar maintenance over a wide range of subsequent wave climates.

9.5.4 MULTIPLE BREAKPOINT MODEL

Another possibility is that waves breaking first over the offshore bar reform over the deeper bar trough to break again over the next shoreward bar. This process is then repeated all the way inshore. Dolan and Dean (1985) developed a numerical model to predict this behavior with reasonable success for bar fields in the Chesapeake Bay. The model predicts wave breaking over each of the bars for representative wave conditions, although no field tests of the model were carried out.

9.5.5 OVERTAKE MODEL

The final possibility posed here is due to the superposition of free and "locked" harmonics of the fundamental wave period. The locked harmonics are those associated with a constant-form nonlinear Stokes wave. These different waves are observed in wave tanks, especially for large waves, unless great care is taken in prescribing the wavemaker motion (Buhr Hansen and Svendsen 1974). Because the free and locked harmonics travel at different speeds, they alternately reinforce and cancel at different locations along the wave tank. At locations of cancellation, sediment accumulations have been observed (Bijker, van Hijum, and Vellinga 1976; Hulsbergen 1974), leading to the supposition that this could be a cause for multiple bars. However, in addition to requiring a mechanism for the free harmonics in nature, it can be shown that the bar spacing would increase with shallower water with this model, which is contrary to what is observed.

9.6 BEACH CUSPS

Beach cusps are rhythmic features in the alongshore direction located on the beach face. Beach cusps appear as longshore-regular planforms with an individual cusp composed of rather pointed features (the horns) aimed toward the sea and an embayment between adjacent horns. At a given tide stage, cusps may be located above the water level if they were formed at a higher tide stage, or they may be actively worked on by the swash, running up and down in the embayments. A set of cusps may extend for a kilometer or more along the shoreline with little apparent variation in their dimensions. The most obvious physical dimension of cusp systems is their spacing, which ranges from 1 m or less to over 60 m. Larger alongshore rhythmic features exist; however, as discussed in Section 9.3, they are believed to have different causes.

There is no general agreement as to the mechanisms for the formation and maintenance of beach cusp systems. Like other coastal processes, the likelihood is that there are numerous valid causes, some of which are more prevalent and others of which occur less often or only under special circumstances. The following conditions are generally conducive, but not universally necessary, for their formation: steep beach face slope, surging waves approaching more or less normally to the shore, a fairly well-defined dominant wave period (narrow wave spectrum), and sediment sizes ranging from medium sand to cobbles. High tide seems to be conducive to the formation of cusps because the water level is relatively stationary for a time sufficient for cusps to be generated; further, high-tide cusps are not erased during other tide levels (Seymour and Aubrey 1985). Beach cusps can form within a few minutes (Komar 1973); however, this is not likely to be the general case (Holland 1998 indicates that the development can take hours). Sometimes the beach cusp spacings are not very regular, and it is possible that the spacings are adjusting to a change in wave characteristics or other controlling parameters. As discussed later, once beach cusps are well-formed, they seem to be self-maintaining and may be able to preserve their spacing even under conditions that would not result in formation at that particular spacing.

9.6.1 POSSIBLE MECHANISMS

Numerous mechanisms have been proposed for the generation and maintenance of beach cusps. Among the hypotheses are inhomogeneities of the beach face material (Johnson 1919, Kuenen 1948), hydrodynamic instabilities in the wave swash (Gorycki 1973), the breaking waves themselves (Cloud 1966), the presence of rip currents (Shepard 1963, Komar 1971, Hino 1974), the breaching of dune ridges and berms (Jefferson 1899, Timmermans 1935, Evans 1938), edge waves (Escher 1937, Komar 1973), intersecting synchronous wave trains (Dalrymple and Lanan 1976), the instability of the littoral drift (Schwartz 1972), swash processes (Dean and Maurmeyer 1980), and finally self-organization (Werner and Fink 1993). Although many of these theories may have validity to some degree, we will focus on just four: edge waves, synchronous waves, self-organization, and swash processes.

9.6.1.1 Edge Waves

As discussed in Chapter 5, edge waves are waves trapped along the shore by refraction and may be present as propagating or standing waves. When present as standing waves, there is a fixed longshore periodicity in water surface elevation and water particle kinematics. For a planar beach of slope β and any given edge wave period T_e, many longshore wave modes are possible, as given by the following dispersion relationship:

$$L_e = \frac{gT_e^2}{2\pi} \sin(2n+1)\beta \tag{9.2}$$

in which $n = 0, 1, 2, \ldots$ is the mode of the edge wave (which is also equal to the number of zero crossings of the water surface elevation in the offshore direction).

A feature that is both attractive and raises questions for an edge wave mechanism for cusps is that there are so many modes that can be used to compare with the measured spacings from the field, it is generally possible to find one or more modes that will agree reasonably well with observed cusp spacings. Secondly, for field measurements, there is always uncertainty as to the appropriate period to utilize in Eq. (9.2), and, because the edge wavelength varies as the square of the period, the predicted length varies substantially. Finally the beach slope m is not uniform across the surf zone, usually varying from a maximum at the beach face and decreasing in an offshore direction.

To provide an indication of the possible range of spacings using Eq. (9.2), consider the following conditions:

$$\beta = 2.86° \pm 10\%$$

$$T_e = 8s \pm 10\%$$

$$n = 0, 1, 2, 3, 4, 5$$

The results are presented by the solid lines in Figure 9.8, where it is seen that even the preceding small ranges of variability encompass the entire range of possible beach cusp spacings from 5.5 m and larger. Thus, it is generally possible to select combinations of modal numbers, wave period, and effective beach slope that will agree well with the observed beach cusp spacings regardless of whether edge waves had a role in generating the cusps.

As was briefly mention in Chapter 5, Holman and Bowen (1979), and later, Kirby et al. (1981), using a simpler method, have developed numerical methods for determining the edge wavelengths for the case of arbitrary nearshore beach profiles, thereby eliminating the need to define an effective planar slope. Holman and Bowen pointed out that the choice of an average planar slope on natural profiles and the use of Eq. (9.2) can lead to errors in edge wavelength of well over 100 percent – even for simple profiles.

Guza and Inman (1975) conducted laboratory and theoretical studies of edge waves and the formation of beach cusps, which indicated that friction due to the turbulent surf zone conditions will heavily dampen edge waves with higher modal numbers. They suggested that a relevant edge wave period for cusp generation T_e, is a subharmonic of the incident wave period T_w and thus that T_e is equal to $2T_w$, the subharmonic wave being the result of nonlinear wave interactions. Guza and Inman further found that the zeroth mode edge wave is most likely to survive and be a causative mechanism for beach cusp formation. The edge wave spacing for this condition is

$$L_e = \frac{g(2T_w)^2}{2\pi} \sin \beta$$

The associated range of edge wave spacings has been added to Figure 9.8 as the dashed line. Note that the cusp spacing is one-half of the edge wavelength.

In their laboratory experiments, Guza and Inman tuned the incident waves to obtain a standing subharmonic edge wave along their sloping beach. The beach was covered with a thin veneer of sand. The standing edge waves removed the sand in the area wetted by the run-up of these waves, leaving the bare concrete. A photograph of their results is shown in Figure 9.9.

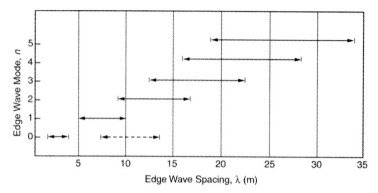

Figure 9.8 Range in beach cusp spacing explainable by various modes of edge waves with 10-percent variability in wave period and beach slope. The dashed line is for a subharmonic edge wave.

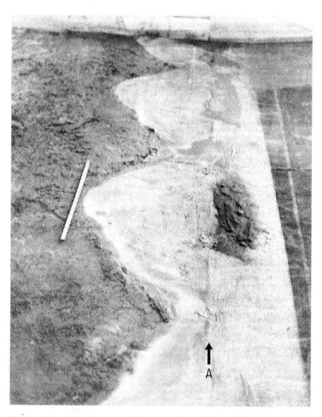

Figure 9.9 Cuspate morphology produced by subharmonic edge waves. Black is sand, and white is exposed concrete. A meter stick is shown for scale, and A denotes the still water level (Guza and Inman 1975. Copyright © by the American Geophysical Union).

One notable feature of the subharmonic edge wave mechanism is that the edge wave motion requires the incident water wave to run up *in every other embayment* simultaneously, which is not often observed in nature. In the field, usually all the embayments are washed by the incident waves each wave period. Also, the general edge wave mechanism predicts that the dominant run-up occurs in the embayments contrary to most observations in nature.

9.6.3 SYNCHRONOUS WAVES

Shaler (1895) was one of the first to present a beach cusp theory. He proposed that the uprushing wave swash is indented because of the presence of intersecting waves; therefore, at periodic locations along the beach, there is more water than at others. This led to the scouring of the cusp embayments. Branner (1900) presented the first theory for intersecting waves of the same period. More recently, Dalrymple and Lanan (1976) examined empirically the generation of beach cusps caused by two incident wave trains with different directions but of the same wave period. The wave trains superimpose and cancel each other out spatially, resulting in periodically spaced nodal lines along the shoreline. At these locations, rip currents occur, as do the horns of cusps. The spacing of the cusps coincides with the spacing of the nodal lines, which is easily obtained from the offshore wave data,

$$L_c = \frac{gT^2}{2\pi(\sin\theta_1 - \sin\theta_2)},$$

where θ_1 and θ_2 are the offshore wave angles of the two wave trains. This theory has been verified in the laboratory and results in water circulations within the cusps that agree with the observations of Kuenen (1948) and Komar (1971) but are contrary to what are observed generally in nature. In this model, the water rushes up the center of the embayment as the wave moves ashore and flows seaward at the horns of the cusp.

9.6.4 SELF-ORGANIZATION

Werner and Fink (1993) described a numerical model for the generation of cusps based on simple laws for particle motion and sediment transport including Newton's laws for motion with gravity and a cubic sediment transport law. Beginning with an initially planar beach, the fluid flow moves the sediment into cusp patterns that are quite robust; that is, they are not strongly affected by changes in model parameters and assumptions. The cusps form through a combination of positive and negative feedbacks: when incipient changes occur on the beach face, more deposition takes place to form the cusps, whereas more fully formed cusps tend to retard deposition and erosion.

The water circulation in these model beach cusps is such that the waves divide at the cusp horns and return seaward at the embayment; the motions in adjacent embayments occur simultaneously rather than once every other wave, as would be required for the edge wave mechanism.

The spacing of these cusps is proportional to the horizontal swash excursion ξ_x, as in Dean and Maurmeyer (1980), which is presented next.

9.6.5 SWASH MECHANISM FOR BEACH CUSP INITIATION AND MAINTENANCE

A similar mechanism for beach cusps is based on an extension of the swash mechanism discussed earlier in Chapter 5. The basic concept of the swash mechanism is that the interaction of the cusp topography and the swash causes circulation cells that tend to construct and then perpetuate the cusp features. A critical requirement is that the wave period T_w be approximately equal to the natural swash period T_n, thereby establishing a condition of (near) resonance. Consider for simplicity the initial condition to be waves approaching directly toward a planar beach face. Swash mechanisms on this planar beach face have been discussed previously with a natural period given approximately by

$$T_n = \frac{2V_0}{g \sin \beta},$$

(9.3)

where V_0 is the swash velocity at the origin and β is the beach face slope. We now consider small perturbations of elevation on the beach face such that, in some places and some times, the swash would be predominantly onshore and predominantly offshore elsewhere. When the swash flows are predominantly offshore, because the wave and swash periods are equal, the outflow tends to nullify the next uprush at those locations. However, at the location of predominant uprush, the next uprush is unimpeded, and the sediment-laden water proceeds upslope; in the process, gravity and friction reduce the velocity, causing deposition and contributing to the cusp horns and the lateral flow, which reduces uprush in the areas of predominant outflow (the embayments) and allows the uprush on the horns to occur with vigor. When the cusps are constructed to their so-called equilibrium geometry, the uprush occurs with substantial force and provides sediment to maintain the cusp elevation. The water flows laterally to join with the lateral flow from the adjacent horn and then returns seaward in the middle of the embayment, sometimes as a fairly rapidly flowing stream with considerable momentum, contributing to scour of the embayment and to nullification of the next uprush. The flow patterns are portrayed in Figure 9.10.

The approach to developing an analytical representation of the swash zone mechanism of beach cusps follows Dean and Maurmeyer (1980) and is based on the rather simple swash mechanics discussed previously but extended to allow lateral flow and calibrated with one empirical coefficient.

An approximate expression for the elevation z of the cusps above a datum is

$$z(x, y) = A\, y(1 + \epsilon \sin kx),$$

(9.4)

where A is the average beach slope ($A = \tan \beta$), k is the alongshore wave number associated with the cusp spacing, $k = 2\pi/\lambda$ (where λ is the cusp spacing), and ϵ is a parameter defining the relief of the cusp feature; the x- and y-coordinate system is shown in Figure 9.10. For example, the elevations at some arbitrary distance y_* up the beach face at the cusp horn and embayment follow from Eq. (9.4) and are

$$z(x_H, y_*) = Ay_*(1 + \epsilon)$$
$$z(x_E, y_*) = Ay_*(1 - \epsilon)$$

(9.5)

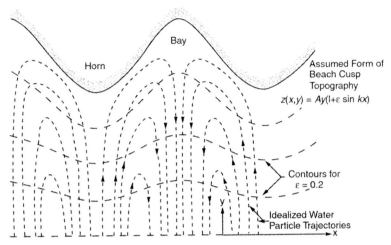

Figure 9.10 Assumed beach cusp topography and observed mean flow patterns. The uprush is concentrated on horns, and seaward return flow is concentrated in bays (from Dean and Maurmeyer 1980).

Thus, the parameter ϵ can be determined as

$$\epsilon = \frac{z_H - z_E}{2Ay_*}, \tag{9.6}$$

that is, ϵ is relative vertical relief expressed as the ratio of the amplitude of the vertical relief between the cusp and horn to the average vertical elevation of the cusp above the reference datum ($z = 0$). Figure 9.10 presents an example of cusp contours calculated from Eq. (9.4) for $\epsilon = 0.2$.

With reference again to Figure 9.10, the differential equations for the water particle trajectories in the x and y directions, with frictional effects excluded, are

$$\frac{d^2\xi_x}{dt^2} = -g\frac{\partial z}{\partial x} = -gA\epsilon ky\cos kx$$

$$\frac{d^2\xi_y}{dt^2} = -g\frac{\partial z}{\partial y} = -gA(1 + \epsilon\,\sin kx) \approx -gA, \tag{9.7}$$

where ξ_x is the longshore displacement of the water particle and ξ_y is the uprush coordinate. Solutions to Eqs. (9.7) are

$$\xi_x(t) = \xi_{x_0} + V_{0x}t - \frac{1}{2}t^2 gAyk\epsilon\cos kx$$

$$\xi_y(t) = \xi_{y_0} + V_{0y}t - \frac{gAt^2}{2}, \tag{9.8}$$

where ξ_{x_0} and ξ_{y_0} are the initial particle positions, and V_{0x}, V_{0y} are the initial swash velocities at $y = 0$. The longshore displacement ξ_x depends on x, y, and ϵ. For our purposes here we will set $\xi_{y0} = 0$ and $V_{0x} = 0$ and take the following as representative values of x and y:

$$y \approx \overline{y} = \frac{(\xi_y)_{max}}{2} \tag{9.9}$$

$$\cos kx \to \overline{\cos kx} = \frac{2}{\pi} \tag{9.10}$$

The maximum uprush of the water particle, Eq. (9.8), is

$$(\xi_y)_{\max} = \xi_y (T_n/2) = \frac{V_0^2}{2gA} \tag{9.11}$$

after using the definition for T_n, Eq. (9.3). Then, we find that the length of the beach cusp is

$$\lambda = 2(\xi_x)_{\max} = 2\left[\frac{T_n^2}{2} g A \overline{y} \frac{2\pi}{\lambda} \epsilon \frac{2}{\pi}\right] \tag{9.12}$$

or, noting that

$$T_n^2 = \frac{8(\xi_y)_{\max}}{gA}, \tag{9.13}$$

we have

$$\lambda^2 = \left[16\epsilon(\xi_y)_{\max}^2\right] \tag{9.14}$$

Therefore, the final representation for the cusp spacing is

$$\lambda = 4\sqrt{\epsilon}\,(\xi_y)_{\max}, \tag{9.15}$$

which indicates that the spacing of the beach cusps should be proportional to the magnitude of the swash.

9.6.5.1 Field Experiments of Longuet-Higgins and Parkin

Longuet-Higgins and Parkin (1962) carried out field studies of beach cusps at Chesil Bank, England. The conditions for studying initiation of beach cusps were ideal because each high tide obliterated the forms that had developed at a previous tide level. Figure 9.11 presents the variation of beach cusp spacing with wave height, swash length, and wave period. With reference to Eq. (9.2), if edge waves are the responsible mechanism, the cusp spacing should vary with the square of the wave period. From Figure 9.11(a), no such dependence is evident. By far, the best dependence occurs between spacing and swash length, as shown in Figure 9.11(c). This is in general accordance with Eq. (9.15). It is noted for later purposes that the expression between spacing, λ, and excursion, $(\xi_y)_{\max}$ (both in feet), is approximately

$$\lambda = 0.45(\xi_y)_{\max} + 10.0 \tag{9.16}$$

that is, there is an offset at $(\xi_y)_{\max} = 0$.

9.6.5.2 Field Measurements of Dean and Maurmeyer

Dean and Maurmeyer (1980) reported on a fairly extensive set of field measurements of beach cusps. The measurements included plane table surveys of beach cusps, grain size analyses, and wave observations but no measurements of wave conditions. The measurements were made at Point Reyes and Drakes Bay, California. Point Reyes beaches are directly exposed to the Pacific Ocean and are composed of fairly coarse

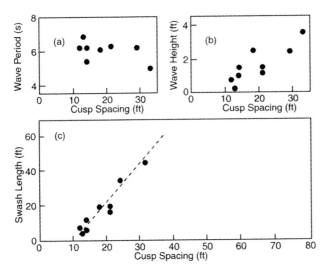

Figure 9.11 Beach cusp spacing related to wave height, swash length, and wave period. Measurements obtained by Longuet-Higgins and Parkin (1962) at Chesil Beach, England.

material; beach cusps are nearly always present. At certain times, up to four berm elevations were present with relict beach cusps present on two of the three inactive berms and active beach cusps on the lower berm. Table 9.1 summarizes the various sets of measurements during the 2 years that visits were made to these beaches. It is of interest that, during 1977, the beach cusp spacings were approximately 60, 40, and 30 m on the upper, middle, and lower (active) berms, respectively. Because greater swash excursions were required to build the higher berm elevations, these results are qualitatively consistent with the hypothesis that larger swash excursions are associated with greater beach cusp spacings (i.e., Eq. (9.15)). The field data also included eight plane table surveys of beach cusps, and from these the values of ϵ were extracted, as summarized in Table 9.2.

Table 9.1 Spacing Characteristics of Beach Cusps at Point Reyes and Drakes Bay Beaches

Location	Berm Level	Year	No. of Cusps	Average Spacing (m)	Standard Deviation of Spacing (m)
Point Reyes South Beach	Upper	1977	11	62.0	8.8
	Upper	1979	10	61.6	9.5
	Mid	1977	15	41.4	7.2
	Mid (Active)	1979	21	42.8	7.9
	Lower (Active)	1977	30	27.9	2.5
Drakes Beach	Only One Berm	1977	31	29.8	5.2
	Present	1979	26	23.2	4.5

Source: (Dean and Maurmeyer 1980).

Table 9.2   Summary of Beach Cusp Parameters

Date and Location of Survey	ϵ
Drakes Bay Beach August 25, 1977	0.17
Point Reyes Beach, North July 22, 1978	0.17
Point Reyes Beach, South July 23, 1978	0.10
Drakes Bay Beach July 23, 1978	0.08
Drakes Bay Beach July 23, 1979	0.11
Point Reyes Beach, South July 24, 1979	0.15
Point Reyes Beach, South July 27, 1979	0.22
Point Reyes Beach, South July 29, 1979	0.25

Source: Dean and Maurmeyer (1980).

With the results available in Table 9.2 and Eq. (9.15), we are in a position to make a crude comparison with the results found by Longuet-Higgins and Parkin, as presented in Figure 9.11 and Eq. (9.16). Using an average value of $\epsilon = 0.16$, we find from Eq. (9.15)

$$\lambda = 1.6 \, (\xi_y)_{max} \tag{9.17}$$

compared with the best fit from Longuet-Higgins and Parkin of

$$\lambda = 0.45 \, (\xi_y)_{max} + 10.0 \tag{9.18}$$

These two equations are compared with the data in Figure 9.12, where it is seen that Eq. (9.17) only agrees with the data for the smaller cusp spacings.

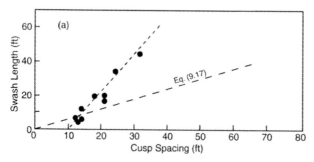

Figure 9.12 Comparison of field data by Longuet-Higgins and Parkin (1962) with Eq. (9.17) based on a swash model.

Table 9.3 Summary of Uprush Excursion and Beach Cusp Measurements on July 28, 1979

Location	Uprush Excursions		Beach Cusp Spacings		
	No. of Obs.	Average Value, $(\xi_y)_{max},$ (m)	No. of Obs.	Average Value, $\lambda,$ (m)	$\lambda/(\xi_x)_{max}$
Drakes Bay[a]	46	12.5	26	23.2	1.8
Point Reyes	11	14.6	4	46.0	3.2

[a]Cusps in process of formation.

Dean and Maurmeyer also reported on measurements of a series of uprush excursions $(\xi_y)_{max}$ and associated beach cusp spacings. These results are reported in Table 9.3 and present ratios of $\lambda/(\xi_y)_{max}$ of 1.8 and 3.2 for Drakes Bay and Point Reyes, respectively. On the basis of the data in Table 9.2 and Eq. (9.15), the predicted ratio is in the range of 1.2 to 2.0.

9.6.5.3 Field Measurements of Smith and Bodie

Smith (1973) and Bodie (1974) conducted field measurements at Del Monte Beach within Monterey Bay, California. Smith's measurements were carried out daily during the winter of 1972–73 at three sites and included documentation of cusp geometry and wave conditions. The time scale required for cusp formation was found to be several hours to several days. Additionally, he noted that the cusps formed serially, not simultaneously; that the swash motion observed was as depicted in Figure 9.10; and that the wave height was the most important determinant of cusp spacing. Large waves obliterated the cusps, and cusp formation was primarily a depositional process.

Bodie's studies were conducted from January to March of 1974 on the same beach studied by Smith with similar results. During the study period, four cycles of beach cusp formation were documented with intervening periods of no cusps. The time required for cusp formation ranged from 6 hours to 3 days, but one day was a typical time scale. All cusps were observed to begin from an initial "embryo horn," and the remainder formed sequentially from this initial perturbation. The spacing was found to be primarily a function of wave height and period. Although swash measurements were conducted in both studies, swash was not a parameter identified by the regression analysis as being strongly correlated with cusp spacing. This may be due to the difficulty of discriminating the effects of swash excursion and wave length.

9.6.5.4 Field Measurements by Holland and Holman

Holland and Holman (1996) have reported on a well-instrumented field experiment (Duck94) in which wave gauges and video camera technology were employed during two cusp formation events at Duck, North Carolina. The data were analyzed specifically to determine whether the model of cusp formation by synchronous or subharmonic edge waves was consistent with the measured cusp spacing and hydrodynamics.

It was concluded that there was no statistical support for the formation of the cusps by the edge wave mechanisms proposed. However, some evidence was found for a mechanism involving edge waves that had not been identified previously. It was proposed that swash motions over well-developed cusps at the incident wave frequency could excite edge waves at the incident wave frequency, which then could result in a second set of cusps at a spacing consistent with the edge wave length.

9.6.5.5 Field Measurements of Takeda and Sunamura

Takeda and Sunamura (1983) measured a total of 52 combinations of swash excursion $(\xi_x)_{max}$ and beach cusp spacing λ. They did not report the parameter ϵ, and thus a direct comparison with Eq. (9.15) is not possible. The range of ξ_x/λ was 0.6 to 5.6 with a peak in the data of 1.7 and a mean of 2.0. This average is in reasonable agreement with Eq. (9.17), for which the ratio is 1.6 and is based on the average ϵ values from Point Reyes and Drakes Beach, California.

9.6.5.6 Field Measurements of Masselink and Pattiaratchi

Masselink and Pattiaratchi (1998) conducted detailed measurements of beach cusps at two sites on the southwestern coast of Australia. Additionally, they solved Eq. (9.7) numerically, thereby obtaining accurate solutions. They found three types of cusp morphology that could be segregated by the parameter $\epsilon\,(\xi_m/\lambda)^2$. For a range of this parameter between 0.015 and 0.15, the typical flows depicted in Figure 9.10 occurred. For values of the parameter outside this range, the swash motion was not "in resonance" with the waves.

9.6.5.7 Field Observations by Holland

Holland (1998) analyzed almost 9 years of video imagery taken at the Field Research Facility in Duck, North Carolina. Fifty-seven cusp formation events were observed during that time, and almost all of the cusp generation occurred within 1 to 3 days after major storms. The mean cusp spacing at Duck was about 26 m in length. The incident wave direction for 98 percent of the cusp events was within $\pm 12°$ of the beach normal, and most often the incident wave spectrum was narrow banded, implying a dominant incident wave train. Additionally, he found that the beach had to be reflective; that is, with a steep slope.

9.7 SUMMARY

There are several ubiquitous coastal features that still defy positive identification of the mechanisms that form them or, at least, will provoke a good argument among coastal scientists. These features are reminders of our lack of complete knowledge of the forces and processes that shape the nearshore. Further advances in understanding and identifying these controlling processes will require additional focused and definitive nearshore measurements supplemented with advances in analytical and numerical representation of the processes. By arriving at better descriptions of these processes, we have tools for an improved understanding of other processes that are more germane to coastal engineers.

REFERENCES

Bagnold, R.A., "Sand movement by Waves: Some Small Scale Experiments with Sand of Low Density," *J. Institution of Civil Engineers*, 27, 4, Paper 5554, 447–469, 1947.

Bakker, W.T., "Mathematical Theory about Sand Waves and Its Application on the Dutch Wadden Isle of Vlieland," *Shore and Beach*, 5–14, October 1968.

Bijker, E.W., E. Van Hijum, and P. Vellinga, "Sand Transport by Waves," *Proc. 15th Intl. Conf. Coastal Eng.*, ASCE, Honolulu, 1,149–1,167, 1976.

Bodie, J.G., "Formation and Development of Beach Cusps on Del Monte Beach, Monterey, California," M.S. thesis, U.S. Naval Postgraduate School, Monterey, CA, 66 pp., 1974.

Bowen, A.J., and D.L. Inman, "Edge Waves and Crescentic Bars," *J. Geophys. Res.*, 76, 36, 8662–8671, 1971.

Branner, J.C., "The Origin of Beach Cusps," *J. Geol.*, 8, 481–484, 1900.

Bruun, P., "Migrating Sand Waves or Sand Humps, with Special Reference to Investigations Carried Out on the Danish North Sea Coast," *Proc. 5th Intl. Conf. Coastal Eng.*, ASCE, New York, 269–295, 1954.

Buhr Hansen, J., and I.A. Svendsen, "Laboratory Generation of Waves of Constant Form," *Proc. 14th Intl. Conf. Coastal Eng.*, ASCE, Copenhagen, 321–339, 1974.

Carter, T.G., P.L.-F. Liu, and C.C. Mei, "Mass Transport by Waves and Offshore Sand Bedforms," *J. Waterways, Harbors, Coastal Eng. Div.*, ASCE, 99, 165–184, 1973.

Cloud, P.E., "Beach Cusps: Response to Plateau's Rule," *Science*, 154, 890–891, 1966.

Dalrymple, R.A., and G.A. Lanan, "Beach Cusps Formed by Intersecting Waves," *Bull. Geol. Soc. America*, 87, 57–60, 1976.

Dean, R.G., and E.M. Maurmeyer, "Beach Cusps at Point Reyes and Drakes Bay Beaches, California," *Proc. 17th Intl. Conf. Coastal Eng.*, ASCE, Sydney, 863–885, 1980.

Dolan, T.J., and R.G. Dean, "Multiple Longshore Sand Bars in the Upper Chesapeake Bay," *Est. Coastal and Shelf Sci.*, 21, 727–743, 1985.

Escher, B.G., "Experiment on the Formation of Beach Cusps," *Leidse Geol. Meded.*, 9, 79–104, 1937.

Evans, O.F., Classification and Origin of Beach Cusps," *J. Geol.*, 46, 615–617, 1938.

Evans, O.F., "The Low and Ball of the East Shore of Lake Michigan," *J. Geol.*, 48, 476–511, 1940.

Gorycki, M.A., "Sheetflood Structure: Mechanism for Beach Cusp Formation," *J. Geol.*, 81, 109–117, 1973.

Grove, R.S., C.J. Sonu, and D.H. Dykstra, "Fate of Massive Sediment Injection on a Smooth Shoreline at San Onofre, California," *Proc. Coastal Sediments '87*, ASCE, New York, 531–538, 1987.

Guza, R.T., and D.L. Inman, "Edge Waves and Beach Cusps," *J. Geophys. Res.*, 80, 21, 2997–3012, 1975.

Hino, M., "Theory on Formation of Rip-Current and the Cuspidal Coast," *Proc. 14th Intl. Conf. Coastal Eng.*, 901–919, 1974.

Holland, K.T., "Beach Cusp Formation and Spacing at Duck, USA," *Continental Research*, 18, 1081–1098, 1998.

Holland, K.T., and R.A. Holman, "Field Observations of Beach Cusps and Swash Motions," *Marine Geology*, 134, 77–93, 1996.

Holman, R.A., and A.J. Bowen, "Edge Waves on Complex Beach Profiles," *J. Geophys. Res.*, 84, C10, 6339–6346, 1979.

Holman, R.A., and A.J. Bowen, "Bars, Bumps, and Holes: Models for the Generation of Complex Beach Topography," *J. Geophys. Res.*, 87, 457–468, 1982.

Hulsbergen, C.H., *Proc. 14th Intl. Conf. Coastal Eng.*, ASCE, Copenhagen, ASCE, 392–411.

Jefferson, M.S.W., "Beach Cusps," *J. Geol.*, 7, 237–246, 1899.

Johnson, D.W., *Shore Processes and Shoreline Development*, New York: J.W. Wiley, 584 pp., 1919.

King, C.A.M., and W.W. Williams, "The Formation and Movement of Sand Bars by Wave Action," *Geog. Journal*, 113, 70–84, 1949.

Kirby, J.T., R.A. Dalrymple, and P.L.-F. Liu, "Modification of Edge Waves by Barred-Beach Topography," *Coastal Eng.*, Vol. 5, 1981.

Komar, P.D., "Nearshore Cell Circulation and Formation of Giant Beach Cusps," *Bull. Geol. Soc. America*, 82, 2643–2650, 1971.

Komar, P.D., "Observations of Beach Cusps at Mono Lake, California," *Bull. Geol. Soc. America*, 84, 3593–3600, 1973.

Kuenen, Ph.H., "The Formation of Beach Cusps,"*J. Geol.*, 56, 34–40, 1948.

Lau, J., and B. Travis, "Slow Varying Stokes Waves and Submarine Long-Shore Bars," *J. Geophys. Res.*, 78, 4489–4498, 1973.

Lettau, H., "Stehende Wellen als Ursache und Gestatender Vorgänge in Seen," *Ann. Hydrogr. Mar. Meterol.*, 60, 385, 1932.

Lippman, T.C., and R.A. Holman, "Quantification of Sand Bar Morphology: A Video Technique Based on Wave Dissipation," *J. Geophys. Res.*, 94, C1, 995–1011, 1989.

Lippman, T.C., and R.A. Holman, "The Spatial and Temporal Variability of Sand Bar Morphology," *J. Geophys. Res.*, 95, C7, 11575–11590, 1990.

Longuet-Higgins, M.S., and D.W. Parkin, "Sea Waves and Beach Cusps," *Geog. Journal*, 128, 194–201, 1962.

Masselink, G., and C.B. Pattiaratchi, "Morphological Evolution of Beach Cusps and Associated Swash Circulation Patterns," *Marine Geology*, 146, 93–113, 1998.

Noda, H., "A Study on Mass Transport in Boundary Layers in Standing Waves," *Proc. 11th Intl. Conf. Coastal Eng.*, ASCE, 227–235, 1968.

Schwartz, M.L., "Theoretical Approach to the Origin of Beach Cusps," *Bull. Geol. Soc. America*, 83, 1115–1116, 1972.

Seymour, R.J., and D.G. Aubrey, "Rhythmic Beach Cusp Formation: A Conceptual Synthesis," *Marine Geology*, 65, 289–304, 1985.

Shaler, N.S., "Beaches and Tidal Marshes of the Atlantic Coast," *Natl. Geogr. Monog.*, I, 137–168, 1895.

Shepard, F.P., *Submarine Geology*, 2nd ed., New York: Harper and Row, 557 pp., 1963.

Short, A.D., "Multiple Offshore Bars and Standing Waves," *J. Geophys. Res.* 80, 3838–3840, 1975.

Silvester, R., "Growth of Crenulate-shaped Bays to Equilibrium," *J. Waterways, Port, Coastal, and Ocean Eng.*, ASCE, 96, WW2, 275–287, 1970.

Silvester, R., and S. Ho, "Use of Crenulate Shaped Bays to Stabilize Coasts," *Proc. 13th Intl. Conf. Coastal Eng.*, ASCE, Vancouver, 1347–1365, 1972.

Smith, D.H., "Origin and Development of Beach Cusps at Monterey Bay, California," M.S. thesis, U.S. Naval Postgraduate School, Monterey, CA, 63 pp., 1973.

Takeda, I., and T. Sunamura, "Formation and Spacing of Beach Cusps," *Coastal Eng. in Japan*, 26, 121–135, 1983.

Terpstra, P.D., and M.J. Chrzastowski, "Geometric Trends in the Evolution of a Small Log-Spiral Embayment on the Illinois Shore of Lake Michigan," *J. Coastal Res.*, 8, 3, 603–617, 1992.

Thevenot, M.M., and N.C. Kraus, "Longshore Sand Waves of Southhampton Beach, New York – Observation and Numerical Simulation of their Movement," *Marine Geology*, 126 (1–4), 249–269, 1995.

Timmermans, P.D., "Proeven ove den Invloed Golven op een Strand," *Leidse Geol. Meded.*, 6, 231–386, 1935.

Verhagen, H.J., "Sand Waves Along the Dutch Coast," *Coastal Eng.*, 13, 129–147, 1989.

Vichetpan, M., "Development of Crenulate Shaped Bays," Thesis No. 280, Asian Institute of Technology, Bangkok, 1971.

Werner, B.T., and T.M. Fink, "Beach Cusps as Self-Organized Patterns," *Science*, 260, American Association for the Advancement of Science, 968–971, 1993.

Wind, H.G., "An Analytical Model of Crenulate Shaped Beaches," *Coastal Eng.*, 23, 243–253, 1994.

Wright, L.D., and A.D. Short, "Morphodynamic Variability of Surf Zones and Beaches: A Synthesis," *Mar. Geology*, 56, 93–118, 1984.

Wright, L.D., A.D. Short, J.D. Boon III, B. Hayden, S. Kimball, and J.H. List, "The Morphodynamic Effects of Incident Wave Groupiness and Tide Range on an Energetic Beach," *J. Mar. Geology*, 74, 1–20, 1987.

Yasso, W., "Plan Geometry of Headland-Bay Beaches," *J. Geol.*, 73, 702–714, 1965.

EXERCISES

9.1 Suppose that you visit a beach where beach cusps are present and the run-up (or swash) occurs alternatively in adjacent embayments. What does this imply with regard to the formation mechanism for the cusps? Alternatively, the swash occurs simultaneously in adjacent embayments. What mechanism(s) does this imply?

9.2 Two rocky headlands, situated 3 km apart, are located on a shoreline that runs in a north–south direction and faces west. Between the headlands is a sandy beach. The predominant wave direction is from the north–northwest. What is the likely indentation of the sandy shoreline? Sketch the shoreline.

9.3 A spiral bay is bounded by two erosion-resistant headlands, as shown in Figure 9.13. The updrift supply of sediment to the bay Q_{so} is one-half the transport capacity of the waves, Q_{s*}. At the location indicated as B, a sand mining operation removes sand at a rate of Q_{sm}, which is smaller than Q_{so}. This sand mining begins at time $t = 0$.

   (a) Qualitatively sketch the bay planform after it has equilibrated to the sand removal. Discuss the reasons for the characteristics of the altered planform.

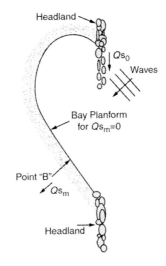

Figure 9.13 Sketch for Problem 9.3.

(b) Qualitatively sketch the bay planform after equilibration if the rate of sand mining were equal to Q_{so}. Discuss the reasons for the altered planform.

(c) Provide a sketch for part **(b)** of the distribution of $Q(s)$, where s is the longshore coordinate from the headland.

9.4 Consider the two possibilities (edge waves and swash processes) as causative mechanisms for beach cusps. Fill out the table below, indicating the effect (increase, decrease, or none) that an increase in each variable would have on beach cusp spacing. In addition, provide a brief explanation for each choice.

	Variables					
Theory	Wave Period	Beach Slope	Wave Height	Sediment Size	Swash Length	Cusp Relief[a]
Edge Wave						
Swash Processes						

[a]Cusp relief refers to the elevation difference between the horns and the embayments.

CHAPTER TEN

Modeling of Beaches and Shorelines

SUPERTANK may conjure up in your mind the title of a grade B movie but actually refers to one of the largest cooperative physical model studies ever carried out in coastal engineering. It was organized by the U.S. Army Corps of Engineers for the purpose of determining storm-induced erosion effects on coastlines in a wave tank large enough that modeling scale effects were not important.

The 8-week experiment, involving investigators from numerous universities, the Corps, and other institutions began in July 1991 in the O.H. Hinsdale Wave Research Laboratory at Oregon State University. The tank is 104 m long, 3.7 m wide, and 4.6 m deep with a hydraulic hinged wave paddle at one end capable of generating random or monochromatic waves up to 1.5 m and absorbing reflected waves. A sand beach of 0.22 mm diameter was constructed over 76 m of the tank, utilizing approximately 1700 metric tons of beach sand trucked in from the Oregon coast.

The instrumentation in the tank included wave gages, current meters, pressure gauges, video cameras, and sediment concentration gauges to measure the hydrodynamics, and beach and dune response. More details about the experiments are given in Kraus and Smith (1994) and Smith and Kraus (1995).

10.1 INTRODUCTION

A major goal of coastal engineering is to develop models for the reliable prediction of short- and long-term nearshore evolution. Ideally, these models, with appropriate wave and sediment information, would allow prediction of the behavior of the shoreline and the offshore bathymetry over some given period of time. Further, if coastal structures or beach nourishment were placed along this shoreline, the model would be able to evaluate the effectiveness and impacts of such modifications over any period of time. This coastal model would also be used by coastal planners for shoreline management.

There are two types of coastal models: physical models, which are real models of the region of interest that are physically smaller than the prototype, and equation-based models involving the solution of the equations governing the physical phenomena in the coastal zone. The latter type includes numerical models, solved on a computer, and analytical models which, although simpler, often provide conceptual tools for analysis and understanding.

No fully satisfactory shoreline model presently exists of either type owing to several factors, including our meager understanding of shoreline sediment transport, the scaling requirements for sediment, the uncertainties of representing the wave environment over short and long periods, and our inability to prescribe fully the hydrodynamics of the surf zone. However, models of both types have been developed partly to determine how well our present state of knowledge permits us to model the coastal environment, partly to identify topics about which our knowledge is weak, and partly to provide the best possible design tools for coastal projects.

In the future, it is likely that the most successful coastal models will be numerical models, which allow flexibility in the choice of initial and boundary conditions and the representation of arbitrary forcing; such models would permit evolutionary growth by being readily modified as our understanding of coastal processes increases. Even now the equations governing the behavior of the shoreline are too complicated to be fully captured in an analytic model, and physical models are hampered by the necessary scaling constraints and relatively high costs. Numerical models, complemented by good judgment developed by field experience, are anticipated to provide the user with reliable predictions.

In this chapter, physical, analytic, and numerical models of beach profile and shoreline planforms will be discussed.

10.2 PHYSICAL MODELING OF COASTAL PROCESSES

A model of a beach profile or of a segment of the shoreline can be built in a laboratory wave tank or basin. Typically, we simply scale down the prototype geometrically, multiplying all dimensions by a given fraction, which is the model scale. Then waves are generated that impinge on the coastal model, and the results are measured. However, determining the full-scale values corresponding to the model results is difficult, partly owing to the problems of scaling phenomena from the laboratory size to the prototype and partly because we do not utilize the same wave climate as occurs in nature. (Much of the following is from Dalrymple 1989(b). For a fuller discussion of coastal modeling, see Hughes 1993.)

The scaling problem arises because, although desirable, it is often not practical to carry out tests at prototype scale either because of limitations in the size of the laboratory or to the costs involved in the testing. Therefore, the prototype is reduced in size to a model that can be studied conveniently. The horizontal length scale of the model ℓ_r is the ratio of the length of a characteristic feature in the model ℓ_m, to the length of the same feature in the prototype, ℓ_p. Typical model scales are $\ell_r = \ell_m/\ell_p = 1/10$ to $1/100$.

Scale effect problems can arise in small-scale models when some of the forces in the model become larger (or smaller) in importance than they are in the prototype. For example, we know in the prototype that gravity is the important restoring force that permits waves to propagate. In the laboratory, a model using extremely short waves may encounter the problem that surface tension becomes more important than gravity, leading to unrealistic wave effects.

To counter scale effects, some models are distorted in such a way that the horizontal length scale is different from the vertical depth scale. For example, a model may have a horizontal length scale of 1/100 with a vertical length scale of 1/10. This means that the model water depths are deeper than in a nondistorted model. This model distortion may ensure that the frictional effects of the bottom do not inappropriately dominate the model flows.

10.2.1 WAVE BASINS AND WAVE TANKS

Another important aspect of laboratory model testing is the artificiality introduced into any testing by the test facilities themselves. Of primary importance is the appropriate wave climate to use in the laboratory. (An overview of wavemaker theory is provided by Svendsen 1985.)

Historically, a single-frequency (*monochromatic*) wave train has been used for testing. The generation of this type of wave train must be done with great care, for it has long been known that the sinusoidal motion of a wave paddle at one end of a test tank does not produce the desired wave train owing to nonlinear processes. For finite amplitude waves with $kh > 1.36$, Benjamin–Feir instabilities occur. Waves with slightly different frequencies than those generated grow and exchange energy with the desired wave train (see, for example, Mei 1983).

The free second harmonics of the wave motion introduced by the nonlinear motion of the wavemaker were explained by Madsen (1971) and Mei and Ünlüata (1972). A finite amplitude wave can be considered to have a fundamental wave, with angular frequency σ, plus higher harmonics (with frequencies $n\sigma, n = 2, 3, \ldots$). All of these harmonics travel at the speed of the fundamental wave. A wave with twice the frequency of the fundamental wave is created by the wavemaker that satisfies the dispersion relationship

$$(2\sigma)^2 = gk_f \tanh k_f h \tag{10.1}$$

This free wave travels with speed $C_f = 2\sigma/k_f$, which is different than the speed of the primary wave. This means that these two waves will be in and out of phase with each other at various locations down the tank and that, at different locations along a wave tank with a horizontal bottom, measured wave heights will vary considerably. (This phenomenon was discussed in the preceding chapter with regard to the "overtake" distance Section 9.5.5.)

Other wave tank problems include the generation of long bounded waves under wave groups (Ottesen Hansen et al. 1980), wave reflection, and side wall energy losses due to friction and roughness.

Presently, many laboratories are carrying out tests in facilities equipped for the generation of random waves with energy spectra corresponding to those measured in nature. Numerous methods have been developed to ensure that the wave spectrum in the wave tank is approximately the same as in nature; some methods involve filtering white noise to shape a spectrum, whereas others rely on summing numerous wave trains with different amplitudes and phases to form the spectrum. In addition, several facilities have the ability to absorb waves reflected from the beach at the wavemaker by using absorbing wavemakers.

In the field, however, we know that a directional spectrum of waves exists composed of waves from many directions and frequencies; therefore, there has been a development of directional wave basins to provide realistic wave climates in the laboratory. These rectangular basins have a multielement wavemaker at one end typically comprising many smaller wavemakers that are individually computer controlled. However, these directional basins have the additional problems of wave reflection from the side walls. Some facilities design elaborate wave absorption at the sidewalls to prevent this. For reflective side walls, if a wavemaker creates a single wave train at an angle to the basin centerline, this wave train will both reflect from one of the sidewalls, creating a pattern of crossing waves, and diffract into the other sidewall. The net result is that, away from the wavemaker, the sea state in the basin is no longer a single wave train. Dalrymple (1989(a)) introduced "designer waves" to cope with this problem by using reflective basin sidewalls to "image" the wavemaker to an infinite length. In this fashion, a long-crested wave train can be created at any location in the basin. This methodology was verified in tank tests by Mansard, Miles, and Dalrymple (1992).

Wave basins can also create undesirable wave motions owing either to the presence of reflected waves from the side walls, the wavemaker, or the beach, or the generation of waves not appropriate for the prototype situations. Examples of such waves include the following:

> *Cross waves*, which are nonlinear waves generated at wave paddles that can propagate into the wave basin;
> *Resonant basin oscillations*, which occur when the wavemaker frequency corresponds to a basin seiching frequency;
> *Resonantly generated long waves*, such as standing edge waves in a basin with a beach; or
> *Spurious waves* that are short waves generated by directional wavemakers as the result of aliasing problems.

Spurious waves arise because of the 2π ambiguity that can occur from wave paddle to paddle when the wavelengths are short compared with two-paddle widths (Sand 1973). These undesirable waves can build up in magnitude sufficiently to remove all validity of the model testing. For a multielement wavemaker lying along the y-axis, an oblique wave train can be made by oscillating each of the paddles in a sinusoidal motion with a phase difference between paddles so that the wavemaker appears to be moving in a snakelike motion described by

$$\zeta(y,t) = Y\cos(k\sin\theta y - \sigma t),$$

where Y is the maximum paddle displacement and is related by wave theory to the wave height in the basin (Dean and Dalrymple 1991), k is the wave number of the desired wave (from the dispersion relationship, Eq. (5.2)), and θ is the wave direction (in radians) from the x-axis. For a multielement wavemaker with paddles of width

W, the phase of the paddle displacement is different for neighboring paddles by an amount $kW \sin \theta$ because of the distance between paddles. Now consider another paddle motion creating a wave train with a different direction θ_s:

$$\zeta_s(y, t) = Y \cos(k \sin \theta_s\, y - \sigma t)$$

The phase difference between adjacent paddles is now $kW \sin \theta_s$. If there is an angle θ_s for which

$$kW \sin \theta_s = kW \sin \theta \pm 2m\pi, \tag{10.2}$$

the motion of the wavemaker is the same for both cases. Therefore, the wavemaker actually creates both wave motions even if you only wanted the first.

Solving for the spurious wave direction, we obtain

$$\theta_s = \sin^{-1}\left(\frac{kW \sin \theta \pm 2m\pi}{kW}\right) \quad \text{for } m = \pm 1, \pm 2, \ldots \tag{10.3}$$

For $kW < \pi$, there is no spurious wave motion, for the argument of the arcsine in Eq. (10.3) exceeds unity. This means that, for desired waves with lengths greater than two paddle widths, there are no spurious waves. It is only for short wave trains that spurious waves can be generated. Figure 10.1 shows the possible spurious wave directions as a function of the desired wave direction θ and the dimensionless paddle width kW. To use this graph, the wave number of the desired wave is computed; then kW is found. From the abscissa, a line is drawn vertically to a line corresponding to the desired wave direction (if there is one or more). Moving horizontally to the

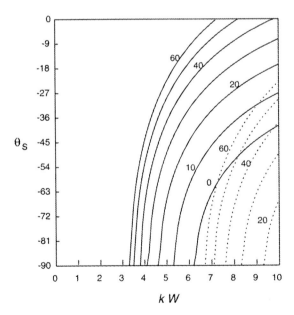

Figure 10.1 Spurious wave direction versus dimensional paddle width for a multielement wavemaker. The solid line denotes the first spurious mode; the dotted line denotes the second possible mode. The values on the isolines are wave directions in degrees.

ordinate, the direction of the spurious wave is obtained. As an example, for $kW = 5$ and a desired wave angle of $20°$, there is a single spurious wave with angle $\theta_s = -66°$ corresponding to $m = -1$ in Eq. (10.3). For $kW = 10$, there would be two spurious wave trains.

10.2.2 FIXED-BED MODELS

For the study of waves in the vicinity of a coastal structure or harbor, or to examine wave-induced flows, for example, a fixed-bed model can be conveniently used because sediment transport is not the focus of the testing. The bathymetry in this model is often molded of concrete or high-density polyurethane foam to a shape geometrically similar to the prototype although smaller in size.

The vertical length scale h_r is defined in the same way as the horizontal length scale presented previously; it is the ratio of the vertical lengths in the model to the prototype values,

$$h_r \equiv \frac{h_m}{h_p}$$

The vertical length scale can be larger than ℓ_r if the model is *distorted*.

Fixed-bed models often are required to reproduce wave shoaling, refraction, diffraction, reflection, and wave breaking correctly. This means that the model should be undistorted ($\ell_r = h_r$) when short waves are used in the testing. The reason for this follows. Froude scaling is used for water waves; the Froude number is defined as

$$\mathcal{F} \equiv \frac{U}{\sqrt{gh}}, \tag{10.4}$$

where the velocity U is a wave-related velocity, say, particle velocities or celerity, g is the acceleration of gravity, and h is a characteristic depth. Froude scaling means that the Froude number in the model is the same as that in the prototype, which in turn means that the ratios of inertial forces to gravity forces are the same in both. The length scales in a model are usually specified; we need to determine the other scale relationships such as the velocity, time, and force scales. For the Froude modeling, the relationship for the velocity and time scales is

$$\mathcal{F}_r = 1 = \frac{U_r}{\sqrt{g_r h_r}} \tag{10.5}$$

Therefore, we have $U_r = \sqrt{h_r}$ because gravity will be the same in both model and prototype. If we write the velocity in terms of a horizontal length and a time, we have

$$t_r = \frac{\ell_r}{\sqrt{h_r}} \tag{10.6}$$

On the other hand, we know that the wavelength of a water wave is dependent on the wave period and water depth from the dispersion relationship (Eq. (5.3))

$$L = \frac{gT^2}{2\pi} \tanh kh, \tag{10.7}$$

where $k = 2\pi/L$, the wave number. For deep water, the wavelength ratio is $L_r = T_r^2$. Because this is a length scale, we have $L_r = \ell_r = T_r^2$. Solving for T_r, we have $T_r = t_r = \sqrt{\ell_r}$. Comparing with our previous expression for the t_r, we require that $h_r = L_r = \ell_r$. This requirement for an undistorted model also holds for intermediate depth because Eq. (10.7) requires that $(kh)_r = 1$, or, substituting for the wave number, we obtain $h_r = L_r = \ell_r$. Thus, for short waves, the model must be undistorted.

However, for long waves, this is not a requirement because the wave phase speed is determined by the local water depth rather than Eq. (10.7), the wave length is determined by $L = \sqrt{gh}\, T$, and Eq. (10.6) is used to determine the time scale. For tidal models of large estuaries, it is of utmost importance that the models be distorted. For example, a model of the Chesapeake Bay that was placed in a building that covered 14 acres ($56,700\ \text{m}^2$) had a length scale of $1/1000$ and a vertical (depth) scale of $1/100$. This put the greatest depth in the model at about 0.4 m. If the model had not been distorted, the greatest depth would have been 0.04 m, and most of the model would have been dominated by viscous effects and bottom friction that would have been unrealistically large. The model distortion is needed to provide for realistic flows. A consequence of the distortion is that bottom friction is often too small in the model, and detailed calibrations with field measurements of flow velocities and water surface elevations are needed to verify the performance of the model. Often metal tabs, attached to the bottom and extending up into the water column, are used to increase the friction and calibrate the model. Although this calibration procedure is an expensive endeavor, there are no inherent difficulties, and these models have been constructed for several estuaries and tidal inlets.

10.2.3 MOVABLE BED MODELS

To model the behavior of a beach profile, a tidal inlet, or a section of shoreline, for example, a movable bed model provides information concerning the directions and volumes of sediment transport.

The use of movable bed models is still an art rather than a science, owing to the large number of parameters that dominate the problem. A traditional tool for determining modeling criteria has been the Buckingham Pi theorem, which involves determining all the dimensionless numbers that exist for a particular problem. For a movable bed model, in which wave and sediment parameters are important, we have these dimensionless parameters (Kamphuis 1985):

$$\Pi_s = \left\{ \frac{v_* d}{\nu},\ \frac{\rho v_*^2}{(\rho_s - \rho)gd},\ \frac{\rho_s}{\rho},\ \frac{H_0}{wT} \right\}, \tag{10.8}$$

which are, respectively, the grain size Reynolds number $\mathbf{Re}_*$, the Shields parameter, Ψ (which is sometimes called the densimetric Froude number $\mathbf{F}_*$ or the *mobility number*), the specific gravity of the sediment, and finally the Dean number. In Chapter 8, these parameters were shown to be important to sediment transport. The velocity v_* is the shear velocity, which is defined as $v_* = \sqrt{\tau_b/\rho}$, and ρ_s is the density of the sediment.

Now, for equivalence of model and prototype, all of these dimensionless numbers should be the same in model and prototype. This would guarantee that the motion of the sediment and water would be the same. However, unless the scale ratio is unity, this equivalence is precluded by a variety of factors. First, the vertical and length scales must be the same to ensure that the wave field (for short waves) is correctly modeled. Then, the scales must not be so small that the sediment to be used in the model becomes cohesive. To illustrate, a 1/100 model would mean that 0.4-mm sand in the prototype would become 0.004 mm in the model, which is well into the clay and cohesive silt range. This implies that either the model must be large (so that cohesionless sand is used in the model) or that the sand size must be distorted. Further, the scaling criteria themselves lead to conflicting requirements. For grain size Reynolds and Froude similitude, we require $(\mathbf{Re}_*)_r = 1$ and $(\mathbf{F}_*)_r = 1$. From the first relationship (and using the same fluid in model and prototype so that $v_r = 1$), we have that $d_r = 1/\sqrt{h_r}$, for the shear velocity is assumed to scale with the square root of the vertical scale (which will be the same as the length scale). This permits the grain sizes in the model to be proportionately larger than in the prototype. Alternatively, the densimetric Froude number indicates that the grain size ratio should be proportional to the depth scale, $d_r = h_r$, which may lead to a model sediment that is too fine. Owing to this irreconcilable conflict (and a similar one for Reynolds and Froude similitude for the waves), we must judiciously choose the scale relationships in such a way that the most important parameters are used and the less important are neglected.

In the following two subsections we will separate the movable bed models into two classes: the beach profile model and the general coastal model. The profile model will be a two-dimensional model of profile evolution conducted in a wave tank, and the coastal model will be a three-dimensional model carried out in a wave basin.

10.2.3.1 Beach Profile Models

One method to determine scale relationships between prototype and full scale is to compare prototype scale beach profiles and their evolution with time empirically to model results, using sand or another material to represent sand. Noda (1972) has carried out such an empirical analysis, which related the grain size scale ratio d_r to the depth scale h_r and the submerged specific weight ratio $s_r = [(\rho_s - \rho)/\rho]_r\, g_r$. He concluded that the same material should not be used in model and prototype and that the scales should be related as follows:

$$d_r(s_r)^{1.85} = h_r^{0.55} \tag{10.9}$$
$$\ell_r = h_r^{1.32} s_r^{-0.386} \tag{10.10}$$

where, again, ℓ_r is the horizontal length scale. If sand is used in the model, $s_r = 1$, and we have

$$d_r = h_r^{0.55} \tag{10.11}$$
$$\ell_r = h_r^{1.32} \tag{10.12}$$

Noda's relationships imply that the models must be distorted and have a smaller sand size than in prototype but are not scaled by the vertical scale, which is scaled by a larger ratio, and thus the model sand is larger than that required by the

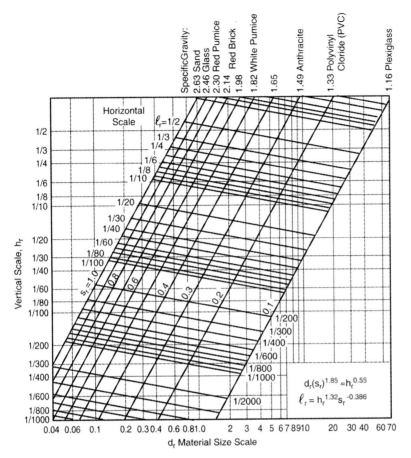

Figure 10.2 Graphical representation of Noda's model relationships (modified from Whalen and Chatham 1979).

densimetric Froude number. Figure 10.2 shows Noda's relationships graphically. The diagonal line farthest to the left is the one to be used for sand in both the model and prototype.

EXAMPLE

A model is being constructed at a horizontal scale of 1/20. Two different modeling materials are available: sand and crushed anthracite coal. According to Noda, what are the appropriate scales for the size of the material and the resulting vertical scale ratios?

For the sand, we use the leftmost of the diagonal lines in Figure 10.2, corresponding to a specific gravity of 2.65. Descending the curve until we encounter the horizontal scale of 1/20, we find that the size ratio of the sand is 0.3, meaning that the model sand size should be 3/10 that of the prototype, and that the vertical length scale is 1/10, with a resulting model distortion of $\ell_r/h_r = 1/2$.

For the coal, we follow the 1.49 specific gravity line to the intersection with 1/20 horizontal scale. Here we have a size ratio of 2, so that the coal particles will be twice the size of those in nature, and the vertical length scale is about 1/14, permitting a smaller scale model.

Alternative model scalings have been developed based on the Dean number, for Dean (1973) showed that the type of beach profile obtained in a wave tank is dependent on the $\mathbf{D}$ parameter. For values of $\mathbf{D}$ exceeding a given constant, a storm profile results, whereas for smaller values, a normal profile exists (see Section 8.4.1). Dalrymple and Thompson (1976), and Hughes (1983) have developed model laws based on the supposition that $\mathbf{D}_r = 1$.

Dean (1985) argued that a model law should have the following features: Froude scaling should be possible for the waves, the model is undistorted, the $\mathbf{D}$ value should be the same in model and prototype, and the model should be large enough that surface tension and viscous effects are not overly represented. Therefore, we have

$$\mathbf{F}_r = 1 \quad \text{or} \quad t_r = \sqrt{\ell_r} \tag{10.13}$$

$$\mathbf{D}_r = 1 \quad \text{or} \quad w_r = \ell_r/t_r, \tag{10.14}$$

which lead to the requirement that the fall velocity ratio $w_r = \sqrt{\ell_r}$. Therefore, the same sized sand cannot be used in model as in prototype, for the fall velocity of the sediment scales as the square root of the length scale (as do all velocities in a Froude model).

Kriebel, Dally, and Dean (1986) carried out validation tests of Dean's modeling criteria by using the large-scale wave tank tests of Saville (1957) for the prototype. Their procedure was to select a model sediment of a smaller size than the prototype. The fall velocity for both sands was determined, and then, from the criterion that $\ell_r = w_r^2$, the length scale was determined by the fall velocity scale. Figure 10.3 presents an example of the comparison. This model allows a comparison with Noda's empirical relationships. For example, Dean (1985) pointed out that if the fall velocity is proportional to the grain size, say, $w \approx d^{1.02}$, then $d_r = h_r^{0.49}$, which is nearly the same as Noda's first relationship (where the exponent was 0.55). The validation tests showed that for eroding profiles, where a storm profile results, good agreement was found with the prototype. For an accreting profile, recovery was not as clearly verified.

Hughes and Fowler (1990) carried out a comprehensive series of wave tank tests with medium-sized sand and combined these results with those from a large-scale program of tests in Hannover, Germany. They basically carried out a test program parallel to that described above by Kriebel et al. (1986); however, they limited their tests to conditions that produced erosion. In addition to regular waves, their tests included irregular waves. The effect of an impermeable sloping base underlying a layer of sand was also investigated. In general, the conclusions were the same as those of Kriebel et al.; there was correspondence between the model and "prototype" data using the deep water Dean number H_0/wT as a basis for the scaling. Figure 10.4 presents an example of their results for the case of an impermeable layer under the sand.

Vellinga (1986) has also developed a model criterion based partly on the Froude scaling (for the waves) but permitting distortion. His model is based on the concept

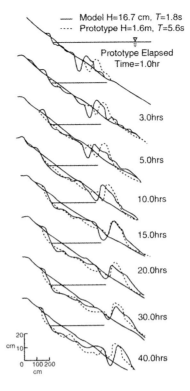

Figure 10.3 Comparison of beach profiles from medium- and large-scale wave tanks with scaling according to $(H_0/wT)_r = 1$ (from Kriebel et al. 1986). Prototype data are from Saville (1957). The initial profile slope is 1:15. Model dimensions are shown. Scale ratio = 1:10.

that the suspended sediment be scaled, which leads to a distortion of the model, defined as $\Omega = \ell_r/h_r$, of

$$\Omega = \left(\frac{h_r}{w_r^2}\right)^{0.25} \tag{10.15}$$

The time scale for the development of the profile is $t_r = \sqrt{h_r}$, which is consistent with Froude scaling. For an undistorted model, $\Omega = 1$, he has $h_r = L_r = w_r^2$, as does

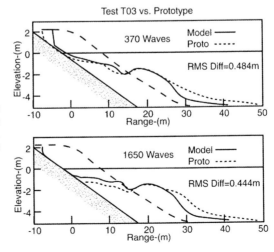

Figure 10.4 Comparison of beach profiles from medium- and large-scale wave tanks with scaling according to $(H_0/wT)_r = 1$. Large scale dimensions are shown (from Hughes and Fowler 1990).

Dean (1973). Dette and Uliczka (1986) have made comparisons of Vellinga's model, using the same sediment in the model and the prototype, and have shown good agreement in the surf zone, but discrepancies offshore of the breaker line, where the suspended sediments do not follow the precepts of the model.

10.2.3.2 Planform Physical Modeling

There are two ways to model a large three-dimensional problem. The simplest is to make a fixed-bed model out of a hard substance such as concrete and then to use a (sediment) tracer within the model to determine the locations of potential shoaling or erosion problems. These models are called *fixed-bed tracer models* (Whalen and Chatham 1979). The initial stages of sedimentation can be determined by these models, and there is less of a need to worry about scale-model relations for the sediment because the rates of erosion and morphological changes are not being measured.

The second method is the full-sediment model, which evolves under the action of the waves to provide the experimenter with the long-term effects of the sediment transport. The model must be designed with a sediment-scaling relationship in mind and the results interpreted based on the model law or by calibration with historical field data. Kamphuis (1985) has examined the choice of modeling relationships for a coastal model. He has derived the so-called Best Model based on requiring the similitude of the Froude number and the Shields parameter, which leads to an undistorted model. His validation does not stress cross-shore transport, and he indicates that perhaps the Dean number may be a more important parameter when the cross-shore transport is important. Other models include the Lightweight Model, which involves similitude of the grain Reynolds number and Shields parameter. This requirement entails using a material other than sand in the model.

A major problem to be addressed by planform modelers is the continuity of the longshore sediment transport. Most modeling basins are rectangular in planform with waves obliquely incident on a beach at one end. Only a section of the shoreline is modeled, and the basin sidewalls represent total littoral as well as hydraulic barriers. The sediment moving along the beach is impounded at one end and eroded from the other end of the beach. Therefore, backpassing arrangements have to be devised to move the sand from the downdrift end of the beach back to the updrift end.

One novel approach to the problem of sidewalls was the development of the spiral wave tank by the authors. This approach, using a spiral wavemaker (Dalrymple and Dean 1972; Mei 1973; Trowbridge, Dalrymple, and Suh 1986) in the middle of a circular wave basin with a sand beach lining the inner circumference of the basin, represents an "infinitely long beach," for the sand transport on the circular beach is continuous around the basin. Although this device has attractiveness for coastal studies, there is the drawback that spiral wavemakers cannot create a large range of wave directions.

The most important task of a modeler is the verification of the model. Once a model scale is chosen, experiments must be carried out to verify that the model will reproduce known field results. If this is not done, there is no way to prove that the model behaves in the same manner as the prototype despite the choice of the optimum model.

In conclusion, to quote Kamphuis (1975), "Owing to the variety and magnitude of scale effects, which can only be fully understood by an experimenter with experience, modeling coastal areas will continue to appear an art." But until we are able to predict the behavior of sediments in simple model basins, we will have difficulty estimating the sand transport in the field.

10.3 ANALYTICAL MODELING

10.3.1 AN ANALYTICAL TIME-VARYING PROFILE MODEL

The response of an equilibrium profile to a rise in mean water level is shoreline recession. As shown in Chapter 7, the equilibrium recession is proportional to the water level rise (Bruun's rule). For a hurricane or other fast-moving storm, the sea level rise is a function of time, and the conditions for shoreline recession may not last long enough to reach the value predicted by Eq. (7.44), which we will label as R_∞. On the basis of numerous numerical modeling results, Kriebel and Dean (1993) noted that the recession of a beach approached R_∞ exponentially:

$$R(t) = R_\infty \left(1 - e^{-t/T_s}\right),$$ (10.16)

where T_s is the time scale of the erosion (on the order of tens of hours). Differentiating this expression and assuming R_∞ is constant, we can obtain

$$\frac{dR}{dt} = \frac{R_\infty - R(t)}{T_s},$$ (10.17)

after substituting for the exponential function using Eq. (10.16).

During a hurricane, the water level changes with time; therefore, the associated maximum erosion R_∞ also changes with time. Kriebel and Dean assumed that this time dependency could be represented by replacing R_∞ (the recession due to the maximum water level rise) with $R_\infty f(t)$, where $f(t)$ is a dimensionless function of time that increases with time to a maximum of unity and then decreases back to zero. Introducing this into Eq. (10.17) yields

$$\frac{dR(t)}{dt} + \frac{R(t)}{T_s} = \frac{R_\infty f(t)}{T_s}$$ (10.18)

This equation can be solved by Laplace transforms to yield a convolution for $R(t)$:

$$R(t) = \frac{R_\infty}{T_s} \int_0^t f(t) \, e^{-(t-\tau)/T_s} \, d\tau$$ (10.19)

A reasonable representation for the time-varying function is $f(t) = \sin^2(\sigma t)$, where $\sigma = \pi/T_D$, with T_D being the duration of the storm. For this case, Eq. (10.19) can be solved as

$$R(t) = \frac{R_\infty}{2} \left[1 - \frac{1}{1 + \beta^2} \left\{ \beta e^{-2\sigma t/\beta} - [\cos(2\sigma t) + \beta \sin(2\sigma t)] \right\} \right]$$ (10.20)

Here, $\beta = 2\sigma T_s = 2\pi T_s/T_D$. This solution is only a function of the dimensionless time parameter β and σt, the dimensionless time.

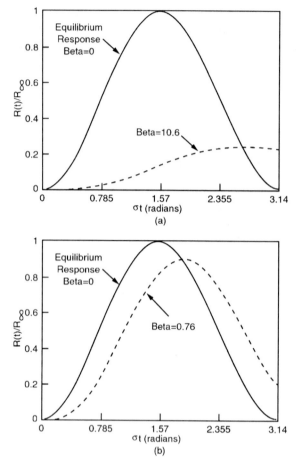

Figure 10.5 Shoreline recession due to idealized storm surge.
(a) Short duration storm. (b) Long duration storm (from Kriebel
and Dean 1993).

Figure 10.5 shows the response of the shoreline to two different types of storms
compared with an instantaneous response ($\beta = 0$). A fast-moving storm (hurricane)
is shown in the top panel and a slower moving storm (northeaster) is shown at the
bottom. Clearly, the faster the storm crosses a shoreline, the less time for shoreline
recession to occur.

The time of the maximum recession for a given value of β is easily obtained and
so is the corresponding value of the recession. These were obtained by Kriebel and
Dean and are shown in Figure 10.6. The phase lag is the dimensionless time of the
maximum erosion after the peak of the surge (at $\sigma t = \pi/2$).

An estimate of T_s was also provided as follows based on a numerical model:

$$T_s = 320 \frac{H_b^{3/2}}{\sqrt{g} A^3} \left(1 + \frac{h_b}{B} + \frac{m W_b}{h_b} \right), \tag{10.21}$$

where W_b is the width of the surf zone (to depth of breaking h_b), B is the berm height,
A is the profile scale parameter, and m is the slope of the offshore transition from
the storm to the original profile.

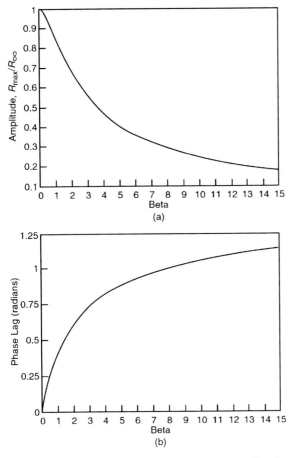

Figure 10.6 Maximum recession and phase lag as a function of β (from Kriebel and Dean 1993).

10.3.2 A ONE-LINE PLANFORM MODEL

The one-line model is the simplest contour model, and it describes the time history of the shoreline position along a shoreline. The first one-line model was presented by Pelnard-Considère (1956), who examined the behavior of groins on a beach. Since that time, the "diffusion" equation, which he developed, has been applied to many different situations. We will derive his equation and also apply it to several situations of engineering relevance.

The shoreline is shown in Figure 10.7(a), with the x-axis oriented alongshore and the y-axis offshore. The equation of the shoreline position is

$$y = y(x, t)$$

Owing to the irregularities in the shoreline, the shoreline normal does not point directly offshore at most locations. The local normal can be determined as

$$\boldsymbol{n} = \frac{-\frac{\partial y}{\partial x}\boldsymbol{i} + 1\boldsymbol{j}}{\sqrt{1 + \left(\frac{\partial y}{\partial x}\right)^2}}$$

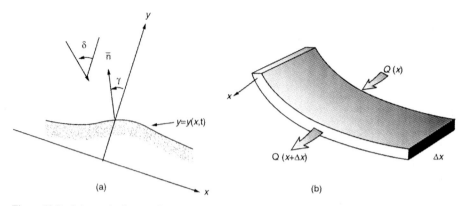

Figure 10.7 Schematic diagram for the one-line model. (a) Plan; (b) profile.

The vectors, $\boldsymbol{i}$, $\boldsymbol{j}$ are the unit vectors in the x and y directions. The angle made by the local beach normal to the y-axis is, by definition of the dot product,

$$\gamma = \cos^{-1}(\boldsymbol{n} \cdot \boldsymbol{j})$$

The first equation used in the development of the one-line model is the alongshore sediment transport formula from Chapter 8:

$$Q = \frac{K\rho H_{\mathrm{b}}^{5/2}\sqrt{g/\kappa}\sin 2(\delta_{\mathrm{b}} - \gamma)}{16(\rho_{\mathrm{s}} - \rho)(1 - p)} = C_q \sin 2(\delta_{\mathrm{b}} - \gamma), \tag{10.22}$$

where C_q is defined here for convenience and $(\delta_{\mathrm{b}} - \gamma)$ measures the angle of wave incidence relative to the shoreline normal. The breaking wave angle of incidence δ_{b} is measured from the fixed y-axis. (The relationship between δ_{b} and γ to the angles measured from north, as in Chapter 8, is found by introducing an angle μ that measures the angle between the x-axis and north or $\epsilon = \mu - \pi = \beta + \gamma$, as shown in Figure 10.8. The angle to the beach normal from north is determined by $\beta = \epsilon - \gamma$, and $\alpha = \epsilon - \delta$. Therefore, the angle of wave attack on the beach is $\beta - \alpha = \delta - \gamma$).

AQ-is this where parens go?

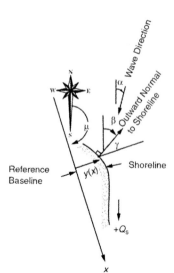

Figure 10.8 Definition sketch for the one-line model utilizing compass directions.

The second and last equation is the conservation of sand equation. Consider a section of the coastline that is Δx long, as shown in Figure 10.7(b). We will assume that the profile is in equilibrium in the cross-shore direction. This is a key assumption; regardless of the shoreline orientation, it is assumed that the equilibrium profile applies at all x locations. Now, if the rate at which sand is entering the cross section owing to alongshore transport is greater than the rate at which it is leaving from the other side of the profile, there must be an accumulation of sand with time in the profile, causing the shoreline to advance offshore. The trick used in the sea level rise and shoreline retreat calculations in Chapter 7 will be repeated here. The volume of sand necessary to move a profile seaward (shoreward) is the shoreline accretion (recession) times the height of the active profile, $(h_* + B)$, where h_* is the depth of closure and B is the berm height. Restating this in mathematical terms,

$$\Delta t[Q(x) - Q(x + \Delta x)] = [y(t + \Delta t) - y(t)](h_* + B)\,\Delta x \tag{10.23}$$

or, through the use of Taylor series and the argument that Δx and Δt become very small, we obtain

$$\frac{\partial y}{\partial t} + \frac{1}{(h_* + B)}\frac{\partial Q}{\partial x} = 0 \tag{10.24}$$

Now, by substituting the expression for transport rate, Eq. (10.22), into this equation, we obtain the final equation. For computer calculation, this is sufficient, as shown in the next section. For analytical solutions, some further approximations are needed. Expanding the trigonometric terms in Q, we have

$$Q = C_q \sin 2(\delta_b - \gamma)$$
$$= C_q[\sin 2\delta_b(\cos^2 \gamma - \sin^2 \gamma) - 2\cos 2\delta_b \sin \gamma \cos \gamma] \tag{10.25}$$

From the definition of the shoreline normal, n, $\sin \gamma = -n \cdot i$ and $\cos \gamma = n \cdot j$. This leads us to

$$\sin \gamma = \frac{\frac{\partial y}{\partial x}}{\sqrt{1 + \left(\frac{\partial y}{\partial x}\right)^2}}; \quad \cos \gamma = \frac{1}{\sqrt{1 + \left(\frac{\partial y}{\partial x}\right)^2}} \tag{10.26}$$

and

$$\tan \gamma = \frac{\partial y}{\partial x} \tag{10.27}$$

These definitions for $\sin \gamma$ and $\cos \gamma$ are substituted into the definition of Q, which yields, for small values of $\partial y/\partial x$,

$$Q = C_q \sin 2\delta_b - 2C_q \cos 2\delta_b \frac{\partial y}{\partial x} = Q_0 - G(h_* + B)\frac{\partial y}{\partial x}, \tag{10.28}$$

where $G = 2C_q \cos 2\delta_b/(h_* + B)$. The term Q_0 is the background transport rate for a shoreline parallel to the x-axis, and the second term represents the transport induced by the alongshore shoreline slope due to the shoreline deviation from the x-axis. We can now carry out the derivatives required in Eq. (10.24); again, we assume that

$\partial y/\partial x \ll 1$ with the final expression

$$\frac{\partial Q}{\partial x} \simeq -G(h_* + B)\frac{\partial^2 y}{\partial x^2} \tag{10.29}$$

Introducing this simplified form for the derivative of Q into the sand conservation equation, Eq. (10.24), we obtain the final version of the Pelnard-Considere equation,

$$\frac{\partial y}{\partial t} = G\frac{\partial^2 y}{\partial x^2} \tag{10.30}$$

This equation is the classical one-dimensional diffusion equation, which is well known in mathematical physics, and a variety of solutions exist for this equation, as we will see. The parameter G, defined after Eq. (10.28), is referred to as the *longshore diffusivity*. For small angles of wave incidence with respect to the shore normal, $\delta_b \ll 1$ and $G = 2C_q/(h_* + B)$. The diffusivity parameter has units of length squared per unit of time; for the shoreline of Florida, $0.02 < G < 0.14\,\mathrm{ft^2/s}$ $(0.002 < G < 0.014\,\mathrm{m^2/s})$.

We will now examine some of the solutions of this equation. Many of these are available for different coastal situations; collections of solutions appear in Le Mehaute and Soldate (1977), Walton and Chiu (1979), and Larson, Hanson, and Kraus (1987).

10.3.2.1 Steady Solution

For a steady-state solution, $\partial y/\partial t = 0$, and the shoreline position y is only a function of x. The diffusion equation reduces to

$$\frac{\partial^2 y}{\partial x^2} = 0, \tag{10.31}$$

which has as a solution

$$y(x) = ax + b, \tag{10.32}$$

the equation of a straight line, with two arbitrary constants, a and b, to be determined. The constant a can be related to the orientation angle of the beach, γ, for a is $\tan \gamma$, from Eq. (10.27). The b is the location of the shoreline at $x = 0$. This indicates that any straight shoreline is in equilibrium with the incident wave field, not because the amount of material moved on the shoreline is zero but because this amount is uniform: the amount of material transported into a beach section by waves is equal to the amount leaving the section. This is a dynamic equilibrium as opposed to the equilibrium achieved when the shoreline is oriented perpendicularly to the wave direction $\delta_b = \gamma$. The reason for the dynamic equilibrium is that, as long as there is no gradient in the longshore transport along a beach, there is no beach change.

10.3.2.2 Periodic Beach

For this special case, the beach has a sinusoidal variation in shoreline position, as might occur with *spatially periodic* sand waves (as discussed in Chapter 9). The diffusion equation is used to predict the subsequent behavior of the beach (Le Mehaute and Brebner 1961).

The initial shoreline orientation is assumed to be

$$y(x, 0) = B \cos \lambda x,$$

where B is the initial amplitude of the shoreline perturbations and λ is their longshore wave number. The spacing of these perturbations, L_p, is equal to $2\pi/\lambda$. The mean beach position is located at $y = 0$.

If we assume that the solution to the one-line equation, Eq. (10.30), is periodic in x, then we have

$$y(x, t) = f(t) \cos \lambda x.$$

Substituting into the equation, we have

$$\frac{\partial f}{\partial t} = -G\lambda^2 f, \tag{10.33}$$

where G, again, is the longshore diffusivity. Solving this equation leads to the following total solution:

$$y(x, t) = Be^{-G\lambda^2 t} \cos \lambda x \tag{10.34}$$

The amplitude clearly approaches zero exponentially with time. The smaller the value of λ, or the longer the spacing between the shoreline features, the longer the time for the beach to evolve to a straight beach.

A surprising result occurs when the wave angle of approach δ_b is greater than 45° as the sign of G changes. This results in an *exponential increase* of the periodic shoreline features. This might apply to the situation of cuspate features on long lagoons, as discussed earlier in Chapter 3. For these cases, the largest waves would be generated along the axis of the lagoon consistent with our discussions in Chapter 8 about unstable littoral drift roses.

10.3.2.3 Sand Waves

The periodic bypassing of sand past an inlet can create a temporally periodic and spatially damped sand wave that propagates along the shoreline. Bakker (1968a) examined the one-line model for a progressive wave solution. The initial condition (at $x = 0$) is that the shoreline oscillates landward and seaward a distance A with a period $T = 2\pi/\sigma$. His solution is

$$y(x, t) = Ae^{-\sqrt{\sigma/2G}\, x} \cos\left(\sqrt{\frac{\sigma}{2G}}x - \sigma t\right) \tag{10.35}$$

The wavelength of the motion increases with the period of the motion and the longshore diffusivity G. The wavelength is given by $L = \sqrt{4\pi GT}$, and the speed of the sand waveform is $C = \sqrt{2G\sigma}$.

The exponential damping of the wave with distance x is very strong. After propagating only one-half a wave length, the wave amplitude has decreased to 4 percent of its original size A. After one wave length, the sand wave essentially disappears.

Bakker found good agreement between the simple one-line model solution with field data in the Netherlands. For his case, six profiles, monitored for 100 years by the Rijkswaterstaat, showed a 60-year periodicity in the shoreline position. From the one-line model, the sand waves have a wavelength of 17.3 km, propagating at a speed of 0.29 km per year, based on a G value of 0.4×10^6 m^2/yr. (Interestingly, the range of the longshore diffusivity for the shoreline of Florida is similar in magnitude: $0.06 \times 10^6 < G < 0.4 \times 10^6$ m^2/yr.)

10.3.2.4 Point Application of Fill

As an idealized example of the performance of beach fill, which will be discussed in greater detail in Chapter 11, the fate of a point application of a beach fill is determined here.

The beach fill will be placed at the point $x = 0$ in such a way that the planform area of the placed fill is M; however, it is located only at one point. Mathematically, we will use a Kronecker delta function for this. Needless to say, this is not likely to be representative of a real beach fill! Thus, the initial shoreline is

$$y(x, 0) = M\delta(x),$$

where the delta function is unity at $x = 0$ and zero elsewhere. The solution to Eq. (10.30) can be obtained through the use of Fourier transforms.

The result is

$$y(x, t) = \frac{M}{\sqrt{4\pi G t}} \, e^{-(x/\sqrt{4Gt})^2} \tag{10.36}$$

The point beach fill diffuses with time in the longshore direction symmetrically, regardless of the angle of wave incidence. This is shown in Figure 10.9. The shoreline displacement at the origin, $x = 0$, is simply

$$y(0, t) = \frac{M}{\sqrt{4\pi G t}},$$

decreasing with a rate proportional to $t^{-1/2}$. It is also important to note that the total

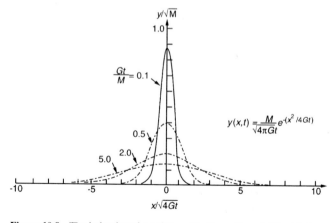

Figure 10.9 The behavior of a point application of beach fill with time.

area of the beach fill remains constant,

$$\int_{-\infty}^{\infty} y(x,t)\,dx = M$$

Two useful characteristics of the beach fill are the location of the *centroid* of the fill, x_c, indicating the geometric center of the fill volume, and the fill *variance*, σ^2, which provides an indication of the spreading of the fill. The centroid is defined as

$$x_c(t) = \frac{\int_{-\infty}^{\infty} xy(x,t)\,dx}{\int_{-\infty}^{\infty} y(x,t)\,dx} \tag{10.37}$$

For this case, $x_c(t) = 0$ because $y(x,t)$ is symmetric about the y-axis. The time rate of change of the centroid, or the velocity with which the majority of the fill is moving, is defined as $U_c = dx_c/dt$ (zero for this case). The volumetric variance is defined as

$$\sigma^2 = \frac{\int_{-\infty}^{\infty} (x - x_c)^2\, y(x,t)\,dx}{\int_{-\infty}^{\infty} y(x,t)\,dx} \tag{10.38}$$

The square root of the variance is defined as the *volumetric standard deviation* σ.

The time rate of change of σ^2 is

$$\frac{d\sigma^2}{dt} = \frac{\frac{\partial}{\partial t}\int_{-\infty}^{\infty} (x - x_c)^2 y(x,t)\,dx}{\int_{-\infty}^{\infty} y(x,t)\,dx} \tag{10.39}$$

Integrating by parts, and using Eqs. (10.24) and (10.28), we can show that

$$\frac{d\sigma^2}{dt} = 2G \tag{10.40}$$

regardless of the shape of the initial fill. Integrating with respect to time yields

$$\sigma^2 = \sigma_0^2 + 2Gt,$$

where σ_0^2 is the variance at time $t = 0$. The variance increases linearly with time and with longshore diffusivity G. For the point fill project, the initial variance is zero.

10.3.2.5 Rectangular Beach Fill

Most beach nourishment projects can be approximated as a rectangular planform. Perhaps not surprisingly, because Eq. (10.30) is linear, the formal solution to this problem is simply the superposition of many point fills with each point fill displaced alongshore to represent the rectangular planform. The solution is

$$y(x,t) = \frac{Y}{2}\left\{ \mathrm{erf}\left[\frac{\ell}{4\sqrt{Gt}}\left(\frac{2x}{\ell} + 1\right)\right] - \mathrm{erf}\left[\frac{\ell}{4\sqrt{Gt}}\left(\frac{2x}{\ell} - 1\right)\right] \right\}, \tag{10.41}$$

where ℓ and Y are the initial project length and width, respectively, and the error function, "erf ( )" is defined as

$$\mathrm{erf}(z) = \frac{2}{\sqrt{\pi}} \int_0^z e^{-u^2}\,du \tag{10.42}$$

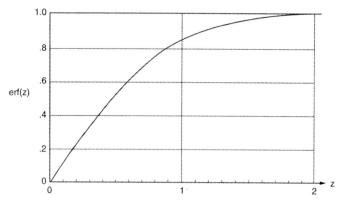

Figure 10.10 The error function.

in which u is a dummy of integration. The error function is tabulated in most books of tables (e.g., Abramowitz and Stegun 1964) and is available in most scientific software libraries. The function $erf(z)$ is antisymmetric about its origin and is presented in Figure 10.10 for positive values of its argument.

Equation (10.41) is presented in Figure 10.11 in nondimensional form. The characteristics of this solution are discussed in detail in Chapter 11. Like the point application example, the rectangular beach fill solution behaves symmetrically; Figure 10.11 shows just half of the evolution of the fill with dimensionless time. Regardless of the wave direction, the evolution of the fill is the same both updrift and downdrift, and the centroid remains at $x = 0$ for all time. An explanation of this paradoxical situation is that the updrift behavior is the result of the impoundment of longshore transport by the presence of the fill, which acts somewhat like a groin. On the downdrift side, the sand spreads out exactly symmetrically to the updrift side. An alternate consideration is, because the Pelnard-Considere solution (Eq. (10.30)) is linear, the background transport due to oblique waves is superposed with the transport from the fill caused by normally incident waves.

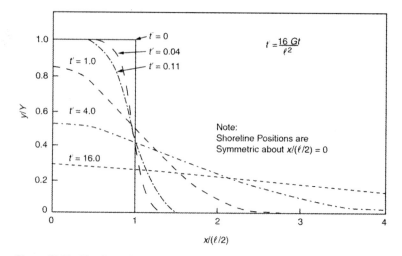

Figure 10.11 Nondimensional shoreline evolution for rectangular planform.

The shoreline displacement at the center of the fill is

$$y(0, t) = Y \operatorname{erf}\left[\frac{\ell}{4\sqrt{Gt}}\right],$$

which is a slower erosion rate than for a point fill. Because the erf function is approximately unity for arguments greater than 2, the center shoreline does not significantly decrease (except for profile equilibration) until $t > \ell^2/(64G)$.

The initial variance of the rectangular fill is easily shown to be $\sigma_0 = \ell^2/12$. The amount of fill within $\pm\sigma$ of the origin is $2\sigma Y = 2\ell Y/\sqrt{12}$, or 57.7 percent of the fill is within one standard deviation of the origin. With time, the percentage of the fill encompassed by $\pm\sigma$ tends to 68.3 percent as the shape of the profile evolves to a normal probability distribution (compare Eq. (10.36)).

10.3.2.6 Littoral Barriers

For a headland or groin, which does not permit the bypassing of sand, we must establish the boundary conditions that apply. At the barrier, taken to have length ℓ, the shoreline must have an orientation that allows for no sediment transport because no material can pass the structure:

$$\frac{\partial y(0, t)}{\partial x} = -\tan \delta_b$$

ensures that the shoreline normal is in the direction of the waves. Far from the barrier, the derivative goes to zero as the beach becomes straight. The solution for this case is found most conveniently by using Laplace transforms:

$$y(x, t) = \pm \left[\sqrt{\frac{4Gt}{\pi}}\, e^{-x^2/(4Gt)} - |x|\operatorname{erfc}(|x|/\sqrt{4Gt})\right] \tan \delta_b, \tag{10.43}$$

where the complementary error function, $\operatorname{erfc}(\ ) = 1 - \operatorname{erf}(\ )$, and the plus and minus solutions apply for the updrift and downdrift sides of the barrier: $x < 0$ and $x > 0$, respectively. This solution is presented in dimensionless form in Figure 10.12, which

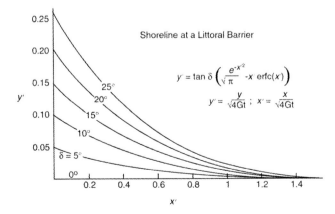

Figure 10.12 Nondimensional shoreline evolution for littoral barrier, Eq. (10.43).

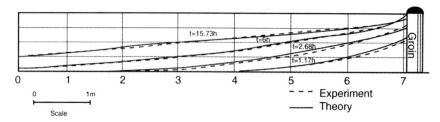

Figure 10.13 Comparisons between experimental and theoretical results (Le Mehaute and Brebner 1961).

shows that the barrier impounds sand with time in an accretionary fillet. Laboratory experiments by Le Mehaute and Brebner (1961) showed good agreement with this theoretical result (see Figure 10.13).

At the barrier, the growth of the beach, $y(0, t)$, seaward with time is easily determined for $x < 0$ as

$$y(0, t) = \sqrt{\frac{4Gt}{\pi}} \tan \delta_b \tag{10.44a}$$

The area of beach accreted at the barrier is given by

$$A(t) = \int_0^\infty y \, dx = Gt \tan \delta_b, \tag{10.44}$$

showing a linear increase of area with time. The volume of sand accreted corresponding to this area is found by multiplying A by $(B + h_*)$.

As the beach next to the barrier grows seaward, it eventually equals the length of the structure, and bypassing of sand will commence. The time of incipient bypassing t^* is easily found from Eq. (10.44a) by setting the width of the beach to ℓ, which yields

$$t^* = \frac{\pi \ell^2}{4G \tan^2 \delta_b}$$

At this time, the accretional fillet has an area of

$$A(t^*) = Gt^* \tan \delta_b = \frac{\pi \ell^2}{4 \tan \delta_b}$$

This area is just $\pi/2$ larger than the area determined by connecting the seaward tip of the barrier to the shoreline with a line normal to the incident wave direction. The impounded volume at incipient bypassing $V(t_*)$ is

$$V(t_*) = (B + h_*) \frac{\pi \ell^2}{4 \tan \delta_b} \tag{10.45}$$

After bypassing commences, the analytical solution we have just used is no longer valid, for the boundary condition at the barrier changes to $y(0, t) = \ell$. Pelnard-Considere found the solution to this problem to be

$$y(x, t) = \ell \, \text{erfc}(x/\sqrt{4Gt'}) \tag{10.46}$$

The $y(x, t^*)$ given by Eqs. (10.43) and (10.46) only match exactly at $x = 0$ and $x = \infty$, and thus we have two different solutions that have to be matched consistently.

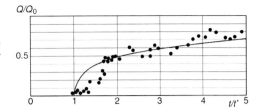

Figure 10.14 Temporal variation of the transport past a groin (from Le Mehaute and Brebner 1961).

This has been done in the past by matching the sand volumes impounded by Eqs. (10.43) and (10.46). This matching provides a determination of how the time scale in the second equation (10.46) is related to t^* of the first. This comparison shows that Eq. (10.46) can be viewed as the subsequent evolution of the profile given by Eq. (10.43) when the time in the last equation is replaced by $t' = t - (1 - \pi^2/16)t^*$.

The amount of sand bypassing the barrier can be computed by the equation for sediment transport using the beach planform slope at the barrier. Le Mehaute and Brebner developed Figure 10.14, which shows how the transport rate increases as the barrier begins bypassing. Note that it takes an infinite amount of time for the original transport rate on a beach, Q_0, to be recovered, which implies that there is always sand impounding in the updrift fillet.

A measure of the updrift influence of the sand fillet at the barrier can be taken arbitrarily as a shoreline displacement $y(x_m, t)$ of a given fraction of the groin length, occurring a distance x_m updrift of the barrier. If we take $y(x_m, t) = \ell/100$, then we can find that $x_m = 2.85\ell/\tan \delta_b$, which is almost three times larger than the point of intersection of a line drawn from the seaward tip of the barrier, perpendicular to the wave direction, to the beach.

After bypassing of the barrier begins, the rate of impoundment of sand by the barrier slows. The area accreted is now

$$A = \frac{\ell \sqrt{4Gt'}}{\sqrt{\pi}}$$

The updrift influence of the barrier, again based on $y(x_m, t')/\ell = 1/100$, is $x_m = 1.82\sqrt{4Gt'}$. According to the one-line theory, this updrift influence increases nonstop until the entire shoreline updrift of the barrier is displaced seaward a distance equal to the offshore length of the barrier. (This is presumably the result of assuming the same equilibrium profile everywhere.)

Downdrift (or on the other side of the impounding barrier), Eqs. (10.43) and (10.46) can also be used to provide the erosion planform as an antisymmetric version of the updrift planform subject to the assumption that diffraction is not important in the shoreline modification.

10.3.2.7 River Deltas

The one-line model has been applied to river deltas, where a source of sand is supplied by the river, by Grijm (1960, 1964) and by Bakker and Edelman (1964), using a two-line model. For a one-line model, Eq. (10.43) is used symmetrically on both sides of the source ($x = 0$), with δ_b chosen to transport half of the sediment discharge, Q_r:

$$\delta = \sin^{-1}\left[\frac{Q_r}{2G(h_* + B)}\right] \tag{10.47}$$

10.3.3 THREE-DIMENSIONAL MODELING

Bakker (1968(b)) introduced a two-line model based on the argument that the beach profile is not always in equilibrium along a beach. For example, the beach updrift of a groin is steeper than that downdrift of the barrier simply because of the impounding of material on one side and erosion on the other.

The profile is divided into two sections, a shallow section that extends offshore to a depth D_1, and a deeper section, which continues offshore to a depth $D_1 + D_2$, which is the depth of closure. The distances of the contours from the baseline are y_1, which is the distance of the shoreline from the y-axis, and y_2', which is the distance that the contour D_1 is from the axis.

The cross-shore sediment transport is based on an equilibrium profile concept. For a profile in equilibrium, the distance between the contours is fixed, $y_2' - y_1 = w$. The cross-shore transport rate is assumed to be proportional to the amount that the actual contour spacing differs from w, or

$$q_y = C_y(y_1 - y_2' - w), \tag{10.48}$$

where C_y is a dimensional transport constant. The direction of the cross-shore transport is positive in the seaward direction, occurring when the right-hand side of Eq. (10.48) is positive, which implies that the shallow portion of the beach is narrower than equilibrium and thus that sand must be moved offshore to widen the beach and approach cross-shore equilibrium. Alternatively, a negative argument means that the beach is too wide and the transport is onshore to narrow the beach. This equation can be simplified with the definition of $y_2 = y_2' - w$, which is a measure of how much the contour spacing differs from the equilibrium value.

The cross-shore transport equation must be coupled with two conservation of mass equations – one for each region:

$$D_1\frac{\partial y_1}{\partial t} = -q_y - \frac{\partial Q_1}{\partial x}$$

$$D_2\frac{\partial y_2}{\partial t} = q_y - \frac{\partial Q_2}{\partial x}, \tag{10.49}$$

where Q_1 and Q_2 are the longshore transports at each contour taken to be proportional to the contour planform slope, as in Eq. (10.28):

$$Q_1 = Q_{1_0} - q_1\frac{\partial y_1}{\partial x}$$

$$Q_2 = Q_{2_0} - q_2\frac{\partial y_2}{\partial x}$$

Figure 10.15 Filling of a groin by the two-line model (from Bakker 1968).

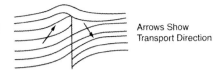

Arrows Show
Transport Direction

Substituting for the Qs gives two equations:

$$D_1 \frac{\partial y_1}{\partial t} = -C_y(y_1 - y_2) + q_1 \frac{\partial^2 y_1}{\partial x^2}$$

$$D_2 \frac{\partial y_2}{\partial t} = C_y(y_1 - y_2) + q_2 \frac{\partial^2 y_2}{\partial x^2} \qquad (10.50)$$

Bakker then rearranged these equations by introducing three new variables: $y = (D_1 y_1 + D_2 y_2)/D$, where $D = D_1 + D_2$ and

$$y_e = (y_1 - y_2)e^{\left(\frac{C_y D t}{D_1 D_2}\right)}$$

With these new variables and the assumption that $q_1/D_1 = q_2/D_2$, which implies that contours with the same curvature will fill at the same rate, he obtained two simple diffusion equations:

$$\frac{\partial y}{\partial t} = G \frac{\partial^2 y}{\partial x^2}$$

$$\frac{\partial y_e}{\partial t} = G \frac{\partial^2 y_e}{\partial x^2}, \qquad (10.51)$$

where G, the diffusivity, is defined here as $G = (q_1 + q_2)/D$.

Given an appropriate value of G and initial and boundary conditions, these equations provide a means to calculate y_1 and y_2'. Bakker has provided an example for a single groin, filling as a function of time, which is shown in Figure 10.15.

Hulsbergen, Bakker, and van Bochove (1976) carried out laboratory tests of groins on long, straight beaches and concluded that the model worked well for the simple case of a long, straight beach with a simple longshore current system; however, for more complicated cases, the model was not as accurate.

The two-line model has been applied to sand waves on the coastline of the Netherlands by Dijkman, Bakker, and de Vroeg (1990).

Because the two-line model (Eq. (10.50)) has linear equations, we can eliminate one of the variables from the equations to make a single equation. The final result is

$$D_1 D_2 \frac{\partial^2 y_1}{\partial t^2} + C_y D \frac{\partial y_1}{\partial t} - (D_2 q_1 - D_1 q_2) \frac{\partial^3 y_1}{\partial x^2 t} - C_y q \frac{\partial^2 y_1}{\partial x^2} + q_1 q_2 \frac{\partial^4 y_1}{\partial x^4} = 0,$$

where $q = q_1 + q_2$. This equation can be used to calculate the damping of sand waves just as was done for the one-line model. Substituting a periodic solution for the sand wave,

$$y_1 = Ae^{i(kx - \sigma t)},$$

where A is the excursion of the inner contour, we obtain a dispersion relationship for the sand wave that relates the (real valued) alongshore wavenumber k of the sand wave to the angular frequency σ:

$$-D_1 D_2 \sigma^2 - [C_y D + (D_2 q_1 + D_1 q_2)k^2]i\sigma + C_y q k^2 + q_1 q_2 k^4 = 0 \tag{10.52}$$

This quadratic equation has two solutions for σ that are purely complex numbers for small k values, indicating that no (damped) propagating solutions exist – only solutions that decay exponentially with time.

10.4 NUMERICAL MODELING

10.4.1 PROFILE MODELING

Beach profile modeling is most often used to evaluate the impact of a major storm with elevated storm tides and wave heights on the beach and the dunes. Another application is to predict the profile evolution of a beach nourishment project that is placed at a slope initially steeper than the equilibrium profile. A third application is in a three-dimensional model that simulates more realistic situations by cross-shore sediment transport. In this latter application, longshore sediment transport may interact with a barrier to cause profile steepening and flattening on the updrift and downdrift sides of a structure, respectively, thus shunting sand seaward and landward. In this section, we will concentrate on the more recent models that have been developed with only brief mention of earlier models.

On the basis of published data and results of a series of wave tank tests using fairly coarse sediments, Swart (1974) developed an empirical method for predicting beach profile evolution. The profile was schematized in three vertical zones and the equilibrium profile established in terms of empirical nondimensional relationships. The method recognizes that an equilibrium profile will be approached under the sustained action of steady waves and tides. The procedure is quite complex and involves numerous empirical expressions; the reader is referred to the original article for the details. The only known numerical applications of Swart's theory to field conditions is by Swain and Houston (1984) for storm erosion at Santa Barbara, California, and near Oregon Inlet, North Carolina. They modified Swart's approach to allow for a time-varying wave height and water level. They demonstrated that tides are important in predicting the magnitude of shoreline erosion during a storm.

Kriebel (1982), Moore (1982), and Kriebel and Dean (1985) have proposed a dynamic equation for cross-shore sediment transport,

$$q = K(\mathcal{D} - \mathcal{D}_*) \tag{10.53}$$

in which $\mathcal{D}$ and $\mathcal{D}_*$ are the actual and equilibrium values of wave energy dissipation per unit water volume and K is an empirical constant (different from the longshore transport coefficient of the same name). Equation (10.53) is combined with the sand conservation equation to obtain a solution; however, rather than using an equation for depth of the form $h = f(y)$, they inverted the expression to $y = f(h)$ and wrote

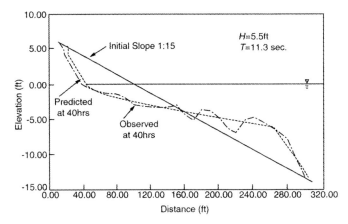

Figure 10.16 Comparison of calibrated profile response model with large wave tank data by Saville (1957) (from Kriebel 1986).

the conservation of sand as

$$\frac{\partial y}{\partial t} = -\frac{\partial q}{\partial h} \tag{10.54}$$

This equation is solved numerically by an implicit method. The profile is represented by several specified elevation increments with y the distance from some fixed baseline to a profile contour h. The constant K was first determined based on the large wave tank data of Saville (1957) and was later calibrated with Hurricane Eloise profile data (1977), resulting in a modified K value of 0.0045 ft^4/lb or 8.7×10^{-6} m^4/N. Figures 10.16 and 10.17 present examples of applications of the method for wave tank and full-scale data, respectively. This numerical model is called EDUNE.

Vellinga (1986) developed a profile response model based on a series of small-scale model tests at various scales and later confirmed the model predictions with large-scale laboratory data and prototype results. The eroded profile was found to depend on wave height and grain size. On the basis of laboratory tests, scale relationships were developed that form the foundation for extrapolation of the test results to conditions not included in the laboratory test results. Laboratory results were

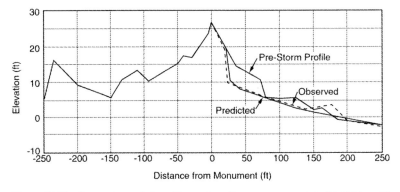

Figure 10.17 Comparison of the calibrated profile response model with field profile affected by Hurricane Eloise as reported by Chiu (1977) (from Kriebel 1986).

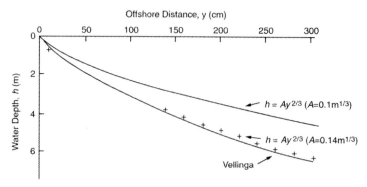

Figure 10.18 Comparison of Vellinga profile with $h = Ay^{2/3}$ for $A = 0.1$ m$^{1/3}$ ($d = 0.2$ mm) and 0.14 m$^{1/3}$ ($d = 0.4$ mm).

obtained for a reference erosion profile associated with the following conditions:

$$H_{0_s} = 7.6 \text{ m}$$
$$T = 12 \text{ s}$$
$$d_{50} = 0.225 \text{ mm}$$
$$t = 5 \text{ hours after start of tests}$$

(where all expressions above are in terms of prototype dimensions, and t is a reference time for the erosion profile). The recommended reference profile with origin at the intersection of the elevated storm surge is

$$h = 0.47(y + 18)^{0.5} - 2.00, \tag{10.55}$$

where variables are in the metric system. Figure 10.18 presents a comparison of the profile represented by Eqs. (7.4) and (10.55) for $A = 0.1$ m$^{1/3}$ (Figure 7.6, $d = 0.2$ mm) and for $A = 0.14$ m$^{1/3}$ ($d = 0.4$ mm); for the latter case, the differences are relatively small. Vellinga recommended that, for durations in excess of 5 hours, the erosion quantity above surge level be increased by 10 percent for each hour that a level 1 m below maximum storm surge level is exceeded by 5 hours with a maximum of about 50 percent.

10.4.2 PLANFORM MODELING

There are two tacks that have been taken to develop numerical models to describe the behavior of the shoreline. The first is the use of contour line models, which keep track of the position of the offshore contours with time, and the second is to make a grid of the offshore region and to determine how the water depth within each grid changes with time. On the basis of historical precedence, we will discuss the contour models first.

In general, planform models (as contrasted to three-dimensional models) imply the use of a one-line model because the offshore profile is usually considered to remain constant in form and to move landward or seaward with the monitored contour and vertically with the water level. These models can be based on the primitive equations of motion using the equations of transport (Eq. (8.17)) and

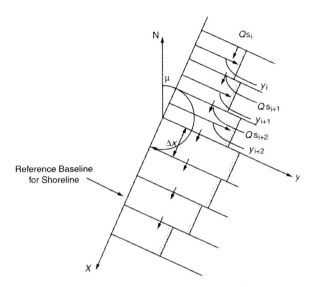

Figure 10.19 Schematic diagram for numerical model.

continuity (Eq. (10.24)), or they can employ the diffusion equation (Eq. (10.30)), which is the combined form of the two. The use of the primitive forms allows greater flexibility in applying boundary conditions and internal transport restraints such as groins. In the discussion to follow, we will first examine applications of the primitive equations followed by the combined equations.

10.4.2.1 One-Line Model Based on Primitive Equations

This model is based on numerical solution of the transport equation and the continuity equation. The most appropriate grid is one in which the transport quantities Q_i are specified at the grid lines and the y_i values apply at the grid midpoints. As shown in Figure 10.19, the subscript on Q denotes the grid on which positive transport (to the right of an offshore-looking observer) would flow. These equations may be solved by explicit or implicit numerical methods. The explicit method is the more direct and is the one that will be illustrated here. In the explicit method, the equations of transport and continuity are solved sequentially. The procedure can be illustrated as follows. The y values are held fixed, whereas the Q values are calculated according to

$$Q_i^{n+1} = C_{q_i} \sin 2\left(\beta_i^n - \alpha_{b_i}^n\right), \tag{10.56}$$

where

$$C_{q_i} = \frac{K H_{b_i}^{5/2} \sqrt{g/\kappa}}{16(s-1)(1-p)}$$

$$\beta_i = \mu - \frac{\pi}{2} - \tan^{-1}\left(\frac{y_i - y_{i-1}}{\Delta x}\right), \tag{10.57}$$

where β_i is the azimuth for the shoreline normal and $\alpha_{b_i}^n$ is the azimuth from which the refracted breaking wave originates at the ith grid line. The superscripts denote the time level, corresponding to $t = n\Delta t$. The Q_i^{n+1} are calculated using the angles determined at time level n. The updated Q^{n+1} values are held, whereas the y values are changed in accordance with the continuity equation:

$$y_i^{n+1} = y_i^n - \frac{\Delta t}{\Delta x(h_* + B)}(Q_{i+1}^{n+1} - Q_i^{n+1}) \tag{10.58}$$

The basis of the term "explicit" is now more apparent and is due to the manner of treating the two variables (Q and y) at any time level as independent rather as interdependent. The explicit method has a stability criterion that limits the time step Δt, approximately, to

$$\Delta t \le \frac{\Delta x^2}{2G}, \tag{10.59}$$

where G is defined as before:

$$G = \frac{K H_b^{5/2} \sqrt{g/\kappa}}{8(s - 1)(1 - p)(h_* + B)} \tag{10.60}$$

Appropriate boundary and possibly internal conditions must be posed to complete problem formulation. Various types of boundary conditions are possible and are implemented fairly directly for the explicit model. The simplest boundary condition might be to fix the displacement of the shoreline at the ends of a represented shoreline segment, say, $y(0)$ and $y(\ell)$, where ℓ is the length of the shoreline segment represented by the model. With these values fixed, the adjacent Qs (i.e., Q_2 and Q_{IMAX}) are computed using these specified values and the adjacent y values in accordance with Eq. (10.58). The second type of boundary condition would be to specify a transport Q. This could apply, for example, downdrift of a jetty where the only transport was that due to a sand bypassing operation. In this case it would be best to locate the updrift grid line at the jetty. Internal boundary conditions may arise, for example, where a groin is placed. Again, the groin should be placed at a grid line, and the transport past the groin would be zero if the y values on either side of the groin were less than the groin length; transport past that grid line would occur only if the y displacement on one or both sides of the grid line were such that a portion of the total transport was bypassed. With changes in the wave direction, the boundary condition at an internal boundary could change from one of no transport to one with a finite transport. Figure 10.20 illustrates possible boundary and internal transport boundary conditions.

The implicit method has been mentioned briefly earlier and will be described. The term *implicit* derives from the simultaneous solution nature of the method. The unknowns are regarded as Q_i^{n+1} and Δy_i^{n+1}. A set of simultaneous equations (in tridiagonal form) is established by expanding the transport equation in terms of the Δy^{n+1} values adjacent to Q_i, that is,

$$Q_i^{n+1} = Q_i^n + \frac{\partial Q_i}{\partial \beta_i}\left(\frac{\partial \beta_i}{\partial(\Delta y_{i-1}^{n+1})}\right)\Delta y_{i-1}^{n+1} + \frac{\partial Q_i}{\partial \beta_i}\frac{\partial \beta_i}{\partial \Delta y_i^{n+1}}\Delta y_i^{n+1}, \tag{10.61}$$

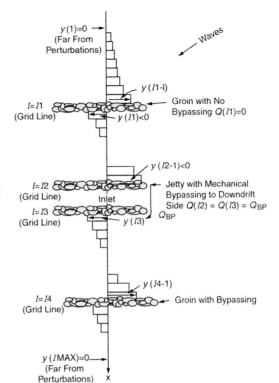

Figure 10.20 Boundary conditions for the numerical model.

which can be rearranged to the form

$$A_i \, \Delta y_{i-1}^{n+1} + B_i Q_i^{n+1} + C_i \, \Delta y_i^{n+1} = D_i \tag{10.62}$$

in which

$$A_i = -\frac{\partial Q}{\partial \beta_i} \frac{\partial \beta_i}{\partial \Delta y_i} = -2C_i \cos 2(\beta_i^n - \alpha_{b_i}^n) \frac{\Delta x}{\sqrt{(\Delta x)^2 + (y_i^n - y_{i-1}^n)^2}}$$

$$A_i \approx -2C_i \cos 2(\beta_i^n - \alpha_{b_i}^n)$$

$$B_i = 1$$

$$C_i = -A_i$$

$$D_i = Q_i^n \tag{10.63}$$

The continuity equation can be expressed in a similar form

$$A_i' Q_i^{n+1} + B_i' \Delta y_i^{n+1} + C_i' Q_{i+1}^{n+1} = D_i' \tag{10.64}$$

and

$$A_i' = -1$$

$$B_i' = \frac{2\Delta x(h_* + B)}{\Delta t}$$

$$C_i' = 1$$

$$D_i' = Q_i^n - Q_{i+1}^n - \frac{2\Delta x(h_* + B)y_i^n}{\Delta t} \tag{10.65}$$

Equations (10.62) and (10.64) form a set of tri-diagonal simultaneous equations, which, when provided with appropriate boundary conditions, can be solved by a variety of schemes very quickly. One approach is the so-called double-sweep method in which the first sweep relates the adjacent unknowns, say Δy_i^{n+1} and Q_i^{n+1}, and the second sweep quantifies the unknowns.

The most elaborate one-line model (GENESIS) was developed by Hanson and Kraus and is described in Hanson (1989) and Hanson and Kraus (1989), which is the culmination of a long series of works, such as Hanson and Larson (1987) and Kraus, Hanson, and Harikai (1984). This model, for use in design, includes wave refraction and diffraction of the wave field, and thus offshore breakwaters, for example, can be studied.

10.4.2.2 N-Line Contour Model

Perlin and Dean (1983) developed the N-line model, which is the extension of the one- and two-line contour models to an arbitrary number of contour lines. The cross-shore sediment transport is based on an equilibrium profile argument and the resulting transport if the profile is out of equilibrium. At each time step, a large set of simultaneous equations is solved.

10.4.2.3 Coastal Models

Rather than examining the behavior of the contour lines, coastal models or, as they are sometimes called, coastal morphodynamics models, examine the behavior of the bathymetry by calculating the sediment transport locally and using the conservation of sand to determine the local depth change over some unit of time. These types of models have the advantage of being able to deal easily with nonmonotonic beach profiles and will likely be the major type of nearshore bathymetric model in the future.

A variety of models have been, and are presently being developed, for coastal modeling (see the review by De Vriend, Zyserman et al. 1993 and their comparison of four European models). However, this is a dynamically changing field because of our increasing knowledge of sediment transport in the nearshore region and also as the result of increasing computational power, which permits modeling of complex processes now that were unthinkable decades ago.

Typically these coastal models require several important components, as shown in Figure 10.21.* The first model component is the wave module, which provides wave heights, directions, and perhaps, radiation stresses for each of the model grid locations. This wave module could be a wave-averaged model or a time-dependent model, such as a Boussinesq wave model. This module may dictate the size of the grids, for the wave models require certain sizes of computational grids for convergence. The second component of the model is the currents module, which computes the wave-induced currents for each grid point. These currents could simply be the depth-averaged

---

* Ideally, these components are implemented as removable modules, and thus, as our understanding or modeling of one particular component improves, the coastal model can be improved.

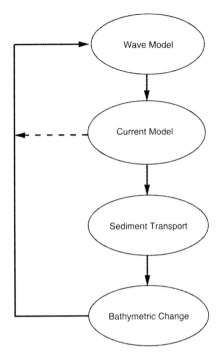

Figure 10.21 Modular diagram for generic three-dimensional coastal model.

velocities, or they could include the undertow. These current results are then used in the sediment transport component that calculates the transport rates. Finally, the conservation of sand module (based on Eq. (6.13)) is used to determine the bathymetric change with time and to modify the bathymetry. As shown by the arrows, the process is then repeated for the next time step, Δt, which must be chosen to provide for stability of the hydrodynamic module as well as the sediment change module. In general, the time scale for the bottom evolution is longer than for the hydrodynamics, and the models can be iterated on the sediment modules more often than over all the modules, as shown in the figure. De Vriend (1987) has discussed the nuances of the morphodynamic time scales and their relationship to the stability of the model.

Because of the complexity of the nonlinear wave and sediment dynamics and their interactions, different approaches have been taken to simplify the modeling. There are short-term models to predict the behavior of a beach over time scales on the order of hours. Ideally these models can predict the evolution of the coast during a storm. Long-term models are designed to predict coastal behavior over scales of a year or more. These types of models would enable a coastal engineer or a planner to show the natural evolution of a beach and, if desired, to compare that to its evolution using hard or soft erosion mitigation devices.

Some of the important issues are still to be resolved for both types of models. How is the randomness of the wave field (because of the randomness in our weather patterns) to be included? Should one representative wave train or a stochastic approach with different wave scenarios be used and the results ensemble averaged? How are the storms to be included, particularly because they play such a major role in the beach profile? How are the tides to be included? De Vriend, Capobianco et al. 1993 have discussed many of these issues.

As examples of models that use the Boussinesq equations for the hydrodynamics, the reader is referred to Sato and Kabiling (1994), who used the Bailard equations for the sediment transport.

REFERENCES

Abramowitz, M., and I.A. Stegun, *Handbook of Mathematical Functions*, New York: Dover Press, 1964.

Bakker, W.T., "Mathematical Theory About Sand Waves and its Application on the Dutch Wadden Isle of Vlieland," *Shore and Beach*, 36, 2, 5–14, 1968a.

Bakker, W.T., "The Dynamics of a Coast with a Groyne System," *Proc. 11th Intl. Conf. Coastal Eng.*, 492–517, London, 1968b.

Bakker, W.T., and T. Edelman, "The Coastline of River-Deltas," *Proc. 9th Intl. Conf. Coastal Eng.*, ASCE, Lisbon, 199–218, 1964.

Chio. T.Y., "Beach and Dune Response to Huricane Eloise of September, 1975," *Proc. Coastal Sediments '77*, New York: ASCE, 116–134, 1977.

Dalrymple, R.A., "Directional Wavemaker Theory with Sidewall Reflection," *J. Hyd. Res.*, 27, 1, 23–34, 1989a.

Dalrymple, R.A., "Physical Modelling of Coastal Processes," in *Recent Advances in Hydraulic Physical Modelling*, R. Martins, ed., NATO ASI, Martinus Nijhof Publishers, Rotterdam, 567–588, 1989b.

Dalrymple, R.A., and R.G. Dean, "The Spiral Wavemaker: Theory and Experiment," *Proc. Intl. Conf. Coastal Eng.*, ASCE, Vancouver, 689–706, 1972.

Dalrymple, R.A., and W.W. Thompson, "A Study of Equilibrium Profiles," *Proc. 15th Intl. Conf. Coastal Eng.*, ASCE, Honolulu, 1277–1296, 1976.

Dean, R.G., "Heuristic Models of Sand Transport in the Surf Zone," *Proc. Conf. Eng. Dynamics in the Surf Zone*, Sydney, 208–214, 1973.

Dean, R.G., "Physical Modelling of Littoral Processes," in *Physical Modelling in Coastal Eng.*, R.A. Dalrymple, ed., Rotterdam, A.A. Balkema, 119–139, 1985.

Dette, H.H., and K. Uliczka, "Prototype and Model Evolution of Beach Profile," *Symposium on Scale Effects in Modelling Sediment Transport Phenomena*, Intl. Assoc. Hyd. Res., Toronto, 35–48, 1986.

De Vriend, H.J., "Analysis of 2DH Morphological Evolutions in Shallow Water," *J. Geophys. Res.*, 92, C4, 3877–3893, 1987.

De Vriend, H.J., M. Capobianco, T. Chesher, H.E. de Swart, B. Latteux, and M.J.F. Stive, "Approaches to Long-term Modelling of Coastal Morphology: A Review," *Coastal Eng.*, 21, 225–269, 1993.

De Vriend, H.J., J. Zyserman, J. Nicholson, J.A. Roelvink, P. Péchon, and H.N. Southgate, "Medium-term 2DH Coastal Area Modelling," *Coastal Eng.*, 21, 193–224, 1993.

Dijkman, M.J., W.T. Bakker, and J.H. de Vroeg, "Prediction of Coastline Evolution for the Holland Coast," *Proc. 22nd Intl. Conf. Coastal Eng.*, ASCE, Delft, 1935–1947, 1990.

Grijm, W., "Theoretical Forms of Shorelines," *Proc. 7th Intl. Conf. Coastal Eng.*, ASCE, The Hague, 197–202, 1960.

Grijm, W., "Theoretical Forms of Shorelines," *Proc. 9th Intl. Conf. Coastal Eng.*, ASCE, Lisbon, 219–235, 1964.

Hanson, H., "Genesis – A Generalized Shoreline Change Numerical Model," *J. Coastal Res.*, 5, 1, 1–27, 1989.

Hanson, H., and N.C. Kraus, "Genesis: Generalized Model for Simulating Shoreline Change," U.S. Army Corps of Engineers, Coastal Engineering Research Center, CERC-MP-89-19, 1989.

Hanson, H., and M. Larson, "Comparison of Analytic and Numerical Solutions of the One-Line Model of Shoreline Change," *Proc. Coastal Sediments '87*, ASCE, 500–514, 1987.

Hughes, S.A., *Physical Models and Laboratory Techniques in Coastal Engineering*, Singapore: World Scientific, 568 pp., 1993.

Hughes, S.A., "Movable-bed Modelling Law for Coastal Dune Erosion," *J. Waterway, Port, Coastal, and Ocean Eng.*, ASCE, 109, 2, 164–179, 1983.

Hughes, S.A., and Fowler, J.E., "Validation of Movable-bed Modelling Guidance," *Proc. 22nd Intl. Conf. Coastal Eng.*, ASCE, 2457–2470, Delft, 1990.

Hulsbergen, C.H., W.T. Bakker, and G. van Bochove, "Experimental Verification of Groyne Theory," *Proc. 15th Intl. Conf. Coastal Eng.*, ASCE, Honolulu, 1439–1458, 1976.

Kamphuis, J.W., "The Coastal Mobile Bed Model – Does It Work?" *Proc. Modelling '75*, ASCE, San Francisco, 993–1009, 1975.

Kamphuis, J.W., "On Understanding Scale Effect in Coastal Mobile Bed Models," in *Physical Modelling in Coastal Engineering*, R.A. Dalrymple, ed., Rotterdam, A.A. Balkema, 141–162, 1985.

Kraus, N.C., and J.M. Smith, "SUPERTANK Laboratory Data Collection Project, Vol. 1. Main Text," Tech. Rept. CERC-94-3, January, 1994.

Kraus, N.C., H. Hanson, and S. Harikai, "A Shoreline Change at Oarai Beach Past, Present and Future," *Proc. 19th Intl. Conf. Coastal Eng.*, ASCE, Houston, 2107–2123, 1984.

Kriebel, D.L., "Beach and Dune Response to Hurricanes," MCE Thesis, University of Delaware, 334 pp., 1982.

Kriebel, D.L., "Verification Study of a Dune Erosion Model," *Shore and Beach*, 54, 3, 13–21, 1986.

Kriebel, D.L., W.R. Dally, and R.G. Dean, "Undistorted Froude Model for Surf Zone Sediment Transport," *Proc. 20th Intl. Conf. Coastal Eng.*, ASCE, Taipei, 1296–1310, 1986.

Kriebel, D.L., and R.G. Dean, "Numerical Simulation of Time-Dependent Beach and Dune Erosion," *Coastal Eng.*, 9, 3, 221–245, 1985.

Kriebel, D.L., and R.G. Dean, "Convolution Method for Time-Dependent Beach-Profile Response," *J. Waterway, Port, Coastal, Ocean Eng.*, ASCE, 119, 2, 204–226, 1993.

Larson, M., H. Hanson, and N.C. Kraus, "Analytical Solutions of the One-Line Model of Shoreline Change," U.S. Army Corps of Engineers Coastal Engineering Research Center, Tech. Report CERC-87-15, 1987.

Le Mehaute, B., and A. Brebner, "An Introduction to Coastal Morphology and Littoral Processes," Civil Engineering Dept., Rpt. 14, Queen's University, Canada, 1961.

Le Mehaute, B., and M. Soldate, "Mathematical Modelling of Shoreline Evolution," U.S. Army Corps of Engineers, Coastal Engineering Research Center, Misc. Rpt. 77-10, 56 pp., 1977.

Madsen, O.S., "On the Generation of Long Waves," *J. Geophys. Res.*, 76, 36, 8672–8683, 1971.

Mansard, E.P.D., M.D. Miles, and R.A. Dalrymple, "Numerical Validation of Directional Wavemaker Theory with Sidewall Reflection," *Proc. 23rd Intl. Cont. Coastal Eng.*, ASCE, Venice, 3468–3481, 1992.

Mei, C.C., *The Applied Dynamics of Ocean Surface Waves*, New York: John Wiley & Sons, 740 pp., 1983.

Mei, C.C., "Shoaling of Spiral Waves in a Circular Basin," *J. Geophys. Res.*, 78, 977–980, 1973.

Mei, C.C., and Ü. Ünlüata, "Harmonic Generation in Shallow Water Waves," in *Waves on Beaches*, R.E. Meyer, ed., New York: Academic Press, 181–202, 1972.

Moore, B.D., "Beach Profile Evolution in Response to Changes in Water Level and Wave Height," MCE, University of Delaware, 164 pp., 1982.

Noda, E.K., "Equilibrium Beach Profile Scale Model Relationships," *J. Waterway, Harbors and Coastal Eng. Div.*, ASCE, 98, 4, 511–528, 1972.

Ottesen Hansen, N.E., S.E. Sand, H. Lundgren, T. Sorensen, and H. Gravesen, "Correct Reproduction of Group-Induced Long Waves," *Proc. 17th Intl. Conf. Coastal Eng.*, ASCE, Sydney, 784–800, 1980.

Pelnard-Considere, R., "Essai de Théorie de l'Evolution des Formes de Rivage en Plages de Sable et de Galets," *4th Journees de l'Hydraulique, Les Energies de la Mer*, Question III, Rapport No. 1, 1956.

Perlin, M., and R.G. Dean, "A Numerical Model to Simulate Sediment Transport in the Vicinity of Coastal Structures," U.S. Army Corps of Engineers Coastal Engineering Research Center, Tech. Report CERC-83-10, 1983.

Sand, S.E. "Three-dimensional Deterministic Structure of Ocean Waves," Series Paper 24, Institute of Hydrodynamics and Hydraulic Engineering, Technical University of Denmark, 1979.

Sato, S., and M.B. Kabiling, "A Numerical Simulation of Beach Evolution Based on a Non-linear Dispersive Wave-Current Model," *Proc. 24th Intl. Conf. Coastal Eng.*, ASCE, Kobe, 2552–2570, 1994.

Saville, T., "Scale Effects in Two Dimensional Beach Studies," *Trans. 7th Gen. Mtg.*, Intl. Assoc. Hyd. Res., 1, A3-1–A3-10, 1957.

Smith, J.M., and N.C. Kraus, "SUPERTANK Laboratory Data Collection Project, Vol. II. Appendices A–I," Tech. Rept. CERC-94-3, Sept., 1995.

Svendsen, I.A., "Physical Modelling of Water Waves," in *Physical Modelling in Coastal Engineering*, R.A. Dalrymple, ed., Rotterdam, A.A. Balkema, 13–47, 1985.

Swain, A., and J.R. Houston, "Onshore–Offshore Sediment Transport Numerical Model," *Proc. 19th Intl. Conf. Coastal Eng.*, ASCE, Houston, 1244–1251, 1984.

Swart, H., "Schematization of Onshore–Offshore Transport," *Proc. 14th Intl. Conf. Coastal Eng.*, ASCE, Copenhagen, 884–900, 1974.

Trowbridge, J., R.A. Dalrymple, and K. Suh, "A Simplified Second-Order Solution for a Spiral Wave Maker," *J. Geophys. Res.*, 91, C10, 11,783–11,789, 1986.

Vellinga, P., "Beach and Dune Erosion During Storm Surges," Delft Hydraulics Laboratory, Comm. No. 372, 1986.

Walton, T.L., and T.Y. Chiu, "A Review of Analytical Techniques to Solve the Sand Transport Equation and Some Simplified Solutions," *Proc. Coastal Structures '79*, ASCE, 809–837, 1979.

Whalen, R.W., and C.E. Chatham, Jr., "Coastal Erosion," in *Coastal Hydraulic Models*, U.S. Army Corps of Engineers, Coastal Engineering Research Center Special Rept. 5, May 1979.

EXERCISES

10.1 The multielement wavemaker at the University of Delaware consists of 34 paddles with a width of 0.52 m. For a water depth of 0.8 m, determine the shortest wave period that can be used without the appearance of spurious waves.

10.2 Given a wave paddle width of 0.52 m and a desired wave train of 0.8 s in 0.8 m of water at a direction of 20°, determine if spurious waves will occur. If so, what spurious wave direction will occur?

10.3 A laboratory study is to be carried out to determine the equilibrium beach profile. The median grain size in the prototype is 0.4 mm with wave conditions of 1 m height and a 4 s period. Develop scale relationships for the following cases, based on Dean number similitude:

(a) If the sand used in the model has a grain size of 0.2 mm.

(b) A length scale of 1/25 is dictated by the space limitations of the laboratory. What sand size should be used?

10.4 Suppose sand is extracted from the midpoint of a surf zone at a rate that is equal to one-half of the longshore sediment transport. Describe the effects on the contours, including the shoreline contour, and discuss the sediment transport mechanisms. Consider unidirectional transport.

10.5   Consider a long, straight coastline with waves advancing directly toward shore. At time $t = 0$, a sand mining program is initiated in which the rate of sand removed from the beach face is Q_0. Assume that the active depth of the profile is h_*.

(a) Sketch the approximate planform of the beach with increasing time.

(b) For given Q_0, how will the shape of the planform vary with h_* and H_b?

(c) The sand removal will cause an indentation in the shoreline that increases with time. How does the maximum indentation increase with time?

(d) In particular, if the maximum indentation after 1 year is 100 ft, what will it be after 4 years?

10.6   Sand is removed from the surf zone at a rate of 100 m³/day.

(a) In Case A, the waves approach the shoreline normally. Qualitatively sketch the planform variation with time.

(b) In Case B, the waves approach the shoreline such that along the straight shoreline the transport is 50 m³/day. Qualitatively sketch the beach planform variation with time.

10.7   Consider the construction of a groin of length ℓ perpendicular to the shoreline. Using a one-line model,

(a) consider two possible structures of lengths ℓ and 2ℓ and compare the volumes of downdrift erosion that would occur due to the two structures

(a-1) prior to bypassing of the shorter structure,

(a-2) after bypassing of the longer structure.

(b) After a very long time, how would the two erosion volumes compare?

(c) According to the Pelnard-Considere equation, would the results differ if the structure were constructed at an angle to the shoreline? If so, why?

10.8   Sand is being added at a constant rate as a point source to the shoreline. The wave height is constant. Observations indicate that the shoreline advanced 100 m seaward after 1 year. How many years will be required for the shoreline to advance 250 m?

10.9   For a shoreline position that is oscillated at the point $x = 0$, the one-line analytical solution is given in Eq. (10.35). Describe in words the activities that could explain the shoreline behavior at $x = 0$. Develop an equation for the sand supply and removal, $Q(0, t)$ at $x = 0$, which would be required for this situation.

10.10   For the initial shoreline position of $y(x, 0) = B \cos \lambda x$, the shoreline position is slowly returned to a straight line according to the Pelnard-Considere solution given by Eq. (10.34).

(a) Sketch the shoreline position and the alongshore sediment transport for normal wave incidence at time $t = 0$.

(b) Sketch the sediment transport for oblique wave incidence.

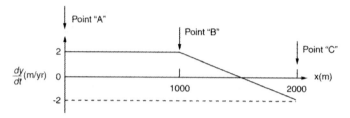

Figure 10.22   Shoreline recession rates for Problem 10.11.

    (c) Qualitatively explain why the shoreline oscillations damp out faster for the shorter wavelength shoreline perturbations. There are two reasons.

10.11   The net longshore sediment transport rate at Point A in Figure 10.22 is 80,000 m³/yr. The annual shoreline changes between points A and C are shown in the figure. What is the net annual transport rate at Points B and C? Assume $B = 2$ m, $h_* = 6$ m between A and B and $B = 1$ m, $h_* = 4$ m between B and C. Assume no cross-shore or sand source contributions to the sediment budget.

PART FOUR

SHORELINE MODIFICATION AND ANALYSIS

CHAPTER ELEVEN

Beach Fill and Soft Engineering Structures

In the early 1970s, much of Miami Beach, Florida, was badly eroded, lined with seawalls, and compartmented by long steel groins, making it impossible for tourists to walk along the beach. The numbers of visitors to Miami Beach and hotel occupancy rates were declining and, at high tide during storms, the impact of large waves against the seawalls and the resulting vibrations were reported to be worrisome to guests in the upper (more expensive) rooms.

From 1976 to 1981, the Miami Beach Nourishment Project was undertaken, widening the beach by 100 m over a length of 16 km. The project extends from Bakers Haulover Inlet at the north to Government Cut Inlet at the south. Both of these termini have large, relatively impermeable jetty structures to prevent end losses from the project. The project required in excess of 10 million m^3 of sand obtained from offshore borrow areas by large hydraulic dredges, and it cost $64 million (1980 USD).

The construction of the nourishment project took place in the midst of rising concern over the effects and effectiveness of shoreline projects. The project attracted negative publicity by receiving the Golden Fleece Award that was given annually by a popular senator to the project deemed to be the greatest waste of public funds. There were other outspoken claims that the project would wash away with the first major storm and was a waste of money. However, several years later, the project received an award from the American Shore and Beach Preservation Society for its effectiveness.

Now with over 20 years' experience by which to judge the project, it must be considered a great success – so much so that international arrivals at the Miami International Airport are greeted with the words "Welcome to Miami and its beaches!" The Miami Beach economy, so dependent on a viable recreational beach resource, has been revitalized and hosts 21 million visitors per year, which is more than twice the combined number of the three most heavily visited U.S. national parks: Yosemite, Yellowstone, and the Grand Canyon (Houston 1996).

11.1 INTRODUCTION

The protection of beaches against erosion has always been an important aspect of civil engineering works. History is replete with the loss of valuable coastal structures such as homes, harbors, lighthouses, and other valuable coastal property to erosion induced by an encroaching sea.*

This chapter and the next examine various methods that have been used (and misused) to protect shorelines and coastal properties from coastal erosion. Here we will address the "softer" structures, such as beach nourishment, whereas the next chapter examines "hard" structures: offshore breakwaters, groins, seawalls, and revetments. We will endeavor to explain the advantages and disadvantages of each method and indicate why one method should be favored over others in a particular design.

Erosion control measures should incorporate a reduction in the cause of the beach erosion where possible when, for example, the erosion is caused by human activities along the shoreline such as at inlets or harbors. However, pervasive natural sea level rise, for example, is not a cause that can be reduced by engineering design, although its effects can be.

For each coastal erosion mitigation measure, knowing the mechanism of its effectiveness is important: to know, for example, that some of the mitigation schemes are able to reduce the wave field at the shoreline, or simply provide a sacrificial beach, whereas others try to impede the longshore transport of sand. In a particular situation, one mitigation scheme will work better than some others owing to this difference in operation. In fact, some methods will fail in one situation and do very well in others. Also, some approaches can provide a net benefit to adjacent shorelines, whereas in other cases, if misapplied, can cause severe erosion. Above all, it is important to remember that, although it is an obvious fact, none of the mitigation measures discussed in these two chapters make sand!

11.2 BEACH NOURISHMENT (BEACH FILL)

The placement of sand on a beach to restore (or to build) it is referred to as *beach nourishment* or *beach fill* and is the most nonintrusive technique available to the coastal engineer. Typically, sand from offshore or onshore sources is placed on the eroding beach. The sources, amounts, qualities, placement, and fate of the beach fill are all problems addressed in this section.

Beach nourishment, with its attendant widening of the beach, is used to accomplish several goals as follows:

* To build additional recreational area,
* To offer storm protection (both by reducing the wave energy nearshore and creating a sacrificial beach to be eroded during a storm), and
* To provide, in some cases, environmental habitat for endangered species.

The second goal of storm protection is often underappreciated. If a beach fill is "lost" through a major storm, this "lost" sand will have been eroded instead of the sand of the unnourished beach, thus protecting the upland. Further, the sand is not lost. It

---

* On the other hand, there are also numerous historical cases of harbors being abandoned because of infilling by sediments, which is quite a different coastal engineering problem but a significant one.

is, most likely, either transported offshore into the bar system to be returned during conditions of milder waves, or it is displaced alongshore.

Of the many remedial measures for beach erosion, beach nourishment is the only approach that introduces additional sand into the system. All others seek to rearrange the existing sand in a manner that will benefit a portion of the beach. Proponents of some of these approaches have claimed that these methods will cause sand to be brought from offshore, thereby building beaches without impacting adjacent beaches; however, there has been no documented proof of this claim. It is clear that an eroding shoreline has a deficit of sand, and the most effective solution is one that will restore a beach and the benefits that a wide beach provides. Other than beach nourishment, there are few approaches that address this basic need of supplying additional sand without adversely affecting adjacent beaches.* In summary, because beach nourishment is the only approach that directly addresses the deficit of sand in the system without at least the potential of causing adverse effects on adjacent property, it is the most benign and acceptable approach to beach erosion mitigation. This conclusion is supported by the National Research Council (1996), which has strongly endorsed beach nourishment and has issued substantial design guidelines.

A significant problem associated with beach nourishment design is predicting the lifetime of a fill project. Usually, a beach slated for nourishment is eroding, and the likelihood is that the replacement of the originally eroding beach is simply a means of turning back time, for the erosion mechanisms are still in place; hence, the beach will eventually erode back to its original state and continue to erode further. The question is, How long will this process take?

In beach nourishment, sand is usually placed on the beach at a slope steeper than its equilibrium profile. Also, inherently the beach represents a planform that is out of equilibrium. Once sand is replaced, waves begin to restore equilibrium both in profile and in planform. The process of subsequent shoreline change can be considered in three stages (see Figure 1.4):

1. The profile equilibration, which generally results in a cross-shore transfer of sand from the upper to lower portions of the profile and, thus, a shoreline recession, but not a transfer of sand out of the profile;
2. A transfer of sand along the beach from a "spreading out" of sand resulting from the planform anomaly created by the placed sand;
3. Background shoreline erosion due to ongoing processes before the project was emplaced.

These three components of change are operative simultaneously; however, they usually have somewhat different time scales. Profile equilibration typically dominates on time scales on the order of years, whereas, as we will see for the longer projects, longshore diffusional losses generally occur on the order of decades. Usually it is assumed that the background erosion losses continue at the same rate as before the

---

* One exception is the prevention of sand losses at sinks (or almost sinks) such as at inlets where the sand is dispersed over large areas, perhaps to ebb and flood tidal shoals. It has been shown (Dean 1993) that terminal structures constructed at the ends of barrier islands have been effective in stabilizing these areas even to the extent of resulting in a widening of the beach.

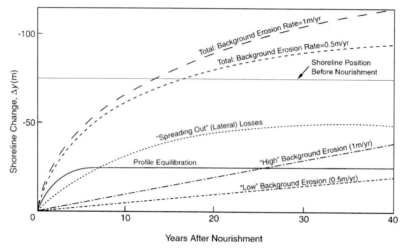

Figure 11.1 Qualitative illustration of three components of shoreline recession following a beach nourishment project shown for two background erosion rates and initial nourished width of 75 m.

project; thus, their effect is the same for each year. Figure 11.1 qualitatively illustrates the shoreline changes for each of the three effects noted and their sum for two background erosion rates.

11.2.1 MECHANICS OF BEACH FILL CONSTRUCTION

The placement of sand on the beach can be done mechanically or hydraulically. The mechanical transportation of sand from a borrow area to the fill area implies that the material is transported by truck or other land-based means of transport to the beach. At the borrow area, the sand is placed in the truck by normal earthmoving equipment. In general, this method is used for small beach fills owing to the high cost of the transportation and the impact on traffic and road surfaces from the heavy loads. In the United States, at least 98 percent of the sand placed in current beach nourishment projects is by hydraulic means. At the borrow sites, which are usually offshore, the sand is lifted from the bottom by a hydraulic dredge and then is pumped via floating pipelines to the fill site, where it is discharged onto the beach. A third method, coupled with offshore sources, uses hopper barges or hopper dredges. These are filled with the borrow material and then taken to the fill site, where the cargo of sand is discharged into shallow water or pumped directly to the beach.

11.2.2 PLACEMENT OF THE FILL

At the fill site, the borrow sand can be placed in a variety of positions. Early experiments in beach nourishment involved placing the material offshore of the beach with the intention that the wave action would move the material naturally onto the beach. The advantage of this scheme is that the material, say obtained by hopper dredges, could be dumped in deeper water with attendant safety and economic benefits. The disadvantage is that the material might not be moved by natural processes toward the shore and would remain unavailable to the beach (this is discussed further in Section 11.3). Other sites for the fill are on the back beach, perhaps the dunes, the foreshore, or distributed over the profile. Planform options are to spread the fill material over the eroding beach, to place the material primarily at the updrift end of the project area, or to place the material in a pile, or a groinlike planform (Führböter 1974).

11.2.3 COMPATIBILITY OF THE BORROW MATERIAL

The original sand on the beach will be referred to as the native material. As discussed in Chapter 2, it has certain characteristics, including mineralogy, composition, and size. Its presence (despite the eroding nature of the beach), rather than another sand, indicates that it is, in some measure, the stable, or nearly the stable, sand size for this beach. Should finer sand be placed on the beach, it would not be in equilibrium, and we expect that much of the material would be moved offshore by the waves.

An ideal sediment source for fill is one that has compatible material, which means the borrow material has the same characteristics as the native material. Of course, this is not likely to be the case, for the beach sand and the borrow sand were deposited under different environmental conditions and may have originated from different sources.

Krumbein and James (1965) proposed a method that considered the grain size distributions $f(\phi)$ of the borrow and native materials to be each represented by lognormal distributions, as discussed in Chapter 2. This method defines compatibility of the borrow material on the basis of the proportion of borrow material distribution common with the native sand size distribution. This approach appears somewhat reasonable in discounting the finer fraction of the borrow material, which is assumed to be transported offshore, but less reasonable in discounting similarly the proportion of coarse material that is in excess relative to the native sand.

James (1974) developed a complex method addressing the relative renourishment frequency for different sand characteristics; however, this procedure addresses longshore sediment transport and considers the nourishment project to be located in an area where the ambient longshore sediment transport had been interrupted completely.

Dean (1974) presented a method that attempted to address the deficiency noted in the Krumbein and James method just discussed. The borrow material was only discounted for the excessive proportion of fines present; excessive proportions of coarser material were included in the compatible fraction. However, it was considered that all the fine fraction smaller than a critical value was lost. This method resulted in a considerably higher compatibility index for the fill material than the approach of Krumbein and James (1965).

The *overfill factor* is denoted as K, which specifies the number of cubic meters (or yards) of material to be placed on a beach to retain one cubic meter (or cubic yard). The procedure used by Dean was to examine the size distribution of the fill material, which was assumed to be log normally distributed: $f_F(\phi)$, using ϕ units,

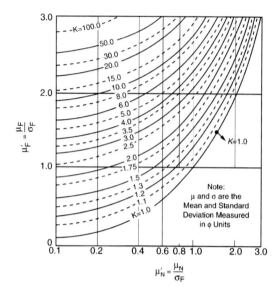

Figure 11.2 Required replacement volume K of borrow material to obtain one unit of compatible beach material (from Dean 1974).

defined in Chapter 2. Because the finer fractions of the fill will be lost, the (altered) size distribution for the material that remains is

$$f_{b_a} = \begin{cases} K f_F(\phi), & \phi \le \phi_* \\ 0, & \phi > \phi_*, \end{cases} \tag{11.1}$$

indicating that all sand finer than ϕ_* is lost. The K factor is chosen so that

$$1 = K \int_{-\infty}^{\phi_*} f_F(\phi)\, d\phi = \frac{K}{\sqrt{2\pi}\,\sigma_F} \int_{-\infty}^{\phi_*} e^{-\frac{(\phi-\mu_F)^2}{2\sigma_F^2}}\, d\phi \tag{11.2}$$

Because we expect the size of the material on the beach to be in equilibrium with the natural processes, the mean diameter of the fill material is to be the same as the native material. Therefore, we have

$$\frac{K}{\sqrt{2\pi}\,\sigma_F} \int_{-\infty}^{\phi_*} \phi\, e^{-\frac{(\phi-\mu_F)^2}{2\sigma_F^2}}\, d\phi = \mu_N \tag{11.3}$$

or, substituting for K:

$$\frac{\int_{-\infty}^{\phi_*} \phi\, e^{-\frac{(\phi-\mu_F)^2}{2\sigma_F^2}}\, d\phi}{\int_{-\infty}^{\phi_*} e^{-\frac{(\phi-\mu_F)^2}{2\sigma_F^2}}\, d\phi} = \mu_N \tag{11.4}$$

This last equation can be solved for ϕ_*, and Eq. (11.2) provides the value of K. Dean has provided a convenient figure for the K factor in terms of the sand characteristics (see Figure 11.2).

EXAMPLE 1

An eroding beach has a sand characterized by a mean diameter of 1.5ϕ and a standard deviation of $\sigma = 0.91$. The beach fill available for use has a mean diameter of $2.96\ \phi$ and a standard deviation of 1.76. Determine the overfill ratio necessary for this finer-sized fill material.

To use Figure 11.2, we need to determine the dimensionless ratios, $\mu'_N = \mu_N/\sigma_F$, and $\mu'_F = \mu_F/\sigma_F$. From the given data, we have $\mu'_N = 0.85$, and $\mu'_F = 1.68$. From the figure, the overfill ratio is 2.05, meaning roughly we need to place twice as much fill as ultimately desired for the fill project, due to the high losses of fine material. (Finding a more compatible fill may reduce project costs due to a smaller required volume of sand; however, this must be balanced against increased transportation costs from a possibly more distant borrow site.)

James (1975) developed overfill and renourishment factors based on his 1974 publication that considers the relative characteristics of the borrow and native sand characteristics. Similar to earlier methods, this procedure was based on the size distribution of the sands rather than their associated equilibrium profiles. Compared with the method of Dean (1974), this method considers that only the portion of the fines in excess of the native distribution is lost; the method thus predicts a more compatible borrow material. The renourishment factor is the ratio of the required renourishment frequencies of the nourished beach if constructed of a particular fill sand compared with a compatible sand. The James method was adopted for use in the 1984 edition of the *Shore Protection Manual* of the U.S. Army Corps of Engineers.

11.2.3.1 Profile Types

By using the equilibrium beach profile concepts described in Chapter 7, we can develop an alternative method for determining the compatibility of the fill. This method has the advantage of including the effects of the forces that shape the natural and altered beach profiles. The Corps of Engineers has adopted this approach as a measure of compatibility (U.S.A.C.E. 1994).

The dependence of sediment scale factor A on diameter d, as shown in Figure 7.6, and Table 7.2 leads to some interesting consequences relative to beach nourishment. When a volume V of fill sand per unit length is added to the native beach profile, it is assumed that it will equilibrate eventually to form $h = A_F y^{2/3}$. Depending on the fill and native sediment scale parameters, A_F and A_N, respectively, and the volume added, the nourished beach profile can be intersecting, nonintersecting, or submerged, as presented in Figure 11.3.

A necessary but insufficient requirement for profiles to intersect is that the placed material be coarser than the native. Similarly, nonintersecting or submerged profiles will always occur if the placed sediment is the same size as, or finer than, the native. However, nonintersecting profiles can occur if the placed sediment is coarser than the native. Figure 11.4 illustrates the effect of placing the same volume of four different-sized sands. In Figure 11.4(a), sand coarser than the native is used, an intersecting profile results, and a relatively wide beach Δy_0 is obtained. In Figure 11.4(b), the same volume of sand of the same size as the native is used, a nonintersecting profile results, and the dry beach width gained is less. More of the sand volume is required to fill out the more mildly sloped underwater profile. In Figure 11.4(c), the placed sand is finer than the native, and much of the fill sand is utilized in satisfying the requirements of the more mildly sloped underwater profile. In a limiting case, shown

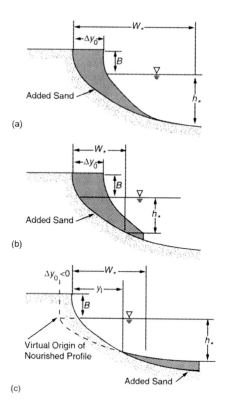

Figure 11.3 Three generic types of nourished beach profiles (from Dean 1991). (a) Intersecting profile $A_F > A_N$; (b) nonintersecting profile; (c) submerged profile $A_F < A_N$.

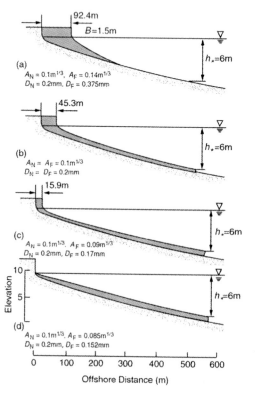

Figure 11.4 Effect of nourishment scale parameter, A_F on width of resulting dry beach. Four examples of decreasing A_F with same added volume per unit beach length (from Dean 1991). (a) Intersecting profiles; (b) nonintersecting profiles with coarser fill; (c) nonintersecting profile with finer fill; (d) limiting case of nourishment advancement with nonintersecting profiles.

in Figure 11.4(d), no dry beach results because all the sand is used to satisfy the underwater profile requirements. For still finer sands, the location of the equilibration and the landward intersection points would both move seaward, resulting in another type of submerged profile.

Figures 11.5(a) through 11.5(d) illustrate the effects of nourishing with greater and greater quantities of a sand that is considerably finer than the native. With increasing volumes, the landward intersection of the native and placed profiles occurs closer to shore, and the seaward limit of the placed profile moves seaward. Figure 11.5(d) is the case of incipient dry beach formation, as in Figure 11.4(d).

11.2.4 VOLUME CALCULATIONS

The Dutch method of beach nourishment (Verhagen 1992) is simple, direct, and completely empirically based. The method comprises five steps:

- Step 1: Perform coastal measurements (for at least 10 years).
- Step 2: Calculate the loss of sand in cubic meters per year per coastal section.
- Step 3: Add 40 percent loss.
- Step 4: Multiply this quantity by a convenient lifetime (for example, 5 years).
- Step 5: Put this quantity somewhere on the beach between the low-water-minus 1-m line and the dune foot.

Interpreted in the framework of processes discussed previously, the additional 40 percent can account for several effects, including spreading out losses, sand that is finer than the native, and the possibility of fines in the borrow material that will be winnowed out and not be stable when mobilized by the waves. Verhagen also

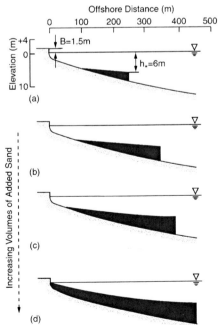

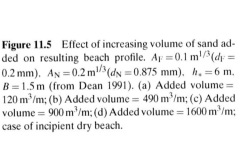

Figure 11.5 Effect of increasing volume of sand added on resulting beach profile. $A_F = 0.1 \text{ m}^{1/3} (d_F = 0.2 \text{ mm})$, $A_N = 0.2 \text{ m}^{1/3} (d_N = 0.875 \text{ mm})$, $h_* = 6 \text{ m}$, $B = 1.5 \text{ m}$ (from Dean 1991). (a) Added volume $= 120 \text{ m}^3/\text{m}$; (b) Added volume $= 490 \text{ m}^3/\text{m}$; (c) Added volume $= 900 \text{ m}^3/\text{m}$; (d) Added volume $= 1600 \text{ m}^3/\text{m}$; case of incipient dry beach.

notes that, in the Netherlands, the spreading losses are expected to be relatively small because the project length-to-width is on the order of 20 to 40.* Monitoring and modification of the approach for the first renourishment are recommended based on the monitoring results. Verhagen states that (as of 1992), in the Netherlands, all beach nourishment projects except one have been designed using this method.

We can quantify the results presented in Figures 11.3–11.5 by utilizing equilibrium profile concepts. It is necessary to distinguish the three cases noted in Figure 11.3. The first is with *intersecting profiles*, as indicated in Figure 11.3(a) and requires $A_F > A_N$. For this case, we need to determine the volume placed per unit shoreline length V_1 that will yield a shoreline advancement Δy_0. The calculation is

$$V_1 = B \Delta y_0 + \int_0^{y_i} A_N y^{2/3} \, dy - \int_0^{(y_i - \Delta y_0)} A_F y^{2/3} \, dy, \tag{11.5}$$

where y_i is the distance from the origin of the original shoreline to the position where the two profiles intersect. Integrating, we obtain

$$V_1 = B \Delta y_0 + \frac{3}{5} A_N y_i^{5/3} - \frac{3}{5} A_F (y_i - \Delta y_0)^{5/3} \tag{11.6}$$

The distance y_i is determined by equating the two profiles at the depth of intersection,

$$A_N y_i^{2/3} = A_F (y_i - \Delta y_0)^{2/3},$$

which results in

$$y_i = \frac{\Delta y_0}{1 - \left(\frac{A_N}{A_F}\right)^{3/2}} \quad \text{or, for later use, } \Delta y_0 = y_i \left[1 - \left(\frac{A_N}{A_F}\right)^{3/2} \right] \tag{11.7}$$

Substituting yields

$$V_1 = B \Delta y_0 + \frac{3}{5} A_N \frac{\Delta y_0^{5/3}}{\left[1 - \left(\frac{A_N}{A_F}\right)^{3/2} \right]^{2/3}} \tag{11.8}$$

Introducing the dimensionless parameters $V' = V/BW_*$, $\Delta y_0' = \Delta y_0/W_*$, $B' = B/h_*$, we can rewrite the final result in nondimensional form as

$$V_1' = \Delta y_0' + \frac{3}{5B'} (\Delta y_0')^{5/3} \frac{1}{\left[1 - \left(\frac{A_N}{A_F}\right)^{3/2} \right]^{2/3}}, \tag{11.9}$$

where W_* is the offshore distance associated with the breaking depth h_* on the native (unnourished) profile, that is

$$W_* = \left(\frac{h_*}{A_N}\right)^{3/2}, \tag{11.10}$$

---

* This range, however, appears to be small enough to cause considerable spreading-out or diffusive losses to the beaches adjacent to the project. Additionally, we will see later in this chapter that this ratio of project length to width is not the relevant parameter for longevity.

and the breaking depth h_*, and breaking wave height H_b are related by

$$h_* = H_b/\kappa$$

with $\kappa (\approx 0.78)$, the spilling breaker index.

For *nonintersecting but emergent profiles* (Figure 11.3(b)), the corresponding volume V_2 in nondimensional form is

$$V_2' = \Delta y_0' + \frac{3}{5B'} \left\{ \left[\Delta y_0' + \left(\frac{A_N}{A_F} \right)^{3/2} \right]^{5/3} - \left(\frac{A_N}{A_F} \right)^{3/2} \right\} \tag{11.11}$$

For $\Delta y_0' \ll 1$,

$$V_2' \simeq \frac{(B'+1)\Delta y_0'}{B'},$$

which, in dimensional form, is simply $V_2 = (B + h_*)\Delta y$.

It can be shown from Eq. (11.7), with y_i set to W_*, that the critical value of $(\Delta y_0')$ for distinguishing between intersecting and nonintersecting of profiles is given by

$$\Delta y_0' + \left(\frac{A_N}{A_F} \right)^{3/2} - 1 \begin{cases} <0, & \text{Intersecting Profiles} \\ >0, & \text{Nonintersecting Profiles} \end{cases} \tag{11.12}$$

The critical *volume* associated with intersecting and nonintersecting profiles is

$$(V')_{c1} = \left(1 + \frac{3}{5B'} \right) \left[1 - \left(\frac{A_N}{A_F} \right)^{3/2} \right] \tag{11.13}$$

and applies only for $(A_F/A_N) > 1$. Also of interest is that the critical volume of sand that will just yield a finite shoreline displacement for nonintersecting profiles $(A_F/A_N) < 1)$, is

$$(V')_{c2} = \frac{3}{5B'} \left(\frac{A_N}{A_F} \right)^{3/2} \left\{ \frac{A_N}{A_F} - 1 \right\} \tag{11.14}$$

Figure 11.6 presents these two critical volumes versus the scale parameter ratio (A_F/A_N) for the special case $h_*/B = 4.0$, that is, $B' = 0.25$.

Figure 11.6 Volumetric requirement for finite shoreline advancement (Curve 1, Eq. (11.14)); volumetric criterion for intersecting profiles (Curve 2, Eq. (11.13)); results for $h_*/B = 4$; from Dean (1991).

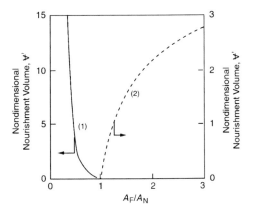

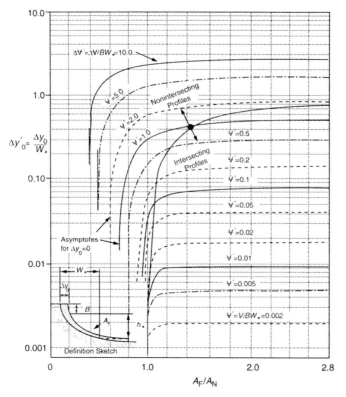

Figure 11.7 Variation of dimensionless shoreline advancement $\Delta y_0/W_*$ with A_F/A_N and V' for $h_*/B = 2.0$ (from Dean 1991).

The results from Eqs. (11.9), (11.11), and (11.12) are presented in graphical form in Figures 11.7 and 11.8 for cases of $h_*/B = 2$ and 4. Plotted is the nondimensional shoreline advancement ($\Delta y_0'$) versus the ratio of fill to native sediment scale parameters A_F/A_N for various isolines of dimensionless fill volume $V'(= V/(W_*B))$ per unit length of beach. It is interesting that the shoreline advancement increases only slightly for $A_F/A_N > 1.2$; for smaller values the additional shoreline width Δy_0 decreases rapidly with decreasing A_F/A_N. For A_F/A_N values slightly smaller than plotted, there is no shoreline advancement; that is, as in Figure 11.4(d).

With reference to Figure 11.4(d) for submerged profiles, it can be shown that

$$\frac{\Delta y_0}{y_I} = 1 - \left(\frac{A_N}{A_F}\right)^{3/2} \tag{11.15}$$

where $\Delta y_0 < 0$ and the nondimensional volume of added sediment V' can be expressed as

$$V' = \frac{3}{5B'}\left\{\left[\left(\frac{A_N}{A_F}\right)^{3/2} + \Delta y_0'\right]^{5/3} + \frac{(-\Delta y_0')^{5/3}}{\left[\left(\frac{A_N}{A_F}\right)^{3/2} - 1\right]^{2/3}} - \left(\frac{A_N}{A_F}\right)^{3/2}\right\} \tag{11.16}$$

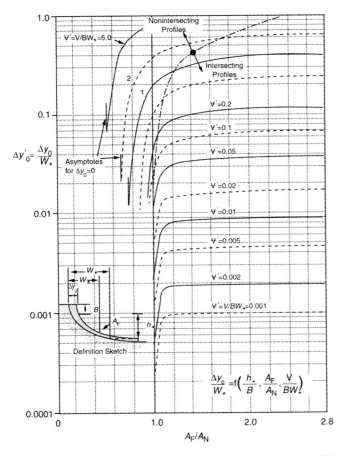

Figure 11.8 Variation of dimensionless shoreline advancement $\Delta y_0/W_*$ with A_F/A_N and V' for $h_*/B = 4.0$ (from Dean 1991).

The amount of additional dry beach width Δy_0 due to a given added volume of sand per unit beach length V for three nourishment sand sizes and a native sand size of 0.2 mm is shown in Figure 11.9, which is developed from the equations given in this section. For nourishment sand coarser than the native size, the additional dry beach width first increases rapidly with fill volume V owing to the intersecting nature of the profiles and the shallower depths being filled. For larger V, the profile becomes nonintersecting, and the relationship becomes approximately parallel to the curve for the fill grain size equal to the native size. For nourishment sand smaller than the native, the fill profile is submerged for V less than the threshold; then, the relationship is once again nearly parallel to the line for the same fill size as the native.

From this figure, we can infer, for the same material in the fill as the native beach, that the dry beach planform area of a beach nourishment project on a long straight beach is conserved with time, as is shown in Chapter 10. For coarser fill material, the planform actually increases with time because V decreases with time owing to the spreading of the fill material, and the additional dry beach width per unit volume is greater for smaller volumes than for larger volumes, as discussed

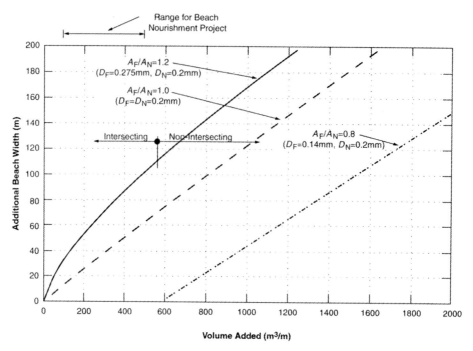

Figure 11.9 Additional dry beach width versus volume of sand added per unit length of beach for three different fill sands; $h_* = 6$ m, $B = 2.0$ m.

earlier in this section. In fact, the asymptotic beach plan area is $V_0 \ell / B$, where ℓ is the length of the fill and B is the berm height. Finally, for finer fill material, the planform diminishes with time, approaching zero as the largest value of V approaches the threshold for finite shoreline displacement.

These trends illustrate the need to represent the nourishment grain size distributions by more than one representative size. For example, for the finer fill material, portions of the nourishment and native grains size distributions would overlap, moderating the results predicted for a single nourishment grain size. Nevertheless, the preceding results are considered to be qualitatively correct and may explain some of the reported anomalous performance behavior of some nourishment projects.

11.2.4.1 Beach Fill Volume at a Seawalled Beach

In some cases, coastal armoring may have been constructed before nourishment is carried out. If the erosion continues to the degree that the beach profile no longer extends above the water line, as shown in Figure 11.10, a threshold volume will be required to be placed before any dry beach results and before the methods described in the previous sections apply. If not taken into consideration, the neglect of this additional volume will result in an underestimation of the required volumes and hence an underdesigned project with a shorter lifespan. For illustration purposes, we start by developing the threshold amount for the simplest case in which the nourishment and native sand are compatible.

With reference to Figure 11.10, where the initial depth at the wall is h_w and the depth of closure is h_*, the required nondimensional threshold volume per unit beach

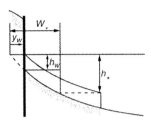

Figure 11.10 Definition sketch for beach nourishment fronting a sea-walled beach for nonintersecting profiles.

length, $V'_{\text{threshold}}$, to yield an incipient dry beach for compatible sediment is

$$V'_{\text{threshold}} = \frac{3}{5B'}[(y'_w + 1)^{5/3} - y'^{5/3}_w - 1], \tag{11.17}$$

where $y'_w = (h_w/h_*)^{3/2} = y_w/W_*$.

For the case of noncompatible nourishment and native material, it is necessary, as before, to separate the conditions into nonintersecting and intersecting profiles.

For nonintersecting profiles, the nondimensional volume is determined as before:

$$V'_{\text{threshold}} = \frac{3}{5}\frac{1}{B'}\left\{\left[y'_w + \left(\frac{A_F}{A_N}\right)^{-3/2}\right]^{5/3} - (y'_w)^{5/3} - \left(\frac{A_F}{A_N}\right)^{-3/2}\right\}, \tag{11.18}$$

and the corresponding results for intersecting profiles are

$$V'_{\text{threshold}} = \frac{3}{5B'}y'^{5/3}_w\left[\frac{1}{\left[1 - \left(\frac{A_F}{A_N}\right)^{-3/2}\right]^{2/3}} - 1\right] \tag{11.19}$$

The criterion separating intersecting and nonintersecting profiles is

$$y'_w = 1 - \left(\frac{A_F}{A_N}\right)^{-3/2}$$

If the left-hand side exceeds the right side, the profiles are nonintersecting and vice versa.

11.2.5 BEACH PLANFORM RESPONSE

The one-line Pelnard-Considere model discussed in Chapter 10 can provide some useful information concerning the placement of beach fill. If the fill is placed on a beach in a rectangular planform with alongshore length ℓ and an offshore dimension Δy_0, then the one-line model can be used to predict the subsequent behavior of the fill.

The analytic solution for this case was expressed in terms of two error functions:

$$y(x, t) = \frac{\Delta y_0}{2}\left\{\text{erf}\left[\frac{\ell}{4\sqrt{Gt}}\left(\frac{2x}{\ell} + 1\right)\right] - \text{erf}\left[\frac{\ell}{4\sqrt{Gt}}\left(\frac{2x}{\ell} - 1\right)\right]\right\} \tag{11.20}$$

The dimensional parameter G is the longshore diffusivity defined in the previous chapter as

$$G = \frac{K H_b^{5/2}\sqrt{g/\kappa}\,\cos 2\delta_b}{8(s - 1)(1 - p)(h_* + B)} \tag{11.21}$$

In Figure 10.10, this solution was plotted as a function of time; because it is symmetric, only half the solution was shown. Initially, the two ends of the planform begin to spread out, but as the diffusional effects move toward the middle, the planform shape begins to look more like a normal distribution.

Many interesting results can be obtained by examining Eq. (11.20). First, an important parameter in the solution is $\sqrt{Gt}/\ell$, where, again, ℓ is the length of the fill project and G is the longshore diffusivity parameter in the diffusion equation. The larger the value of this time parameter, the more the planform has changed. If the quantity $\sqrt{Gt}/\ell$ is the same for two different situations, then it is clear that the planform evolutions are also geometrically similar. For example, if two nourishment projects are exposed to the same wave climate but have different lengths, the project with the greater length will last longer. In fact, the longevity of a project varies as the square of the length; thus, if Project A, with a shoreline length of 1 km, "loses" 50 percent of its material in a period of 2 years, Project B, subjected to the same wave climate but with a length of 4 km, would be expected to lose 50 percent of its material from the region where it was placed in a period of 32 years. Thus, the project length is very significant to its performance. The presence of a background erosion rate would reduce the differences between the longevities of these two projects.

Consider next the case in which two fill projects are of the same length but are exposed to different wave climates. The coefficient in the one-line model, G (Eq. (11.21)) varies with the wave height to the 5/2 power, and thus if Project A is located where the wave height is 1 m and loses 50 percent of its material in a period of 2 years, Project B, with a similarly configured planform located where the wave height is 0.25 m, would be expected to last a period of 64 years.

In Figure 11.11, a graphical depiction of Eq. (11.20) is presented for various combinations of wave height and project length for the case of no background erosion.

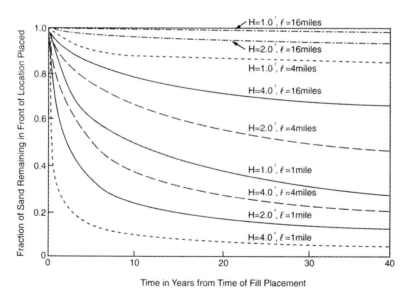

Figure 11.11 Fraction of material remaining in front of location placed for several wave heights and project lengths ℓ.

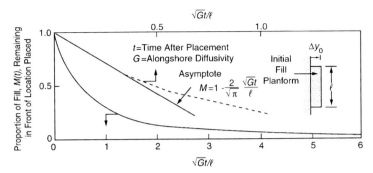

Figure 11.12 Proportion of material remaining in fill area versus dimensionless time.

As an example of the use of this figure, a project 4 mi in length at a location where the wave height is 4 ft would lose 60 percent of its material in 7 years, and a second project in a location where the wave height is 2 ft and the project length is 16 mi would lose only 10 percent of its material in a period of 40 years.

It is possible to obtain an analytical expression for the proportion of sand $M(t)$ remaining at the placement site, where

$$M(t) = \frac{1}{\Delta y_0 \ell} \int_{-\ell/2}^{\ell/2} y(x,t)\, dy, \tag{11.22}$$

or

$$M(t) = \frac{\sqrt{4Gt}}{\ell\sqrt{\pi}} \left(e^{-(\ell/\sqrt{4Gt})^2} - 1\right) + \operatorname{erf}(\ell/\sqrt{4Gt}), \tag{11.23}$$

which is plotted in Figure 11.12 along with the asymptote

$$M(t) \simeq 1 - \frac{\sqrt{4Gt}}{\ell\sqrt{\pi}}, \tag{11.24}$$

which is valid for small t; more specifically, for

$$\sqrt{Gt}/\ell < \frac{1}{2}$$

The quantity Gt/ℓ^2 can be referred to as a dimensionless time. For large values of this parameter,

$$M(t) \simeq \frac{1}{2\sqrt{\pi}} \frac{\ell}{\sqrt{Gt}}$$

A useful yardstick (meterstick, now?) for estimating the half-life of a project is obtained by noting that $M = 0.5$ for $\sqrt{Gt}/\ell \simeq 0.46$. Therefore, we can define a half-life t_{50} as

$$t_{50} = (0.46)^2 \frac{\ell^2}{G} = 0.21 \frac{\ell^2}{G} \tag{11.25}$$

Note that the half-life does not depend on the volume of fill emplaced.

EXAMPLE 2

The half-life of a project depends on the wave climate that it experiences through the G parameter. Determine how long a given beach fill will last if the parameters in G (see Eq. (11.21)) are

$$h_* + B = 8\,\text{m}; \quad \kappa = 0.78$$
$$s - 1 = 1.65; \quad p = 0.35$$

Substituting into G, we get $G = 0.052 H_b^{\frac{5}{2}}$. We can now write t_{50} as

$$t_{50} = 4.1 \frac{\ell^2}{H_b^{\frac{5}{2}}}\ \text{s} \tag{11.26}$$

The project will be 10 km in length; therefore, we have $t_{50} = 4.1 \times 10^8 / H_b^{5/2}$ s, or $t_{50} = 13/H_b^{5/2}$ yrs. For a wave height of 0.5 m, the half-life is 73 years, whereas for a 1-m wave, the half-life is only 13 years. (Note that if background erosion were present, the half-life would be reduced. The effect of background erosion will be considered in detail later.)

11.2.5.1 Longevity of Beach Fills

The longevity of the beach fill depends on the geometry of the project and the nature of the fill material, as discussed earlier in this section. In addition, longevity depends on the wave climate to which the project will be exposed during its lifetime. The wave climate cannot, of course, be predicted over the lifetime of the project exactly; therefore, other approaches must be taken. One approach is the design wave, which would represent all of the waves occurring during the lifetime of the fill. Another approach would be a probabilistic one that generates numerous possible sequences of wave heights that the beach fill might experience, which are then used to predict the beach fill behavior. All of the subsequent possible fill behaviors can then be averaged (or ranked) to determine the probability of the beach fill lifetime. From these numerous realizations of the beach behavior, such statements as "there is a 50-percent probability that the project will last 10 years" can be made. Strine and Dalrymple (1989) have used this approach for the Delaware coastline in conjunction with the analytical model of beach fill evolution.

Perhaps surprisingly, the longevity of a nourishment project constructed on a long, straight shoreline is independent of the sequence of possible storm events but rather depends on their cumulative effect, which is characterized by $\int_0^t G(t')\,dt'$. This implies that the long-term behavior of a beach fill impacted immediately by a major storm is the same as a beach fill that experiences the same storm but at a later time in its lifespan. To prove this result, we use the governing equation for the shoreline position, which was derived in Chapter 10 as the diffusion equation, Eq. (10.30). Restating it for convenience, we have

$$\frac{\partial y}{\partial t} = G \frac{\partial^2 y}{\partial x^2}, \tag{11.27}$$

where $y(x)$ is the shoreline position and G is the longshore diffusivity, which is assumed to be a function of time t.

Consider a Fourier decomposition of the shoreline position such that the shoreline can be mathematically represented as

$$y(x, t) = \sum_{n=0}^{\infty} y_n(x, t) = \sum_{n=0}^{\infty} a_n(t) \cos(k_n x), \tag{11.28}$$

where the $a_n(t)$s represent the time-varying amplitude of each Fourier mode. The nth component of the shoreline fluctuation is substituted into Eq. (11.27), yielding the following differential equation for the time function $a_n(t)$:

$$\frac{da_n}{dt} = -G(t)k_n^2 a_n, \tag{11.29}$$

where the time dependence of G has been indicated explicitly. Equation (11.29) can be rewritten as

$$\int \frac{da_n}{a_n} = -k_n^2 \int G(t) \, dt \tag{11.30}$$

and integrated to

$$a_n(t) = a_n(0) e^{-k_n^2 \int_0^t G(t') \, dt'} \tag{11.31}$$

The final form of the nth component of the shoreline is

$$y_n(x, t) = a_n(0) e^{-k_n^2 \int_0^t G(t') \, dt'} \cos(k_n x) \tag{11.32}$$

Because the differential equation is linear and we have shown the independence of the result on the storm sequence for one of the components, the equation applies for all components. Thus, it is proven that the evolution of a beach nourishment project depends only on the integral in the exponent in Eq. (11.32) and not on the sequencing of the events (storms) that result in this integrated value. Secondary effects not included in the linear diffusion equation may cause some deviation from this result, but these are expected to be relatively small.

11.2.5.2 The Sea Level Rise Effect on Beach Fills

Sea level rise affects the costs and viability of beach nourishment as a long-term remedial measure. The full evaluation in design requires consideration of the effects of both longshore and cross-shore effects. Considering only cross-shore effects and invoking the Bruun rule (Eq. (7.44)), one can show that the rate of shoreline retreat is

$$\frac{\partial R}{\partial t} = \frac{1}{h_* + B} \left[\frac{\partial S}{\partial t} W_* - \frac{\partial V}{\partial t} \right], \tag{11.33}$$

where $\partial V / \partial t$ is the volumetric rate of nourishment addition per unit length of beach (all other variables have been defined previously). The rate of nourishment required to keep the beach stable is found by setting $\partial R / \partial t = 0$, resulting in

$$\frac{\partial V}{\partial t} = W_* \frac{\partial S}{\partial t} \tag{11.34}$$

According to Eq. (11.34) (which considers only cross-shore effects), the required nourishment rate for shoreline stability is proportional to the rate of sea level rise. Also to be recognized in design, the background erosion rate is a reflection, in part, of the present rate of sea level rise. However, at most locations, the long-term erosion rates are *not* in accord with the Bruun rule, and it is appropriate to describe a simple method that is applicable in such cases. Denoting the existing background erosion and sea level rise rates as

$$\frac{\partial R_0}{\partial t}, \quad \text{and} \quad \frac{\partial S_0}{\partial t},$$

respectively, the effect of an increased sea level rise rate is to require a volumetric rate of nourishment with compatible sediment:

$$\frac{\partial V}{\partial t} = (h_* + B)\frac{\partial R_0}{\partial t} + W_* \left(\frac{\partial S}{\partial t} - \frac{\partial S_0}{\partial t} \right), \tag{11.35}$$

where the first term on the right-hand side represents the volume required for the present rate of sea level rise, $\partial S_0/\partial t$, and the second term is the amount due to the increased sea level rise rate. This equation has the advantage that it includes, in a lumped manner, all of the causes contributing to the present shoreline retreat, whether due to gradients of longshore sediment transport or to sea level rise. In addition, any limitations in the Bruun rule will be present only in the *increase* in sea level rise rates.

Stive, Nicholls, and de Vriend (1991) have stressed the need to consider the effects of sea level rise on beach nourishment requirements. Owing to the long time scales associated with sea level rise, a combination of models is recommended, including the deductive models that are normally used in predicting the short-term performance of beach nourishment projects and a second large-scale coastal evolution model based more on concepts and a general understanding of the prevailing long-term coastal processes.

11.2.5.3 Effects of Combined Spreading and Background Erosion

After placement of a nourishment project on a long, straight beach, the volume within the nourished region will diminish over time because of the combined effects of diffusive losses and background erosion. Relying on the superimposability associated with the Pelnard-Considère heat conduction equation (Eq. (11.23)), the amount remaining, $M(t)$ at any time t due to the combined effects of diffusive losses and background erosion, requires a new term to be added to Eq. (11.23):

$$M(t) = \frac{\sqrt{4Gt}}{\ell\sqrt{\pi}} \left(e^{-(\ell/\sqrt{4Gt})^2} - 1 \right) + \text{erf}\,(\ell/\sqrt{4Gt}) - \frac{\partial R}{\partial t}\frac{t}{\Delta y_0} \tag{11.36}$$

where $\Delta y_0 = V_0/[(h_* + B)\ell]$ is the initial width of the beach (the nourishment sand is considered to be compatible with the native and the profile has equilibrated) and V_0 is the total initial volume of placed sand. For this case the values of $M(t)$ can be negative owing to the effects of the background erosion. Figure 11.13 presents isolines of the proportion remaining, $M(t)$, in terms of two nondimensional terms, $\sqrt{Gt}/\ell$, which we have encountered previously, and $(dR/dt)t/\Delta y_0$. (Note that the values of $M(t)$ presented in Figure 11.12 are found along the abscissa, for which $dR/dt = 0$.)

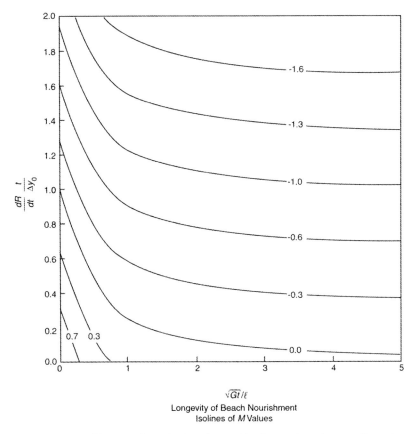

$$\frac{\sqrt{Gt}/\ell}{}$$

Longevity of Beach Nourishment
Isolines of M Values

Figure 11.13 $M(t)$ versus dimensionless time and background erosion rate. For values of the ordinate greater than 0.2, most of the M values are negative after a short time.

11.2.5.4 Effects of Setting Back the Fill Boundaries

It is often desirable to retain as much of the fill material within a given boundary, which might, in fact, correspond to the property boundaries of the community funding the project. One approach to this problem is to make the length of the fill less than the length of the community, and thus the material that moves out of the fill area benefits community beaches first and is not moved away early in the life of the project. If the distance from the end of the fill project to the end of the community is defined as $\Delta\ell$, then the amount of fill retained within the community can be obtained as a function of the dimensionless parameter $\Delta\ell/\ell$, as shown in Figure 11.14 for three different relative setbacks. It can be seen that the effects of the setbacks are greatest early in the project life (say $\sqrt{Gt}/\ell < 0.6$), when a relative setback of $\Delta\ell/\ell = 0.5$ would increase the percentage of material retained from 42 percent for no setback to 73 percent with the setback. On the downside, the beach within the setback region is narrower than those in the fill area and is narrower than the case of no setback.

11.2.5.5 Effects of Tapered Ends on Beach Fill

Because the sharp gradients at the ends of the fill planform lead to increased long-shore sediment transport rates, a logical question is, Does the use of tapered ends

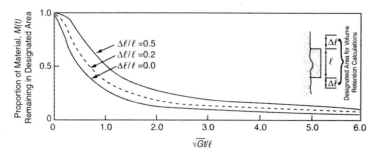

Figure 11.14 Proportion of fill remaining in beach length $\ell + 2\Delta\ell$.

on an otherwise rectangular beach fill affect the longevity of a project? If longevity is based on the retention of sand within the placement area, tapered planforms have a substantially longer life than a rectangular planform, for with tapered ends, the transport of material out of the project area is less initially. Figures 11.15(a) and (b) show the evolution of representative rectangular and tapered fill areas. Table 11.1 summarizes the cumulative losses from the placement area over the first 5 years. The tapered project loses 33 percent less than a rectangular project. This question of tapered ends has been treated in detail by Hanson and Kraus (1993) and Walton (1994). However, Dean (1996) has shown that if the same volumes are placed as rectangular or trapezoidal planforms and the retention percentage is based on the placement length of the trapezoidal fill, the greatest proportional retention *always* occurs for the initial rectangular planform. The reason is, in part, that the fill boundaries have been set back from the region for which the retention calculations are conducted as

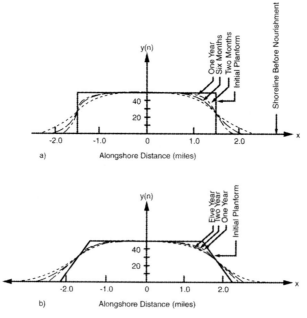

Figure 11.15 Representative evolutions of (a) an initially rectangular beach planform on an otherwise straight beach and (b) a tapered fill project of the same area.

Table 11.1   Comparison of Cumulative Percentage Losses from Rectangular and Tapered Fill Planforms ($G = 0.02$ ft^2/s; $\ell = 3$ mi; $\Delta y_0 = 55$ ft)

Years after Placement	Cumulative Percentage Losses with	
	Rectangular Planform	Rectangular Planform with Triangular Fillets
1	5.7	2.4
2	9.5	4.6
3	11.8	6.6
4	13.8	8.3
5	15.5	9.8

discussed earlier in this section. Also, an initially rectangular planform soon evolves into one with tapered ends (see Figures 10.11 and 11.15).

11.2.5.6 Multiple Nourishments

The fill material in a nourishment project will eventually erode away as the result of diffusional losses and background erosion. To maintain the beach, it will need to be renourished periodically. A design consideration, then, is to determine the most appropriate renourishment period. Many factors bear on this determination including mobilization and demobilization costs, total sand requirement over the project lifetime, environmental effects, and public perception.

This problem has been addressed by Dette, Führböter, and Raudkivi (1994) by assuming that the proportionate volume remaining, $M(t)$, is represented by an exponentially decreasing function of time:

$$M(t) = e^{-kt} \tag{11.37}$$

By introducing the half-life T_{50} such that $M(T_{50}) = 0.5 = \exp(-kT_{50})$, we can determine k and rewrite Eq. (11.37) as

$$M(t) = 2^{-t/T_{50}}, \tag{11.38}$$

where T_{50} can be determined from data obtained from the first, subsequent, or both renourishments. Before the results obtained by Dette et al., are described it is appropriate to note that the assumed model for volume decrease (Eq. (11.37)) differs substantially from that determined from a Pelnard-Considère analysis presented as Eq. (11.23). The differences in the shapes of these two relationships are shown in Figure 11.16, where the two curves are matched at their respective half lives. For long times, the exponential relationship approaches zero much more rapidly than the result based on the Pelnard-Considère solution, making its use very conservative.

In the model considered by Dette et al., the volume per unit length of beach within the longshore limits of a beach nourishment project is allowed to decrease to

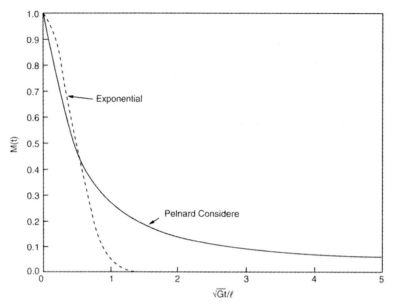

Figure 11.16   Comparison of Pelnard-Considère and exponential decay for $M(t)$.

$\overline{V}_{\min}$, at which time the beach is renourished to the full volumetric limit $\overline{V}_0\,\ell$, where $\overline{V}_0$ is the average volume per unit length of beach for complete nourishment given by $\overline{V}_0 = \overline{V}_{\min} + \overline{V}_w$, where $\overline{V}_w$ is the amount of sand required to renourish the beach. The time interval between renourishments is denoted as T_w and can be shown to be constant from renourishment to renourishment by

$$T_w = \frac{T_{50}}{\ln 2}\ln\left(1 + \frac{\overline{V}_w}{\overline{V}_{\min}}\right), \tag{11.39}$$

which follows from setting $M(T_w) = \overline{V}_{\min}/(\overline{V}_{\min} + \overline{V}_w)$ in Eq. (11.38).

An additional result of engineering interest is that the average rate of required sand replacement by renourishment, Q_w, is

$$Q_w = \frac{\overline{V}_{\min}\ell}{T_w}\left(2^{T_w/T_{50}} - 1\right), \tag{11.40}$$

from which it can be shown by application of the L'Hospital rule that the minimum average rate is that associated with a continuous rate of renourishment as given by

$$Q_{\min} = \frac{\overline{V}_{\min}}{T_{50}}\ln 2, \tag{11.41}$$

and, because T_{50} is a known constant, the average minimum supply rate is known.

Finally, for a project with a renourishment rate Q_w, the ratio of the actual supply rate to the minimum is defined as α and is given by

$$\alpha = \frac{2^{T_w/T_{50}} - 1}{(T_w/T_{50})\ln 2}, \tag{11.42}$$

which is always greater than or equal to unity. Dette et al. also present data from several nourishment projects, including one at Norderney, Germany, spanning 25 years. Some of the results from these projects indicate a loss rate more in accordance with the Pelnard-Considere solution than Eq. (11.38).

EXAMPLE 3

The Dette et al. (1994) results relating to the performance of beach nourishment projects are of interest and useful; however, as might be expected from comparison of the two relationships in Figure 11.16, the behavior of beach renourishment projects according to the Pelnard-Considère solution may be quite different. First we consider a case in which there is no background erosion, a project length ℓ of 6 km, an effective wave height of 0.5 m, an active profile depth h_* of 9 m, and maximum and minimum volumes of 3 million m^3 and 2 million m^3, respectively. The results are shown in Figure 11.17 over a period of 200 years, where it is seen that 19 renourishments are required and that the renourishment interval increases from 1.25 years for the first renourishment to 20.2 years for the last renourishment.

For situations in which background erosion exists, the renourishment interval characteristics are affected significantly. Figure 11.18 presents the same case as before, with the exception that a background erosion rate of 0.2 m per year is included. For this case, many more (33) renourishments over the 200-year period are necessary, and the first renourishment interval is 1.2 years

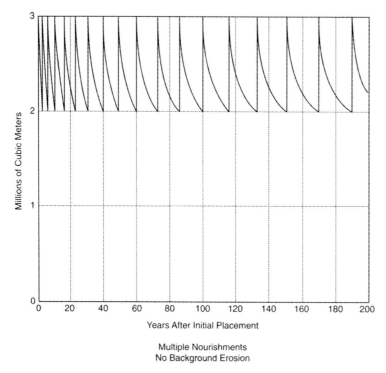

Multiple Nourishments
No Background Erosion

Figure 11.17 Multiple nourishments as a function of time (no background erosion); $H_b = 0.5$ m, project length $= 6$ km, $h_* + B = 9$ m.

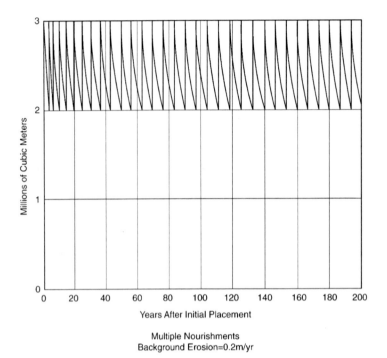

Figure 11.18 Multiple nourishments as a function of time (background erosion $= 0.2$ m/yr). Other conditions are the same as in previous figure.

and the last (27th), 6.7 years, a significantly smaller difference compared to the preceding example of no background erosion. For comparison purposes, the renourishment interval, T_{w_o}, associated with background erosion alone (no spreading "losses") is

$$T_{w_o} = \frac{V_w}{(h_* + B)(e\ell)},$$

in which e represents the background erosion rate. For the example here, the renourishment interval due to background erosion alone is 17.4 years. Thus, the longshore diffusional losses are thus much greater than those due to background erosion.

11.2.5.7 Migration of Beach Fills

As mentioned in Section 10.3, the centroid of a beach nourishment planform placed on a shoreline composed of compatible sand will remain nearly fixed even under the action of oblique waves. With the transport equation linearized, the reason can be shown to be a result of the material transported from the downdrift end of the nourishment being replaced by transport from the updrift side. If the beach fill is composed of sand that is coarser than the native material, the planform centroid can be shown to migrate updrift because the fill material leaves the project area more slowly than the finer native material is transported into the fill area. Conversely, the centroid of a fill composed of finer material will move downdrift, for the updrift transport into the fill area is less than that leaving (Dean and Yoo 1992).

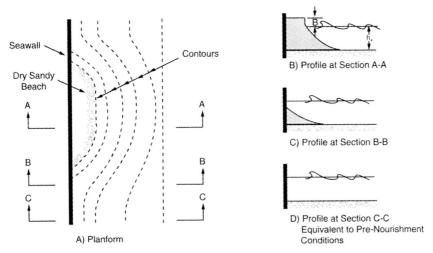

Figure 11.19 Beach nourishment profile in front of a seawall (from Dean and Yoo 1994). (a) Planform; (b) profile at section A-A; (c) profile at section B-B; (d) profile at section C-C equivalent to prenourishment conditions.

The limit of differences in the transport rates of the in-place and nourishment material is that of nourishment on a seawalled shoreline where there is no transport of native material. This situation may be approximated where there is a rock bottom offshore; for our purposes here, we will assume that the depth in this region is uniform, as shown in Figure 11.19. Treating this problem in an idealized manner and assuming that, at each profile where sand is present the cross-sectional area of sand A_c is uniform, Dean and Yoo (1994) have shown that, under the action of oblique waves, the centroid translates down the coast at a rate that increases with time and that the rate of increase of the shoreline variance is $2G$, the same rate of increase that occurs on a shoreline composed of compatible material. The reason for the migration of the centroid is that, because there is no inflow of sediment on the updrift side of the nourishment, the sand present there is cannibalized and transported along the nourished region to the downdrift side, where it is deposited. An approximate relationship for the rate of centroid migration can be developed as follows. The linearized transport rate is Eq. (10.28):

$$Q = Q_0 - G(h_* + B)\frac{\partial y}{\partial x} \tag{11.43}$$

If $\partial y/\partial x$ is small, the effective transport can be considered as Q_0 along the entire nourished shoreline. Thus, if the nourishment is considered to be of uniform width Δy_0, the updrift planform boundary is moved downdrift at a rate

$$\Delta x_{\text{updrift}} = \frac{Q_0 \Delta t}{A_c},$$

and the downdrift planform limit is increased at the same rate. Thus, the translational rate of the centroid can be expressed as

$$\frac{\Delta x}{\Delta t} = \frac{Q_0}{A_c} \tag{11.44}$$

The cross-sectional area is related to the total placed volume by $A_c = V/\ell_e$, where ℓ_e is the effective length of the nourishment planform. Because, as the result of spreading, the effective length of the project increases with time, resulting in a decrease in A, the speed of centroid migration also increases.

In a series of laboratory tests, Dean and Yoo (1994) and Yoo (1993) have shown the significance of the boundary condition near the ends of the nourishment. At these locations, where there is very little sand in the cross section and most of it is predominantly underwater, the transport rates are reduced significantly. For oblique waves, the transport at the downdrift end was much more efficient in extending the nourishment length than the transport at the updrift end was in reducing the nourishment length. This resulted in a more rapid increase the variance of the planform than either predicted by theory ($2G$) or as found experimentally for the case of normal wave incidence.

11.2.5.8 Erosional Hot Spots and Related Phenomena

Erosional hot spots can occur on either nourished or natural beaches and are localized sections of beach that erode more rapidly than the neighboring sections. On nourished beaches, the erosion may occur sporadically, or it may be chronic. The presence of erosional hot spots is of concern to completed beach nourishment projects owing to public perception and to an early need to address the problem that could require expensive dredge mobilization or the use of earth moving equipment to relocate sand within the project limits. Physical relocation of sand may not be popular among property owners fronting that portion of the beach from which the sand would be removed. It has been stated that erosional hot spots have occurred to some degree on all nourished beaches (Campbell 1994). Figure 11.20 shows an example of an erosional hot spot (near DNR monument R185) that occurred at Delray Beach, Florida.

Bridges and Dean (1996) have identified several possible mechanisms causing erosional hot spots as follows:

- Wave refraction due to offshore bathymetry, either natural or fill-created, including breaks in offshore bars,
- Wave refraction, diffraction, dissipation and reflection by borrow pits,
- Grain size variation due to different sand sources,
- Variation in placement methods, for hydraulic dredging results in a different beach than mechanically placed fill,
- Structures such as seawalls can lower the profile locally, or groins can induce rip currents, and
- Headland effects.

In some cases, it is evident on natural beaches that breaks in natural bars or reefs can cause areas of intensified erosion of the subaerial beach. Raichle, Ellsworth, and Bodge (1998) postulated offshore flows of sand through these depressions in the offshore features. These features may persist for months and may be associated with storms (see Section 9.2). A second cause predominantly associated with nourished beaches is the refraction of the incident waves over irregularly placed contours. This irregularity, which is associated with variable density of placed material in the

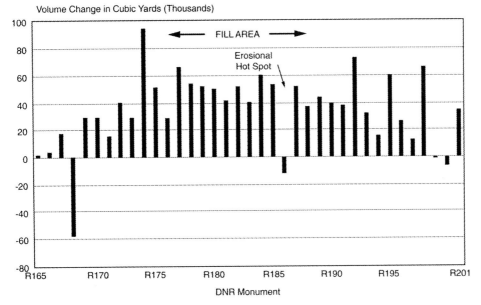

Figure 11.20 Erosional hot spot on Delray Beach, Florida (from Beachler 1993).

longshore direction, could result from various causes, including the irregular placement of different volumes of sand during the fill construction. Dean and Yoo (1993)
have discussed this phenomenon in terms of the "residual" contours that are placed
at depths greater than those activated by the waves. These contours then remain in
place and cause waves to refract and imprint their shape on the shoreline. For an
offshore contour with a displacement about the mean contour alignment of Δy_R, the
displacement of the shoreline Δy_S about its mean alignment is

$$\Delta y_S = \left[1 - \frac{C_*}{C_1} \right] \Delta y_R \tag{11.45}$$

in which C_1 and C_* are the wave celerities at the outer limit of the placed material
and the celerity at the depth of closure, respectively.

   The refraction mechanism described in the preceding paragraphs can also be
viewed in a more intuitive way: the shallower bathymetric contours tend to conform
to the lower contours. This effect occurred at Tybee Island, Georgia, where beach
nourishment was placed along a seawalled shoreline such that a more or less uniform
beach width from the seawall was established. The 2-m contour bulged out near the
southern one-third of this 4.5-km-long island. In retrospect, it is not surprising that the
nourished sand was reorganized by the natural refractive processes into a very wide
beach landward of the seaward bulging 2 m contour and other areas were stripped of
the placed sand. In the area of the wide beach, two rows of fairly high dunes formed;
see Figure 11.21.

   Wave refraction and possibly diffraction and reflection effects from bathymetry
created by beach nourishment can be manifested in different ways. Grand Isle,
Louisiana, a 7.5-mile-long barrier island on the Gulf of Mexico has been nourished

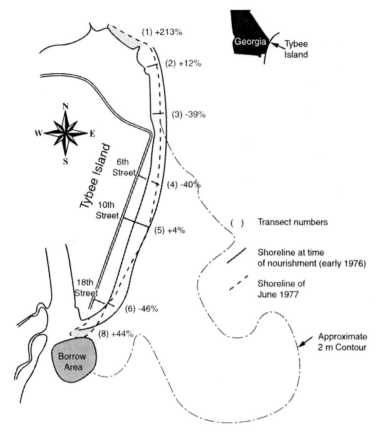

Figure 11.21 Tybee Island, Georgia, showing area of stability after 1976 nourishment (modified from Oertel et al. 1977).

with over 2 million yd^3 of material since 1954 (Gravens and Rosati 1994). Combe and Soileau (1987) have reported that, as part of the nourishment of Grand Isle, Louisiana, two borrow pits located about 0.8 km seaward of the shoreline, caused two large cuspate features to form along the nourished beach, as shown in Figure 11.22. The sand in the features was drawn from the adjacent beaches, leaving the beach between the salients quite narrow. This effect was probably due to the combined effects of refraction by the greater depths in the borrow pits, deflecting waves away from the leeward shoreline, forward scattering of the waves by the pit, and diffraction of the waves shoreward of the pits, into the "shadow" region, causing sand to be transported and deposited there. As an example of this, McDougal, Williams, and Furukawa (1996) have shown, for waves in shallow water, that if the cross-shore dimensions of the borrow pit are equal to the wavelength, substantial wave scattering can occur. Figure 11.23 compares the wave fields that result from a solid parallelipiped extending through the full water column (Figure 11.23(a)) and a rectangular pit (Figure 11.23(b)) that is dredged to a depth three times the water depth.

The accumulation of sand behind a borrow pit could come about through several other possible mechanisms. The sheltering of the shoreline leeward of the pit could cause sand, drawn from the neighboring beaches, to be deposited – particularly if the

Figure 11.22 Salients on Grand Isle, Louisiana, due to offshore borrow pits (Combe and Soileau 1987).

Figure 11.23 Diffraction coefficients for rectangular breakwater (a) and a dredged pit (b). Length of the breakwater and pit: $L/2$, width of breakwater and pit: L, pit depth: $3h$, $h/L = 0.027$, Normal Wave Incidence. Note the strong reflection from the pit (from McDougal, Williams, and Furukawa 1996).

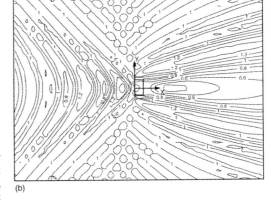

(b)

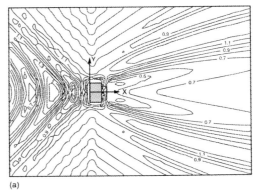

(a)

transport direction changes regularly. Additionally, the high wave set-up that would occur on the adjacent beaches and the low to negligible setup behind the pit could drive currents into the sheltered area. Alternatively, if the refractive effects of the pit cause a change in wave direction at the adjacent beaches, sand could be taken away from these beaches, leaving a sheltered area landward of the pit. Gravens and Rosati (1994), based on a numerical model study, concluded that the salients at Grand Isle are due to higher waves adjacent to the pits.

Horikawa, Sasaki, and Sakuramoto (1977) evaluated the effects of dredged holes on the shoreline with numerical and physical models. The numerical model (which included refraction but not diffraction or scattering of the waves) showed that salients tended to occur behind holes, presumably because of the reduced wave heights. The physical experiments, conducted with unidirectional waves, showed more pronounced salients than predicted by the numerical model. The difference was attributed to scale effects.

Dalrymple (1970), investigating the use of a drag scraper to pull offshore sand onto a beach in Florida, examined the behavior of the borrow pit that was created about 50 m offshore. The drag scraper excavated a total of approximately 23,000 m^3 from about the 3-m contour and placed it on the beach. Offshore surveys showed that the borrow pits filled from the side rather than from the beach face. No obvious effect of the borrow pit was found on the beach face. The explanation is that the annual littoral drift in this area was estimated to be more than seven times the amount excavated by the drag scraper, and the pit created by the scraping operation filled rapidly; measurements showed it filling at 6 cm/day.

11.3 SUBMERGED BERMS

Sand placement as underwater berms can occur simply as a means of dredge material disposal for beach stabilization or nourishment purposes, or a combination of these factors. A major cost savings often accrues if beach fill material can be placed offshore rather than on the beach in the expectation that natural processes will move the material to the beach.

Although the interest in, and construction of, underwater berms has intensified in recent years, the first effort appears to have occurred in 1935 at Santa Barbara, California, and many such projects have been conducted subsequently. Table 11.2 (Otay 1994) presents a summary of underwater berms and their characteristics, including whether they were judged to be stable or migrated. Of the berms placed to benefit the landward beaches, possible designs could be a feeder berm, in which case sand would be transported to the beach from an active berm or as a stable berm that causes damping of the waves and thus sheltering of the landward beach. The performance of underwater berms has been investigated both in the laboratory and through field monitoring programs. Hands (1991) has provided a thorough review of the behavior of 11 berms and their performance. Several examples will be discussed below followed by an assessment of the capabilities of available methods for predicting their stability and performance characteristics.

Table 11.2 Summary of Underwater Berms, Otay, 1994

Location	Date	Placed Volume [m³]	Water Depth [m]	Mound Relief [m]	Sand Size [mm]	Wave Height [m]	Wave Period [s]	On/Off-Motion	Shore Protection	Reference
Santa Barbara, CA	1935	154,000	6.1	1.5	0.18	—	—	stable	none	Hall and Herron (1950)
Atlantic City, NJ	1942	2,700,000	4.6–7.6	—	0.32	—	—	stable	none	"
Long Branch, NJ	1948	460,000	11.5	2.1	0.34	1–2	7–9	stable	none	"
Durban, South Africa	1970	2,500,000	7–16	0–8.3	0.35	1–2	—	both	indirect	Zwamborn et al. (1970)
Copacabana Beach, Brazil	1970	2,000,000	4–6	—	0.4–0.5	0.7	10–14	onshore	direct	Vera-Cruz (1972)
Long Island Sound, CT	1974	1,170,000	18.3	9.1	silt	0.1	—	stable	none	Bokuniewicz et al. (1977)
Lake Erie, OH	1975	18,000	17	0.36	silt	0.1	—	stable	none	Danek et al. (1978)
New River Inlet, NC	1976	26,750	2–4	1–8	0.49	0.55	7.3	onshore	direct	Schwartz and Musialowski (1977)
Limfjord Barriers, Denmark	1976	22,000	4–5	2.1	0.25–0.3	—		onshore	direct	Mikkelsen (1977)
Tauranga Bay, New Zealand	1976	2,000,000	11–17	9	—	—		stable	—	Healy et al. (1991)
Dam Neck, VA	1982	650,000	10–11	3.3	0.08	—	—	stable	—	Hands and DeLoach (1984)
Sand Island, AL	1987	350,000	5.8	1.8–2.1	0.22	—	—	onshore	direct	Hands and Bradley (1990)
Fire Island, NY	1987	320,000	4.9	2	—	—	—	—	—	McLellan et al. (1988)
Jones Inlet, NY	1987	300,000	4.9	2	—	—	—	—	—	"
Mobile, AL (outer mound)	1988	14,300,000	10.7–13.7	6.6	fine	0.3–0.8	3.4–4.6	stable	indirect	"
Coos Bay, OR	1988	4,000,000	20–26	4.6–7.6	0.25–0.3	2.7	11.5	loss	none	Hartman et al. (1991)
Silver Strand, CA	1988	113,000	4.6–5.5	2.1	0.2	0.62	13.1	onshore	direct	Andrassy (1991)
Kira Beach, Australia	1988	1,500,000	7–10	2	—	4	8	onshore	direct	Smith and Jackson (1990)
Mt. Maunganui, New Zealand	1990	80,000	4–7	2	—	0.5–1.5	5.9	—	direct	Foster et al. (1994)
Port Canaveral, FL	1992	120,000	5.3–6.8	1.65	—	1.2	6.3	onshore	direct	Bodge (1994)
Perdido Key, FL	1992	3,000,000	5–6	1.75	0.3	0.45	5.7	stable	indirect	Otay (1994)

The purpose of the underwater berm placed downdrift of Santa Barbara, California, in 1935 was to serve as a source of material to the eroded beaches downdrift of the Santa Barbara Harbor, which had been constructed in the late 1920s (see the preface to Chapter 8). The resulting mound was approximately 700-m long and 1.5-m high and was placed approximately 300 m offshore in a water depth of 6 m. The mound was monitored 2 years after placement and was determined to be relatively stable; thus, it was not providing fill sand to the beaches.

Another early berm placement was located off Atlantic City, New Jersey, in 1945 by the U. S. Army Corps of Engineers. Over a period of 7 years, a total of 2.7 million m^3 was placed in water depths of 4.6 to 7.6 m with the expectation that sand would move ashore, nourishing the beaches. Subsequent monitoring demonstrated no measurable movement of the material.

Zwamborn, Fromme, and FitzPatrick (1970) reported on a berm placement off Durban, South Africa, in which 2.5 million m^3 were deposited in water depths ranging between 7 and 16 m, forming an underwater mound 1.6-km in length. The results of 4 years of monitoring demonstrated that the berm had remained essentially stable and had provided considerable protection to the leeward beaches. Physical model tests were conducted before berm placement to quantify its effectiveness in reducing wave energy. The conclusion was that the stability of the leeward beaches was the result of wave height reduction to "nonerosion" levels.

Copacabana Beach near Rio de Janeiro, Brazil, was nourished starting in 1969 with 1.5 million m^3 of sand, and concurrently an offshore berm of approximately 2 million m^3 was placed in water depths of 4 to 6 m (Vera-Cruz 1972). The results of an extensive field monitoring program confirmed that the offshore berm had been of considerable benefit to the overall performance of the beach nourishment project. Evaluations were based on measurements of beach width, and these were compared with the results from a hydraulic model study conducted before construction of this project to investigate the effectiveness of the method.

Otay (1994) has described the monitoring results of an underwater berm placed off Perdido Key, Florida. The material was a byproduct of a channel-deepening project into Pensacola Bay and comprised approximately 3 million m^3 of generally clean sand with a median diameter of about 0.33 mm deposited in water depths ranging from 5 to 6.5 m. Additionally, about 4.5 million m^3 of sand were placed as direct beach nourishment. A profile from the monitoring program is shown in Figure 11.24. Monitoring included repetitive beach profiles and wave measurements. It was found that, although considerable smoothing of the internal topographic relief and perimeter of the berm occurred, no translation of the centroid of the berm could be detected by methods considered to provide horizontal accuracy within ±3 m. The data suggested that the berm had exerted a stabilizing effect on the beach leeward of the berm.

In a summary of 11 berm projects, Hands (1991) compared their stability with the inner and outer limits as predicted by Hallermeier's approach (1980, 1981a,b) and as discussed in Chapter 8. All 11 berm installations were judged to be consistent with these limits. If the berm crest elevation was above Hallermeir's inner limit, it was found to be active, and if the crest elevation was in a depth greater than the outer depth limit, the berm was judged, on the basis of field measurements, to be stable.

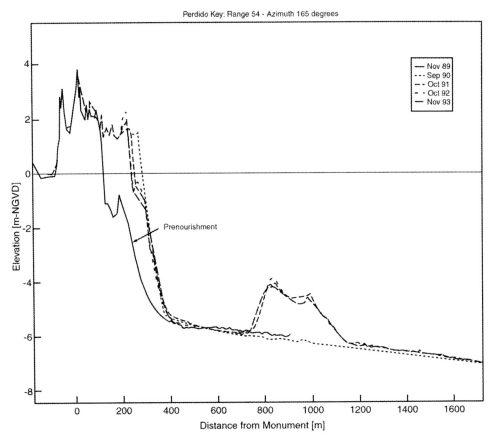

Figure 11.24 Pre- and postnourishment beach profiles at Perdido Key, Florida, showing underwater berm placement.

Worthy of note is that the Perdido Key berm was predicted by this methodology to be active contrary to the field results. Considering a TMA (Texel-Marsden-Arsloe) wave spectrum, Hands also compared the probability that near-bottom velocities would exceed particular values and concluded that berms are stable if the 75-percent velocity value exceeds 40 cm/s; if the 95-percent velocity value exceeds 70 cm/s, the berms are active.

Larson and Kraus (1992) have evaluated methods for predicting whether berms will move onshore or offshore. The database for their development was the offshore bar at Duck, North Carolina. In examining the behavior of this bar, a base profile with the following form was used to establish the bar volume above the reference profile:

$$h = A_* \left[y + \frac{1}{\lambda} \left(\frac{D_0}{D_{\text{inf}}} - 1 \right) (1 - e^{-\lambda x}) \right]^{\frac{2}{3}}, \tag{11.46}$$

which was developed by Larson (1991) to allow for variable sediment size across the profile. In Eq. (11.46), A_* represents the sediment scale parameter in the offshore region, D_0 and D_{inf} are the equilibrium energy dissipation values in the inshore and offshore, respectively, and λ^{-1} is a characteristic length scale governing the rate at which D_0 approaches D_{inf}.

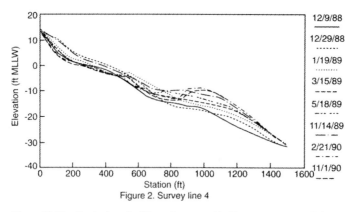

Figure 11.25 Evolution of offshore berm profile (from Andrassy 1991). The first postfill survey was conducted on 12/29/88.

In analyzing the data, Larson and Kraus assumed that onshore sediment transport is associated with decreasing volumes of the bar and vice versa for increasing bar volumes. Several criteria evaluated earlier by Kraus, Larson, and Kriebel (1991) for bar formation were tested, including the fall velocity parameter H_0/wT, deep water wave steepness and ratio of deep water wave height to $d_{50\%}$, and the dimensionless fall velocity $w/\sqrt{gH_0}$, where H_0 is the deep water wave height and w is the sediment fall velocity. Different coefficients were found to be more appropriate than those determined in earlier correlations by Kraus et al. (1991) for nearshore bars. For example, the critical parameter for the Dean number, H_0/wT, was found to be 7.2 rather than 3.2, as found earlier in Chapter 8. The method was tested for the case of a nearshore berm placed at Silver Strand, California, in which approximately 110,000 m^3 of sand were placed in a water depth of 5 to 8 m. The average height of the placed material was approximately 2 m. Wave measurements and profile monitoring were conducted for approximately the first 5 months of the project. It was found that the profile was quite active as evidenced by the sand spreading shoreward over this relatively short period. Figure 11.25 from Andrassy (1991) presents the evolution of one of the profiles. All parameters were in agreement with the observed onshore transport. For example, the average value of the fall velocity parameter was 2.3 compared with the previously cited critical value of 7.2 above which the investigators found that seaward transport would occur.

Douglass (1995) has applied the Bailard bedload transport equation in conjunction with Stokes wave theory and the equation of sand conservation to investigate the cross-shore transport of sand mounds. This approach has resulted in a convection–diffusion equation in which the speed of mound movement and the diffusion coefficient are expressed in terms of wave and sediment parameters and a single calibration coefficient. Douglass has recommended that the calibration coefficient be determined from field observations in order to predict future behavior. It appears that because of the velocity asymmetry in the Stokes wave theory, onshore mound migration would always be predicted, except in the case of extremely steep mound slopes.

Hands (1991), considering the design objective of an offshore berm to be either as a feeder beach to the adjacent shoreline or as a stable berm to shelter the beach through wave energy reduction, has accepted the profile zonation approach by Hallermeier as appropriate. It appears that several features of this approach should be considered with reservation. First, Hallermeir's method was developed for the case of offshore regions where the sediment sizes are more narrowly grouped than those that will be placed in berms. Thus, it may be necessary to reconsider the role of sediment size in berm stability. Second, because berms will remain offshore for many years, and the Hallermeier method has an annual wave basis, it may be worthwhile to consider a different period as the basis for evaluation. Finally, for those berms that are active, only the Douglass method is able to predict the rate of transport, but this method requires evaluating a constant prior to use. Also, at least one set of fairly complete measurements is not consistent with the available methodology (Perdido Key). It therefore appears prudent to utilize the available methodology with a degree of caution and to recognize the uncertainties in the prediction methodology. Finally, additional, carefully conducted field monitoring of this important problem should be carried out.

11.4 BEACH DRAINS

A novel method for reducing the erosion of a beach is the beach drain system, which consists of burying a pipeline along the beach to lower the water table on the beach face by pumping or gravity. Successful demonstrations have been carried out in the laboratory and apparently in the field, although the field data are more ambiguous.

In the laboratory, Machemehl, French and Huang (1975), using well pipes placed perpendicularly to the surf zone and pumping 65 gpm/(ft of beach), showed that accretion occurred on the beach face but not farther offshore in the surf zone. Kawata and Tsuchiya (1986) examined the case for solitary and regular waves on the beach face and concluded that the method worked. For the solitary wave tests, it was clear that the beach face was much steeper with pumping compared with the nonpumping cases. (It is difficult to determine the flow rates for their tests.) More recently, Weisman et al. (1995) showed that the presence of a tide range is helpful to the drain system and that it works for both erosive and accretionary conditions.

Additional laboratory tests have encompassed passive gravitational systems (no pumping) involving the installation of a permeable layer below the beach, reducing the water table and returning the captured flow to the surf zone at the seaward end of the installed drain. Kanazawa et al. (1996) have discussed three types of permeable underlayers, all of which demonstrated some increased beach stability.

There are several reports of sand drains in the field. In Denmark, two sand drains have been installed. The first at Hirthals was designed to obtain sea water for the aquaria of the North Sea Research Center and to supply the heat pump system. The beach is purported to have accreted at the site when the pumps were started. To determine if the results were applicable elsewhere, the Danish Geotechnical Institute (1986) carried out a test at Thorsminde, where, in 1985, a 500-m installation began operating. From their analysis of periodic beach profiles, they concluded that the method worked. This is not entirely clear from their data, for they do not explain well how their volume data were computed; nor does there appear to be a strong signal because over the course of the year the north section of the project eroded. Bruun (1989) pointed out that the effect of the drain is noticed on the beach face; offshore erosion will not be noticeably affected by the drain.

At Sailfish Point, Florida, a sand drain was installed along 600 ft of beach (Terchunian 1989). Beach profiles were taken monthly to determine the efficacy of the method. Dean examined early profiles (Dean 1989) but was unable to draw a definite conclusion as to the performance of the system; however, a analysis of more data (Dean 1990) involving surveys over 18 months concluded that the system appeared to result in some limited stabilization of the beach. Nevertheless in 1997, the system was turned off.

A field experiment of a permeable layer passive drain was carried out at the Hazaki Research Pier in Japan (Katoh and Yanagishima 1996). A permeable layer (88 m in the cross-shore direction, 7.8 m wide, and 0.2 m thick) was buried at a depth of approximately 3 m in the beach face with instrumentation to measure the total flow drained by the system. The sand at this site is 0.18 mm in diameter. Surveys were conducted before, during, and after typhoon storm events, which caused substantial infragravity wave motion in the nearshore. Measurements documented that the system lowered the water table relative to the adjacent beach (up to 1 m), resulting in less erosion during the storm and more rapid recovery after the storms. Maximum flow rates in the system were approximately 3 l/s, which is actually quite small.

11.4.1 PRINCIPLE OF OPERATION

It is not obvious how this method works, which it clearly does in the laboratory. Several hypotheses have been put forward, which will be discussed: Pumping

- Increases the deposition of sand on the beach face by lowering the water table such that the uprush water volume is greater than the downrush volume (Chappel et al. 1979, Vesterby and Parks 1988). Weisman et al. (1992) measured more wave uprushes with time at the onset on pumping due to the reduced backwash than for nonpumping conditions.
- Increases the effective sediment fall velocity by inducing a small vertical component of velocity in the fluid.
- Increases the effective sand size (fall velocity) by increasing the local pressure gradient.
- Reduces the outflow from the beach face during the ebbing tide (Danish Geotechnical Institute 1986).
- Filters out the sand present in the surf zone; that is, flow is withdrawn from the surf zone, which has a given sand concentration due to wave breaking (Cheng 1990).

The first of these mechanisms argues that the wave uprush on the beach face carries sand; as the water then rushes down the beach under the action of gravity, it carries sand back into the surf zone. This argument is appealing because highly permeable beaches tend to have a greater slope than those of lesser permeability. By using a drain, more of the water sinks into the beach, and therefore the sediment load is deposited on the beach face.

By inducing a downward flow into the beach, the effective fall velocity is increased. The steepness of the beach face is inversely related to the fall velocity (Dalrymple and Thompson 1976), and thus the beach face is steeper with the pumping. As the tide falls at the beach, the water table in the beach is higher than the ocean level, resulting in seepage from the beach. This seepage causes erosion of the beach, for there is a lift force on the sand grains on the beach (Grant 1948).

Another way to view the action of pumping is that the water withdrawn from the pump is clear; the beach acts as a large filter for the sediment-laden water being withdrawn from the surf zone. By multiplying the average concentration of the water in the surf zone by the flow rate at the well C_s, the maximum deposition rate can be estimated by

$$Q_s = C_s Q_w, \tag{11.47}$$

where Q_w is the pumping rate and Q_s is the deposition rate.

One way to get an idea of the average volumetric concentration is to follow the method of Dean (1973), which was discussed in Chapter 8. If the dissipation of energy by a single sand grain is (Eq. (8.31)),

$$\mathcal{D} = (\rho_s - \rho)g\frac{\pi d^3}{6}w, \tag{11.48}$$

(where w is the fall velocity, ρ_s and ρ are the densities of the sand and water, respectively, and d is the diameter of the sand grain), then the number of sand grains in the surf zone can be estimated by dividing the energy flux in the surf zone per unit length of beach by the dissipation of one grain:

$$N_s = \frac{\epsilon E C_g \cos\theta}{\mathcal{D}} = \epsilon\frac{\rho H_b^2}{8}\frac{C_g \cos\theta_b}{(\rho_s - \rho)g(\pi d^3/6)w}$$

The parameter ϵ is the fraction of energy dissipated by the falling sand grains. Now, multiplying N_s by the volume of a sand grain and then dividing by volume of the surf zone per unit length of beach A_c, we obtain the average volumetric concentration of sand in suspension.

$$C_s = \frac{N_s(\pi d^3/6)}{A_C} \tag{11.49}$$

Substituting Eq. (11.49) into (Eq. (11.47)), we find that

$$Q = \epsilon\frac{\rho H_b^2}{8}\frac{C_g \cos\theta_b}{(\rho_s - \rho)gwA_c}Q_w \tag{11.50}$$

If, for example, $C_s = 0.05$ in Eq. (11.47), then $Q_s = 0.05Q_w$, or a deposition rate of about 0.00075 yd³/ft of beach per minute per gallon, or 1 yd³ per day per foot of beach based on a 1 gpm pumping rate.

One of the puzzling factors about almost all the explanations is the extremely small flow rates into the beach. Pumping rates in the field are on the order of 1–2 gpm/ft of beach. For a pipe that is buried 2 m under the beach face, this creates a very slow induced flow within the surf zone. The ratio of the seepage velocities within the sand to the velocities within the jet of fluid rushing up the beach face is about 1/1000, according to Kawata and Tsuchiya. Bruun (1989) pointed out that the method ought to be more effective in mild conditions than storm conditions because the velocities (and energy levels) are far higher in the surf zone during a storm.

From the laboratory studies, the influence of the drain is to steepen the beach by retaining more sand on the upper part of the beach face at the expense of the lower part. In a wave tank, there is no net increase of sand on the profile (owing to conservation of sand); the profile is only reshaped. It may be that, in a three-dimensional test, the sand accumulated on the upper part of the beach would come at the expense of the downdrift beach. (There is no free lunch.) There have been, however, no three-dimensional tests in a laboratory.

REFERENCES

Andrassy, C.J., "Monitoring of a Disposal Mound at Silver Strand State Park," *Proc. Coastal Sediments '91*, ASCE, 1970–1984, 1991.

Beachler, K.E., "The Positive Impacts to Neighboring Beaches from the Delray Beach Nourishment Program," *Proc. 6th Ann. Natl. Conf. on Beach Preservation Tech.*, Florida Shore and Beach Preservation Assoc., Tallahassee, 223–238, 1993.

Bridges, M., and R.G. Dean, "Erosional Hot Spots: Characteristics and Causes," *Proc. 10th National Conference on Beach Preservation Technology*, Florida Shore and Beach Preservation Assoc., 1996.

Bruun, P., "The Coastal Drain: What Can It Do or Not Do?" *J. Coastal Res.*, 5, 1, 123–126, 1989.

Campbell, Thomas J., personal communication, 1994.

Chappel, J., I.G. Eliot, M.P. Bradshaw, and E. Lonsdale, "Experimental Control of Beach Face Dynamics by Water Table Pumping," *Eng. Geology*, 14, 29–41, 1979.

Cheng, A.H.-D., personal communication, 1990.

Combe, A.J., and C.W. Soileau, "Behavior of Man-made Beach and Dune, Grand Isle, Louisiana," *Proc. Coastal Sediments '87*, ASCE, 1232–1242, 1987.

Dalrymple, R.A., "An Offshore Beach Nourishment Scheme," *Proc. 12th Intl. Conf. Coastal Eng.*, ASCE, Washington, DC 955–966, 1970.

Dalrymple, R.A., and W.W. Thompson, "A Study of Equilibrium Beach Profiles," *Proc. 15th Intl. Conf. Coastal Eng.*, ASCE, Honolulu, 1277–1296, 1976.

Danish Geotechnical Institute, "Coastal Drain System: Full Scale Test – 1985 Tormindetangen," June 1986.

Dean, R.G., "Heuristic Models of Sand Transport in the Surf Zone," *Proc. Conf. Eng. Dynamics in the Surf Zone*, Sydney, 208–214, 1973.

Dean, R.G., "Compatibility of Borrow Material for Beach Fills," *Proc. 14th Intl. Conf. Coastal Eng.*, ASCE, Copenhagen, 1319–1333, 1974.

Dean, R.G., "Principles of Beach Nourishment," in *CRC Handbook of Coastal Processes and Erosion*, P.D. Komar, ed., Boca Raton: CRC Press Inc., 217–232, 1983.

Dean, R.G., Independent Analysis of Beach Changes in the Vicinity of the Stabeach System at Sailfish Point, Florida, Consulting Report Prepared for Coastal Stabilization, Inc., 1989, 15 pp.

Dean, R.G., Independent Analysis of Beach Changes in the Vicinity of the Stabeach System at Sailfish Point, Florida, Consulting Report to Coastal Stabilization, Inc., Rockaway, NJ, September 1990, 16 pp.

Dean, R.G., "Equilibrium Beach Profiles: Characteristics and Application," *J. Coastal Res.*, 7, 1, 53–83, 1991.

Dean, R.G., "Terminal Structures at Ends of Littoral Systems," *J. Coastal Res.*, Spec. Issue, 18, A.J. Mehta, ed., 195–210, 1993.

Dean, R.G., "Beach Nourishment Performance: Planform Considerations," *Shore 2nd Beach*, 64, 3, 36–39, 1996.

Dean, R.G., and C.-H. Yoo, "Beach Nourishment Performance Predictions," *J. Waterway, Port, Coastal, and Ocean Eng.*, ASCE, 118, 6, 567–586, 1992.

Dean, R.G., and C.-H. Yoo, "Beach Nourishment in Presence of Seawall," *J. Waterway, Port, Coastal, and Ocean Eng.*, 120, 3, 302–316, 1994.

Dette, H.H., A. Führböter, and A.J. Raudkivi, "Interdependence of Beach Fill Volumes and Repetition Intervals," *J. Waterway, Port, Coastal, and Ocean Eng.*, ASCE, 120, 6, 580–593, 1994.

Douglass, S.L., "Estimating Landward Migration of Nearshore, Constructed Sand Mounds," *J. Waterway, Port, Coastal, and Ocean Eng.*, ASCE, 121, 5, 247–250, 1995.

Führböter, A., "A Refraction Groyne Built by Sand," *Proc. 14th Intl. Conf. Coastal Eng.*, ASCE, Hamburg, 1451–1469, 1974.

Grant, U.S., "Influence of the Water Table on Beach Aggradation and Degradation," *J. Marine Res.*, 7, 655–660, 1948.

Gravens, M.B., and J.D. Rosati, "Numerical Model Study of Breakwaters at Grand Isle, Louisiana," Misc. Pap. CERC-94-16, U.S. Army Corps of Engineers, 75 pp., 1994.

Hallermeier, R.J., "Sand Motion Initiation by Water Waves," *J. Waterway, Port, Coastal, Ocean Div.*, ASCE, 106, WW3, 299–318, 1980.

Hallermeier, R.J. "A Profile Zonation for Seasonal Sand Bars from Wave Climate," *Coastal Engineering*, 4, 253–277, 1989.

Hallermeier, R.J., "Seaward Limit of Significant Sediment Transport by Waves: An Annual Zonation for Seasonal Profiles," CETA 81-2, U.S. Army Corps of Engineers, Coastal Engineering Research Center, 1981b.

Hands, E.B., "Unprecedented Migration of a Submerged Mound Off the Alabama Coast," *Proc. 12th Ann. Conf. Western Dredging Assoc.*, 1–25, May 1991.

Hanson, H., and N.C. Kraus, "Optimization of Beach Fill Transitions," in "Beach Nourishment Engineering and Management Considerations," D.K. Stauble and N.C. Kraus, eds., *Coastal Zone '93*, ASCE, 103–117, 1993.

Horikawa, K., T. Sasaki, and H. Sakuramoto, "Mathematical and Laboratory Models of Shoreline Changes Due to Dredged Holes," *J. Faculty of Engineering*, University of Tokyo, XXXIV, 1, 49–57, 1977.

Houston, J.R., "International Tourism and Beaches," *Shore & Beach*, 64, 2, 3–4, 1996.

James, W.R., "Beach Fill Stability and Borrow Material Texture," *Proc. 14th Intl. Conf. Coastal Eng.*, ASCE, Copenhagen, 1334–1349, 1974.

James, W.R., "Technique for Evaluating Suitability of Borrow Material for Beach Nourishment," U.S. Army Coastal Engineering Research Center, Tech. Memo No. 60, 1975.

Kanazawa, H., F. Matsukawa, K. Katoh, and I. Hazegawa, "Experimental Study of the Effect of Gravity Drainage System on Beach Stabilization," *Proc. 25th Intl. Conf. Coastal Eng.*, ASCE, Orlando, 2640–2651, 1996.

Katoh, K., and S. Yanagishima, "Field Experiment of Gravity Drainage System on Beach Stabilization," *25th Proc. Intl. Conf. Coastal Eng.*, ASCE, Orlando, 2564–2665, 1996.

Kawata, Y., and Y. Tsuchiya, "Applicability of Sub-sand System to Beach Erosion Control," *Proc. Intl. Conf. Coastal Eng.*, ASCE, Taiwan, 1255–1267, 1986.

Kraus, N.C., M. Larson, and D.L. Kriebel, "Evaluation of Beach Erosion and Accretion Predictors," *Proc. Coastal Sediments '91*, ASCE, 572–587, 1991.

Krumbein, W.C., and W.R. James, "A Lognormal Size Distribution Model for Estimating Stability of Beach Fill Material," U.S. Army Coastal Engineering Research Center, Tech. Memo. No. 16, 1965, 17 pages.

Larson, M., "Equilibrium Profile of a Beach with Varying Grain Size," *Proc. Coastal Sediments '91*, ASCE, 905–919, 1991.

Larson, M., and N.C. Kraus, "Analysis of Cross-shore Movement of Natural Longshore Bars and Material Placed to Create Longshore Bars," Tech. Rpt. DRP-92-5, U.S. Army Coastal Engineering Res. Center, 1992.

Machemehl, J.L., J.T. French, and N.E. Huang, "New Method for Beach Erosion Control," Proc. Civil Engineering in the Oceans, III, ASCE, University of Delaware, 142–160, 1975.

McDougal, W.G., A.N. Williams, and K. Furukawa, "Multiple Pit Breakwaters," *J. Waterway, Port, Coastal and Ocean Eng.*, ASCE, 122, 27–33, 1996.

Oertel, G.F., C.F. Chamberlain, M. Larsen, and W. Schaaf, "Monitoring the Tybee Beach Nourishment Project," *Proc. Coastal Sediments '77*, ASCE, Charleston, 1049–1056, 1977.

Otay, E.N., Long-Term Evolution of Nearshore Disposal Berms, Ph.D. dissertation, Dept. of Coastal and Oceanographic Engineering, University of Florida, 1994.

Raichle, A.W., G.L. Elsworth, and K.R. Bodge, "Identification and Diagnosis of Beach Fill Hot Spots, Broward County, Florida," *Proc. Beach Pres. Technology '98*, Florida Shore and Beach Preservation Assoc., 61–76, 1998.

Stive, M.J.F., R.J. Nicholls, and H.J. de Vriend, "Sea-Level Rise and Shore Nourishment: A Discussion," *Coastal Eng.*, 16, 147–163, 1991.

Strine, M.A., and R.A. Dalrymple, "Beach Fill at Fenwick Island, Delaware," Center for Applied Coastal Research, Res. Rpt. 89–1, 1989.

Terchunian, A.V., "Performance of the Stabeach©System at Hutchinson Island, Florida," *2nd Ann. Nation. Conf. Beach Pres. Tech.*, 229–238, 1989.

U.S. Army Corps of Engineers, "Beach-Fill Volumes Required to Produce Specified Dry Beach Width," Coastal Eng. Tech. Note II-32, Coastal Engineering Research Center, 6 pp., 1994.

Vera-Cruz, D., "Artificial Nourishment of Copacabana Beach," *Proc. 13th Intl. Conf. Coastal Eng.*, ASCE, Vancouver, 1451–1463, 1972.

Verhagen, H.J., "Method for Artificial Beach Nourishment," *Proc. 23rd Intl. Conf. Coastal Eng.*, ASCE, Venice, 2474–2485, 1992.

Vesterby, H., and J. Parks, "Beach Management with the 'Coastal Drain System'," Florida Shore and Beach Preservation Assoc., Gainesville, 1988.

Weisman, R.N, G.S. Seidel, and M.R. Ogden, "Effect of Water-Table Manipulation on Beach Profiles," *J. Waterway, Port, Coastal, and Ocean Eng.*, ASCE, 121, 2, 134–142, 1995.

Walton, T.L, Jr., "Shoreline Solution for Tapered Beach Fill," *J. Waterway, Port, Coastal, and Ocean Eng.*, ASCE, 120, 6, 651–655, 1994.

Yoo, C.H., *Realistic Performance of Beach Nourishment*, Ph.D. Dissertation, Dept. Coastal and Oceanographic Eng., Univ. Florida, 150 pp., 1993.

Zwamborn, J.A., G.A.W. Fromme, and J.B. FitzPatrick, "Underwater Mound for the Protection of Durban's Beaches," *Proc. 12th Intl. Conf. Coastal Eng.*, ASCE, Washington, DC, 975–1000, 1970.

EXERCISES

11.1   List and discuss at least four variables that affect the longevity of a beach nourishment project.

11.2   Suppose that a beach is nourished with sand smaller than that originally present. The breaking height is known. There are two possible configuration types of the final equilibrium profile depending on the volume of nourishment material added, its stability characteristics, and so forth. Illustrate these two configurations by sketches providing a comparison with the original profile.

11.3   You are asked to evaluate two sediment sources with different sizes for use as beach nourishment material. The sizes are 0.2 and 0.4 mm.

   (a) If the costs per unit volume for the two sources are the same, which would you choose?

   (b) On the assumption of normal wave conditions, state the main advantage of the material that you chose.

(c) From the standpoint of storm wave conditions, list two substantial advantages of the material that you chose.

11.4 Suppose you are considering beach nourishment at two sites where the wave climate is the same but the sand sizes differ. Fine sand is located at Site A, and coarse sand at Site B.

(a) Consideration is being given to obtaining sand by dredging from a particular distance from shore (within 300 m). At which site would you believe this approach to be most appropriate? Explain your answer.

(b) Consideration is being given to dredging from a particular depth contour (within 4 m). At which site would you believe this approach to be most appropriate? Explain your answer.

11.5 Some early attempts at beach nourishment involved an arrangement by which sand was removed from regions within the surf zone and placed on the beach. Consider a beach on the East Coast of the United States with the beach facing due east.

(a) For longshore sediment transport from north to south, discuss qualitatively the effect that you would expect this operation to have on the beaches to the north and to the south of the operation.

(b) For the same area, it has been found that the winter bar is located 200 m offshore. Discuss the relevance of this location relative to the "borrow area."

11.6 Verify the fill volume required for a fill in front of a seawall $V_{\text{threshold}}$ is as given in Eq. (11.17) for compatible sand.

11.7 A beach nourishment project is being considered in which the native sand size is 0.2 mm, three sizes of nourishment material are available, and the costs of the various sand sizes differ. The sizes are $d_{F_1} = 0.15$ mm, $d_{F_2} = 0.20$ mm, $d_{F_3} = 0.30$ mm. Values of $h_* = 6$ m and $B = 1.5$ m may be considered for the original beach. As part of an economic analysis to determine the optimum sand source, you are asked to plot the additional dry beach widths Δy_0s versus the volume V for the three sand sizes. Consider the range of volumes added to be between 0 and 1500 m^3/m. Explain why the three curves for Δy_0 in Figure 11.9 are all nearly parallel for the larger values of the volume.

11.8 Discuss the effect of sediment size on the volume of sand that must be added to advance a beach by 10 m for the following conditions: $H_b = 2$ m, $S = 0$, $B = 2$ m, $h_* = 8$ m and $d_N = 0.2$ mm. Calculate the volumes for $d_{F_1} = 0.1$ mm and $d_{F_2} = 0.3$ mm.

11.9 Consider a seawalled shoreline at which the offshore bottom is rocky and horizontal with a depth of 2 m at the toe of the wall. Sand of 0.3 mm diameter is to be added.

(a) How much sand is required to just yield a dry beach (zero width)?

(b) How much sand is required to yield a dry beach with width of 20 m and a berm height of 2 m?

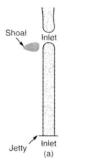

Figure 11.26 Sketch for Problem 11.11.

(c) Suppose that the rocky bottom is inclined seaward with a slope m of 1:100. Formulate a method relating sand volume added to shoreline displacement.

(d) For the conditions in **(c)**, what is the volume of sand required to just yield a dry beach?

(e) What is the volume of sand required to yield a dry beach width of 20 m with a berm height of 2 m?

11.10   Discuss in greater detail each of the reasons for the presence of hot spots in a beach fill.

11.11   The island shown in Figure 11.26 is bounded by an inlet at each end and is 4.5 km long. The original equilibrium planform is shown in Figure 11.26(a) and was controlled by a jetty at the south end such that no sand was lost but the jetty was holding all the sand that it could. Extensive dredging of the northern ebb tidal shoal caused the equilibrium planform to rotate clockwise, as shown in Figure 11.26(b), with a recession of 20 m at the north end of the island. It is desired to reinstate the original island width at the north end of the island.

(a) Assume that the beach planform is linear. How much sand was lost in the shoreline rotation? Consider that $B = 2$ m, $h_* = 6$ m.

(b) Suppose that a beach nourishment project is carried out at the north end of the island. Describe the effects. Would the nourishment change the equilibrium planform?

(c) What engineering recommendations would you make to reestablish the equilibrium beach planform such that the beach will advance 20 m at the north end? Your recommendation should not include reestablishment of the shoal at the northern inlet. The only structural modifications allowed are to the jetty.

11.12   Sand of diameter 0.22 mm is placed at an initial planar slope m_i of 1:80. The breaking wave height H_B is 4 m, and it may be assumed that the depth of closure h_* equals H_B/κ, where $\kappa = 0.78$. The berm height is 1.5 m.

(a) What type of profile is this?

(b) What is the equilibrium shoreline change Δy?

(c) Draw both the initial and equilibrated profiles.

(d) Repeat these steps for an initial slope of 1:20.

CHAPTER TWELVE

Hard Engineering Structures

Following the devastating 1962 northeaster along the East Coast of the United States, a groin field was constructed at Westhampton Beach, Long Island, New York, with a very deleterious effect on the downdrift (west) beaches. This shore protection project was planned to include a beach fill and to extend to the west to Moriches Inlet. For various reasons, the western part of the groin field was never completed, nor was the beach fill placed. A total of 11 groins were constructed in 1966 and 4 more in 1970 with the most westerly of these located approximately 2.3 km east of Moriches Inlet. The rather substantial net westerly longshore sediment transport began filling the easterly groins with an associated deficit of sand experienced in the 2.3-km gap between the groin field and the eastern jetty of Moriches Inlet. (Excellent reviews of the background for this area are provided by Heikoff 1976 and Nersesian, Kraus, and Carson 1992.) As a result of the erosional stress downdrift of the groins, the barrier island was breached in 1980. By the time the breach was repaired in 1981, it had grown to 1000 m in width and required 980,000 m^3 of sand to close (Schmeltz et al. 1982). A second breach occurred during a severe storm in November 1992. This breach also grew, destroying houses while decisions were pending for closure.

Since the construction of the groins, the average shoreline change within the groin field has been a seaward advancement of approximately 80 m; the greatest advancement of 112 m has occurred in the eastern compartment, and the smallest advancement of approximately 15 m within the westernmost compartment. Between 1962 and 1991, the 14 groin compartments trapped more than 1.8 million m^3 of sand above National Geodetic Vertical Datum (NGVD). It is estimated that the total volume trapped easily exceeded 3 million m^3 and that the shoreline immediately downdrift of the westernmost groin receded approximately 75 m. In summary, the Westhampton Beach groin field is a well-documented example of both the effectiveness of groins in trapping sand from the littoral system and the associated impacts on the downdrift shoreline.

12.1 INTRODUCTION

Hard structures are built to prevent the further erosion of a beach or to impede the motion of sand along a beach. Structures designed to prevent erosion of the upland fall into the general category of coastal armoring and include seawalls and revetments. The second type that impedes sand movement includes groins, jetties, and detached breakwaters and, unless this type of structure is filled artificially to its capacity concurrent with construction, there will almost certainly be adverse effects on the adjacent shorelines (particularly on the downdrift side), for any sand they impound is taken from the littoral system. Although difficult to estimate, the potential impact of jetties, groins, and coastal armoring on an adjacent beach is on the approximate order of 100:10:1, respectively.

Conditions most suitable for armoring include a shoreline that is stable or has a moderate erosion rate and is set back from the normally active surf zone such that, under ordinary conditions, the armoring will be buried under dunes and will be active only during severe storms. In this way, the armoring performs much like an insurance policy – out of sight until needed but present to provide the needed protection when required.

Dean (1986) has recommended a mitigative approach for armoring on an eroding coastline that calls for the placement of sand annually in the amount that has been prevented from entering the system by the armoring structure.* This approach maintains a more natural littoral system; however, from a regulatory point of view, monitoring and enforcement of the mitigative sand placement is required.

In this chapter a variety of coastal structures are presented; some better than others. For a given site, with its unique environment and causes of erosion, different structures may be useful. It is clear, however, that an inappropriate structure can exacerbate the situation and harm adjacent beaches as well. Therefore, the design of a hard structure solution requires a good understanding of the causes of erosion and the characteristics and performance of the numerous hard structure options.

12.2 PERCHED BEACH

One method of beach nourishment that has been proposed to reduce the amount of fill required is an offshore sill, as depicted in Figure 12.1. For this to function well, it is assumed that there is no loss of material to the offshore over the sill, and further, unless the project is extremely long, groins are present at each end of the project to ensure that there will be no significant alongshore losses. The first assumption is likely to be violated in practice, limiting effectiveness of this method of beach nourishment.

To determine the amount of material necessary to fill the perched beach, we consider the placement of a sill in a water depth of h_1 on the original equilibrium profile. The height of the sill is $\Delta h = h_1 - h_2$, where h_2 is the water depth over the sill. Fill sand will be placed shoreward of the sill, completely filling the profile out to the new equilibrium shape. The amount of fill necessary is determined by considering the volume balance

$$\int_0^{y_1} h_N(y)\,dy + By_1 = \Psi + \int_0^{y_2} h_F(y)\,dy + By_2, \tag{12.1}$$

---

* This is one of the tenets of sand rights, which are discussed at the end of Chapter 14.

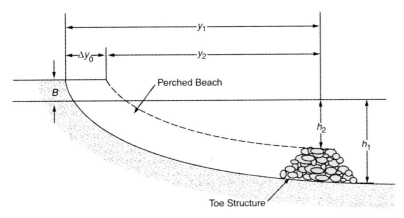

Figure 12.1 Schematic of a perched beach.

which is the volume per unit beach length below the berm height B. Integrating and replacing y_1 with $(h_1/A_N)^{3/2}$ and a similar replacement for y_2 leads us to the following equations for the shoreline advancement, Δy, and the required fill volume:

$$\Delta y_0 = y_1 - y_2 = \left(\frac{h_1}{A_N}\right)^{3/2} - \left(\frac{h_2}{A_F}\right)^{3/2} \tag{12.2}$$

$$V = \left(\frac{h_1}{A_N}\right)^{3/2}\left(B + \frac{3h_1}{5}\right) - \left(\frac{h_2}{A_F}\right)^{3/2}\left(B + \frac{3h_2}{5}\right) \tag{12.3}$$

Experience has shown that a sill is unable to hold this volume owing to scour immediately landward of the sill, violating the first assumption.

Sorenson and Beil (1988) conducted two-dimensional wave tank studies to evaluate the effectiveness of a perched beach. For the erosive conditions tested and with four different sill locations, sediment was carried seaward over the sill and scour occurred immediately landward of the sill, reducing the retention capacity of the perched beach system. The total shoreline recession on the silled beaches was greater than on a natural beach; however, Sorenson and Beil did find that, as time passed, the erosion from the beaches with sills was eventually slower than for a natural beach. Figure 12.2 compares the initial and final beach profiles for a natural and a silled beach.

Observations in nature show that broad reefs are effective in retaining a wider beach. Additionally, in practice, it is necessary to consider longshore effects, for the problem is not two-dimensional.

12.3 GROINS

Groins have served traditionally to prevent shoreline erosion on shorelines with significant alongshore transport. A groin, usually built as a vertical barrier extending directly offshore, impounds a fillet of sand on its updrift side, resulting in erosion on the downdrift side. This filling behavior of a single groin was examined in Chapter 10 through the use of the one-line model, which showed, on a noneroding beach, that there was as much downdrift erosion as there was updrift accretion, although predicting the distribution of this erosion and accretion along the beach

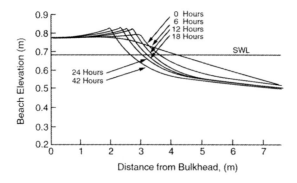

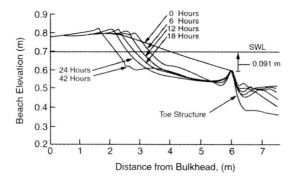

Figure 12.2 Model study of the perched beach concept comparing natural beach (top) to a perched beach (bottom) (from Sorenson and Beil 1988).

is more problematic. This balance of accretion and erosion can be convincingly argued from a conservation of sand perspective. On a long, straight beach, the updrift transport rate is Q. Downdrift of the groin the transport rate is again Q. Therefore, the rate of deposition against the groin must be equal to the rate of erosion. For this reason, groins must be used carefully.

The purpose of the groin is to provide a template for the desired beach profile. This means that the groin should not be too high or too long, which would create a complete barrier to waves and swash, completely disrupting the nearshore processes. A danger is that materials can be deflected offshore into bars and lost locally to the nearshore.

To ameliorate the downdrift erosion associated with a single groin, groins are often used to form a groin field, which means that many groins are used spaced along the beach. Ideally, the spacing of the groins in the field is chosen so the impoundment of sand on the updrift side of a groin will extend to the next updrift groin. Thus, when the predominant wave angle is small, the spacing can be large. This use of multiple, correctly spaced groins forces the erosion to be displaced to the last groin in the groin field. An extreme example of erosion downdrift of a groin field is at Westhampton Beach, New York, where the field of 15 large groins caused devastating downdrift results, as discussed earlier. See Figure 12.3. However, when an eroding beach terminates at an inlet or the end of an island, the downdrift erosion is often not a problem because downdrift effects equate to less sand carried into the inlet

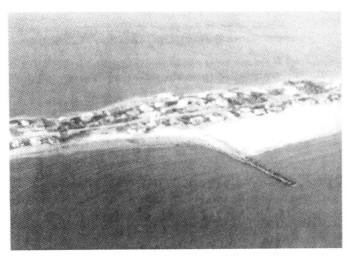

Figure 12.3 Groin field at Westhampton Beach, New York (from Nersesian, Kraus, and Carson 1992).

channel or lost to offshore bars. (An exception to this might be when significant bar bypassing exists at an inlet.) A terminal groin then functions as a jetty. (This means that care must be taken in designing the footings of the structure.)

In Florida, adjustable groins have been tried. The concept grew out of the seasonal nature of the littoral transport. When the drift is high in one direction, the groins are made high to slow the drift. Then, when the drift reverses, the groin heights are lowered to permit the material to move back updrift. This maximizes the "updrift" deposition of material. The problem is the high cost of maintaining the adjustable groins. Biological fouling, jamming by sand, and high labor costs usually force the abandonment of the seasonal adjustments.

Groins, either singly or in a field, work efficiently where there is a significant amount of alongshore transport. If the groins are located in regions of little longshore transport, such as a nodal region of drift reversal, they do not work well at all.

Groins enhance the behavior of beach nourishment projects. As explained in Chapters 10 and 11, the major losses from a beach fill occur through the diffusive losses at the project ends by longshore sediment transport. By "templating" the beach fill planform at the ends, which means that the groins must be built far enough seaward to ensure that the fill material does not easily pass around them, the losses are reduced greatly.

When emplacing a groin field, it is important to consider filling the groins with sand while they are being constructed. Otherwise, any impoundment of sand within the groin field comes at the expense of the downdrift beaches.

The landward ends of groins need to extend landward into the duneline to prevent flanking during storms or the winter season. The landward extent can be estimated using two indicators of beach change: (1) the seasonal shoreline changes and (2) the annual erosion rate (including the effect of a relative sea level rise). For example, multiplying the annual erosion rate by the presumed lifetime of the structure gives one measure of how far landward the groin should extend.

12.3.1 GROIN PLANFORM

To determine the behavior of groin systems, we can idealize the shoreline with a one-line model, as discussed in Chapter 10. In actuality, the presence of the groin over a portion of the beach profile will curtail longshore transport there, whereas offshore, transport will occur unimpeded. However, in the following discussion, a one-line model will be used to provide guidance.

Dean (1984) developed a solution of the one-line model for two groins, showing the planform evolution of the fill between them. The groins have an offshore length of L and a spacing of W. For the updrift groin, no sand transport was assumed, whereas at the other boundary, the offshore extent of the beach was fixed at $x = L$. Note that these boundary conditions are not the same, and thus Dean's solution is restricted to two groins; it is not valid for a full groin field. In dimensionless form, the solution is

$$\frac{y(x,t)}{W \tan \alpha} = \frac{L}{W \tan \alpha} + \frac{x}{W} - 1 + 2 \sum_{n=0}^{\infty} \frac{1}{\mu_n^2} e^{-\mu_n^2 G \, t/W^2} \cos\left(\frac{\mu_n x}{W}\right)$$

and

$$\mu_n = \frac{2}{(2n+1)\pi},$$

and α is the angle the waves make with the y-axis. In Figure 12.4, this solution is shown for the special case of $L/W \tan \alpha = 2.0$. After a long time the solution becomes asymptotic to a plane beach perpendicular to the wave direction.

$$\frac{y(x)}{W \tan \alpha} = \frac{L}{W \tan \alpha} + \frac{x}{W} - 1$$

The ratio of remaining fill to the initial fill (WL) is easily found by geometric considerations to be

$$M = 1 - \frac{W \tan \alpha}{2L}$$

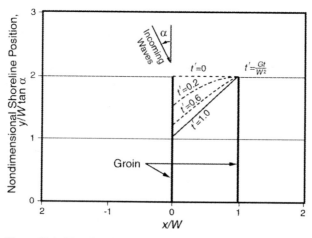

Figure 12.4 The shoreline change between two filled groins with time. Here, $L/(W \tan \alpha) = 2.0$.

The final shoreline configuration from this model indicates that the beaches between the groins will eventually be oriented in a direction normal to the incident
wave angle so that no alongshore transport occurs. For sites where the incident wave
direction is predominantly in one direction, the spacing between the groins can easily
be related to the active length of the groins; that is

$$\frac{L}{W} > \tan \alpha \tag{12.4}$$

This guarantees that the shoreline at the updrift groin does not flank the groin. For
shorelines characterized by seasonally varying incident wave directions, the beaches
within the groin fields will be continually shifting orientation, always attempting to
become perpendicular to the constantly changing wave direction.

For a groin field consisting of many groins in equilibrium with longshore drift and
bypassing a quantity Q_0, the shoreline orientation between the groins is

$$y(x) = A(x - W) + L, \tag{12.5}$$

where the parameter A depends on Q_0, but it is less than $\tan \alpha$. The amount of
material in the groin compartment divided by the maximum possible volume is

$$M = 1 - \frac{AW}{2L}$$

A method to mitigate the effects of a groin field is the tapering of groin lengths with
the longest groins in the center of the field and the shortest at the ends. Coupled
with initial filling of the groin field, this should result in a reduced downdrift erosion
pattern because the last groin is not so long as to cause significant downdrift erosion.
Tapering the end groins will also promote the transport past the downdrift groin
occurring along, rather than seaward of, the shoreline. In some cases of a single groin,
or an abrupt terminus of a groin field, the transport can be observed to occur along
a submerged bar, thereby not providing the maximum benefit to the neighboring
downdrift beaches.

12.3.2 GROIN TYPES

"Template" groins with the upper elevation only slightly above the sand profile to be
maintained also reduce sand and water diverted seaward by the waves and currents.
Further they can potentially be raised and lowered seasonally, depending on the
structure. Permeable groins allow sand and water to flow through, rather than around,
the groin. Raudkivi (1996) and Trampeneau, Göricke, and Raudkivi (1996) have
discussed fields of permeable groins that were constructed with closely spaced piles.
These permeable structures were found to be effective in retaining sand without
apparent deleterious downdrift effects.

There are numerous other shapes of groins besides a simple straight groin.
T-head groins are used to provide some diffractive shelter to the beaches at their
base in order to mitigate the downdrift erosion. Z-groins and round-headed groins
are other configurations. Appropriate use of these shapes can lead to the formation
of crenulate impoundments similar to the crenulate bays of Chapter 9.

12.4 OFFSHORE BREAKWATERS

Rather than physically impeding the alongshore transport of sand, as do groins, the offshore breakwater works on the principle of reducing the amount of wave energy that reaches a coastline – either to limit shoreline erosion or to provide a safe haven for boats, dredging equipment, or bathers. Typically, offshore (or detached) breakwaters are built seaward of the breaker line parallel to the shoreline. Often, numerous breakwaters are used in a long line along the shore. The use of multiple offshore breakwaters instead of one long, continuous breakwater is motivated by several factors: (1) it is cheaper to build multiple breakwaters because less construction material is required, and (2) the openings between breakwater segments permit water exchange between the nearshore and offshore and allow for some wave action behind the structures, providing recreational advantages.

Offshore breakwaters can be emergent (preventing overtopping), submerged (with the structure crest below the wave trough level), or partially submerged. Here, we will discuss emergent breakwaters and, in the next section, submerged ones.

In the United States, offshore breakwaters have been used at several sites; for example, single breakwaters are located at Santa Monica, Venice, and Channel Islands Harbor, California, and multiple offshore breakwaters are at Lakeview Park, Lorain, Ohio; Presque Isle, Pennsylvania; Holly Beach, Louisiana and Colonial Beach, Virginia. In other countries, offshore breakwaters have been used more frequently. Although construction of offshore breakwaters in the United States has increased recently from less than 20 two decades ago to more than 150 today, this is far less than the more than 4000 in Japan (Seiji, Uda, and Tanaka 1987). Figure 12.5 shows multiple offshore breakwaters along the Kaike coast of Japan. The small gap widths have created circular beaches, parallel to the diffracted wave crests in the embayments, and tombolos.

There is no doubt that a properly constructed emergent, long offshore breakwater will protect a shoreline. If no waves reach the coast, then there is no beach erosion! A very long multiple breakwater system with small gaps between the breakwaters will also protect the shoreline. The question becomes, How large can the gaps be and what level of shore protection is afforded by the breakwater system? Also of concern in offshore breakwater design is whether these structures will accumulate sand from the adjacent beaches owing either to ambient longshore sediment transport or, as discussed below, diffraction-induced transport for the case of normal wave incidence. Further, will they bypass sand or cause continued updrift impoundment?

12.4.1 SINGLE OFFSHORE BREAKWATER

A single offshore breakwater, even on a coastline with no longshore transport, will cause localized longshore and cross-shore sediment transport patterns because of the wave diffraction and diffraction-induced currents behind the structure. A depositional feature behind a breakwater may grow seaward. If it reaches the breakwater, it is referred to as a *tombolo*; otherwise, it is a *salient*.

The salient may grow to an equilibrium planform. Conceivably, the effect of wave diffraction is to reduce the wave heights behind the structure but also to turn the waves into the "shadow zone" behind the structure. The reduction of wave heights

Figure 12.5 Multiple offshore breakwaters on the Kaike coast, Japan (courtesy of CTI Engineering Co., Ltd., and the Ministry of Construction, Chugoku Regional Construction Bureau, Hinogawa Work Office).

there implies that the local transport will be less, resulting in impoundment of sand behind the structure; but as the impoundment increases in size, the angle of wave attack is larger, and thus an equilibrium can be developed whereby the wave angle is such as to balance the change in wave height, ultimately resulting in a constant littoral transport from one side of the structure to the other under the action of oblique waves. This is illustrated in Figure 12.6.

The approximate conditions for the formation of a tombolo can be determined from a simple diffraction analysis. The wave field behind the breakwater for a normally incident wave train can be envisioned as overlapping semicircular wave patterns emanating from the tips of the breakwater. A tombolo will form if the radius y_B, which is the distance from the shoreline to the breakwater, is smaller than half the length of the breakwater, $L_B/2$, or $y_B < L_B/2$. This criterion permits an equilibrium planform to be built by deposition from the action of normally incident waves. Suh and Dalrymple (1987) have conducted some small-scale laboratory experiments that verify this relationship when the offshore breakwater is contained within the surf zone (as measured by the surf zone width y_b). For breakwaters located farther offshore ($y_B > y_b$); then $y_B \sim 1$ to $2L_B$.

An important design parameter is the amplitude of the salient, for it is a measure of the amount of beach created by the offshore breakwater. From an examination

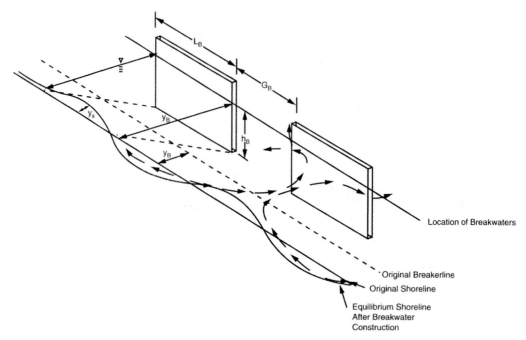

Figure 12.6 Longshore currents and transport in the vicinity of an offshore breakwater (from Suh and Dalrymple 1987).

of their data and those of others, Suh and Dalrymple found that the dimensionless salient amplitude y_s/y_B varies linearly with dimensionless breakwater length L_B/y_B; however, there is a great deal of scatter in the data. For the case where the surf zone width y_b is less than half the distance to the breakwater ($y_b < y_B/2$), a reasonable correlation exists (Suh and Dalrymple 1987):

$$\frac{y_s}{y_B} = 0.16 \frac{L_B}{y_B} \tag{12.6}$$

Larger salients occur when the breakwater is closer to the surf zone or contained within it; however, there is more scatter in the data.

Hsu and Silvester (1990) showed that the dimensionless ratio $(y_B - y_s)/L_B$, which is the dimensional distance from the breakwater to the salient, can be plotted against y_B/L_B with very little scatter in experimental data. A best-fit line through the data gave

$$\frac{(y_B - y_s)}{L_B} = 0.678 \left(\frac{y_B}{L_B} \right)^{1.215} \tag{12.7}$$

Recasting the equation in terms of $(y_B - y_s)/y_B$, we find that the equation becomes

$$\frac{(y_B - y_s)}{y_B} = 0.678 \left(\frac{y_B}{L_B} \right)^{0.215} \tag{12.8}$$

This yields the interesting result that $(y_B - y_s)/y_B = 1$ (that is, there is no salient) when $y_B/L_B = 6.09$; in other words, if the breakwater is located a distance offshore greater than six times its length, it will have no effect on the shoreline. This

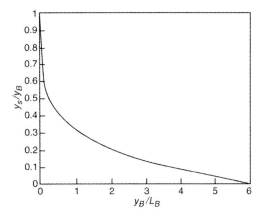

Figure 12.7 Dimensionless salient length versus dimensionless distance offshore.

corroborates the field results of Inman and Frautschy (1965) and Noble (1978). Because the amplitude of the salient y_s is a more useful measure, using the definitions of y_B and $(y_B - y_s)$, one can easily obtain

$$\frac{y_s}{y_B} = 1 - 0.678 \left(\frac{y_B}{L_B} \right)^{0.215} \tag{12.9}$$

This equation is plotted in Figure 12.7. Suh and Dalrymple (1987) pointed out, however, that in the field a tombolo often forms for $y_B/L_B < 1$; therefore, this curve may be unreliable for small values of y_B/L_B.

EXAMPLE

It is desired that the salient behind an offshore breakwater only extend offshore 20 percent of the distance to the breakwater ($y_s/y_B = 0.2$) or less. If the length of breakwater L_B is 600 ft and a single breakwater is to be used, how far offshore should the breakwater be located?

From Figure 12.7, it is clear that the breakwater must be farther offshore than twice the breakwater length ($y_B/L_B > 2$); thus, the breakwater should be located 1200 ft offshore to ensure that the salient (y_s) is not greater than 120 ft. For more accumulation of sand in the salient, the breakwater should be located closer to shore.

12.4.2 MULTIPLE OFFSHORE BREAKWATERS

Multiple offshore breakwaters, if close to the shoreline, can have tombolos at each breakwater, leading to the formation of equilibrium shorelines between each of the breakwaters, as shown in Figure 12.5. The shape of this shoreline planform can be predicted on the basis of diffraction theory (Dean 1978). Alternatively, the breakwaters may be far enough offshore that only salients occur. For either case, unless additional sand is provided to the system, there is a downdrift erosion due to the impounding of sand by the new shoreline features.

Suh and Dalrymple (1987) examined the gap spacing G_B as a parameter for multiple offshore breakwaters. The smaller the gap, the less energy is transmitted behind the structures. For very large spacings, the interaction between breakwaters disappears, and each structure acts as a single offshore breakwater. Analyzing laboratory data and field results, they obtained a curve for the ratio of the dimensionless salient amplitude y_s/y_B versus $G_B y_B/L_B^2$, the dimensionless gap width, where, again, y_B is the offshore distance of the breakwater and L_B is the breakwater length. They indicated that a tombolo forms for values of $G_B y_B/L_B^2 = 0.5$. If $L_B = 2y_B$ for this case, then $G_B = L_B$. As the dimensionless gap width parameter increases, the size of the salient decreases; therefore, to ensure that no tombolos form, the gap width needs to be larger than the length of the breakwater. More work needs to be carried out to examine all of the parameters necessary to design multiple offshore breakwaters.

Sand transport behind offshore breakwaters depends on the amount of energy reflected from the breakwater and the amount transmitted to the downwave side. For long waves, where the water wavelength exceeds the distance between the midpoints of the breakwaters, the reflection coefficients (for normal wave incidence) can be calculated based on the work of Dalrymple and Martin (1990). They examined vertical breakwaters separated by gaps of length G_B. The length of each breakwater is also G_B. They found that the complex reflection coefficient K_R (involving the absolute value of the reflection coefficient and a phase shift due to the reflection process: $|K_R| e^{i\epsilon_R}$) is given by the following:

$$K_R = \frac{S}{1+S},$$

(12.10)

where

$$S = \sum_{n=1}^{\infty} \frac{2k}{\sqrt{k^2 - (n\lambda)^2}} J_0^2(n\pi G_B/L_b)$$

Here, J_0 is the zeroth-order Bessel function $\lambda = 2\pi/L_b$, and k is the incident wave number ($k = 2\pi/L$).

An interesting effect of multiple offshore breakwaters with constant spacing is that it is possible for more than one wave train to exist behind the breakwaters. The row of breakwaters acts like an optical grating and diffracts the wave train behind the structures. If a long wave train encounters the breakwaters, then only one wave train is transmitted through them with a reduced wave height. If the waves are shorter than the breakwater separation distance, then more than one wave train can exist. These wave trains have the same period as the incident waves but propagate in different directions than the incident wave train. These waves have been explained theoretically and have been observed in the laboratory (Dalrymple and Martin 1990), but they have not yet been observed in nature. It is possible that the occurrence of these synchronous wave trains of different directions behind the structures could lead to periodic nearshore circulation and coastal features such as beach cusps.

12.4.3 SUBMERGED BREAKWATERS

Offshore breakwaters need not be emergent to be effective. Submerged, or reef, breakwaters are designed on the concept that the shallow depth over the structure

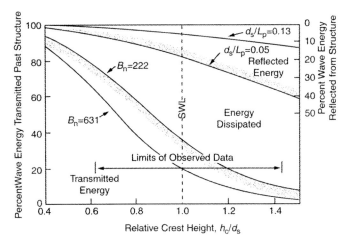

Figure 12.8 Reflection and transmission from breakwaters (from Ahrens 1987).

will induce wave reflection, breaking, and turbulent energy dissipation within the structure, leading to a reduction in wave height shoreward of it.

One advantage of the submerged breakwater is that the structure is not visible and thus does not disturb the view of the sea; however, it may create a navigational hazard unless well marked.

Ahrens (1987) has examined submerged and emergent breakwaters in the laboratory to study the stability of the structures and the amount of reflection and transmission of wave energy by the breakwater. Figure 12.8 shows his results, which involve the following parameters: the relative breakwater height h_c/d_s, where h_c is the elevation of the breakwater above the bottom, d_s is the mean water depth

$$B_n = \frac{A_t}{d_{50}^2},$$

which is the *Bulk number*, or a measure of the number of stones in the structure as A_t is the cross section of the breakwater and d_{50} is the diameter of the median stone size; and finally, the relative water depth, d_s/L_p, where L_p is the Airy wavelength. It is obvious from the figure that the higher the structure, the greater the reduction in wave height behind the structure. This leads to a major disadvantage of these structures: they become less effective during a storm when a storm surge increases the mean depth. However, recall that the sediment transport depends on the wave energy, and thus a reduction in wave height of 10 percent means a reduction in wave energy by 20 percent. Furthermore, sediment deposition occurring during times of low wave height provides a reservoir of sand to be eroded during a storm.

Dean, Chen, and Browder (1997) have documented 35 months of morphological response to a 1200-m long submerged, prefabricated offshore breakwater comprising two separate sections located in 3-m depth off Palm Beach, Florida. They found sand losses occurring *more rapidly behind the structure* than at alongshore sites, indicating that the long breakwater was causing sand to be exported from behind the structure. Dean, Dombrowski, and Browder (1994) proposed that the mass transport of the waves pumps water over the structure but that the normal

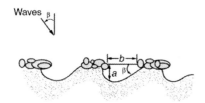

Figure 12.9   Artificial headlands for shoreline stabilization.

seaward return flows are impeded by the presence of the submerged breakwater, leading to alongshore currents that transport sand out the ends of the breakwater. An undistorted 1:16 physical model study by Dean et al. (1994) verified this hypothesis. Using different layouts of offshore breakwaters, including the use of gaps and a staggered* scheme of breakwater segments, can reduce the longshore flows by permitting offshore flow to occur through the gaps. This staggered design was implemented at Vero Beach, Florida. Monitoring has demonstrated that the performance of this installation has been similar to the one at Palm Beach (Stauble and Smith 1999).

12.5 ARTIFICIAL HEADLANDS

Artificial headlands are breakwaters built along the shoreline designed in planform to produce the crenulate bays discussed in Chapter 9. They were championed by R. Silvester of Australia and have been used effectively in several locations. In Singapore, for example, a series of 44 headlands were successful in stabilizing 27 km of beach fill (Silvester and Ho 1972). The purpose of headland control is to force a shoreline into equilibrium planforms that are tied to (eventually or originally) shore-attached structures.

The simplest form of headland control is a series of shore-parallel structures that can be viewed basically as sections of sea walls or shore-attached multiple breakwaters, as shown in Figure 12.9. Erosion between the structures leads to the equilibrium planform often referred to as a spiral bay. As discussed in Chapter 9 in conjunction with natural headlands, the equilibrium planform may be described by a logarithmic spiral. This spiral shoreline tends to orient itself parallel to the local wave crests.

A serious concern for shorelines with modest amounts of sediment transport is that the headland be flanked by erosion and it becomes an offshore breakwater. This may occur as a result of an inadequate elevation of the structure, which is overtopped during high water levels. The likelihood of separation can be minimized by extending a stem, or groin, landward, as shown in Figure 12.10. In Ibaraki Prefecture, Japan, 40 large structures of this shape (150-m long with 100-m T-head structures) with a 1-km spacing have been constructed to stop shoreline erosion there (Saito et al. 1996).

* The staggered scheme consisted of two shore-parallel lines of breakwater segments separated by a distance equal to several breakwater widths and located such that a gap between breakwater segments in one line would be occluded by a breakwater segment in the other line.

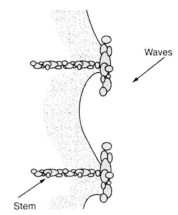

Figure 12.10   The use of stems to ensure the integrity of artificial headlands.

12.5.1 EQUILIBRIUM PLANFORM FOR HEADLAND CONTROL

Hsu, Silvester, and Xia (1986) have carried out laboratory tests on headlands and crenulate planforms. Their results provide a method for determining the planform shape behind artificial headlands. Figure 12.11 is a schematic diagram showing the angle of wave incidence β (defined as the angle made by the wave crest and the control line R_0) and the angle θ, which is drawn with respect to the wave crest. Figure 12.12 shows the experimental results for different angles of wave incidence and the radius ratio for different angles θ measured from the wave crest direction, as shown in the schematic.

To determine the planform of an equilibrium bay between two fixed headlands or to determine if a given bay is in equilibrium, first draw on a sketch of the headlands the control line between the upcoast and downcoast headlands R_0 and measure its length. Then determine the incident wave angle β that best represents the incident wave field or that is parallel with the existing downcoast section of the embayment. Draw the wave crest at the updrift control point (x_u, y_u) as in Figure 12.11. For various values of θ measured from the wave crest line, determine the ratios $R(\theta)/R_0$ and plot them on the sketch. Then connect these points to find the equilibrium planform.

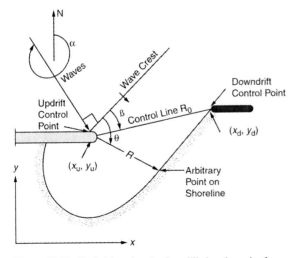

Figure 12.11   Definition sketch of equilibrium bay planform.

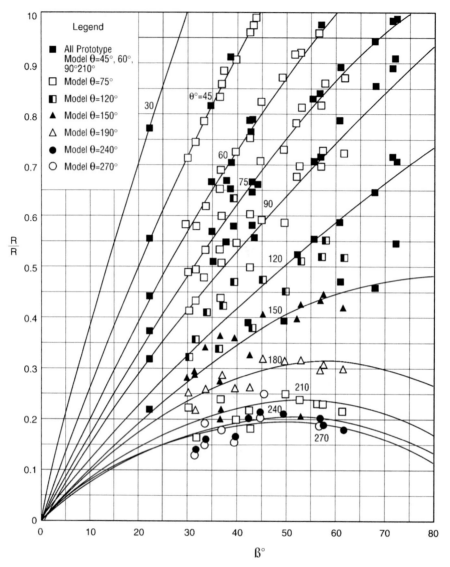

Figure 12.12 Dimensionless radius as a function of angle of wave incidence β for various θ.

12.6 REVETMENTS

Revetments are shore-parallel structures constructed to limit landward erosion. These stone, asphalt, or concrete-block structures are placed on a slope at the foot of bluffs, dunes, or along the beach face. Figure 12.13 shows a typical revetment made of stone. The design of the structures is intended to create wave breaking and loss of energy during the run-up process, limiting the reflection of wave energy from the beach. The design of a stone revetment may consist of two or more layers of stone with the upper, larger stones providing stability against wave attack. The *Shore Protection Manual* (U.S. Army Corps of Engineers 1984) provides the design criteria for stone weights. The stone sizes must be graded to ensure that the lower, smaller stone does not wash out through the upper layers. Typically, the stone is underlain

Figure 12.13 Example of a revetment (from *Shore Protection Manual* 1984).

by a geotextile fabric to prevent the base sand on which the structure is built from washing out.

In designing a revetment, the crest elevation should be sufficiently high to prevent excessive overtopping during high waves, and the toe elevation should be low enough to prevent undermining by scour action. Of special concern should be the construction of a revetment on a shoreline experiencing an erosional trend. One consequence will be the continuing lowering of the profile. If the design wave height was depth limited in front of the structure, then, with increasing depths with time, the wave height capable of reaching the structure during a storm will increase with time as well.

A result of normal profile adjustment to storms is erosion of the beach face and the formation of a shore-parallel storm bar. With the presence of a shore-parallel structure overlaying the beach face, the material needed to build the bar cannot come from the upland; rather, experience has shown that a somewhat smaller volume of sand is eroded near the toe of the structure. Although no precise estimates of the scour depths are available for design, usual guidance suggests scour depths on the order of the wave height at the base of the revetment.

In most cases, the construction of a toe structure consisting of a mound of stone at the base of the structure is advisable. With increasing water depths and scour at the toe of the revetment, this excess toe material will tend to be moved into the scour hole, thereby providing continuing protection for the base of the revetment.

Revetments are effectively used along low-energy coastlines, such as in the Chesapeake Bay, where high erosion rates (on the order of 10 m/yr) dictate shoreline armoring to prevent the loss of upland.

12.7 SEAWALLS

Seawalls protect the upland by preventing wave attack. Seawalls are most often vertical walls fronting a bluff or other highland with the beach or water on the opposite side. The most common structural materials are timber, concrete, or steel-sheet piling. Other seawall designs, usually constructed of concrete, have curved shapes to deflect the incident waves, reducing scour at the toe of the wall, or to reflect the waves. The interaction of seawalls with the beach system is similar in many respects to that of revetments.

One of the more famous seawalls is the Galveston seawall, which was built in 1903 to protect the city of Galveston, Texas, from hurricane storm surges.* In 1900, Galveston was obliterated by a hurricane, resulting in the loss of 6000 lives. This low-lying barrier had developed into a resort community, and the storm simply swept over the island and destroyed all in its path. By 1915, when another storm of similar magnitude hit the island, the seawall had been constructed along the old beach, and fill had been used to raise the land level behind the wall by over 2 m. This time, as the storm struck, the community was higher and safer – only 15 lives were lost. The wall continues to function to this day. In 1983, Hurricane Alicia caused a storm surge that was within 0.1 ft of the peak height of the 1900 hurricane. There were no lives lost in Alicia, again demonstrating the effectiveness of the seawall (and the modern hurricane forecasting and warning systems).

Seawalls usually have a deep foundation for stability. Also, to overcome the earth pressure on the landward side of the structure, "deadmen," or earth anchors buried upland, are often connected to the wall by rods. Structural integrity is frequently provided by a cap. The elevation of the seawall cap and the depth to which the wall extends into the beach are critical to the survival of the wall during storms. If the wall is too low, excessive overtopping may remove a considerable amount of fill from behind the structure and weaken it. Further, the overtopping water saturates and weakens the soil, increasing the pressure against the wall. If the embedment of the wall is insufficient, the toe of the seawall may "kick out" as a result of toe scour and the forces due to the saturated earth behind the wall. Return walls into the upland are necessary at the ends of a seawall. Otherwise, as the shoreline retreats at the end of the wall, erosion will remove some of the fill behind the structure, weakening its resistance to wave attack. As with revetments, it is often useful to place an armor layer of stone at the base of the wall to prevent excessive scouring.

A well-designed seawall, built on a rapidly eroding shoreline, will indeed protect the upland property. On the eastern shore of the Chesapeake Bay in Maryland, seawalls and revetments are used successfully to protect shorelines that have high erosion rates. (The high erosion rates are due to the sediment composing the upland, it is silt and erodes rapidly because there is no lag sand deposit to form a protective beach, as discussed in the preface to Chapter 2.)

12.7.1 EROSION INDUCED BY SEAWALLS

Seawalls are controversial. The statement has been made that "seawalls cause erosion." However, seawalls are almost only built on eroding shorelines, and thus the

* Larson (1999) provides a vivid historical account.

converse of the statement is definitely true: erosion can cause seawalls! There are examples of seawalls in which the beach in front of it has disappeared (sometimes because the seawall was built seaward of the mean water line) or excessive downdrift erosion has occurred. Kraus (1988) has provided a comprehensive review of literature pertaining to beach–seawall interaction.

Several different kinds of erosion are associated with seawalls. Those seawalls that project onto the beach and into the surf zone act as a wide groin with attendant updrift impoundment and corresponding downdrift erosion. There is also the profile response in front of the wall, as discussed in Chapter 7, Section 5.

Barnett (1987) conducted a comprehensive set of wave tank measurements of beach *profile* response to a vertical seawall and found that the additional storm-induced scour volume at the base of the wall was approximately 60 percent of the upland erosion that would have occurred had the seawall not been present. This toe scour is shown in Figure 12.14. These tests are valid for shorelines with little or no alongshore transport.

Griggs et al. (1991) examined seawall behavior in Monterey Bay, California, where the littoral drift rate is high and the seasonal fluctuations in the beach profile are large. The seawalls in this locale are exposed to winter waves as the berm near the wall retreats farther landward than the seaward portion of the wall. On the basis of their beach profile measurements, Griggs et al. did not detect additional scour at the base of the wall. Furthermore, for this study site, they found no net erosion induced by the seawall. The lack of seawall impact at this site is likely due to the extremely large longshore sediment transport that overwhelms any seawall effect.

Walton and Sensabaugh (1979) examined the erosion adjacent to seawalls due to a hurricane. Because of the building of an offshore bar during the storm and the lack of upland material as a source, material will be taken from the adjacent shorelines. This means that there will be additional erosion to adjacent beaches. On the basis of field observations following Hurricane Eloise (1975) in western Florida, Walton and Sensabaugh found that the extent of erosional scour adjacent to the seawalls increased with the length of the wall, as shown in Figure 12.15. See, also, McDougal, Sturtevant, and Komar (1987).

Many other problems associated with seawalls have not been adequately resolved. For example, it is often stated that seawalls can cause the offshore beach profile to

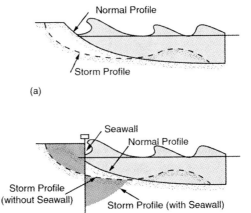

Figure 12.14 Additional scour immediately in front of a seawall due to storms (from Dean 1986). (a) Normal and storm profiles on a natural shoreline; (b) normal and storm profiles on a seawalled shoreline and comparison with profiles on a natural shoreline.

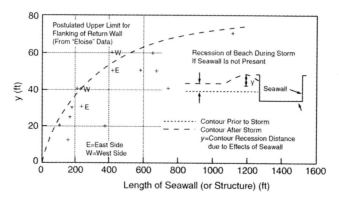

Figure 12.15 Additional shoreline recession adjacent to a seawall y as a function of seawall length (Walton and Sensabaugh 1979).

steepen, implying that greater wave heights will exist near the shoreline. There are no known data to support this claim. Another claim is that seawalls placed well back on the beach are detrimental to the beach and serve no useful purpose. First, to affect the beach, the seawall must be exposed to wave attack. Further, these setback seawalls can protect the upland in the event of a catastrophic erosion event during a severe storm. Clearly, the ideal seawall installation is one that is well set back from the ocean during normal conditions but in place to provide the necessary protection during severe storm events.

12.8 OTHER COASTAL PROTECTION DEVICES

Human beings are naturally inventive, and numerous devices have been patented to prevent beach erosion. Many people are also gullible and eager, sometimes frantic, to find the cure for an eroding beach that may be imperiling their home and often buy such devices without adequate knowledge of their performance. There have been many miraculous claims for such devices, but not one of them can make sand. These devices are often emplaced after a severe erosion event or in the spring following the erosive winter storms but prior to the natural summer recovery of the beach. Then, as the beach evolves to the summer or recovered profile, the structures are often buried, providing the manufacturer further evidence of their effectiveness. Finally, a year after emplacement, at the end of the winter season, the beach is once again stripped to its spring conditions, and the erosion control devices are again exposed, if not destroyed. Before and after photographs (or surveys) are therefore not true indications of the success of an erosion control device. Long-term and large-scale measurements (beach surveys) are necessary.

This same process occurs along the shores of the Great Lakes, which exhibit multiyear lake level fluctuations with erosion and the attendant installation of erosion control devices during the higher lake levels. The return of lower lake levels and the associated natural landward transport of sediment restoring the beaches often result in claims of success for these devices.

The forces that exist in nature are often underestimated. Objects designed to reduce erosion must be able to cope with all the forces present during storms and the normal corrosion and fatigue cycling of parts caused by the oscillatory wave motions. Shore erosion control devices must be designed to withstand the storm sea state.

The U.S. Army Corps of Engineers carried out a series of experimental placements of low-cost shore protection along low-energy coastlines to determine the effectiveness of such devices for use by shoreline homeowners (Housley 1981). Installations of floating offshore breakwaters, revetments, seawalls, and natural vegetation were located along several sites in the United States. Several evaluations were carried out with the conclusion that there was no one low-cost design that was effective for all purposes.* Further, natural vegetation was effective only when coupled with another shore erosion device. Critical to the life of all designs was the proper foundation, materials, and connections.

At the present, it appears that the standard erosion control devices (groins, beach fills, breakwaters, seawalls, and revetments), which have been used for hundreds of years, will continue to be the most widely used devices in the future. With time, we should expect better criteria for their design and use to evolve. We can also hope that, with an increasing understanding of coastal processes through research, new and more efficacious erosion mitigation devices will be developed.

12.9 JETTIES AND INLETS

Jetties are usually stone structures constructed at navigational channels to prevent sand from depositing in the channel and to provide wave protection for vessels. Jetties are so named because their purpose is to constrict the flow at their seaward ends to a velocity sufficient to "jet" sand from the entrance. Jetties can cause several effects on the adjacent shorelines, as discussed below.

By fulfilling its intended function of preventing sand from entering the inlet, the principal effect of a jetty is to obstruct natural sand transport past the inlet, thus starving the downdrift shoreline. In areas of strong, net long-shore sediment transport, this jetty effect can result in substantial downdrift erosion. In fact, if no sand passes the updrift and downdrift jetties, the rates of updrift accretion and downdrift erosion will be equal to the net longshore sediment transport; although the distributions of these effects are not easy to predict precisely. Indeed, many of the more reliable sediment transport estimates are based on monitoring of impoundment rates following jetty construction, which can lead to very dramatic coastal changes (as discussed in Section 6.7). Some previously stable shorelines, now downdrift of jettied and dredged inlets, are eroding at rates in excess of 10 m/yr. Also, some updrift jetties have rapidly become impounded to capacity, resulting in the construction of a jetty extension to further block the sediment transport from causing shoaling in the navigational channel. This, of course, exacerbates the downdrift erosion even more.

The analytical solution for the effect of a littoral barrier for unidirectional sediment transport was presented in Chapter 10. An additional effect of jetty construction can occur if the jetties are either permeable or low, allowing sand to pass through or over them. This is especially important in the presence of bidirectional sand transport. During transport reversals (from the predominant direction), sand will enter the inlet from the downdrift direction, and in this case the downdrift erosion rate is more nearly the gross transport rate Q_G ($Q_G = Q_+ + |Q_-|$) rather than the net Q_N ($Q_N = Q_+ - |Q_-|$). Also, downdrift transport may enter the inlet area over

---

* In coastal protection, you get what you pay for. There are no cheap solutions.

the updrift jetty. The reader may wish to consider whether in this case a net erosion of the updrift shoreline can actually occur. Dean and Perlin (1977) in a study of Ocean City Inlet, Maryland, found that the downdrift jetty was much lower (+4.5 ft) than the natural berm elevation (+8 ft mean sea level) in that location, and, even during periods of northeast waves, sand entered over the south jetty from the south. Jetties that allow substantial quantities of sand to pass over or through them usually have sandy beaches inside and along the inlet channel, potentially causing shoaling problems.

One consequence of sand-tight jetties is that, with no sand entering over or through the jetties, the waves entering the channel will erode the channel banks if they are not protected. Figure 12.16 presents a photograph of Sikes' Cut across St. George Island, Florida. The small crenulate bay features were studied by Dalrymple (1970) and identified to be due to sand transported landward by diffracted waves from the landward ends of the jetties. Indian River Inlet, Delaware, experienced similar bank erosion that was addressed by armoring the eroding area only to result in the shifting of erosion farther landward (Thompson and Dalrymple 1976). As shown in Figure 13.14, this resulted in three sections of progressively wider channel in the landward direction. The lesson for man-made navigational channels is that the channel banks should be protected from wave action for the entire length of the channel (or for as far as the waves are actively moving sediment).

A novel use of artificial headlands to protect inlet channels may be found inside the Ocean City Inlet, Maryland. Dean, Perlin, and Dally (1979) noted that the south jetty was allowing sand to pass northward into the inlet, as just mentioned, and that the unarmored, innermost portion of the inlet's south bank was eroding. They recommended that the south jetty be raised and that several artificial headlands be used there to stabilize the sidewall (Fig. 9.6). Fulford and Bass (1991) reported that, on the basis of several years on postconstruction monitoring, the design has been successful.

Figure 12.16 Inlet through St. George Island, Florida, showing wave–diffraction-induced channel bank erosion.

12.10 MONITORING AND MITIGATION FOR ALL COASTAL STRUCTURES

12.10.1 GENERAL

One of the most important aspects of coastal construction is that all contingencies cannot be predicted or modeled. There is a likelihood that extreme storm conditions or overlooked design factors may cause the structure to behave differently than predicted. This means that the structure may need to be altered or that the maintenance of the structure (meaning also maintenance filling of beach nourishment projects) may have to be implemented more or less often than planned.

The only way to know if the project is working correctly or if modifications are needed is by implementing a monitoring program consisting of periodic surveys at the structure and the nearby beaches coupled with wave and other environmental measurements. This monitoring program not only documents the effectiveness of the project, but it also can provide insight on how to modify the project appropriately, if needed. Knowledge gained through monitoring will benefit future projects through a better understanding of the performance of the structure and its interaction with the adjacent shorelines. *No major structures should be built at a shoreline without an adequate and long-term monitoring program.*

12.10.2 MITIGATION

Armoring the shoreline can result in shoreline erosion of adjacent beaches by denying the material that would have eroded from the upland. Therefore, one mitigation measure is the annual placement of fill equal to that which would have been eroded otherwise.

To estimate the amount of material to provide annually in mitigation on a coast that is eroding with the recession rate R, the following volumetric calculation is made:

$$V = Rd\ell, \tag{12.11}$$

where d is the vertical height of the seawall or revetment from toe to cap and ℓ is the longshore extent of the structure.

REFERENCES

Ahrens, J.P., "Characteristics of Reef Breakwaters," U.S. Army Corps of Engineers, Coastal Engineering Research Center, Tech. Rpt. CERC-87-17, 1987.

Barnett, M.R., "Laboratory Study of the Effects of a Vertical Seawall on Beach Profile Response," M.S. Thesis, Dept. Coastal and Oceanographical Engineering, Univ. of Florida, Gainesville, 1987.

Dalrymple, R.A., "Coastal Engineering Investigation of St. George Island Channel," Dept. Coastal and Oceanographic Engineering, University of Florida, Gainesville, 1970.

Dalrymple, R.A., and P.A. Martin, "Wave Diffraction Through Offshore Breakwaters," *J. Waterways, Port, Coastal and Ocean Eng.*, ASCE, 116, 6, 727–741, 1990.

Dean, R.G., "Diffraction Calculation of Shoreline Planforms," *Proc. 17th Intl. Conf. on Coastal Eng.*, ASCE, Hamburg, 1903–1917, 1978.

Dean, R.G., "Beach Nourishment," in *CRC Handbook of Coastal Processes and Erosion*, P.D. Komar, ed., Boca Raton: CRC Press Inc., 217–232, 1984.

Dean, R.G., "Coastal Armoring: Effects, Principles and Mitigation," *Proc. 20th Intl. Conf. Coastal Eng.*, ASCE, Taipei, 1843–1857, 1986.

Dean, R.G., A.E. Browder, M.S. Goodrich, and D.G. Donaldson, "Model Tests of the Proposed P.E.P. Reef Installation at Vero Beach, Florida," Dept. Coastal & Oceanographic Engineering, Univ. of Florida, UFL/COEL-94/012, August 1994.

Dean, R.G., R. Chen, and A.E. Browder, "Full Scale Monitoring of a Submerged Breakwater, Palm Beach, Florida, USA," *Coastal Eng.*, 29, 291–315, 1997.

Dean, R.G., M.R. Dombrowski, and A.E. Browder, "Performance of the P.E.P. Reef Installation: Town of Palm Beach, Florida: First Six Months Result," Coastal & Oceanographic Engineering Department, Univ. of Florida, UFL/COEL-94/002, February 1994.

Dean, R.G., M. Perlin, and W.R. Dally, "A Coastal Engineering Study of Shoaling in Ocean City Inlet," Dept. of Civil Engineering, Univ. of Delaware, O.E. Rpt. 19, 1979.

Fulford, E.T., and G.P. Bass, "Rehabilitation of the South Jetty, Ocean City, Maryland," *Proc. Coastal Sediments '91*, ASCE, Seattle, 1991.

Griggs, G.B., J.F. Tait, K. Scott, and N. Plant, "The Interaction of Seawalls and Beaches: Four Years of Field Monitoring, Monterey Bay, California," *Proc. Coastal Sediments '91*, ASCE, Seattle, 1871–1885, 1991.

Heikoff, J.M., *Politics of Shore Erosion, Westhampton Beach,* Ann Arbor, MI: Ann Arbor Science, 173 pp., 1976.

Housley, J. "Low-Cost Shore Protection, Final Report," U.S. Army Corps of Engineers, 1981.

Hsu, J.R.C., and R. Silvester, "Accretion Behind Single Offshore Breakwater," *J. Waterway, Port, Coastal, and Ocean Eng.*, ASCE, 116, 3, 362–380, 1990.

Hsu, J.R.C., R. Silvester, and Y.M. Xia, "Static Equilibrium Bays: New Relationships," *J. Waterway, Port, Coastal, and Ocean Eng.*, ASCE, 115, 3, 285–298, 1989.

Inman, D.L., and J.D. Frautschy, "Littoral Processes and the Development of Shorelines," *Coastal Eng.*, Santa Barbara Specialty Conference, ASCE, 511–530, 1965.

Kraus, N.C., "The Effects of Seawalls on the Beach: An Extended Literature Review," *J. Coastal Res.*, Special Issue 4, 1–28, 1988.

Larson, E., *Isaac's Storm: A Man, A Time and the Deadliest Hurricane in History,* New York: Crown Pub, 232 pp., 1999.

McDougal, W.G., M.A. Sturtevant, and P.D. Komar, "Laboratory and Field Investigation of the Impact of Shoreline Stabilization Structures on Adjacent Properties," *Coastal Sediments '87*, ASCE, 961–973, 1987.

Nersesian, G.K., N.C. Kraus, and F. C. Carson, "Functioning of Groins at Westhampton Beach, Long Island, New York," *Proc. 23rd Intl. Conf. Coastal Eng.*, ASCE, Venice, 3357–3370, 1992.

Noble, R.M., "'Coastal Structures' Effects on Shorelines," *Proc. 17th Intl. Conf. Coastal Eng.*, ASCE, Sydney, 2069–2085, 1978.

Raudkivi, A.J., "Permeable Pile Groins," *J. Waterway, Port, Coastal and Ocean Eng.*, 122, 6, 267–272, 1996.

Saito, K., T. Uda, K. Yokota, S. Ohara, Y. Kawanakajima, and K. Uchida, "Observations of Nearshore Currents and Beach Changes Around Headlands Built on the Kashimanada Coast, Japan," *Proc. 25th Int. Conf. Coastal Eng.*, ASCE, Orlando, 4000–4013, 1996.

Schmeltz, E.J., R.M. Sorenson, M.J. McCarthy, and G. Nersesian, "Breach/Inlet Interaction at Moriches Inlet," *Proc. 18th Intl. Conf. Coastal Eng.*, ASCE, Cape Town, 1062–1077, 1982.

Seiji, M., Uda, T., and S. Tanaka, "Statistical Study on the Effect and Stability of Detached Breakwaters," *Coastal Eng. in Japan,* 30, 1, 131–141, 1987.

Silvester, R., and S.K. Ho, "Use of Crenulate Shaped Bays to Stabilize Coasts," *Proc. 13th Intl. Conf. Coastal Eng.*, ASCE, Vancouver, 1347–1365, 1972.

Sorenson, R.M., and N.J. Beil, "Perched Beach Profile Response to Wave Action," *Proc. 21st Intl. Coastal Eng. Conf.,* ASCE, Malaga, 1482–1492, 1988.

Stauble, D.K., and J.B. Smith, "Performance of the P.E.P. Reef Submerged Breakwater Project, Vero Beach, Indian River County, Florida," Second Ann. Rpt. Aug 1996–Sept 1998. U.S. Army Corps of Engineers, Waterway Exp. Station, Vicksburg, 1999.

Suh, K., and R.A. Dalrymple, "Offshore Breakwaters in Laboratory and Field," *J. Waterway, Port, Coastal, and Ocean Eng.*, ASCE, 113, 2, 105–121, 1987.

Thompson, W.W., and R.A. Dalrymple, "A History of Indian River Inlet, Delaware," *Shore and Beach,* 44, 2, 24–31, 1976.

Trampenau, T., F. Göricke, and A.J. Raudkivi, "Permeable Pile Groins," *Proc. 25th Intl. Conf. Coastal Eng.*, ASCE, Orlando, 2142–2151, 1996.

U.S. Army Corps of Engineers, Coastal Engineering Research Center, *Shore Protection Manual*, 3 Vols., 1973.

Walton, T.L., and W.M. Sensabaugh, "Seawall Design on the Open Beach," Florida Sea Grant Report No. 29, University of Florida, 1979.

EXERCISES

12.1 An area was dredged in conjunction with a phosphate loading terminal just inside Gasparilla Island on the west coast of Florida. Following the deepening of this area, an accelerated erosion was noted in the area shown in Figure 12.17.

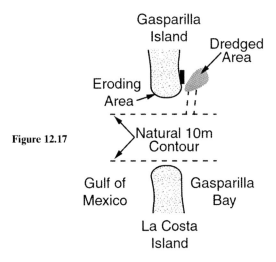

Figure 12.17

(a) Explain, in adequate detail, a plausible cause of the erosion noted above.

(b) On the basis of your explanation in **(a)**, describe a possible remedial measure. Try to define a measure that could be of benefit to the eroding beaches, and in some way, to the loading facility.

12.2 Suppose we have a shoreline that is eroding because of sea level rise (offshore sediment transport) and there is no *net* alongshore transport. A municipality wishes to nourish the beach with large quantities of sand to reduce the erosion rate and losses of sediment by stabilizing the fill with groins. Discuss the following aspects.

(a) Will there be substantial adverse effects to adjacent beaches? Can you see any beneficial effects to adjacent beaches?

(b) Would you recommend nourishing with material of smaller or larger diameter than that originally present on the beach? Why?

(c) Describe the mechanisms of any loss of sediment that you might expect from the nourished area.

(d) What would be the effect of groin length? On what parameters (including cost) would you base the groin length? Provide a justification of your answer.

12.3 The "perched beach" concept has been proposed as a possible method of moving the shoreline a significant distance seaward without requiring prohibitively large volumes of sand to raise the profile over the entire depth range of active motion.

    (a) Describe qualitatively how you would design the toe structure (location and height) to position the shoreline at some desired location.

    (b) Describe the effect of a storm. In particular, would the shoreline return to the prestorm position?

    (c) If any sand is "lost" landward of the toe because of a storm, how would you estimate this volume?

    (d) For given storm characteristics, could you design the toe such that the shoreline would not recede beyond a certain point? Discuss the procedure qualitatively.

12.4 Erosion is occurring in an area where the wave direction can be considered normal to shore. City B (see Figure 12.18) is considering placing segmented offshore breakwaters to reduce erosion of its shoreline.

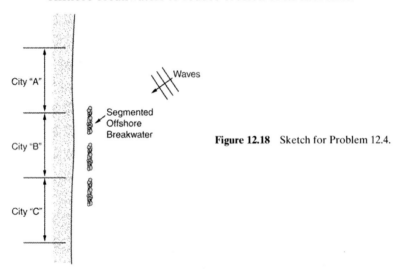

Figure 12.18 Sketch for Problem 12.4.

    (a) You are retained by Cities A and C to advise them of any possible adverse effects of the breakwaters. State any reservations that you would have and why.

    (b) Are there any modifications or additions to the project that would reduce any concerns you might have?

    (c) What type of monitoring project would you recommend to document the effects of the breakwater system on Cities A and C ?

    (d) What characteristics of the offshore breakwaters would control the adverse effects?

12.5 Determine the equilibrium planform for two artificial headlands. The upcoast control point (x_u, y_u) is $(0, 0)$, and the downdrift control point (x_d, y_d) is $(4000\,\text{m}, 1000\,\text{m})$. The angle of wave incidence, measured from North, is $330°$. Calculate the indentation ratio $a/b = a/R_0$ and compare it with the result in Figure 9.5.

CHAPTER THIRTEEN

Tidal Inlets

Inlets are created through barrier islands either naturally by storms or water-courses, or they are man-made. On some coastlines, the spacing between inlets can be quite large, creating public pressure for additional navigational access. This is the case at Santa Rosa Island on the Gulf Coast of Florida. This island extends 80 km from Destin Pass at the east to Pensacola Bay entrance at its western terminus. Both nature and humans have tried to establish a new inlet about midway along the island length at Navarre Beach, Florida, but to no avail. The result was always the same: after several months, the inlet would migrate, shoal, and eventually be sealed off by the waves (see Figure 13.1). In the 1970s, we conducted a study to try to determine the cause of the very brief life of this inlet. Our conclusion was that the tides in this area were insufficient to maintain an inlet. The tides are predominantly diurnal and during certain phases of their cycle have quite small tide ranges (only 10 cm or so). It is during these periods that the scouring capability of the tidal currents is greatly diminished, leaving the inlet at the depositional mercy of the waves. If a storm occurs, sand is deposited in the inlet, contributing to its hydraulic inefficiency and eventual demise. How then do the adjacent inlets on Santa Rosa Sound survive? It appears that, because of their substantially greater cross sections and storage capacity, they can withstand and recover from any storm-induced deposition during these neap tidal periods. Also, the adjacent inlets tend to compete for their share of the tidal flow going into and out of the bay, which provides the scouring capacity necessary for their survival. An additional inlet could actually decrease the stability of the existing inlets.

13.1 INTRODUCTION

Tidal inlets cutting through offshore barrier islands provide many benefits, including navigational access to the lagoon or bay behind the island for commercial shipping, fishing, and recreational boating. Also, the exchange of lagoonal water during a tidal cycle plays a major environmental role, flushing the bay of sediments and pollutants and maintaining water quality and salinity levels for aquatic life.

On the other hand, the presence of a tidal inlet on a shoreline has the drawback that it traps a considerable amount of sand, creating significant potential for beach

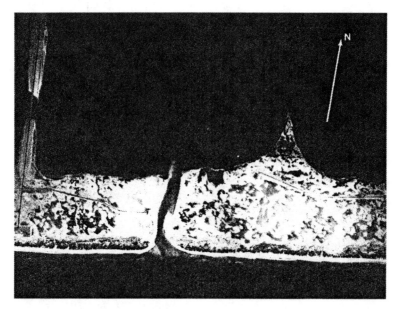

Figure 13.1 Navarre Pass, Florida, circa 1965. The Gulf is at the bottom of the photograph. Note the cuspate foreland in the lagoon. The landform on the left is a causeway constructed on a cuspate foreland to lessen the distance across the lagoon.

erosion on the adjacent beaches. Dean (1988) has indicated that the shoreline erosion on the east coast of Florida can be largely explained by the presence of inlets, which trap sand against updrift jetties and draw in beach sand on flood tides to deposit on flood tidal shoals, and, on the ebb tide, jet some of it far offshore onto ebb tidal shoals. This loss of sand is further exacerbated if sand, removed from the inlet by dredging of navigational channels, is disposed of offshore, outside of the littoral environment. (Marino and Mehta 1988 indicated that dredging of the inlets on the east coast of Florida had resulted in the loss of 41 million m^3 of sand. This is equivalent to about 100 years' worth of the sand transport rate at the Georgia–Florida border, which is the updrift end of the littoral stream in the state.)

This chapter examines both the hydrodynamics of a tidal inlet and the effects of the inlet on the neighboring beaches.

13.2 TIDAL HYDRODYNAMICS

The tidal flows through an inlet are caused by the water level difference between the ocean and the bay. The dynamics of the flow induced by this water level difference have been examined by several investigators; some of the earliest were Lorentz (1925), Brown (1932), and Keulegan (1967). Here we present both the Keulegan method and a linear theory for tides in an idealized inlet and bay system in order to highlight these dynamics. Before these two theories are presented, some common formulation elements will be reviewed.

We will consider a steep-sided bay such that the planform area of the bay A_B is independent of the water level within the bay. There are assumed to be no rivers or streams flowing to the bay. The inlet will have a cross-sectional area A_c, which will

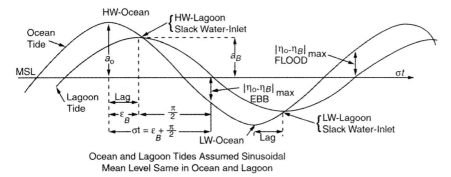

Ocean and Lagoon Tides Assumed Sinusoidal
Mean Level Same in Ocean and Lagoon

Figure 13.2 Time histories of idealized ocean and bay tides.

change as the tide changes, but the width of the inlet will be fixed at b, the mean depth will be h_i, and its length is ℓ. The depth of the bay will be h_B, which is greater than h_i and deep enough so that the propagation time of the tidal wave across the bay is not important (this can be construed to mean either that the bay is very deep, such that the wave speed is very fast, or that the size of the bay is small).

The tide in the ocean will be characterized by a single tidal amplitude a_0, which is half the tide range, and a tidal frequency σ, which will depend on whether the tide is diurnal or semidiurnal (as described in Chapter 4). The expression for the ocean tide will be

$$\eta_0(t) = a_0 \cos \sigma t,$$

and the tide in the bay will be assumed to be

$$\eta_B(t) = a_B \cos(\sigma t - \epsilon_B), \tag{13.1}$$

where a_B is the tidal amplitude in the bay, and ϵ_B is a phase shift, or lag, of the tide in the bay with respect to that in the ocean. This phase shift is depicted in Figure 13.2, which shows a time history of the tides in the ocean and the bay.

The ocean tidal amplitude is larger than that of the bay, leading to tidal heights in the ocean larger (or smaller) than that in the bay at the same instant. When the ocean tide is higher than the bay tide, a flow is induced in the inlet to fill the bay, causing the bay tide to rise. As high tide is reached in the ocean, the bay is still filling because the bay tide lags that of the ocean by the phase angle ϵ_B. As the ocean tide begins to fall, the bay tide is still rising until the two water levels become the same. Notice in the figure how high tide in the bay corresponds to the coincidence of water levels. Then the flow in the inlet ceases (slack tide) and reverses. The same tide lag occurs at low tide, as shown in Figure 13.2. (If the inertial effect of the water in the inlet is significant, the water can continue to flow into the inlet after the two water levels are equal. In fact, it is possible that the bay tidal amplitude will exceed the ocean tidal amplitude, although this is rare in nature.)

Two equations are used to develop these tidal hydrodynamic models. The first is the conservation of mass, based on the argument that the flow into the bay through the inlet will result in an increase in the volume of water in the bay and the opposite for an ebb tide. The second equation, a momentum equation, relates the flow in the inlet to the water level differences.

13.2.1 CONSERVATION OF MASS

The conservation argument, predicated on the assumption that flow rate into the inlet Q fills the bay uniformly over its planform area A_B is

$$Q = Ub\left[h_i + \frac{(\eta_0 + \eta_B)}{2}\right] = \frac{d\left[A_B(h_B + \eta_B)\right]}{dt} = A_B\frac{d\eta_B}{dt}, \tag{13.2}$$

where the water level in the inlet has been taken as the average of those in the ocean and the bay. In words, the inlet flow into the bay must cause a change in the volume of water in the bay. The cross-sectionally averaged velocity in the inlet will be denoted as $U(t)$. If we now linearize this expression to facilitate a mathematical solution, we obtain

$$U A_c = A_B\frac{d\eta_B}{dt}, \tag{13.3}$$

where $A_c = bh_i$ is the mean inlet cross section.

13.2.2 DYNAMIC EQUATION

The dynamic equation can be developed from the Bernoulli equation or the equation of motion expressed along the channel axis. Here we will utilize the equation of motion for the horizontal velocity

$$\frac{du}{dt} = -\frac{1}{\rho}\frac{\partial p}{\partial x} + \frac{1}{\rho}\frac{\partial \tau}{\partial z} \tag{13.4}$$

Considering the pressure distribution to be hydrostatic over the water column $p = \rho g(\eta - z)$, expanding the total differential, and integrating over depth, we obtain

$$h_i\frac{\partial U}{\partial t} + h_i\frac{\partial\left(U^2/2\right)}{\partial x} = -gh_i\frac{\partial \eta}{\partial x} + \frac{1}{\rho}(\tau_\eta - \tau_b), \tag{13.5}$$

where U is the depth-averaged velocity, and τ_η and τ_b are the shear stresses at the water surface and the bottom, respectively. Expressing the bottom shear stress in terms of a Darcy–Weisbach relationship

$$\tau_b = \frac{\rho f|U|U}{8}$$

and neglecting the water surface stress (no wind), we can integrate the equation of motion along the channel to obtain

$$h_i\ell\frac{\partial U}{\partial t} = -gh_i(\eta_\ell - \eta_0) - \frac{f|U|U\ell}{8}, \tag{13.6}$$

where h_i and U have been assumed uniform along the prismatic channel. The water surface elevations are evaluated at $x = 0, \ell$, which are inside the inlet channel and do not have the same value as in the ocean or bay. As shown in Figure 13.3, there is a drawdown as water flows into the inlet from the ocean owing to the presence of a

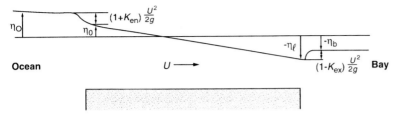

Figure 13.3 Schematic of the water surface along the inlet channel. Not to scale.

velocity head. Therefore, we can relate η_0 to the ocean water level η_o (note carefully the difference between the two subscripts) by

$$\eta_0 = \eta_\mathrm{o} - (1 + K_\mathrm{en})\frac{U^2}{2g}$$

The entrance loss coefficient K_en is included to account for losses due to the flow contraction at the mouth of the inlet.* Also, η_ℓ can be related to η_B:

$$\eta_\ell = \eta_\mathrm{B} - (1 - K_\mathrm{ex})\frac{U^2}{2g},$$

where K_ex is an expansion loss. Estimates of the entrance and exit loss coefficients are

$$K_\mathrm{en} = 0.1 \text{ to } 0.3 \tag{13.7}$$

$$K_\mathrm{ex} \approx 1.0, \tag{13.8}$$

and approximate estimates of the friction coefficient are $0.03 < f < 0.08$.

Substituting the ocean and bay tidal amplitudes into the equation, we now have

$$h_i \ell \frac{\partial U}{\partial t} = -g h_i (\eta_\mathrm{B} - \eta_\mathrm{o}) - h_i \left(K_\mathrm{en} + K_\mathrm{ex} + \frac{f\ell}{4h_i} \right) \frac{|U|U}{2} \tag{13.9}$$

Now, integrating across the channel cross section, we obtain

$$\ell A_\mathrm{c} \frac{\partial U}{\partial t} = -g A_\mathrm{c} (\eta_\mathrm{B} - \eta_\mathrm{o}) - A_\mathrm{c} \left(K_\mathrm{en} + K_\mathrm{ex} + \frac{f\ell}{4h_i} \right) \frac{|U|U}{2} \tag{13.10}$$

The final form of the dynamic equation is

$$(\eta_\mathrm{B} - \eta_\mathrm{o}) + \frac{\ell}{g}\frac{\partial U}{\partial t} + \left(K_\mathrm{en} + K_\mathrm{ex} + \frac{f\ell}{4R_\mathrm{h}} \right) \frac{|U|U}{2g} = 0 \tag{13.11}$$

We have introduced the hydraulic radius, R_h for h_i, but the hydraulic radius is defined in hydraulics as $R_\mathrm{h} = h_i b/(b + 2h_i)$, and will be used as such to include the frictional effects of the channel sidewalls.

Two methods are presented next to determine analytically the hydrodynamic response of the bay–inlet system. The methods differ in the dynamic equations governing the water flow. The Keulegan method (which is more accurate and retains

---

* In hydraulics, these types of loss coefficients are widely used. For example, in pipe flow, loss coefficients are associated with inlets, changes in pipe cross sections, bends, and outlets.

nonlinear terms) will be presented first, and then a linear method, including dynamic effects, will follow.

13.2.3 KEULEGAN METHOD

Keulegan considered the inertial term in Eq. (13.11) to be small, resulting in the reduced form of the dynamic equation

$$(\eta_\text{O} - \eta_\text{B}) = \left(K_\text{en} + K_\text{ex} + \frac{f\ell}{4R_\text{h}} \right) \frac{|U|U}{2g} \tag{13.12}$$

in which the left-hand side represents the potential energy difference between the ocean and the bay and the right-hand side is an expression of the allocation of these energy losses in terms of entrance, exit, and distributed loss components. Solving for U, we have

$$U = \sqrt{\frac{2g|\eta_\text{O} - \eta_\text{B}|}{K_\text{en} + K_\text{ex} + f\ell/4R_\text{h}}} \; \text{sgn}(\eta_\text{O} - \eta_\text{B}) \tag{13.13}$$

Here the signum function, sgn, is $+1$ when its argument is positive, that is, when the ocean tide exceeds the bay tide (flood), and it has the value (-1) when the opposite is true. The conservation of mass equation (Eq. (13.3)) allows us to replace U on the left-hand side, resulting in

$$\frac{A_\text{B}}{A_\text{C}} \frac{d\eta_\text{B}}{dt} = \sqrt{\frac{2g|\eta_\text{O} - \eta_\text{B}|}{(K_\text{en} + K_\text{ex} + f\ell/4R_\text{h})}} \; \text{sgn}(\eta_\text{O} - \eta_\text{B}) \tag{13.14}$$

This governing equation for η_B is made nondimensional by introducing the following nondimensional parameters:

$$t' = \sigma t \tag{13.15}$$

$$\eta'_\text{O}, \; \eta'_\text{B} = \eta_\text{O}/a_\text{O}, \; \eta_\text{B}/a_\text{O}, \tag{13.16}$$

and thus,

$$\frac{d\eta'_\text{B}}{dt'} = K\sqrt{|\eta'_\text{O} - \eta'_\text{B}|} \, \text{sgn}(\eta'_\text{O} - \eta'_\text{B}) \tag{13.17}$$

Surprisingly, Eq. (13.17) depends on only one dimensionless parameter, K, which is termed the *repletion coefficient*, where

$$K = \frac{A_\text{c}}{\sigma a_\text{O} A_\text{B}} \frac{\sqrt{2ga_\text{O}}}{\sqrt{K_\text{en} + K_\text{ex} + f\ell/4R_\text{h}}} \tag{13.18}$$

The French verb for 'to fill' is *replere*, and the repletion coefficient signifies the degree to which the bay will be filled given the particular combination of bay, inlet, and tidal parameters. Equation (13.17) clearly shows that large K values mean that a given water level difference between the ocean and the bay will result in a faster rise in bay level than for a small K value.

All of the parameters in K can be determined from navigational charts, tide tables, and estimates of frictional losses. Examining their influence on the repletion coefficient directly shows that increases in all of the parameters in the numerator (e.g., the inlet cross section) will increase the relative tidal amplitude (a_B/a_O) in the bay, whereas increasing the parameters in the denominator (bay plan area, frictional losses in channel) will decrease the tide in the bay. For large bays, then, the repletion coefficient is likely to be small.

Keulegan solved Eq. (13.17) for a sinusoidal water level variation in the ocean

$$\eta'_O = \frac{\eta_O}{a_O} = \sin t' \qquad (13.19)$$

and developed useful engineering results that we will now discuss.

13.2.3.1 Bay Tidal Range, Maximum Velocity, and Phase Lag

The solid line in Figure 13.4 presents the variation of the ratio of bay to ocean tidal amplitudes, a_B/a_O, with repletion coefficient K. Clearly, for large values of K, the bay fills completely. In a nondimensional sense, it is possible to define bays as "large" or "small," depending on whether the filling is relatively incomplete or complete. Denoting large bays as those with $a_B/a_O < 0.2$, and small bays as those with $a_B/a_O > 0.8$, then the associated K limits are

$$\text{Large Bays: } \frac{a_B}{a_O} < 0.2, \quad K < 0.18 \qquad (13.20)$$

$$\text{Small Bays: } \frac{a_B}{a_O} > 0.8, \quad K > 0.88 \qquad (13.21)$$

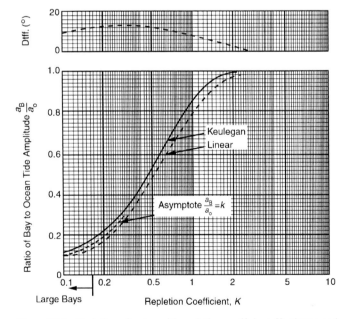

Figure 13.4 Variation of a_B/a_O with repletion coefficient; Keulegan and linear methods (from O'Brien and Dean 1972).

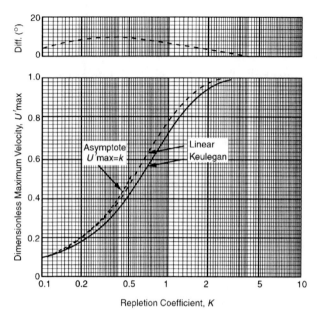

Figure 13.5 Variation of $U'_{\max}$ with repletion coefficient. Keulegan and linear methods (from O'Brien and Dean 1972).

In Figures 13.4–13.6, the Keulegan method results are plotted along with those from the linear method, which will be discussed later. The difference between the two methods is shown at the top of the figures.

The nondimensional maximum velocity $U'_{\max}$ is determined from the continuity equation:

$$Q_{\max} = A_c U_{\max} = A_B \left(\frac{d\eta_B}{dt} \right)_{\max} \tag{13.22}$$

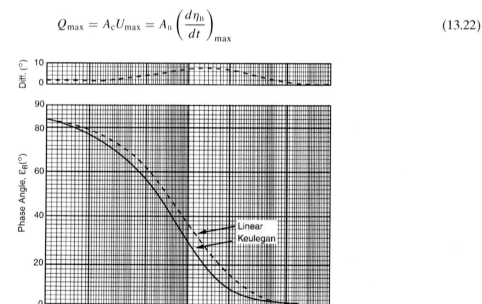

Figure 13.6 Variation of bay phase lag ϵ_B with repletion coefficient. Keulegan and linear methods (from O'Brien and Dean 1972).

and substituting for $\eta_{\scriptscriptstyle B}$ from Eq. (13.1), we obtain

$$\left(\frac{d\eta_{\scriptscriptstyle B}}{dt}\right)_{max} = a_{\scriptscriptstyle B}\sigma; \tag{13.23}$$

thus,

$$U_{max} = \frac{A_{\scriptscriptstyle B}}{A_c}a_{\scriptscriptstyle B}\sigma; \tag{13.24}$$

or in nondimensional form

$$U'_{max} \equiv \frac{U_{max}}{(A_{\scriptscriptstyle B}/A_c)a_{\scriptscriptstyle O}\sigma} = f(K) \tag{13.25}$$

and is shown in Figure 13.5. It is seen that graphical representations of U'_{max} and $a_{\scriptscriptstyle B}/a_{\scriptscriptstyle O}$ are quite similar.

The bay tide lags that in the ocean by a phase, $\epsilon_{\scriptscriptstyle B}$, which is shown graphically in Figure 13.2. The amount of this lag is quite closely related to the relative bay tidal range. In fact, if the flow were linear, it can be shown that

$$\cos\epsilon_{\scriptscriptstyle B} = a_{\scriptscriptstyle B}/a_{\scriptscriptstyle O} \tag{13.26}$$

Figure 13.6 presents the bay phase lag.

13.2.3.2 Other Keulegan Parameters

Keulegan determined several other parameters characterizing tidal flow through inlets. These will be mentioned here, but will not be presented in detail. He showed that, especially for large bays (small K), the bay tidal variation was distinctly nonsinusoidal and could be expressed by

$$\frac{\eta_{\scriptscriptstyle B}}{a_{\scriptscriptstyle O}} = \cos(\sigma t) - a_1\sin(\sigma t + \epsilon_{\scriptscriptstyle B}) - a_1 b_3[\sin(\sigma t + \epsilon_{\scriptscriptstyle B}) - \sin[3(\sigma t + \epsilon_{\scriptscriptstyle B})]]$$
$$-a_1 a_3\sin[3(\sigma t + \epsilon_{\scriptscriptstyle B})] \tag{13.27}$$

Keulegan also presented numerical results for a_1, a_3, and b_3 as a function of K. Usually, we consider the first two terms to represent tidal flows through inlets adequately. However, for small K, the bay tidal displacement is very strongly influenced by friction and tends to be "top hat" in shape (see Figure 13.7) when viewed as a tide gauge record.

13.2.3.3 Examination of Large and Small Bay Asymptotes

Large bays, with small K values, are characterized by a small relative bay amplitude, and it is possible to develop surprisingly simple asymptotes for some of the results. Starting with the expression for velocity from the Bernoulli equation (Eq. (13.13))

$$U = \frac{\sqrt{2g|\eta_{\scriptscriptstyle O} - \eta_{\scriptscriptstyle B}|}}{\sqrt{K_{en} + K_{ex} + f\ell/4R_h}}\ \text{sgn}\,(\eta_{\scriptscriptstyle O} - \eta_{\scriptscriptstyle B}) \tag{13.28}$$

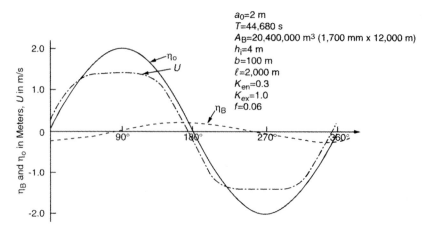

Figure 13.7 Ocean and bay tides for an inlet–bay system with small repletion coefficient.

and, because the bay tide will be small, we approximate it by zero and then note that, for U_{max}, $\eta_0 = a_0$, resulting in

$$U_{max} = \frac{\sqrt{2ga_0}}{\sqrt{K_{en} + K_{ex} + f\ell/4R_h}} \tag{13.29}$$

Casting U_{max} into nondimensional form in accordance with Eq. (13.25), we obtain the surprisingly simple asymptote

$$U'_{max} \approx K \tag{13.30}$$

This asymptote is within 20 percent for $K < 0.5$ (see Figure 13.5).

We can also develop a simple asymptote for relative bay tidal displacement a_B/a_0. The bay tidal amplitude is

$$a_B = \frac{1}{A_B} \int_{\epsilon_U/\sigma}^{\epsilon_U/\sigma + T/4} A_c U \, dt, \tag{13.31}$$

where ϵ_U is the phase shift in the velocity U. Because η_B is small, this velocity is approximated by

$$U = \frac{\sqrt{2g\eta_0}\, \mathrm{sgn}(\eta_0)}{\sqrt{K_{en} + K_{ex} + f\ell/4R_h}} \tag{13.32}$$

and further approximated by

$$U = \frac{\sqrt{2ga_0}\, \cos(\sigma t - \epsilon_U)}{\sqrt{K_{en} + K_{ex} + f\ell/4R_h}}, \tag{13.33}$$

which, when substituted in Eq. (13.31) and integrated, yields for a_B/a_0

$$\frac{a_B}{a_0} = K \tag{13.34}$$

This asymptote is compared with the Keulegan result in Figure 13.4, where it is seen that agreement to within ± 20 percent exists for $K < 1$.

Consideration of our small K results, Eqs. (13.30) and (13.34), each of which contains the head loss factor $\sqrt{K_{en} + K_{ex} + f\ell/4R_h}$, demonstrates that large bays are *friction dominated* because of the relatively large size of the frictional loss term $f\ell/4R_h$.

Small bays, with large K values, by their definition tend to fill completely. In contrast to large bays that are friction dominated, small bays are volume controlled, and the asymptote for a_B/a_O is unity for large K. The velocity asymptote can be developed from the continuity equation

$$Q = A_c U = A_B \frac{d\eta_B}{dt} \approx A_B \frac{d\eta_O}{dt} \tag{13.35}$$

and

$$U_{max} = \frac{A_B}{A_c} a_O \sigma \tag{13.36}$$

In nondimensional terms

$$U'_{max} = 1 \tag{13.37}$$

A result that will be of use later is evident from Eq. (13.36). With increasing inlet cross-sectional area A_c and the same total discharge, the maximum velocity must decrease.

13.2.4 LINEAR METHOD

The dynamic equation for the linear method retains the inertia term in Eq. (13.11) and involves linearizing the shear stress term. The loss terms (entrance, exit, and shear stress) are nonlinear in velocity and include an absolute value sign to ensure a sign change with the velocity. To make the problem tractable, we linearize the loss terms in a manner that will ensure that the maximum discharge into (or out of) the bay will be the same as for the Keulegan method

$$\left(K_{en} + K_{ex} + \frac{f\ell}{4R_h}\right) \frac{|U|U}{2g} \Rightarrow \frac{F\sqrt{2ga_O}\,U}{2g}, \tag{13.38}$$

where F is a dimensionless coefficient that we will quantify later.

We also employ the same expression for U in terms of the water surface elevation in the bay from the conservation of mass argument, Eq. (13.3):

$$U = \frac{A_B}{A_c} \frac{d\eta_B}{dt}$$

The two equations for this problem involve two unknowns: the water surface elevation in the bay η_B and the velocity in the inlet U. We can eliminate one of them; let us select U. Therefore, if we substitute for U in the equation of motion (Eq. (13.11)), then

$$\frac{dU}{dt} = \frac{A_B}{A_c} \frac{d^2\eta_B}{dt^2} = g \frac{(\eta_O - \eta_B)}{\ell} - \frac{F\sqrt{2ga_O}\,A_B}{2\ell A_c} \frac{d\eta_B}{dt} \tag{13.39}$$

or,

$$\frac{d^2\eta_B}{dt^2} + \frac{F\sqrt{2ga_0}}{2\ell}\frac{d\eta_B}{dt} + \frac{gA_c}{\ell A_B}\eta_B = \frac{gA_c}{\ell A_B}\eta_0 \tag{13.40}$$

The form of this equation may be familiar, for it also describes a forced spring-mass-dashpot dynamical system. Here, the mass term is the first term on the left-hand side, the damping term is provided by the shear stress, the spring term is the last on the left-hand side, with gravity providing the restoring force, and the tidal forcing from the ocean appears on the right.

Substituting an assumed solution, $\eta_B = a_B\cos(\sigma t - \epsilon_B)$, into the governing equation and separating into terms dependent on $\cos\sigma t$ and $\sin\sigma t$ yield two equations. From the sine terms, which must equal zero,

$$\tan\epsilon_B = \frac{F\sqrt{2ga_0}}{2\ell\sigma(gA_c/A_B\ell\sigma^2 - 1)} \tag{13.41}$$

or

$$\tan\epsilon_B = \frac{\phi_1}{(\phi_2 - 1)}, \tag{13.42}$$

where the dimensionless parameters are defined as

$$\phi_1 = \frac{F\sqrt{2ga_0}}{2\ell\sigma} \tag{13.43a}$$

$$\phi_2 = \frac{gA_c}{\sigma^2 A_B\ell} \tag{13.43b}$$

These parameters can be viewed as the ratio of frictional forces to inertial forces and the ratio of hydrostatic forces to inertial forces. The ϕ_2 term can also be written as

$$\phi_2 = \left(\frac{\sigma_N}{\sigma}\right)^2,$$

where σ_N is defined as the natural frequency of the inlet–bay system, that is, the frequency of oscillation obtained from Eq. (13.40) in the absence of friction and tidal forcing. This natural frequency is

$$\sigma_N = \sqrt{\frac{gA_c}{A_B\ell}}.$$

Equation (13.42), for the phase shift of the bay tide, shows that the bay tide lags the ocean tide by an amount that depends on the inertia and friction in the inlet.

The second equation, resulting from the terms proportional to $\cos\sigma t$, gives the amplitude of the bay tide in terms of the ocean tide.

$$\frac{a_B}{a_0} = \frac{\phi_2}{\sqrt{(\phi_2 - 1)^2 + \phi_1^2}} \tag{13.44}$$

The tidal velocity, from Eq. (13.3), is

$$U = \frac{A_B}{A_c} \frac{d\eta_B}{dt} = -\frac{A_B \sigma}{A_c} a_B \sin(\sigma t - \epsilon_B)$$

The *tidal prism*, Ω, is the amount of water that flows into the bay from low tide (one of which occurs at $t_1 = -T/2 - \epsilon_B T/2\pi$) to high tide, which occurs at $t_2 = -\epsilon_B T/2\pi$.

$$\Omega = \int_{t_1}^{t_2} U A_c \, dt = \int_{t_1}^{t_2} A_B \frac{d\eta_B}{dt} \, dt = 2 A_B a_B \tag{13.45}$$

13.2.4.1 Comparison of Keulegan and Linear Methods

By comparing these two methods, Keulegan's with its nonlinear friction terms and the linear method with the linearized friction and inertia terms, we can determine the importance of different terms. Furthermore we can establish a method to determine the linear friction term F.

Keulegan's K can be related to the ϕ parameters defined by the linear theory (Eq. (13.43)) by equating the maximum discharges by the two methods, Eqs. (13.13) and (13.3),

$$(Q_{max})_{Keulegan} = (Q_{max})_{Linear} \tag{13.46}$$

$$\frac{\sqrt{2g(\eta_0 - \eta_B)_{max}}}{\sqrt{K_{en} + K_{ex} + f\ell/4R_h}} A_c = A_B \left(\frac{d\eta_B}{dt}\right)_{max}, \tag{13.47}$$

where the right-hand side is obtained from the continuity equation. To do this comparison properly, we have to neglect the inertia term in the linear method. This implies that $\phi_2 \gg 1$; that is, the natural frequency of the basin system should be much larger than the tidal forcing frequency. For most inlets, the inertial term is small and can be neglected.

Using the linear method to determine expressions for $(\eta_0 - \eta_B)_{max}$ and $(d\eta_B/dt)_{max}$, the last equation can be written as

$$K = \frac{\phi_2}{\phi_1^{1/2} \left(\phi_1^2 + \phi_2^2\right)^{1/4}} \tag{13.48}$$

In applying Eq. (13.48), the repletion coefficient K and ϕ_2 are first calculated from their definitions (Eqs. (13.18) and (13.43)) using the known data; then, ϕ_1 is calculated from this equation, which finally allows F to be determined from Eq. (13.43a). With both ϕ parameters determined, the linear method can be used to investigate the problem of interest. For example, the dashed lines in Figures 13.4, 13.5, and 13.6 for the relative bay amplitude, the phase lag, and the nondimensional maximum velocity, respectively, were determined by the linear method, and they compare reasonably well with the Keulegan method. As noted before, the upper panels in the figures represent the percentage difference between the two methods.

EXAMPLE 1

Consider a bay with a plan area $A_B = 2 \times 10^8 \, m^2$, $a_0 = 0.5 \, m$, $f = 0.06$, $A_c = 2000 \, m^2$ ($b = 200$ m and $h_i = 10$ m such that the wetted perimeter is

220 m), $\ell = 1000$ m, $K_{en} = 0.3$, $K_{ex} = 1.0$, and the tide is semidiurnal with period $T = 44{,}640$ s ($\sigma = 2\pi/T = 1.41 \times 10^{-4}$ s^{-1}). Determine the maximum velocity and tidal characteristics of this tidal inlet using the Keulegan and linear methods. Compare results.

Using these values, the Keulegan parameter (Eq. (13.18)) is $K = 0.259$, and ϕ_2 (Eq. (13.43)) is 4.95. Solving Eq. (13.48) for ϕ_1, we obtain $\phi_1 = 18.8$, which then, from its definition (Eq. (13.43)), can be used to find the parameter

$$F = \frac{2\ell\sigma\phi_1}{\sqrt{2ga_o}} \tag{13.49}$$

We are more interested in U_{max}, which is related to F or ϕ_1 through Eq. (13.38) after the cancellation of a U from both sides as follows:

$$U_{max} = \frac{2\phi_1\ell\sigma}{K_{en} + K_{ex} + f\ell/4R_h} = 1.79 \text{ m/s} \tag{13.50}$$

If we compare this result with the Keulegan method for $K = 0.259$, using Figure 13.6, $U'_{max} = 0.237$, and therefore

$$U_{max} = \frac{A_c U'_{max}}{\sigma a_o A_B} = 1.67 \text{ m/s},$$

which is 7 percent different than the linear solution.

Comparing the linear and the Keulegan estimates of the bay amplitude ratio and the phase lag, we have

$$\left(\frac{a_B}{a_o}\right)_{linear} = \frac{\phi_2}{\sqrt{\phi_1^2 + \phi_2^2}} = 0.255$$

and, from the Keulegan method, $(a_B/a_o)_{Keulegan} = 0.29$, which is a 12.2-percent difference.

The phase angle by the linear method is found from Eq. (13.42) modified to neglect the inertia term:

$$(\epsilon_B)_{linear} = \tan^{-1}\left(\frac{\phi_1}{\phi_2}\right) = 75.2°$$

and, from the Keulegan method (see Figure 13.6): $(\epsilon_B)_{Keulegan} = 73.7°$, which is a difference of 1.5°.

In this example, the linear method agrees very well with the Keulegan method, which gives us confidence that we can apply the linear method to problems not treatable by the Keulegan method. (Because, for small bays, the frictional effects are relatively less important, we would expect the two methods to agree better for small bays.)

13.2.4.2 Effect of Inlet Cross-Section A_c on Tidal Prism

The tidal prism Ω is a measure of the amount of tidal flushing of the bay, for there is a substantial amount of mixing with the bay or ocean waters during each change of tide. Increasing the tidal prism can lead to an improvement in water quality; hence, an interest in determining how increasing the cross-sectional area of the inlet would increase the tidal prism.

The tidal prism (Eq. (13.45)) can be written as

$$\Omega = 2 \left(\frac{a_B}{a_O} \right) A_B a_O,$$ (13.51)

which can be rewritten in terms of the linear parameters (Eq. (13.44)) as

$$\Omega = 2 \left(\frac{\phi_2}{\sqrt{(\phi_2 - 1)^2 + \phi_1^2}} \right) A_B a_O$$ (13.52)

Now ϕ_1 does not depend directly on A_c; therefore, we can find the change in Ω due to the change in cross-sectional area by differentiation:

$$\frac{\partial \Omega}{\partial A_c} = 2 A_B a_O \left(\frac{\phi_1^2 - \phi_2 + 1}{[(\phi_2 - 1)^2 + \phi_1^2]^{3/2}} \right) \frac{\partial \phi_2}{\partial A_c}$$ (13.53)

This expression can be recast into the percentage change in tidal prism relative to the percentage change in cross-sectional area (using Eqs. (13.52) and (13.43)).

$$\frac{\partial \Omega / \Omega}{\partial A_c / A_c} = \frac{\phi_1^2 - \phi_2 + 1}{[(\phi_2 - 1)^2 + \phi_1^2]}$$ (13.54)

When the inertia of the inlet can be neglected, the Keulegan parameter can be introduced and used to demonstrate how the prism changes with cross-sectional area, as shown in Figure 13.8. As an example of the use of the figure, if the Keulegan parameter for an inlet system is small, then an X percent change in cross-sectional

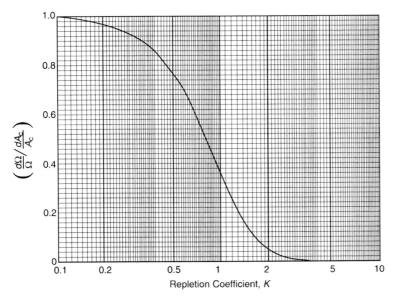

Figure 13.8 Variation of the ratio of the relative change in tidal prism Ω to the relative change in inlet cross-sectional area A_c with the Keulegan parameter K based on the linear method.

area will result in almost an X percent change in tidal prism, although for large K values, there is almost no change in prism with changes in cross-sectional area. Why? Could this have been anticipated?

13.2.4.3 Effect of Changing Bay Planform Area A_B on the Tidal Prism

Although it is not often done now, numerous bays have been subjected to widespread filling for residential and industrial development. Examples include Biscayne Bay, Florida, and there are numerous cases in the Netherlands, where large polder areas have been created. This reduction in bay plan area results in a reduction of tidal prism in part owing to the reduced volume of the bay. Using the linear method, we can examine this effect.

The analysis is the same as in the last section because only ϕ_2 depends on A_B, leading to

$$\frac{\partial \Omega / \Omega}{\partial A_B / A_B} = \frac{\phi_2(\phi_2 - 1)}{(\phi_2 - 1)^2 + \phi_1^2} \tag{13.55}$$

This expression includes inertia. Neglecting inertia leads to a simplification,

$$\frac{\partial \Omega / \Omega}{\partial A_B / A_B} = \frac{\phi_2^2}{\phi_2^2 + \phi_1^2} \tag{13.56}$$

This last result is plotted versus K in Figure 13.9. You can see that there is almost no change in the tidal prism for changes in planform areas of large bays (small K). This is obviously because they do not have time to fill from low to high tide anyway.

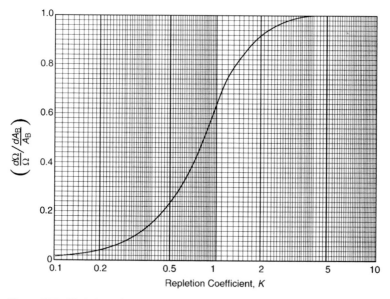

Figure 13.9   Variation of the ratio of the relative change in tidal prism Ω over the relative change in bay area A_B with the Keulegan parameter K based on the linear method.

13.2.4.4 Effect of Changing A_c on the Maximum Velocity

Changes in inlet cross section can change the velocities in the inlet as well as the tidal prism. This has led to interest in deliberately reducing velocities in inlets by increasing their cross-sectional areas. For example, at South Lake Worth Inlet, Florida, which connects the Atlantic Ocean to Lake Worth, the floor of the inlet is rock, and the two sidewalls are concrete, sheet-piling jetties and bridge abutments; thus, the limiting velocity for scouring (approximately 1 m/s) is exceeded substantially and can be hazardous for the less experienced boater. Increasing the cross-sectional area of the inlet appears to be a logical approach; however, we will show that it will not be helpful for this inlet.

In this case we express the maximum velocity in terms of the tidal prism Ω and cross-sectional area,

$$\Omega = \frac{T}{2}\left(\frac{2}{\pi}\right) U_{\max} A_c,$$

where the $2/\pi$ represent the average of $\cos \sigma t$ over half of a tidal cycle. Differentiating this equation, we obtain

$$A_c \frac{dU_{\max}}{dA_c} + U_{\max} = \frac{\sigma}{2}\frac{d\Omega}{dA_c} \tag{13.57}$$

Assuming that inertia effects are small, we can replace $d\Omega/dA_c$ with

$$\frac{d\Omega}{dA_c} = \frac{\phi_1^2}{\left(\phi_1^2 + \phi_2^2\right)}, \tag{13.58}$$

which is Eq. (13.53) for no inertia. This gives us

$$\frac{\partial U_{\max}}{\partial A_c} = \frac{U_{\max}}{A_c}\left(\frac{\phi_1^2}{\left(\phi_1^2 + \phi_2^2\right)} - 1\right) \tag{13.59}$$

after introducing $U_{\max} = \Omega/(\sigma A_c)$. This is finally rewritten in percentage form as

$$\frac{\partial U_{\max}/U_{\max}}{\partial A_c/A_c} = -\left(\frac{\phi_1^2}{\left(\phi_1^2 + \phi_2^2\right)} - 1\right), \tag{13.60}$$

which is plotted versus the repletion coefficient in Figure 13.10. For small repletion coefficients, increasing the cross-sectional area only results in a proportionally small decrease in the maximum velocity. This is the case for South Lake Worth Inlet, but other inlets may differ substantially.

13.2.5 MAXIMUM VELOCITIES IN THE INLET

The maximum velocities that occur in the inlet should not be too large to endanger navigation nor cause excessive channel scour. The inlet velocity is given as

$$U = \frac{A_B}{A_c}\frac{d\eta_B}{dt} \tag{13.61}$$

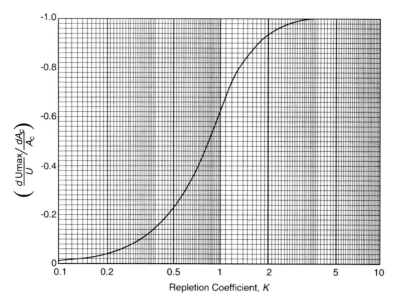

Figure 13.10 Variation of ratio of the relative maximum velocity $U_{\max}$ and cross-sectional area A_c with the Keulegan parameter based on the linear method.

Substituting for η_B, we can show that the maximum velocity $U_{\max}$ is

$$U_{\max} = \frac{\sigma A_B a_B}{A_c}, \tag{13.62}$$

which is very small when the inlet cross-section A_c is small, for a_B is small; $U_{\max}$ increases with A_c, reaching a maximum, and then decreases. The reason for this behavior is that small entrance channels are dominated by friction. As the inlet cross-sectional area increases, the influence of friction is less. However, as the inlet cross section becomes very large, the velocity in the inlet decreases because the bay fills completely during the tidal cycle. This variation of velocity with inlet cross-sectional area is shown as the solid line in Figure 13.11. We will refer back to this variation later.

The velocities predicted in the inlet by the simple linear model and the more complex model of Keulegan do not correctly mirror the realistic nonlinear situation. The flows have a considerable momentum. As they exit the inlet, the flows are similar to jets, traveling large distances, entraining the neighboring fluids, and transporting sediments offshore to ebb tidal shoals. The flows into the inlets, however, are characterized more as sinks with flow from all directions to the inlet. To envision this more graphically, think of blowing out a candle. As you exhale, a jet of air puts out the candle. Now try to inhale the candle out!

13.2.6 TIDAL SETUP IN BAYS

A super elevation of the mean water level in bays can result from winds, fresh water inflows, and net frictional effects (see Figure 13.12). For example, if the bay water is less saline than the ocean water, a mean setup in the bay is required to balance the forces. Also, a net inflow to the bay will result in a setup in order to overcome the

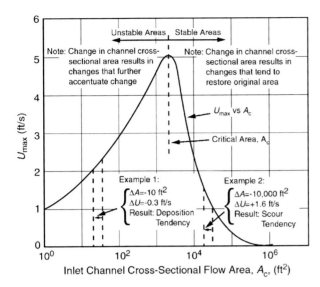

Figure 13.11 Maximum velocity in the inlet as a function of A_c (from O'Brien and Dean 1972). The stability of an inlet depends on this velocity, as discussed in Section 13.4.

friction induced by the net outflow to the ocean. An additional cause of setup in the bay is the time-varying depth in the inlet channel. During times when the water level is greatest, the flow is predominantly bayward and vice versa for the reduced water levels. Thus, this time-varying water level in the inlet acts as a "leaky" check valve, admitting more water to the bay during inflow times and acting to reduce the outflows.

This check valve phenomenon is a nonlinear effect, and thus it is essential to retain some of the nonlinear characteristics in the formulation. However, it is somewhat surprising that a linear representation of the velocity as a function of the ocean and bay water level difference will be adequate to yield an approximate expression for the bay water level setup $\bar{\eta}_B$.

We begin by expressing the ocean and bay water surface displacements and the inlet current as

$$\eta_O = a_O \cos \sigma t = \hat{\eta}_O \tag{13.63}$$

$$\eta_B = a_B \cos(\sigma t - \epsilon_B) + \bar{\eta}_B = \hat{\eta}_B + \bar{\eta}_B \tag{13.64}$$

$$U = \mathcal{C}(\eta_O - \eta_B), \tag{13.65}$$

where $\bar{\eta}_B < a_O$ and $\mathcal{C}$ is a linearized friction coefficient. The water level in the inlet is taken as the average of that in the ocean and the bay. The discharge into the

Bay

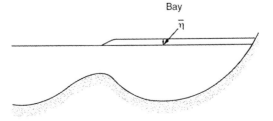

Figure 13.12 Setup within a bay.

bay Q is thus

$$Q = C\left[\hat{\eta}_{\mathrm{O}} - (\hat{\eta}_{\mathrm{B}} + \bar{\eta}_{\mathrm{B}})\right]\left(h_i + \frac{\hat{\eta}_{\mathrm{O}} + \hat{\eta}_{\mathrm{B}} + \bar{\eta}_{B}}{2}\right) \tag{13.66}$$

Requiring the average of Q to be zero over a tidal cycle in order for the long-term bay water level to be steady, and neglecting very small terms of order $\bar{\eta}_{\mathrm{B}}^2$, we find

$$\bar{\eta}_{\mathrm{B}} = \frac{1}{2h_i}[\bar{\hat{\eta}}_{\mathrm{O}}^2 - \bar{\hat{\eta}}_{\mathrm{B}}^2] = \frac{a_{\mathrm{O}}^2}{4h_i}\left[1 - \left(\frac{a_{\mathrm{B}}}{a_{\mathrm{O}}}\right)^2\right] \tag{13.67}$$

It is somewhat surprising that the mean setup can be obtained without a detailed analysis of the problem and also that $\bar{\eta}_{\mathrm{B}}$ is independent of the linearized friction coefficient C.

13.2.7 MULTIPLE BAY SYSTEMS

Although not examined here, it is worthwhile to indicate that the linear method of tidal analysis is quite effective for examining the response of coupled bay systems and the effect of introducing alterations in these inlet systems such as causeways and filled areas.

13.2.8 NONLINEAR TIDAL ANALYSIS

The Keulegan and the linear methods, as presented, only provide the response of the inlet–bay system to a single tidal component such as the M_2 tide. The Keulegan analysis showed that higher harmonics of this forcing can occur owing to the nonlinear nature of the full equations (Eq. (13.27)).

Natural flows in inlets, forced by numerous tidal harmonics and flowing through irregularly shaped tidal channels, often show an asymmetry in velocity over a tidal cycle. For some inlets, the flood tidal flows are faster than those associated with the ebbing tide, whereas for other inlets the opposite is true. This asymmetry in velocities (and the fact that the transport of sediment in an inlet probably depends on the velocity to some power greater than unity) leads to a net importing into, or exporting of, sediment from the inlet–bay system (Boon and Byrne 1981). Speer and Aubrey (1985) and Aubrey and Speer (1985) showed with numerical models that the tidal asymmetry can be brought about through a variation in both the inlet cross-sectional area and the bay cross-sectional area* during the tidal cycle. This variation in bay area often occurs in nature because bays may contain several shoals that wet and dry during the tidal cycle, and they may be surrounded by marsh that fills over part of the tidal cycle. Correspondingly, the inlet cross section (particularly for a natural inlet) may have one or more deeper channels that are always filled and shallower channel banks that emerge during lower tidal stands.

DiLorenzo (1988) carried out an approximate analytical solution to the nonlinear inlet hydrodynamic problem for inlet–bay systems with constant cross-sectional and

---

* Recall that we assumed vertically sided channels and bay so that the cross sections were constant for all stages of the tide.

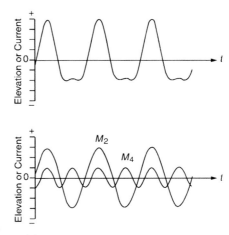

Figure 13.13 Superposition of the M_2 and M_4 tides (from DiLorenzo 1988). (a) Addition of the M_2 and M_4 components (elevation or current) for the case of zero lag. The upper figure is the sum of the two curves in the lower figure. (b) Addition of the M_2 and M_4 components for the case of 90° lag.

(a)

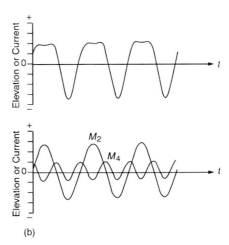

(b)

bay areas. He showed that the inlets can amplify the M_4 tide (with frequency 2σ) much more than the M_2 tide and that large phase shifts between the tides can be introduced. He showed that the two tides can add in phase, leading to a flood-dominated inlet, or they may be out of phase, resulting in an ebb-dominant inlet. These two possibilities are shown in Figure 13.13.

13.3 INLET STABILITY

A tidal inlet is stable if its cross-sectional area and planform location do not change significantly over the years. For the cross section, stability occurs when the scouring ability of the tides coursing through the inlet is balanced by the sediment load carried into the inlet.

Sand is brought into an inlet through a variety of means. On flood tides, the flows into the inlet draw sediment-laden water from the surf zone as well as from offshore shoals. For jettied inlets, sand can be drawn through the jetties from the adjacent

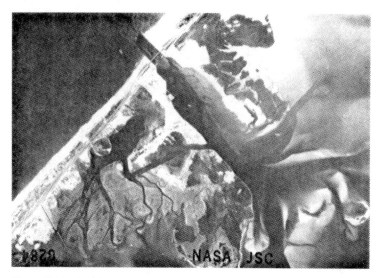

Figure 13.14 Aerial photograph of Indian River Inlet, Delaware, on June 8, 1971 (NASA 1975).

beach directly if they are not sandtight (that is, they do not have an impermeable core) or, for low jetties, the sand can be washed over the top when the waves and tides are high or blown over when the wind is strong. All of the sand carried into the inlet is eventually deposited on shoals within the channel or in the bay. Figure 13.14 shows Indian River Inlet, Delaware, with some of its extensive flood tidal shoals. For long, wide inlets, particularly those that have been widened for navigational purposes, the sand can deposit on the channel banks. On an ebbing tide, the sediment comes from the bay shoreline, bottom, or shoals. This sand is then jettied offshore into the ebb tidal delta. Figure 13.15 shows a bathymetric survey made offshore of Indian River Inlet in 1975 displaying the large asymmetric ebb tidal shoal.* The shoals are at about −13 ft relative to NGVD (National Geodetic Vertical Datum) in an ambient water depth of 35 ft. (Note the scour hole directly in front of the inlet that is probably due to flow concentration during the flood tide.)

O'Brien (1931, 1969) examined a number of tidal inlets on sandy coastlines and found that the size of the inlet cross sections seemed to be constant over time and that larger bays and lagoons had larger inlets. To examine this, he chose those inlets he thought were in equilibrium; that is, the cross-sectional area of the inlet was relatively constant with respect to time. He then plotted these measured cross-sectional inlet areas as a function of the spring tidal prisms for the inlets. He found two families of curves. For natural inlets,

$$A_c = 2.0 \times 10^{-5} \, \Omega, \tag{13.68}$$

where A_c is measured in square feet and Ω has the units of cubic feet. For jettied

---

* This inlet is dominated by the tidal flows at the entrance to Delaware Bay, which is located about 20 km to the north; when Indian River Inlet is ebbing, the tidal currents from Delaware Bay are also ebbing and southerly, forcing the ebb tidal jet from Indian River Inlet to deflect to the south, and, thus, the depositional ebb tidal shoal is located updrift!

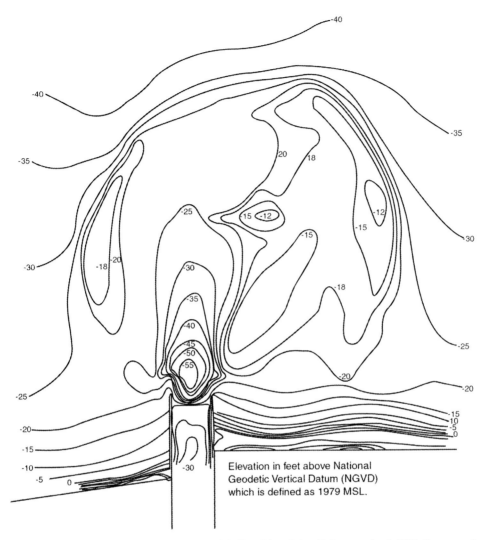

Figure 13.15 Bathymetric survey offshore of Indian River Inlet, Delaware, April 1975 (Lanan and Dalrymple 1977). The contours are in feet.

inlets,

$$A_c = 4.69 \times 10^{-4} \, \Omega^{.85} \tag{13.69}$$

These data are shown in Figure 13.16. The basis of these relationships is that larger tidal prisms mean that more water has to flow into the inlet over the tidal cycle (which is of the same duration, regardless of the inlet's size). This greater flow means more capability to scour the bottom of the inlet to increase its size. Later, Jarrett (1976) examined many more inlets and found, combining all inlets, the results presented in Figure 13.17, which are in essential agreement with the earlier findings of O'Brien. His equation is

$$A_c = 5.74 \times 10^{-5} \, \Omega^{0.95} \tag{13.70}$$

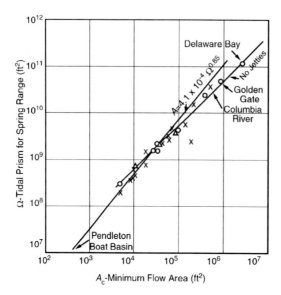

Figure 13.16 Equilibrium tidal inlet relationship (from O'Brien 1969).

His results for stable cross sections, segregated by jettied or nonjettied inlets, are

$$A_c = 1.04 \times 10^{-5} \, \Omega^{1.03} \quad \text{for natural inlets,} \tag{13.71}$$

$$A_c = 3.76 \times 10^{-4} \, \Omega^{0.86} \quad \text{for jettied inlets.} \tag{13.72}$$

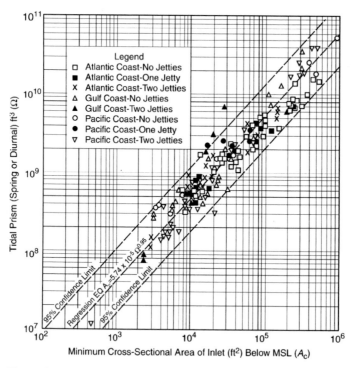

Figure 13.17 Empirical relationship between tidal prism Ω and inlet cross-sectional area A_c (Jarrett 1976).

Beyond providing a method for determining the equilibrium cross section for an inlet in terms of the prism, these relationships can be used to ascertain the change in cross section resulting from a change in tidal prism. For example, reclaiming a portion of an estuary reduces the tidal prism and therefore reduces the equilibrium cross-sectional area of the inlet serving the estuary. Taking the total derivative of Jarrett's equations and then dividing by A_c gives the following percentage change in cross-sectional area:

$$\frac{dA_c}{A_c} = 1.03\frac{d\Omega}{\Omega} \tag{13.73}$$

for natural inlets and

$$\frac{dA_c}{A_c} = 0.86\frac{d\Omega}{\Omega} \tag{13.74}$$

for jettied inlets. These equations show how percentage changes in tidal prism are mirrored directly in changes in equilibrium inlet cross section.

Bruun and Gerritsen (1966) introduced a parameter Ω/Q for examining inlet stability, arguing that the influence of the annual rate of littoral drift Q should play a role in inlet stability because inlets on coasts with high transport rates should have a different size than those serving similarly sized bays on a low-drift coast. Inlets with values of $\Omega/Q > 300$ appear to be stable, whereas those with $\Omega/Q < 100$ are more likely to be unstable and have shifting channels and significant shoals. No equivalent curve to O'Brien's was constructed, however.

Earlier, Bruun and Gerritsen (1959) pointed out that natural tidal inlets are not complete sinks of sediment but often have means to bypass sand from the updrift side of the inlet to the downdrift side. Two of these methods are *bar bypassing* and *tidal bypassing*, which can occur simultaneously at a given inlet. Bar bypassing is the transport of sand by waves and tidal currents along the ebb tidal shoal across the mouth of the inlet. Tidal bypassing occurs primarily through drawing of sand into the inlet by the flood tidal current and the subsequent discharge of the sand back out to sea by the ebb tidal flow. Owing to prevailing wave action or coastal currents over many tidal cycles, some of the sand is moved from one side of the inlet to the other by this mechanism.

The dominant method of sand bypassing was determined by Bruun and Gerritsen to be a function of the tidal prism of the inlet and the amount of littoral drift. They introduced the r factor to distinguish the two cases, where $r = Q/Q_f$, where Q is the littoral drift in, say, cubic yards per year, and Q_f is the maximum spring tide flow rate into the inlet in cubic yards per second. If r is greater than 200–300, then the littoral drift is very large with respect to the flows in the inlet, and bypassing on the ebb tidal bar occurs predominantly. Alternatively, for r values less than 10–20, the material is bypassed by tidal action.

13.4 SEDIMENTARY RELATIONSHIPS AT INLETS

Some of the material scoured out by tidal currents in inlets is transported seaward, where it is deposited in the ebb tidal shoal, and some is transported landward and deposited in the flood tidal shoals. There appears to be no counterbalancing forces that limit the growth of the flood tidal shoal except perhaps the greater currents that occur with the growth of the shoal, reducing the water depth. However, as the flood tidal shoal continues its growth, usually the main channel will migrate to a more hydraulically efficient location, and perhaps the inlet will also migrate. On the other hand, as the ebb tidal shoal increases in volume, it becomes greater in extent and shallower and thus is more susceptible to wave action that tends to transport the sediment back into the nearshore system. On the basis of this argument, it is reasonable to anticipate that the volumes of ebb tidal shoals approach equilibrium sizes.

To evaluate this hypothesis, Walton and Adams (1976) computed the volume of sand in the ebb tidal shoals of 44 inlets and plotted them versus the tidal prisms for the inlets. Shoal volumes were calculated using a procedure developed and applied by Dean and Walton (1975) in which the ideal "no-inlet" bathymetry is constructed based on interpolating the adjacent bathymetric contours under the ebb tidal shoals, and the ebb tidal volume is considered to be that sand volume residing above these no-inlet contours. Walton and Adams categorized their results by wave exposure and showed that there is a relationship between the tidal prism and the ebb tidal shoal volume for each category. The measure of the exposure of the inlet was taken to be the quantity H^2T^2 determined from nearshore (15–20-ft depth) wave gages. This quantity can be interpreted as a measure of the energy per wave. For highly exposed coasts, H^2T^2 was taken to be greater than 300; for mildly exposed coasts, H^2T^2 was less than 30. Their empirical relationships are of the form

$$V = a \, \Omega^{1.23}, \tag{13.75}$$

where V is measured in cubic yards, and the tidal prism Ω is in cubic feet. The parameter $a = 8.7 \times 10^{-5}, 10.5 \times 10^{-5}, 13.8 \times 10^{-5}$, respectively, from high to low wave exposure. Figure 13.18 shows $a = 10.7 \times 10^{-5}$, which is the best fit to all the data that are also shown as well. The larger the tidal prism of the inlet, the larger the ebb tidal flows; hence, the farther offshore sediment is carried by the currents. High wave action will move this material ashore, and thus inlets on mildly exposed coasts can have larger bars than inlets on more highly exposed coasts. An important use of this relationship is to determine, for nonequilibrium inlets or for new inlets (either naturally breached during a storm or created through dredging), the potential volume of sand eventually to be transported offshore into the ebb tidal shoal. (Note that there is also a large volume of sediment lost to flood tidal shoals.) Most of the sand that comprises the shoal is derived from the neighboring beaches and is carried to the inlet by the longshore transport. Therefore, these ebb tidal shoals directly represent the amount of material lost from the beaches.

Escoffier (1940) developed a method for evaluating the stability of inlets by recognizing that, on the basis of inlet hydraulics, the maximum velocity through an inlet increases with cross-sectional area A_c. With increasing A_c, the maximum velocity reaches a peak value for some intermediate value of A_c, denoted $A_{c,c}$, and then decreases for larger cross-sectional areas. Look again at Figure 13.11. He then argued that there was a stable maximum tidal velocity through the inlet that would scour out any excess sand carried into it by wind and waves. This was assumed to be about

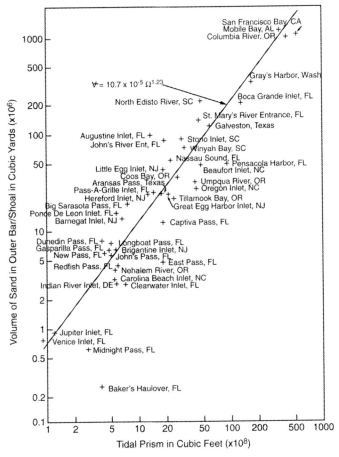

Figure 13.18 Ebb tidal shoal volume versus tidal prism (from Walton and Adams 1976).

3 ft/s based on grain sizes in inlets. Plotting this horizontal line on Figure 13.11 gives two points of intersection: one an unstable equilibrium cross section and the other a stable one; call them A_{c_1} and A_{c_2}. Escoffier contended that an inlet with a cross section of A_{c_1} would have the scouring velocity needed for a stable inlet, but any perturbations in that cross section would set into motion forces causing it either to grow smaller or larger.

To examine this further, O'Brien's stability relationships can be converted into a maximum tidal velocity requirement (O'Brien and Dean 1972) to improve upon Escoffier's assumed 3 ft/s by introducing the following definition for the tidal prism:

$$\Omega = A_c \int U_{max} \cos(\sigma t - \epsilon_U) dt,$$

where the integral is over the half of the tide cycle corresponding to flood flow. This gives

$$\Omega = \frac{A_c U_{max} T}{\pi}$$

Substituting this into O'Brien's relationship for jettied inlets, Eq. (13.68), and re-arranging to solve for U_{max} gives the interesting result for semidiurnal tides ($T = 44,700$ s) that $U_{max} = 3.5$ ft/s. For the natural inlets, the maximum velocity is given by

$$U_{max} = 2.13 \times 10^{-4} A_c^{0.3} \tag{13.76}$$

Therefore, for a stable inlet, the maximum velocities over a tidal cycle should be given by these relationships in order that sand carried into the inlet by waves will be scoured out during a tidal cycle. This sedimentary stability relationship can then be plotted along with the hydraulics relationship to give Escoffier's two solutions A_{c_1} and A_{c_2}.

O'Brien and Dean, following Escoffier's inlet stability argument (Figure 13.11), made the case that inlets on the small A_c side of the peak velocity were unstable and those with A_c values larger than $A_{c,c}$ values were stable. The basis for this stability argument is that if $A_c < A_{c,c}$, and, if a storm transports a large amount of sand into the inlet, the cross section is reduced. Moving to the left on the hydraulic curve in-dicates that a smaller cross section means reduced velocities; thus, the inlet has less capability to flush out the sand. Eventually the inlet can close unless it is dredged. In-lets in this regime are therefore unstable to closure forces. Alternatively, if $A_c > A_{c,c}$, then the reduction in cross section due to a storm increases the velocity in the inlet and hence results in an increased ability of the inlet to flush out the sand. These inlets are therefore stable. In application, this approach to inlet stability is complicated by variations in tidal ranges, uncertainty in the magnitude and sequencing of storms, and other factors.

Van de Kreeke (1992) has pointed out that "stable" inlet cross sections in nature vary with time, and a stable cross section can lie between the two equilibrium points: $A_{c_1} < A < A_{c_2}$. As long as the cross-sectional area of the inlet is not reduced below A_{c_1}, should be stable, for the inlet will have a velocity exceeding that given by the sedimentary stability criterion. This assumption that A is always greater than A_{c_1} implies that no major sand influxes to the inlet can occur.

EXAMPLE 2: THE FRIESIAN ISLANDS

Off the northern coast of the Netherlands and Germany is a barrier island chain separated from the mainland by the Wadden Sea, which is a shallow sea with many shoals and sandbanks. Each of the inlets between these islands provides a portion of the tidal prism for the Wadden Sea. In 1969, the Dutch closed off the Lauwerszee, increasing the arable land area but reducing the tidal prism for the Het Friesche Seegat (the principal inlet serving that section of the Wadden Sea). The corresponding decrease in tidal prism amounted to 105 million m^3 (from 305 to 200 million m^3) according to Steijn (1991). Almost immediately, the Wadden Sea began to shoal locally with 7 million m^3 of new sediment, the ebb tidal shoal began to shrink, the inlet's geometry began to change, and a spit started growing from the eastern side of the inlet (from the island Schiermonnikoog). By 1990, Steijn reported that this spit had grown to a length of 4 km with a width of 100 m. (We thank Ap van Dongeren for pointing out this example.) FitzGerald and Penland (1987) pointed out a similar example of the effect of polders on the Friesian Islands in Germany.

13.5 SAND BYPASSING AT INLETS

As noted before, inlets modified for navigational purposes have a substantially reduced capacity for natural sand bypassing because their cross section may be too wide and sand deposition takes place within the deepened inlet. Maintenance dredging of this material often results in its placement in areas other than downdrift of the inlet, as nature intended. New inlets also have a similar, capacity to interrupt the natural sand transport processes because, in part, the natural pathways for sand have not yet developed and the ebb and flood tidal shoals must be created.

In recognition of the deleterious effects of entrances, the reinstatement of the natural net transport past these entrances has been recognized increasingly as a high coastal engineering priority. Often, sand bypassing is only a partial solution, for the amount of sand available may be inadequate for all interests. Thus, bypassing may restore equity but may still leave inadequate sand resources.

In view of the investment along the nation's shoreline, it is quite surprising that more attention has not been directed toward the technology, improvement, and installation of sand transfer systems at tidal inlets. Sand transfer systems and costs in Florida were reviewed by Jones and Mehta (1980). Several sand transfer system types and individual installations are reviewed in the following subsections.

13.5.1 FIXED SAND TRANSFER SYSTEMS

The earliest U.S. system of this type was installed at South Lake Worth Inlet, Florida (also known as Boynton Inlet), in 1935. The inlet is artificial and was cut in 1927 across a narrow ridge separating Lake Worth from the Atlantic Ocean. Prior to installation of the sand transfer plant, the downdrift beaches eroded rapidly in response to the newly constructed inlet with its short jetties. The sand transfer plant is essentially a fixed dredge pump mounted inside a large housing and draws sand through a nozzle suspended from a movable boom through the pump and across the inlet via a bridge to a downdrift discharge point. Rock underlies the intake area, limiting the depth of available sand. Yeend and Hatheway (1988) have provided a review of the characteristics of this system, which is shown in Figure 13.19.

Because the initial jetty lengths were short and allowed sand to enter and deposit in Lake Worth and did not provide the desired navigational protection, a plan was developed for their extension. The north jetty was extended in 1965, and the residents to the north were effective in relocating the plant approximately 40 m seaward, where it could reach less sand! Recent litigation initiated by the affected downdrift community has resulted in a potentially precedent-setting court order that has brought about the restoration of the downdrift beaches and steps to reinstate the natural sediment flows.

A second fixed sand transfer system is located at Lake Worth Inlet, Florida (also known as Palm Beach Inlet), approximately 25 km to the north of South Lake Worth Inlet. This system was installed in 1948 and is of the same general type as at South Lake Worth Inlet. However, because there is no bridge across this inlet, the discharge line is submerged beneath the inlet, emerges at the south jetty, and continues to a discharge point.

Figure 13.19   South Lake Worth Inlet, Florida. Note the sand transfer plant on the north jetty (from Yeend and Hatheway 1988).

This is one of the inlets noted earlier in which there is insufficient sand to meet the desires of the updrift and downdrift interests, and thus a tug of war over these insufficient sand resources has ensued. Possible causes of the insufficient sand supply are the dredging and disposal in deep water of 2.3 million m³ of sand from the entrance and the dredging and storage of a similar amount to form an island in Lake Worth. At this bypassing plant, the updrift interests were successful in requiring the driving of steel-sheet piling around the plant intake to limit its effectiveness!

In summary, the effectiveness of each of the two fixed bypass plants is not technologically limited but rather has been restricted by politically based action that reduced their capability to fulfill their design objectives.

13.5.2 WEIR JETTY SYSTEMS

Weir jetty systems include (1) a low section built into the updrift jetty that is designed to admit the net longshore sediment transport into (2) a deposition basin within which the sediment is stored until it is later transferred by floating dredge to the downdrift shoreline. A schematic of a weir jetty system is shown in Figure 13.20. Weggel (1983)

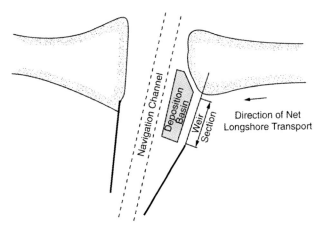

Figure 13.20 Schematic of a weir jetty system.

has presented a review of design considerations of weir jetty systems, and Seabergh (1983) has described the results of a physical model study of Murrells Inlet, South Carolina.

The first weir jetty system was constructed at Hillsboro Inlet, Florida, where the weir section is formed by a natural seaward-dipping limestone reef extending obliquely to the shoreline. An aerial view of the inlet is shown in Figure 13.21. The outer portion of the updrift jetty is a rock structure that rests on the reef. This system is operated by an inlet district authorized legally to levy taxes for its operation; operation has been under way for nearly three decades and has been quite successful.

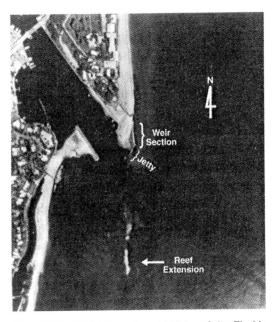

Figure 13.21 Weir jetty system at Hillsboro Inlet, Florida
(from Dean 1988).

A small weir jetty system was built at Boca Raton Inlet, Florida, and is described by Spadoni, Beumel, and Campbell (1983). Initially, the weir section in the updrift (north) jetty was closed by cement-filled sandbags. The updrift and downdrift shorelines responded rapidly by advancing and receding, respectively, threatening a pavilion located at a county park immediately downdrift of the inlet. To alleviate this erosion, the sand bags closing the weir were removed, and a small dedicated dredge now transfers sand to the downdrift beaches. The system appears quite effective, although some sand is swept from the entrance channel and deposition basin to the ebb tidal shoal, which is dredged periodically; the sand is placed on the downdrift beach.

Rudee Inlet, Virginia, located some 20 km south of the entrance to the Chesapeake Bay, also has a weir jetty system. The net transport there is toward the north, and the deposition basin is located between an original south jetty and, to the south, a more recent jetty for which the shoreward section functions as a weir. Sand transfer is effected by a combination of (1) a floating dredge that removes the sediment, bypassing the deposition basin and settling into the inlet channel, and (2) a movable jet pump system operating within the deposition basin.

Other examples of weir jetties that have performed with less success include

1. Masonboro, North Carolina, where, prior to the construction of the south jetty, the main channel migrated through the deposition basin and against the north jetty, threatening its stability.
2. Destin Pass, Florida, where the direction of the longshore transport was assessed incorrectly, leading to the installation of the weir section on the wrong (west) side of the pass.
3. Ponce de Leon Inlet, Florida, where again the channel migrated through the deposition basin and there were reports that the weir jetty had caused a drawdown of the updrift shoreline, leading to the deactivation of the weir section by encasing it in large stones. This inlet has been discussed in Section 1.1.3.
4. Perdido Pass, Alabama, where the discharge of the transferred material is limited to an easement very close proximity to the downdrift (west) jetty.

13.5.3 JET PUMPS

Jet pumps appear to have the potential to contribute effectively to the coastal engineer's goals of sand bypassing. They have no moving parts but are instead driven by a high-pressure jet of water, and they can be buried and restarted without clogging. These devices have been installed in depositional basins and in navigational channels, where shoaling can be critical. Also "swimming" jet pumps have been mounted on a flexible boom with propulsion provided by landward-directed jets and buoyancy controlled by ballast tanks.

The practical nemeses of jet pumps have been (1) the debris in the sand that they entrain, and (2) wear due to sand abrasion. The debris can clog the jet pumps or can lead to an enveloping matrix that decreases the hydraulic efficiency of the intake. To counter the problem of wear, the use of abrasion-resistant materials is being explored, but most jet pumps provide no more than 300 hours of active service before replacement is required.

Figure 13.22 Indian River Inlet, Delaware, bypass system. The jet eductor pump is suspended from the crane and is lowered into the beachface, forming a temporary excavation crater.

The jet pump plant at Indian River Inlet, Delaware, is one of the most successful installations employing this technology (Clausner et al. 1991). In this unique design, the jet pump is movable; it is suspended by a crane-mounted boom and lowered onto the beach near the waterline, where it quickly excavates a hole; in this application the sand–water mixture is transferred by pipeline across the Indian River Inlet bridge to the downdrift (north) shoreline, as shown in Figure 13.22. When not in use, the jet pump is lifted from the excavation pit, and the crane moves behind the dune line – away from the effects of storms and interference with recreational beach users. This system functions extremely well owing, in part, to the easy accessibility to the jet pump and the ability to access wide areas of the beach. If debris concentrates within the removal areas, it can be removed using a clam shell bucket on the same boom.

A jet pump bypassing system has also been installed at Nerang River Entrance in Queensland, Australia (Coughlan and Robinson 1990). This system is quite innovative and also quite elaborate, as seen in Figure 13.23. Ten jet pump eductors are suspended rigidly from a pier constructed as a part of the bypassing system. The eductors elevate the water–sediment mixture to a sloping trough, where it flows by gravity to a collection sump at the landward end of the pier. From this point a standard centrifugal pump is used to transfer the sand under the Nerang River Entrance. Innovative features of the system are that it is completely automatic and operates unattended at night to take advantage of the less expensive electricity rates. The operation is computer driven, and combinations of 4 of the 10 eductors operate simultaneously. If the monitored pressures indicate that a particular eductor has exhausted the sand available to it, it is taken off-line automatically, and another

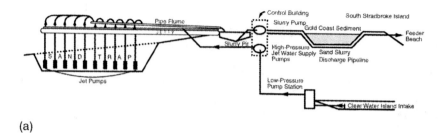

(a)

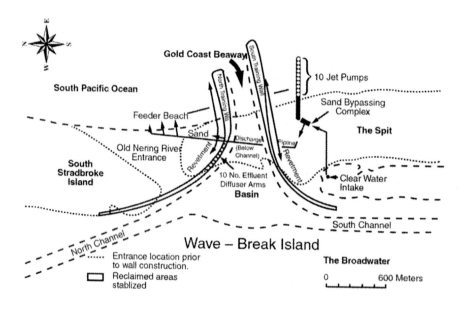

(b)

Figure 13.23 Profile of pier and jet pump bypassing system, Nerang, Australia (from Coughlan and Robinson 1990). (a) Schematic of sand transfer system; (b) layout of Nering River entrance.

eductor is activated. If the monitoring identifies conditions that could be deleterious to the system, the computer shuts it down and dials a responsible party at his or her residence. The Nerang system has been in operation since 1986. Early performance was good; however, debris has collected within the area of influence of the eductors and has resulted in a decrease in operating efficiency with a concomitant increase in the cost of bypassing sand. Also, the eductors were expected to have an operating life of 600 hours, whereas experience has shown it necessary to replace them after 300 to 400 hours of use.

Other areas where jet pumps have been installed include Rudee Inlet, Virginia, where, as just mentioned, jet pumps are used in combination with a floating dredge; Mexico Beach Inlet, Florida, a very small inlet; and Santa Cruz Harbor entrance, California, where this installation is not yet operational despite many years of trying.

In summary of the effectiveness and prognosis for jet pump usage in coastal engineering, it appears that fixed systems have definite limitations imposed by debris clogging and wear. To address both of these problems, it is highly desirable that the system intake be movable and accessible for changing eductors and debris removal.

13.5.4 DREDGING TO MAINTAIN CHANNEL ENTRANCE DEPTHS

Perhaps the most common means of bypassing is simply placing the dredged material removed for navigational maintenance on the downdrift shoreline. This can usually be accomplished readily if a conventional pipeline dredge is employed; however, if a hopper dredge is used, it must be equipped with a pump-out capability to place the sand on the beach.

Historically, material dredged from inlets has been disposed of in deep water in the United States owing to the lower cost of offshore disposal. However, this material almost never returns to the beaches, resulting in a deficit of beach sand, and, consequently, beach erosion. Fortunately, the inherent value of beach sand is recognized now, and the offshore disposal and wasting of sand is being curtailed within the United States.

13.6 INLET DESIGN CONSIDERATIONS

As indicated in earlier sections of this chapter, inlet design requires considerations of inlet stability and impact on the adjacent shorelines.

13.6.1 INLET STABILITY

If an inlet is to be stable against closure without maintenance dredging, it is essential that the characteristics of the bay system and natural ocean tidal range be such that the inlet will satisfy the requirements of the inlet stability curve (Figure 13.11), that is, the inlet must fall on the stable side of the peak, and the associated velocity must be in accordance with O'Brien's finding that the peak velocities be on the order of 3.5 ft/s.

A complicating factor is that tidal ranges vary over a fortnight and may meet those requirements during some, but not all, of the tidal oscillations. In these cases, there has been no definite percentage of time established that the preceding requirements must be satisfied. Rather, the design process should recognize that, during times of low velocities, the inlet is more subject to closure, and, in the event of storms, the inlet must have sufficient capacity to store the sand carried in by the storm without closing and to allow the tidal currents afer the storm to scour the channel to the stable cross-sectional area. It follows that inlets with small cross-sectional areas are more susceptible to storm closure than larger inlets.

If an inlet is subject to closure, or if it is anticipated that the dredging requirements will be excessive, it may be necessary to include one or more jetties in the design.

13.6.2 ADJACENT SHORELINE IMPACT

The impacts of a new inlet on the adjacent shoreline can be due to a reduction in sediment supply (by trapping on shoals and at structures or dredging) or to a redistribution of the sediment present. Inlets with ebb and flood currents and possibly jetty systems will modify the wave energy and direction along the adjacent shorelines. Even if there is no *net* reduction in sediment volume, the shoreline response to modified wave conditions can include local erosion and deposition.

The more serious concern is if a new inlet might result in a net loss of sediment to the nearshore system. For example, the volume of sand required to build the ebb tidal shoal to equilibrium will, of course, be derived from the nearshore system and can cause substantial beach erosion. Material stored in flood tidal shoals likewise represents a loss of sediment to the nearshore system as does material removed permanently from the nearshore system, perhaps by dredging of a shoaled channel by hopper dredge with placement in water depths too great for the sand to return to shore within engineering time frames.

To minimize the impact to the adjacent shorelines, unless a thorough analysis demonstrates the net longshore sediment transport to be negligible, all inlet designs and funding considerations should provide for sand bypassing, monitoring, and some allowance for reasonable modifications of the bypassing design.

13.7 AN EXAMPLE

Oregon Inlet in North Carolina can be considered one of the most dynamic inlets on the East Coast of the United States. The net annual longshore sediment transport at this inlet has been estimated at 700,000 m^3 and, historically, the inlet has migrated southward at an average rate of 75 m per year (Inman and Dolan 1989). One impact of this migration is that more than a third of the 4-km-long bridge spanning the inlet is now over dry land that has been deposited in the inlet since the bridge's construction. This migration has also threatened the south bridge abutment and required lengthening some of the support piles as the deeper portions of the inlet migrated south. Over the past three decades, the Corps of Engineers has pursued the possibility of stabilizing the inlet with jetties. An obstacle to the construction project has been the National Park Service (NPS), which is the custodian of the two adjacent barrier islands and is generally committed to maintaining the property in as natural condition as possible. The NPS has maintained that the jetties are unnecessary and that the channel could be maintained solely through dredging. Additionally, the policy of the state of North Carolina is opposition to coastal structures. However, by 1990, the threat to the south bridge abutment had become so great that the state and the NPS agreed to the construction of a so-called terminal structure on the south island. Coincidentally, the alignment of this bridge abutment would serve as the base of the south jetty if it were constructed. Continuing problems plaguing the jetty construction have been the large costs, disagreements over the benefits, and concerns about the secondary effects of a navigational channel constructed to facilitate fishing boat access on the fishery stocks. Initially estimated at $60 million, current estimates are in excess of $150 million. In response to an NPS concern, the Corps has developed an elaborate sand management plan to be implemented if the jetties are constructed. Previously, most of the sand dredged from the inlet was placed offshore; however, more recently, all of the sand has been placed on the beach or in shallow water – a practice, which, in conjunction with the presence of the south terminal structure, has resulted in an areal increase of 280,000 m^3 over the northerly 3.3 km of shoreline during the period from 1989 to 1992 (Overton et al. 1992). This example illustrates the complex web of technical, economic, social, and political issues with which modern coastal engineers must contend. The question is now, Should the project be constructed? rather than Can the project be constructed?

REFERENCES

Aubrey, D.G., and P.E. Speer, "A Study of Nonlinear Tidal Propagation in Shallow Inlet/ Estuarine Systems. Part I: Observations," *Est. Coast. Shelf Science*, 21, 185–205, 1985.

Boon, J.D., and R.J. Byrne, "On Basin Hypsometry and the Morphodynamic Response of Coastal Inlet Systems," *Marine Geol.*, 40, 27–80, 1981.

Brown, E.I., "Flow of Water in Tidal Canals," *Proc. ASCE*, 96, 747–834, 1932.

Bruun, P., and F. Gerritsen, "By-passing of Sand by Natural Action at Coastal Inlets and Passes," *J. Waterways and Harbors Div.*, ASCE, 85, 1644-1 to 1644-49, 1959.

Bruun, P., and F. Gerritsen, *Stability of Coastal Inlets*, Amsterdam: North-Holland Pub. Co., 1966.

Clausner, J.E., J.A. Gebert, A.T. Rambo, and K.D. Watson, "Sand Bypassing at Indian River Inlet, DE" *Proc. Coastal Sediments '91*, ASCE, Seattle, 1177–1191, 1991.

Coughlan, P.M., and D.A. Robinson, "The Gold Coast Seaway, Queensland, Australia," *Shore and Beach*, 58, 1, 9–16, 1990.

Dean, R.G., "Sediment Interaction at Modified Coastal Inlets: Processes and Policies," in *Hydrodynamics and Sediment Dynamics of Tidal Inlets*, D.G. Aubrey and L. Weishar, eds., New York: Springer-Verlag, 412–439, 1988.

Dean, R.G., and T.L. Walton, "Sediment Transport Processes in the Vicinity of Inlets with Special Reference to Sand Trapping," *Estuarine Research*, II, New York Academic Press, 129–149, 1975.

DiLorenzo, J.L., "The Overtide and Filtering Response of Small Inlet/Bay Systems," in *Hydrodynamics and Sediment Dynamics of Tidal Inlets*, D.G. Aubrey and L. Weishar, eds., New York: Springer-Verlag, 24–53, 1988.

Escoffier, F.F., "The Stability of Tidal Inlets," *Shore and Beach*, 8, 4, 114–115, 1940.

Escoffier, F.F., and T.L. Walton, "Inlet Stability Solutions for Tributary Flow," *J. Waterway, Port, Coastal, and Ocean Eng.*, 104, WW4, 341–355, 1979.

FitzGerald, D.M., and S. Penland, "Backbarrier Dynamics of the East Friesian Islands," *J. Sedim. Petrol.*, 57, 4, 746–754, 1987.

Inman, D.L., and R. Dolan, "The Outer Banks of North Carolina: Budget of Sediment and Inlet Dynamics Along a Migrating Barrier System," *J. Coastal Res.*, 5, 23, 193–238, 1989.

Jarrett, J.T., "Tidal Prism-Inlet Area Relationships," GITI Rpt. 3, U.S. Army Coastal Engineering Research Center, 76 pp., 1976.

Jones, C.P., and A.J. Mehta, "Inlet Sand Bypassing Systems in Florida," *Shore and Beach*, 48, 1, 25–34, 1980.

Keulegan, G.H., "Tidal Flow in Entrances," U.S. Army Corps of Engineers, Committee on Tidal Hydraulics, Tech. Bull. 14, Vicksburg, 1967.

Lorentz, H.A., "Report of the State Commission Zuiderzee, 1918–1925," (in Dutch), 1925.

Marino, J.N., and A.J. Mehta, "Sediment Trapping at Florida's East Coast Inlets," in *Hydrodynamics and Sediment Dynamics of Tidal Inlets*, D.G. Aubrey and L. Weishar, eds., New York: Springer-Verlag, 284–296, 1988.

O'Brien, M.P., "Estuary Tidal Prisms Related to Entrance Areas," *Civil Engineering*, 1, 8, 738–739, 1931.

O'Brien, M.P., "Equilibrium Flow Areas of Inlets on Sandy Coasts," *J. Waterways and Harbors Div.*, ASCE, 95, WW1, 43–52, 1969.

O'Brien, M.P., and R.G. Dean, "Hydraulics and Sedimentary Stability of Coastal Inlets," *Proc. 13th Intl. Conf. Coastal Eng.*, ASCE, Vancouver, 761–780, 1972.

Overton, M.F., J.S. Fisher, W.A. Dennis, and H.C. Miller, "Shoreline Change at Oregon Inlet Terminal Groin," *Proc. 23rd Intl. Conf. Coastal Eng.*, ASCE, Venice, 2332–2343, 1992.

Seabergh, W.C., "Physical Model Study of Weir Jetty Design," *Proc. Specialty Conf. Coastal Structures '83*, ASCE, 876–893, 1983.

Spadoni, R.H., N.H. Beumel, and T.J. Campbell, "The Design and Performance of a Weir at Boca Raton Inlet, Florida," *Proc. Specialty Conf. Coastal Structures '83*, ASCE, 894–901, 1983.

Speer, P.E., and D.G. Aubrey, "A Study of Nonlinear Tidal Propagation in Shallow Inlet/Estuarine Systems," *Est. Coast. Shelf Science*, 21, 207–224, 1985.

Steijn, R.C., "Some Considerations on Tidal Inlets," Literature Survey H 840.45, Delft Hydraulics, May, 130 pp., 1991.

van de Kreeke, J., "Stability of Inlets; Escoffier's Analysis," *Shore and Beach*, 60, 1, 9–12, 1992.

Walton, T.L., Jr., and W.D. Adams, "Capacity of Inlet Outer Bars to Store Sand," *Proc. 15th Intl. Conf. Coastal Eng.*, ASCE, Honolulu, 1919–1937, 1976.

Weggel, J.R., "The Design of Weir Sand By-passing Systems," *Proc. Specialty Conf. on Coastal Structures '83*, ASCE, 860–875, 1983.

Yeend, J.S., and D.J. Hatheway, "Quantification of Sand Transport and Sand Bypassing at South Lake Worth Inlet, Palm Beach County, Florida," *Proc. 21st Intl. Conf. Coastal Eng.*, 2772–2783, 1988.

EXERCISES

13.1 Explain qualitatively why inlets are more widely separated on a coastline with a small tide range than on one with a large tide range.

13.2 Is there a stable inlet that can be constructed on a shoreline with a diurnal tide range of 3 m to connect a lake with the ocean? If so, would it be navigable? The lake, with a plan area of 2×10^6 m², is separated from the ocean by a 1000-m-wide barrier. The littoral transport has been estimated at 100,000 m³ per year. State your assumptions.

13.3 If a new tidal inlet is excavated, it is known that the ebb tidal shoals will grow and eventually achieve an equilibrium.

(a) Describe the processes that result in an equilibrium volume being reached for the shoals; that is, Why do they grow and what limits their growth? For simplicity, assume that the waves approach the shore normally.

(b) Suppose that an inlet with mature ebb tidal shoals is closed. Again, consider normal wave incidence. Discuss the fate of the shoal material. Sketch several stages of the shoreline planform.

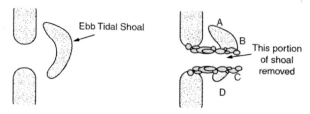

Figure 13.24 Figure for Problem 13.3.

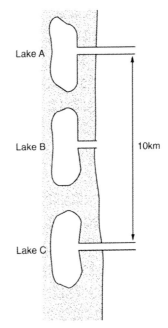

Lake A

Lake B

10km

Figure 13.25 Figure for Problem 13.4.

Lake C

(c) If the ebb tidal shoals appear as shown in Figure 13.24 for natural conditions and two jetties are installed, what will happen to the portions of the shoals labeled AB and CD?

13.4 Lakes A and C in Figure 13.25 have been opened for navigational purposes, and the inlets have stabilized. Plans exist also to open Lake B, with its plan area of 8×10^6 m^2, although no jetties are contemplated. A stability analysis shows that the inlet will be stable with a repletion coefficient of $K = 0.4$. The ocean tidal amplitude is 1.0 m. The tidal period is 12.4 h.

(a) Determine the following for Lake B: bay tidal amplitude, inlet cross section, and maximum inflow velocity.

(b) Would you expect the establishment of an inlet to Lake B to have any adverse effects on the adjacent beaches? Explain.

(c) Would the construction of jetties at Lake B reduce any adverse effects to the adjacent beaches? Explain.

(d) If the inlet without jetties will cause any erosion of the adjacent beaches, calculate this erosion for an effective wave height of 3 m, a berm height of 2 m and a moderately exposed coastline.

CHAPTER FOURTEEN

Shoreline Management

Hurricane Hugo, a Category 4 storm,* made landfall slightly north of the entrance to Charleston Harbor, South Carolina, in September 1989. The storm tide, based on high water marks, peaked in excess of 20 ft, which is second only to the 22.4-ft. storm tide caused by Hurricane Camille (1969) at Pass Christian, Mississippi.

The development along the impacted area of South Carolina had taken place over several decades under generally inadequate structural and setback controls. Moreover, the erosion rates on some portions of the South Carolina shoreline are moderately high, causing older structures to be near the shoreline.

Although no structures built to modern standards suffered major damage, Hurricane Hugo destroyed or caused severe damage to in excess of 15,000 structures. Of these, about 150 were considered to be unbuildable under the management policies in place at the time. (Shore & Beach, 58, 4, 1990, has a number of articles about Hugo's effects.) There has been considerable litigation over whether and under what conditions these structures should be rebuilt. The total damages caused by Hugo exceeded $4.0 billion.

With most of the U.S. shoreline heavily developed, the preceding scenario will recur, and it is important that national and local management policies be in place to regulate construction near the shoreline and reconstruction following storms so that future disaster liabilities and personal hardships will be minimized.

14.1 INTRODUCTION

The current pressure for development along much of the nation's shoreline, coupled with the prognosis for increased rates of sea level rise (Chapter 3) and the possibility of increased storminess due to the Greenhouse effect, have raised questions about the most appropriate policies for coastal zone management. On an eroding coastline, the options are surprisingly few:

1. Retreat by abandoning shoreline development (present and future) or moving it landward at the same rate as the shoreline recession.

---

* The Saffir–Simpson Scale for hurricanes was discussed in Chapter 3.

2. Nourish the beach to stabilize it.
3. Protect the beach using hard structures such as groins, seawalls, or revetments.

Not surprisingly, the long-term background erosion rate and its causes are important determinants in identifying appropriate options. In addition, we must keep in mind that the time scales of development and occupancy along a shoreline are related more to scales of human lifetimes rather than geological time.

Armoring the beach with seawalls or revetments is generally looked on as causing a variety of adverse shoreline impacts. Some of these criticisms are deserved, although the general assessment is probably overly critical. In cases in which the erosion is caused by human activities, such as jetty construction, matters of equity emerge.

Federal legislation in the United States has attempted to reduce losses on an eroding shoreline through insurance. For example, the Upton–Jones Amendment of 1988 provided for reimbursement of up to 40 percent of the value of the structure if it is relocated appropriately once the structure is in imminent danger of collapse. Alternatively, the home owner could collect 110 percent of the structure value by demolishing the property before its falling victim to erosion.

This chapter addresses technical issues of national shoreline management and attempts to provide conceptual guidelines for selection of proper alternatives.

14.2 OPTIONS AND FACTORS

14.2.1 OPTIONS

The effects on an eroding shoreline of the three possible response options mentioned above are illustrated in Figure 14.1. Beach nourishment is the only option that can result in moving the shoreline seaward. Armoring can hold the shoreline at a fixed location, whereas retreat does not affect the shoreline recession rate.

The incorporation of the most appropriate of the three options into a rational shoreline management plan is made more difficult by the probability of an unquantified future increase in the rate of sea level rise.

Each of the options identified above has associated with it a host of technical, economic, and social considerations, all of which are difficult to apply.

14.2.2 FACTORS

The long-term shoreline change rate (LTSCR), including both beach advancement and recession, is the prime factor among those that should be considered in developing a management plan. The LTSCR is the average historical advancement or retreat rate of the shoreline measured in m/yr or ft/yr. Obviously, if the LTSCR is high and negative (recession), the balance is tipped away from shoreline stabilization toward retreat. On the other hand, if the negative LTSCR is small, then, with a relatively small amount of beach nourishment, the shoreline can be kept stable and retreat is a much less attractive option. However, even though the past LTSCR has been small, it is often argued that, with accelerated sea level rise rates, future erosion will be much more rapid; but, from a practical point of view, the uncertainties of these

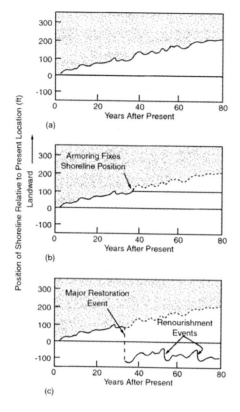

Figure 14.1 The three options available for a property owner on an eroding shoreline. The annual erosion rate is about 2.6 ft/yr. (a) Shoreline position versus time with "retreat" option. (b) Shoreline position versus time with "armoring" option. (c) Shoreline position versus time with "restoration" and renourishment option.

future rates are so great as to render their acceptance unrealistic at present. Additionally, all predictions of increased sea level rise rates suggest that the curve will be concave upward and that the smallest increases will occur in the earliest decades ahead followed by more rapid rates (Figure 3.8) well into the future. Moreover, with the character of the tide gauge results, it may not be possible to confirm conclusively that an acceleration in sea level rise has actually occurred until possibly three or more decades into the future.

Even neglecting *increases* in the rate of sea level rise, the planner is confronted with very real and practical problems in shoreline management. An attempt is made in the following paragraphs to illustrate the manner in which the factors affect the optimum response option. These factors are summarized in Table 14.1.

14.2.2.1 Long-Term Shoreline Change Rates

As mentioned earlier in this section, the LTSCR is the most important factor in identifying the most appropriate option. Although difficult to quantify, if the LTSCR is smaller than, say, -0.7 m/yr and the shoreline has at least a moderate amount of development, then the shoreline should be considered as a reasonable candidate for beach nourishment.

As an example of a situation in which a combination of beach nourishment and structural control can be effective, consider the case of a rapidly eroding area downdrift of an inlet that may have been deepened for navigation. Usually, in such cases,

Table 14.1   Factors Affecting Three Shoreline Management Options

		Option		
Factor	Magnitude	Retreat	Nourish	Armor
Long-Term Erosion Rate	High	X^a		
	Low		X	
Upland Economic	High		X	
Base	Low	X		
Protecting Historic Structures			X	X
Protecting Vital Infrastructure			X	X
Equity Matters Involved			X	X

^{a} Unless reduction in LTSCR possible.

the erosion is due to trapping of sand by the updrift jetty or sand transport into the inlet system or the ebb tidal shoal, where it is lost to the beach. It may be possible to nourish the area and stabilize against losses by constructing a relatively short terminal structure adjacent to the inlet. Such structures have proven very effective along portions of the west coast of Florida where the net longshore transport is minimal and thus the adjacent shorelines are not affected adversely. As a longer-term solution, a sand-bypassing plant could be constructed at the inlet.

Areas with LTSCR greater than −2 or −3 m/yr may not be good candidates for nourishment unless retention structures are used to stabilize the project or there is an unusually high upland economic base that justifies the high cost of stabilization. One other possibility for these eroding shorelines is to identify and remedy, the cause of the erosion and using nourishment to repair the beach.

14.2.2.2 Economic Base

Beaches can serve as a valuable asset to any community. Some cities, such as Miami Beach, Florida, can derive a tremendous source of income from these beaches. Miami Beach attracts 21 million visitors a year who spend on the order of $2 billion USD annually. In fact, in the United States, tourism to beaches provides a significant fraction of the foreign trade ($170 billion annually), and more visitors are attracted to U.S. beaches than its National Parks (Houston 1996). Other countries blessed with beautiful beaches also derive tremendous income from tourism and developing countries often target tourist development of beaches as a means to attract foreign investment and foreign exchange.

If, however, the cost of maintaining the beaches exceeds the ability, willingness, or both, of the community to provide the necessary funding, beach nourishment may not be a viable option. Figure 14.2 presents, qualitatively, the required relationship between upland economic base and long-term background erosion rate for nourishment to be economically practical.

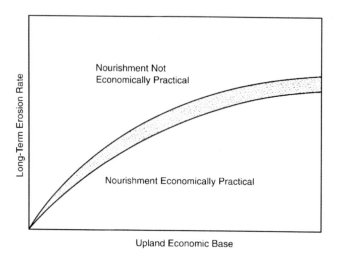

Figure 14.2   Relationship between economical base and LTSCR.

14.2.2.3 Protection of Historic Structures or Vital Infrastructure

When a historic structure or vital infrastructure, such as a storm evacuation route, is jeopardized, economic justification may be a secondary consideration in shoreline maintenance and protection. For example, erosion may threaten an historic monument, possessing an intrinsic value much larger than its economic value. In these cases, it may not be necessary to justify protection on considerations of benefit–cost analyses. Examples include the historic Morro Castle in Puerto Rico, Fort Clinch, Florida, and Fort Fisher, North Carolina. Likewise, it is difficult to place a value on protecting a vital evacuation route that may be essential during times of high erosion potential. In this case, if the threatened area is fairly localized, armoring may be more effective than nourishment, which would require remedial measures over a much longer shoreline segment. As a further specific example of the use of structures, the Cape Hatteras Lighthouse is owned by the National Park Service (NPS). This structure, built in 1870 has been jeopardized off and on for more than 55 years. In recognition of its anticipated short life in 1935, the Lighthouse Commission transferred ownership to the NPS and constructed a functional steel tower replacement located both landward and to the north of the lighthouse. In 1969–70, the U.S. Navy constructed three steel groins to provide protection to a surveillance station located immediately to the north of the lighthouse. The original design was modified slightly by placing the southernmost groin farther south to provide limited protection to the lighthouse. Although great difficulty was experienced in constructing the groins under energetic wave conditions – and the groins have required a moderate amount of repair – they have been very successful in stabilizing the beach owing in large part to the dominant southerly littoral transport at that location. This structure, in jeopardy and essentially abandoned in 1935, is still standing owing, in part, to the three groins constructed for a different purpose than providing stability to the lighthouse.

A National Research Council (1988) study commissioned by the National Park Service to develop recommendations for preserving the lighthouse suggested moving it up to 3700 m inland. This relocation was completed in the summer of 1999. Clearly, on the basis of the available history, an alternate and effective method, which probably would have provided stability for an additional 40 years, would have been to extend the existing groin field farther south by adding another groin. Such a solution was designed and, because there were no structures between the lighthouse and the Cape, the impacts of a new groin would have been minimal. The problem of protecting the lighthouse has, as a common element to other structures, the difficult trade-off between retreat versus structural stabilization or beach nourishment. Additionally, the use of structural measures in a National Park setting poses philosophical and judgmental questions.

14.2.2.4 Equity Considerations

Tidal entrances, modified for navigational purposes, have a substantial capacity to interrupt the natural flow of sand and cause the erosion of neighboring beaches. In practically all cases, these inlets are public works projects either under the jurisdiction of the federal government (U.S. Army Corps of Engineers), the state, or a local public entity. In cases in which property is jeopardized primarily because of erosion caused by a public works project, it seems very reasonable to relax permitting requirements to allow the property to be protected by armoring. Alternatively, a more aesthetic and desirable solution would be to restore the beach with a one-time placement of sand and then to maintain the beach in its near-natural state through bypassing sand around the entrance and placing it on the downdrift beach.

14.3 THE ROLE OF SETBACKS AND CONSTRUCTION STANDARDS

Central goals in shoreline management include (1) reduction of structural and economic losses during storms, and (2) maintenance of a recreational beach for future generations of people and for endangered and other species that nest and live in the beach and dune system. Construction setbacks and construction standards can assist substantially in achieving these goals.

14.3.1 CONSTRUCTION SETBACK

Many coastal states require approval for coastal structure designs. Table 14.2 from the National Research Council (NRC 1990) lists the states with such programs. Some states, for example Florida, regulate the location, size, and structural characteristics of a building allowed on a particular coastal property.

Required construction setbacks generally depend on the beach morphology, the available shore-perpendicular dimensions of the property, and the long-term background shoreline change rate. Many construction setbacks attempt to preserve the integrity and functionality of the beach–dune system. Also, many criteria require that, if the shoreline is eroding, the structure be set back a distance equal to the product of the long-term erosion rate times a prescribed number of years. In most states, beach nourishment is accepted as a means of maintaining the shoreline seaward of a setback line.

Table 14.2 Regulations of Coastal States Concerning Coastal Structures (NRC, 1990)

State/Territory	Recession Rates from Aerial Photos	Recession Rates from Charts	Recession Rates from Ground Surveys	Erosion Setbacks Established*	Reference Feature	Years of Setback	Local Administration	One Foot per Year Standard	Fixed Setback	Floating Setback
Alabama	Y	Y	N	Y	MHW	NA	N	Y	N	NA
Alaska	Y	Y		N	NA	NA	NA	NA	NA	NA
American Samoa	N	N	N	N	NA	NA	NA	NA	NA	NA
California	Y	Y	Y	N	NA	NA	Y	NA	NA	NA
Connecticut	Y	Y		N	NA	NA	NA	NA	NA	NA
Delaware	Y	Y		Y4	TD	NA	Y	N	Y	N
Florida	Y	Y		Y5	NA	30	Y	N	Y	N
Georgia	Y	Y		N	NA	NA	NA	NA	NA	NA
Hawaii	N	N	N	Y	6	N	Y	N	Y	N
Indiana	Y	N	Y	N	NA	NA	NA	Y	NA	NA
Illinois	Y	Y	Y	N	NA	NA	NA	NA	NA	NA
Louisiana	Y	Y	N	N	NA	NA	NA	NA	NA	NA
Maine	N	N	Y	N7	NA	NA	NA	NA	NA	NA
Maryland	Y	Y		N	NA	NA	NA	NA	NA	NA
Massachusetts	Y	Y	N	N	NA	NA	NA	NA	NA	NA
Michigan	Y	N	N	Y	BC2	30	Y	Y	N	Y
Minnesota	Y	N	N	N	NA	NA	NA	Y	N	NA
Mississippi	N	N	N	N	NA	NA	NA	NA	NA	NA

State									
New Hampshire	N	N	N	NA	NA	NA	NA	NA	NA
New Jersey	Y	Y	Y	MHW	50	Y	Y	Y	N
New York	Y	N	Y	BC	30–40	Y	Y	N	Y
North Carolina	Y	N	Y	DC	30–60	NA	NA	NA	NA
N. Mariana's	N	N	N	NA	NA	NA	Y	Y	N
Ohio	Y	Y	N1	BC	30	NA	NA	Y	N
Oregon			N		NA	Y	NA	NA	NA
Pennsylvania	Y	N	Y	BC	50+	Y	Y	Y	Y
Puerto Rico	N	N	N	NA	NA	N	NA	NA	NA
Rhode Island	N	N	Y	CD	30	BL	N7	Y	N
South Carolina	Y	Y	Y	NA	40	NA	Y	Y	NA
Texas	Y	Y	N	NA	NA	NA	NA	NA	NA
Virgin Islands	N	N	N	NA	NA	NA	NA	NA	NA
Virginia	Y	Y	N	MHW	NA	Y	NA	Y	NA
Washington			N	NA	NA	NA	NA	NA	NA
Wisconsin	Y	Y	N3	NA	NA	NA	NA	Y	Y

Note: 1 = setbacks may be established within 2 years; 2 = bluff crest or edge of active erosion; 3 = some counties have setbacks; 4 = has 100 foot setback regulation over new subdivisions and parcels where sufficient room exists landward of setback; 5 = not all counties have coastal construction control lines established; 6 = storm debris line or vegetation line; 7 = 2 feet per year standard.

Y, yes; N, no; NA, not applicable; BC, bluff crest; MHW, mean high water; TD, toe of dune; DC, dune crest, toe of frontal dune or vegetation; BL, base line. A blank means no information was available.

*Most states have setbacks from water line but not based on an erosion hazard.

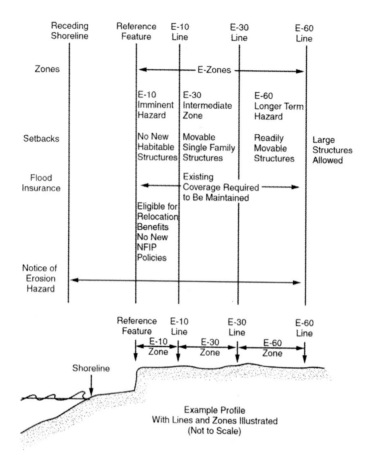

Figure 14.3 National Research Council recommendations (reprinted with permission from NRC 1990. Copyright 1990 by the National Academy of Sciences, courtesy of the National Academy Press, Washington, D.C.), where E-10 denotes a 10-year erosion zone.

North Carolina requires a 30-year setback for smaller structures and a 60-year setback for larger structures. Florida requires a 30-year setback for all structures. The NRC Committee on Coastal Erosion Zone Management (1990) recommended that readily movable smaller structures be allowed behind the 30-year projection line and that large structures be allowed only behind the 60-year projection line. Their recommendations are summarized in Figure 14.3

14.3.2 CONSTRUCTION STANDARDS

Substantial increases in structural integrity during storms can be achieved by structural means at relatively low cost. The three primary factors are as follows:

1. Elevation of the main horizontal shore-parallel structural elements on piling above the maximum wave crests.
2. Use of structural connectors that can resist high wind loads.

Figure 14.4 Damage to structure by Hurricane Opal in Florida.

3. The use of piling with sufficient embedment to withstand wave forces and erosion without loss of foundation integrity.

In the United States, the elevation standards have been required by the Federal Emergency Management Administration for at least 25 years.

Inspection of hurricane-affected areas following storms has demonstrated that the measures described herein are quite effective. Practically all of the major damage occurs to structures that were built before these requirements were imposed. Figure 14.4 presents a photograph of a structure collapsed on the beach as the result of undermining following Hurricane Elena in 1985. Figure 14.5 shows damage wrought by Hurricane Hugo to a restaurant on Folly Island, South Carolina. Neither of these structures was built to modern standards.

14.4 PROTECTIVE VALUE OF A WIDE BEACH

A wide beach serves a very substantial energy-absorbing function for the landward structures. Figure 14.6 shows, for a particular offshore profile, the calculated wave height due to a 100-year wave height with and without a small nourishment project that increased the beach width by only 13 m. Although this wave height is merely reduced by 30 percent (from 1.4 to 0.9 m), the associated damage potential is decreased by a much greater percentage because the damage potential increases with the wave height to a power of 2 or 3.

Figure 14.7 presents the results of a survey of 540 structures following Hurricane Eloise in 1975. The damages to each structure were tabulated as a function of the structure location relative to a control line ("CL" in Figure 14.7) that is approximately

Figure 14.5 Damage to structure by Hurricane Hugo, Folly Island, South Carolina (Wang 1990).

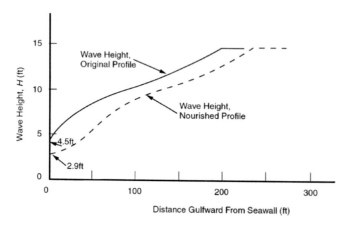

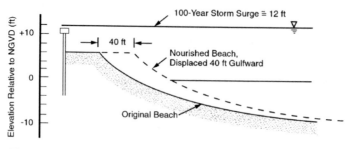

Figure 14.6 Effect of beach nourishment on wave height.

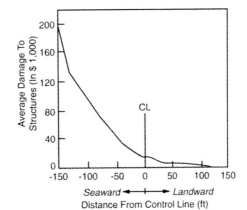

Figure 14.7 Structural damage to structures during Hurricane Eloise in 1978 dollars. Based on data from Shows (1978).

parallel to the shoreline. Two features are evident: (1) the damages were much greater to those structures located near the shoreline and (2) the damage curve is quite steep for locations close to the beach. Figure 14.8 shows the incremental damage reduction for these data by increasing the building setback by 0.3 m. This can be done practically by widening the beach through beach nourishment for this area.

14.5 SAND RIGHTS

Sand rights, an emerging field of law, were first proposed by Stone and Kaufman (1988) as an extension of water rights and the public trust doctrine, which holds that the state has the responsibility to preserve tide lands and navigable waters for the public. Water rights are derived from English common law holding that owners of land adjacent to watercourses could use the water if they put it to beneficial use and did not unduly affect the water quality for downstream users. The fundamental concept of water rights is that the stream or river must be allowed to flow much the same as it did naturally.

Fundamental to the public trust doctrine is that the usage of the resource be beneficial. Sand transported in waterways and the shores of tidal bodies by the action of water is part of the natural system described in the doctrine. Stone and Kaufman

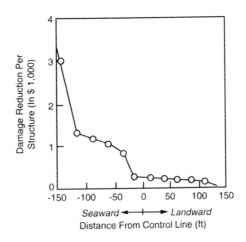

Figure 14.8 Reduction in structural damage due to a 0.3-m wider beach (in 1978 dollars) (from Dean 1988). The figure is based on a seaward shift of the curve in Figure 14.7 by 0.3 m.

quote Roman law: "The shores are not understood to be property of any man, but are compared to the sea itself, and to the sand or ground which is under the sea," Institutes of Justinian (2.1.5).

In California, sand was transported by streams to the sea and then by littoral currents along the coast to littoral traps, such as offshore canyons. Sand mining, damming of the stream, or armoring the shoreline to interrupt this process are then nonbeneficial uses of the resource and are detrimental to downdrift property owners, who must deal with beach erosion and loss of protective beach.

Sand rights imply that human interventions that affect the supply of sand down streams or along the coast must recognize the adverse impacts so the downdrift properties. Magoon and Edge (1999) have argued that the following doctrine should be adopted:

> Human and human-induced actions will not interfere, diminish, modify or impede sand and other sediments or materials from being transported to and along beaches, shores, or any flowing or eolian paths or bodies without appropriate restitution being made.

As an example, if a property on an eroding headland is armored, stopping the erosion and the supply of littoral materials to the beach, the owner of the property should compensate for this loss of erosion products to the littoral system. One method to do this would be to place beach fill in front of the armored area annually, as discussed earlier.

The intent of the sand rights argument is not to restrict the development of the shoreline or stop coastal protection but to require that people evaluate the consequences of their actions and to remediate, or compensate, for downdrift erosion, if it occurs.

REFERENCES

Houston, J.R., "International Tourism and U.S. Beaches," *Shore and Beach*, 64, 2, 3–4, 1996.

Magoon, O.T., and B.L. Edge, "Sand Rights – The Fragile Coastal Balance," *Proc. Canadian Coastal Conf.*, Canadian Coastal Science and Eng. Assoc., Burlington, Ontario, 99–107, 1999.

National Research Council (NRC), *Managing Coastal Erosion*, Water Science and Technology Board and Marine Board, National Academy Press, Washington, D.C., 182 pp., 1990.

National Research Council, *Saving Cape Hatteras Lighthouse from the Sea: Options and Policy Implications*, Board on Environmental Studies and Toxicology, National Academy Press, Washington, DC, 136 pp., 1988.

Shows, E.W., "Florida's Coastal Setback Line – An Effort to Regulate Beachfront Development," *Coastal Management Journal*, 4, 1–2, 151–164, 1978.

Stone, K.E., and B. Kaufman, "Sand Rights: A Legal System to Protect the 'Shores of the Sea'," *Shore and Beach*, 56, 3, 8–14, 1988.

Wang, H., "Water and Erosion Damage to Coastal Structures – South Carolina Coast, Hurricane Hugo, 1989," *Shore and Beach*, 58, 4, 37–47, 1990.

Author Index

Subject Index

Printed in the United Kingdom
by Lightning Source UK Ltd.
133526UK00001B/123-124/A